AF783575

Recent Progress in Acoustics and Vibrations

Volume I

Recent Progress in Acoustics and Vibrations
Volume I

Edited by **Adele Raymond**

NY RESEARCH PRESS

New York

Published by NY Research Press,
23 West, 55th Street, Suite 816,
New York, NY 10019, USA
www.nyresearchpress.com

Recent Progress in Acoustics and Vibrations: Volume I
Edited by Adele Raymond

© 2015 NY Research Press

International Standard Book Number: 978-1-63238-394-5 (Hardback)

This book contains information obtained from authentic and highly regarded sources. Copyright for all individual chapters remain with the respective authors as indicated. A wide variety of references are listed. Permission and sources are indicated; for detailed attributions, please refer to the permissions page. Reasonable efforts have been made to publish reliable data and information, but the authors, editors and publisher cannot assume any responsibility for the validity of all materials or the consequences of their use.

The publisher's policy is to use permanent paper from mills that operate a sustainable forestry policy. Furthermore, the publisher ensures that the text paper and cover boards used have met acceptable environmental accreditation standards.

Trademark Notice: Registered trademark of products or corporate names are used only for explanation and identification without intent to infringe.

Printed in China.

Contents

Preface

Sound is a vibration that propagates as a mechanical wave of pressure and displacement, through some medium. For centuries, sound and its diverse forms of vibrations has been an area of interest for scientists. They have studied these vibrations carefully and derived useful forms of the same that can be helpful in manifold ways. Acoustics is the interdisciplinary science that deals with the study of all mechanical waves in gases, liquids, and solids including vibration, sound, ultrasound and infrasound. On a smaller scale, vibrations are used in measuring instruments; whereas on a larger scale, the applications become varied and more consequential in nature.

The magnitude of contribution of acoustics and vibrations in the advancement of science is immense. Static and dynamic characteristics of composite conoidal shell roofs, biomass from paddy waste fibers as sustainable acoustic material, three-dimensional investigation of the stokes eigenmodes in hollow circular cylinder, the effect of uncertainty in the excitation on the vibration input power to a structure, crack identification method for bridge type structures and side branch interaction with main line standing waves are important propositions for the progression of the field.

This book presents the most innovative and worthy researches and developments in the category. The book owes itself to a number of qualified researchers, scientists and industry veterans across the globe. I thank them for helping me at every step of editing and wish them the best in life. I also wish to thank my family and friends for supporting me at every step.

Editor

1

Structural Vibration Monitoring Using Cumulative Spectral Analysis

Satoru Goto,[1] Yoshinori Takahashi,[2] and Mikio Tohyama[1]

[1] *Wave Science Study, 2-7-8 Fujigaya, Kugenuma, Fujisawa-shi, Kanagawa 251-0031, Japan*
[2] *Tokyo Metropolitan College of Industrial Technology, 8-17-1 Minami Senju, Arakawa-ku, Tokyo 116-0003, Japan*

Correspondence should be addressed to Satoru Goto; ahjinsei@yahoo.co.jp

Academic Editor: K. M. Liew

This paper describes a resonance decay estimation for structural health monitoring in the presence of nonstationary vibrations. In structural health monitoring, the structure's frequency response and resonant decay characteristics are very important for understanding how the structure changes. Cumulative spectral analysis (CSA) estimates the frequency decay by using the impulse response. However, measuring the impulse response of buildings is impractical due to the need to shake the building itself. In a previous study, we reported on system damping monitoring using cumulative harmonic analysis (CHA), which is based on CSA. The current study describes scale model experiments on estimating the hidden resonance decay under non-stationary noise conditions by using CSA for structural condition monitoring.

1. Introduction

Structural health condition diagnostics are needed for ensuring the safety of buildings struck by earthquakes and old buildings. Spectral analysis focuses on the change in the resonant frequency caused by structural degradation, and it yields useful information for health monitoring. However, for structural health monitoring, spectral analysis requires a test signal. Although using a test signal is an easy way to measure structural spectral characteristics, it is impractical to vibrate a structure during every measurement. If the spectral characteristics could be measured using structural vibrations due to wind, ground motion, and so forth, the health of the structure would be able to be monitored all the time. However, this sort of measurement would require a lot of spectral information about external noise characteristics besides the structural spectral characteristics. For this reason, spectral characteristics analysis for structural health monitoring generally requires the external source characteristics to be known at all times [1, 2].

Hirata proposed the short-interval period (SIP) distribution to analyze the dominant spectral components of a structure subject to an unknown vibration or even nonstationary noise [3]. The SIP distribution represents the shape of the frequency response in poor measuring conditions. Recently, Hirata et al. showed that the SIP distribution can be used to get the gradual time variation of the frequency response of a structure subjected to nonstationary vibrations [4]. On the other hand, the SIP distribution requires a sufficiently long signal because the computational process of the SIP distribution uses information on the dominant frequency components taken from a number of short intervals.

Takahashi et al. proposed a health monitoring method that focuses on damping using cumulative harmonic analysis (CHA) to reveal the spectral peaks of hidden structural resonances in short signals [5]. In a recent report, the authors estimated the temporal damping changes in the presence of unknown nonstationary vibrations by using the interpolated power spectrum of CHA [6]. Damping is an important factor in structural diagnostics, and it is related to the resonant decay of transient information in the time domain. For instance, if a pillar has a crack, any resonance decay will be short lived. This means that a way of estimating transient information such as resonant decay would be of value to structural diagnostics. CHA is an extension of cumulative spectral analysis (CSA) proposed by Berman and

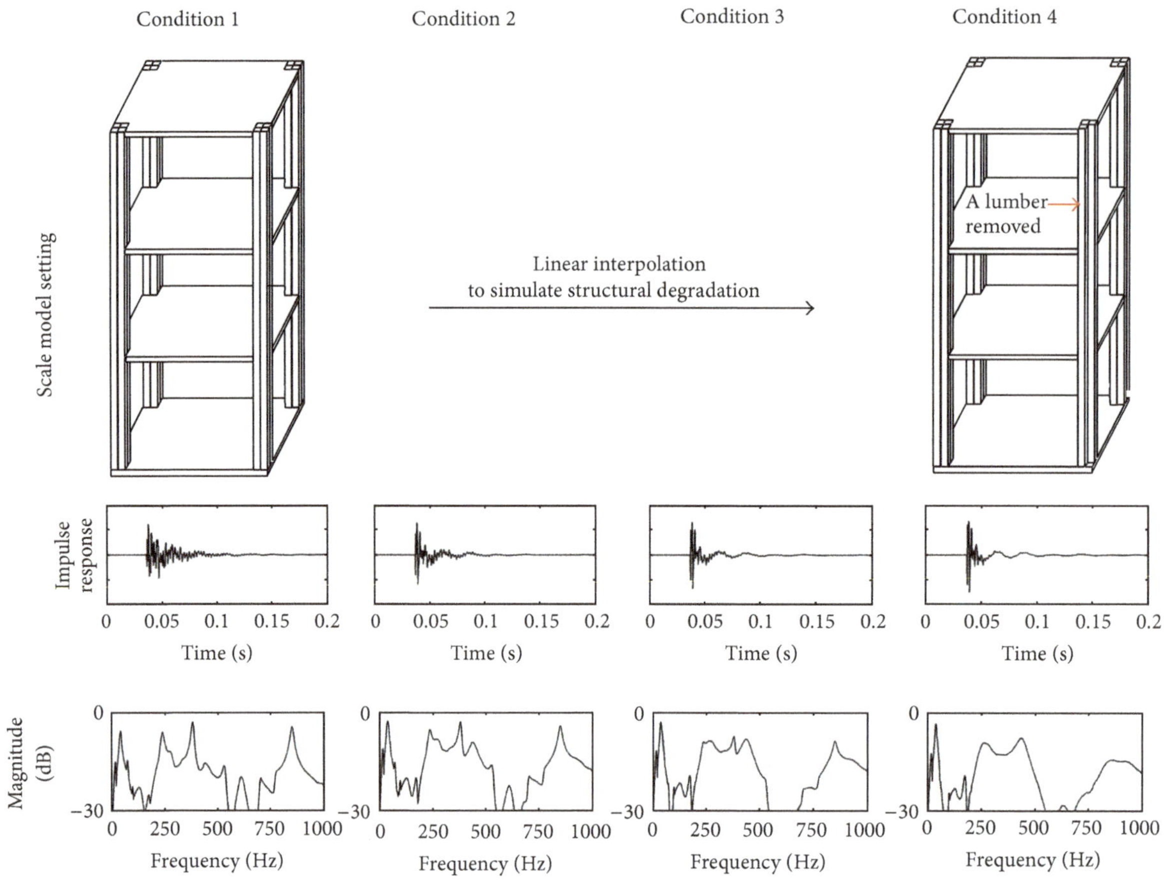

FIGURE 1: Scale models and corresponding measured and interpolated impulse responses and magnitude spectra.

Fincham [7], and CSA determines the transient characteristics of loudspeaker systems. In terms of the relationship between transient information and CHA, Takahashi et al. predicted the howling frequency by using a histogram of CHA power spectrum peaks of the transient decaying portion [8]. Thus, in acoustic systems, spectral analysis with a stepwise time window can accurately detect resonance peaks under the unknown signals and possibly predict the hidden frequency change from the transient decaying portions of the histogram of the CHA peaks in a simple feedback system. Consequently, when using a stepwise time window spectral analysis, we assume that acoustical techniques could be applied to structural analysis in order to reveal the resonance decay characteristics for structural diagnostics under nonstationary noise.

In this paper, we discuss a hidden resonance decay estimation for structural health monitoring that does not need the frequency characteristics of the external noise. We also describe experiments on scale models subjected to simulated nonstationary noise that assessed the effectiveness of resonance decay estimation focused on the transient decay portion.

Section 2 describes the measurement of the scale model's impulse response. Section 3 explains the CSA analysis and impulse response analysis. Section 4 describes the experiments in ideal cases of resonance decay analysis on the decay portions of the spectra. Section 5 discusses the results of the simulation experiment. Section 6 discusses the possibility of quantitative structural diagnostics using experimental CSA results.

2. Scale Model and Impulse Responses for a Simulation Experiment

We made two scale models and measured their impulse responses. Each scale model was a wooden framework (18 cm (W) $*$ 21 cm (D) $*$ 38 cm (H)) of a three-story building. The scale model had four pillars, and each pillar comprised 1 cm square lumber. Impulse responses were recorded at an Fs of 6 kHz using a piezoelectric accelerometer (PV-90B, RION) with an impulsive hammer. Figure 1 shows the scale-model experimental conditions and the linear interpolated impulse response waveforms and magnitude responses. As the upper illustration of Figure 1 shows, we prepared two scale models for condition 1 and condition 4. Conditions 2 and 3 simulated structurally degraded conditions by linear interpolation between the impulse response waveforms of conditions 1 and 4. The lower panels show the impulse response waveforms and magnitude spectra. From these panels, we can see that the magnitude spectra characteristics gradually change from condition 1 to condition 4.

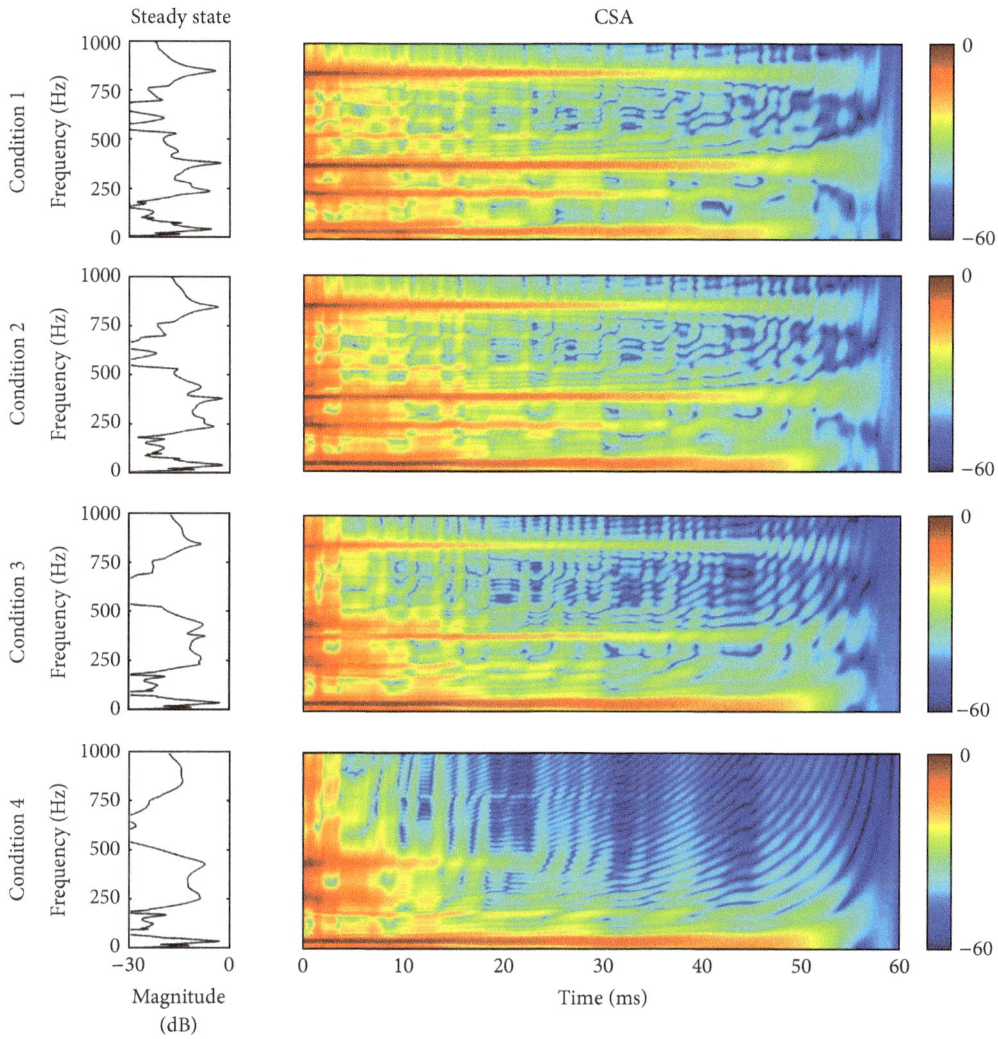

FIGURE 2: CSA of impulse responses.

3. Spectral Analysis with Stepwise Time Window

We used stepwise time window spectral analysis to make the resonance decay estimation. The cumulative spectral analysis (CSA) of a framed input signal $x(n)$ is formulated as

$$\mathrm{CSA}(n,k)=\sum_{m=n}^{N-1}u(m)\,x(m)\,e^{j(2\pi k/N)m}\qquad 0\le n\le N-1. \tag{1}$$

Here, $u(n)$ is a unit-step function corresponding to input signal conditions such as

$$u(n)=\begin{cases}1, & 0\le n\\ 0, & n<0.\end{cases} \tag{2}$$

We set N to include zero padding of 10 times the frequency interpolation of Fs.

Figure 2 shows the CSAs of the impulse responses corresponding to the different structural conditions. We calculated the CSAs by using a 60 ms frame to extract the impulse response decay portion. The left panel shows the magnitude spectrum of the impulse responses in the steady state, whereas the right panel illustrates the results of CSA. This figure shows that the spectral peaks in the magnitude spectra are the same as those in the CSA results. In addition, if there is a peak on the time axis, the changes in the decay of the resonant frequency can be read in the CSAs across conditions. Moreover, the fundamental frequency of 45 Hz does not decay in any of the plots. This is because the structure profile of the scale model did not change in these experiments. Therefore, the fundamental frequency did not decrease much compared with the higher resonances. The effect of CSA can be seen by comparing the peaks above 250 Hz. As for the peak at 850 Hz, it lasts for 40 ms under condition 1. Moreover, the time period of the peak gradually decreases to under 5 ms in going from condition 1 to condition 4. From these results, we supposed that CSA can estimate resonance peaks and their decays even when the decay portion is shorter than the impulse response.

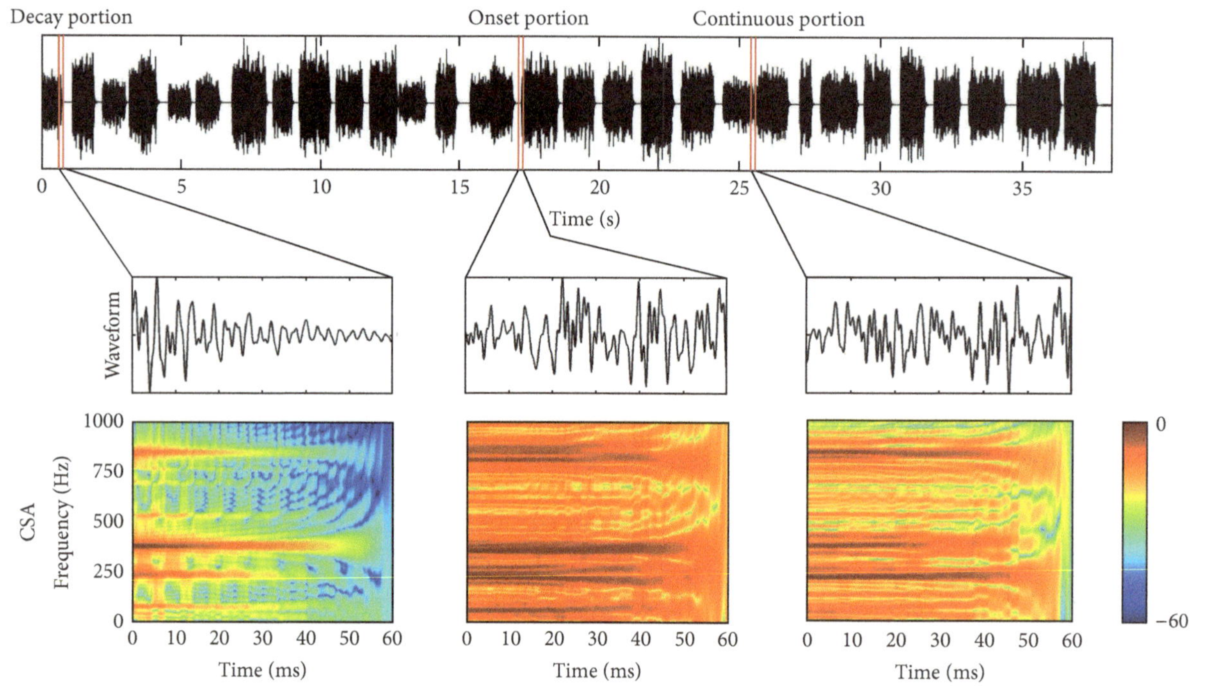

FIGURE 3: Random burst noise and CSA of different three portions.

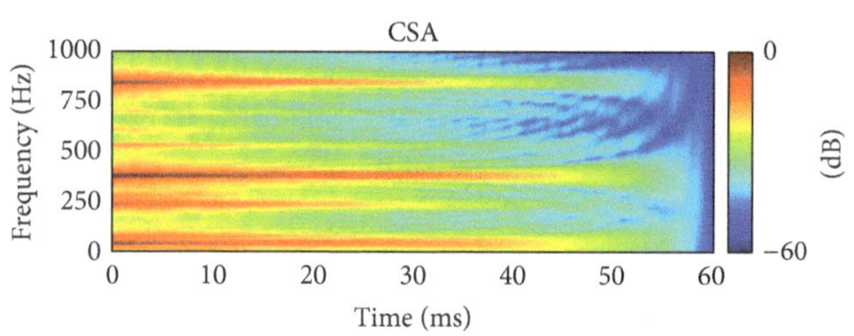

FIGURE 4: Mean of CSAs on all decay portions in the presence of random noise bursts.

4. Random Burst Noise Analysis for the Ideal Case of Resonance Decay

When a vibration stops, a transient decay arises through the system's transfer function. This transient decay includes the effect of the system's impulse response. To simulate this transient decay under ideal circumstances, we prepared random burst white noise having random durations, intervals, and magnitudes.

Figure 3 shows a random burst noise waveform convolved with the impulse response measured under condition 1 and the CSAs of different extracted portions. The upper panel illustrates the random burst noise and highlights of the extracted waveform for different portions. The decay portion on the left side is just after the random noise stopped. The onset portion in the middle is the onset of the convolved random noise. The continuous portion on the right side is the mixture of the decay of the signal and the beginning of the next random noise. The corresponding waveforms are in the second top panels. CSA at the bottom reveals different resonance decays between the decay portion and the other portions. The CSA of the decay portion has similar characteristics to those of the CSA of condition 1's impulse response in Figure 2. This means that the frequency characteristics including spectral decay can be estimated by using CSA on the decaying signal portion. On the other hand, the CSAs of the onset and continuous portions do not have spectral decays. This is because in the onset and continuous portions, the input signal was continuous and an increase in the resonance continued. This means that if the signal is found in the decay potion, CSA will reveal a hidden resonance decay.

Figure 4 shows the mean of the CSAs of the decaying portions of the whole simulated signal. In this figure, we can see that the magnitude spectrum of the impulse response and mean of CSA have the same spectral peaks, even the small peaks between 500 and 750 Hz. Furthermore, comparing this with the decay portion in Figure 3, the decay of the fundamental frequency is close to that of the CSA results of the impulse response in Figure 2. Accordingly, we can see that the mean of CSA is useful for estimating the resonant decay of the structure even in a short-frame signal such as 60 ms. Therefore, if constant monitoring is possible for structural health monitoring, the mean of the decay portions' CSAs could be used to show the structural resonance frequencies and their decays.

5. Structural Degradation Estimation Using Simulated Signals

In most settings, the signal length of the impulse response for sound sources such as musical instruments is short.

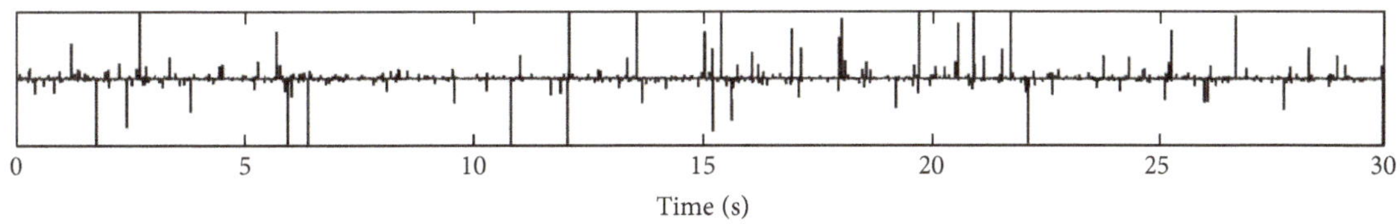

FIGURE 5: Simulated nonstationary noise waveform.

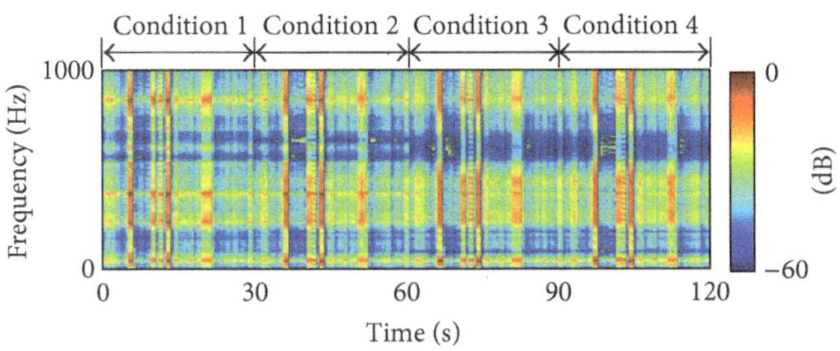

FIGURE 6: Spectrogram of simulated observed signals.

In contrast, structural impulse responses or reverberations of structures or halls are longer because the length of the impulse response depends on the system size and resonance, and the sound source is usually much smaller than the whole system. Therefore, if the observed signals contain a decay portion, it has a lot of information on the structural characteristics related to the impulse response of the system. In this experiment, we prepared the nonstationary noise and convolved it with the simulated structurally degrading impulse responses. The simulated signals were analyzed and the mean of the decay portion CSAs were calculated in order to show the resonance frequencies and their decays.

We prepared simulated nonstationary noise which had a Cauchy distribution. The simulated nonstationary noise $C(n)$ is defined as

$$C(n) = \frac{G_A(n)}{G_B(n)}, \quad (3)$$

where $G_A(n)$ and $G_B(n)$ each mean Gaussian noise.

Figure 5 shows the waveform of the simulated nonstationary noise, and we can see sudden vibrations in it. After the impulse responses were convolved with the nonstationary noise as simulated observed signals, we used the sudden vibrations to extract the transient decay portions. To confirm the spectral changes in a simulated observed signal, we joined 30 seconds of the simulated observed signals for each condition in sequence. Figure 6 shows a spectrogram using a 1-second framed DFT with a 0.5 second overlap of the joined signals. This figure shows the difficulty of revealing spectral changes under nonstationary noise conditions. Moreover, because of the large changes in energy due to the precipitous vibration, it is difficult to discriminate between conditions in the spectrogram. Regarding the resonant decay, such a spectrogram cannot be seen due to the poor time resolution. Putting these facts together, it is clear that a finer analysis is required for estimating the resonance frequency and decay in

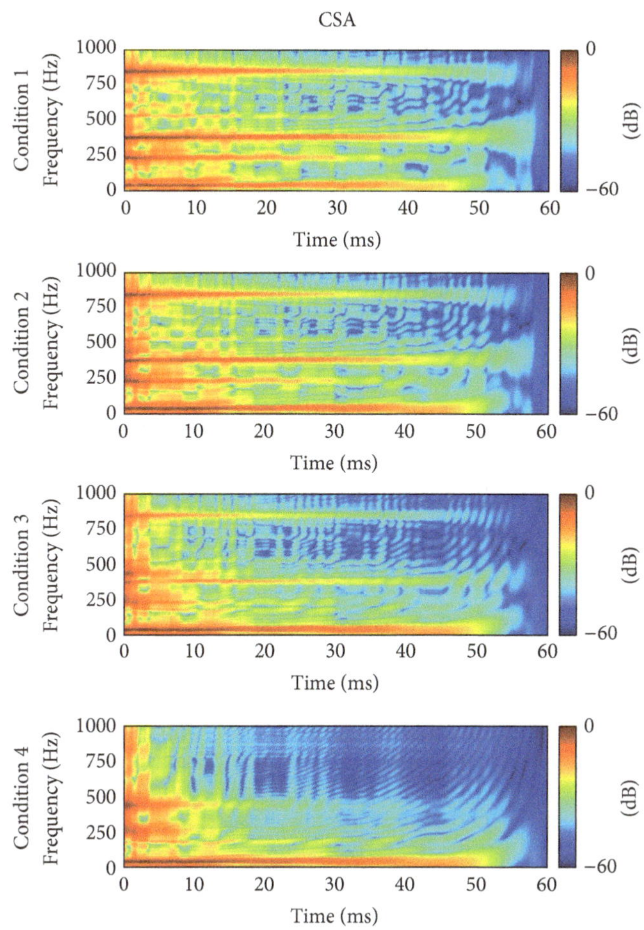

FIGURE 7: Mean of CSAs of decay portions in simulated observed signals.

the presence of nonstationary noise and without knowing the external source characteristics. Accordingly, we conducted resonance frequency and decay estimations using the CSAs of the decay portions.

In this experiment, we extracted 16 decay portions from the simulated observed signals for sudden vibrations in the order of the largest to the sixteenth largest. Figure 7 shows the mean CSAs of the decaying portions of the simulated observed signals for each condition. The resonant frequencies and their decays are very similar to the impulse response CSAs in Figure 2. Moreover, we can see that the durations of the peaks above the 250 Hz gradually decrease from condition 1 to condition 4. Also, the fundamental frequency and the decay duration were the same for all conditions.

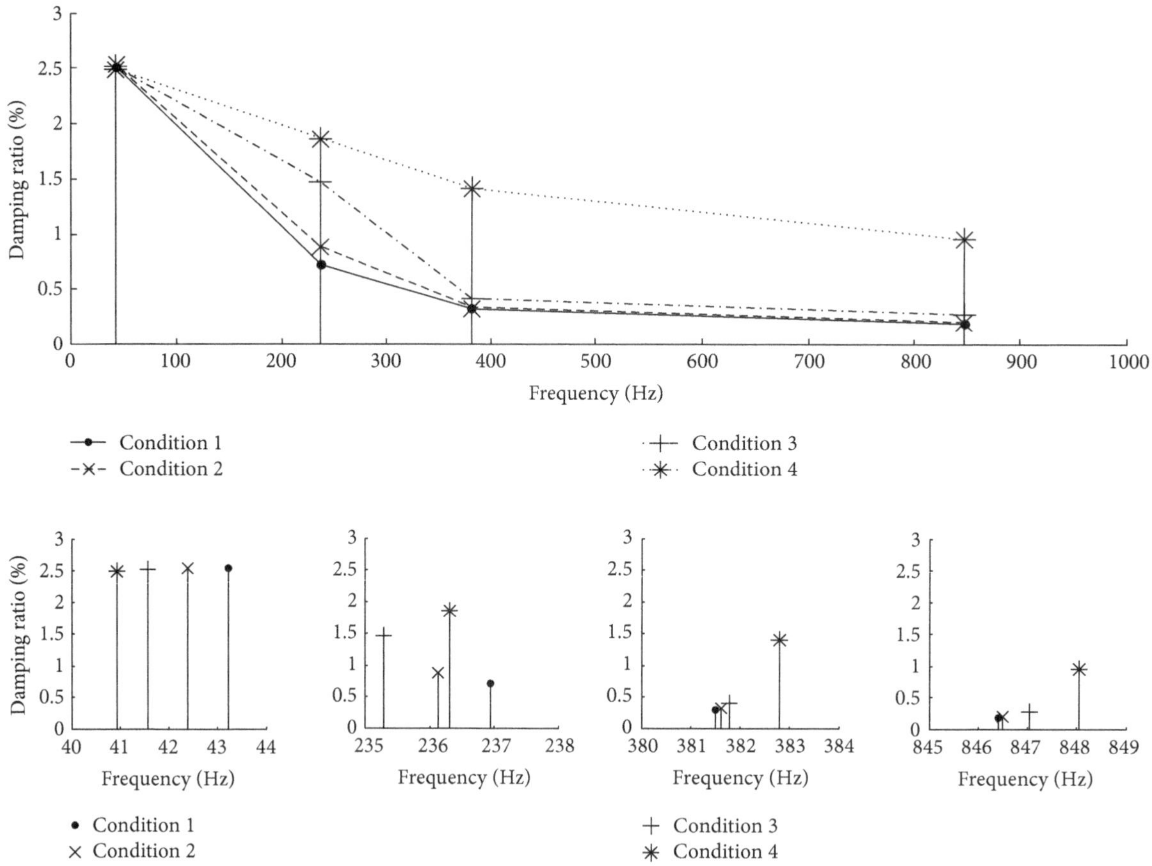

FIGURE 8: Damping ratio calculated from simulated observed signals' CSAs.

This means that the resonance changes for structural health monitoring are reliable even in the presence of nonstationary noise. Consequently, the CSAs of the decay portions can reveal structural changes in a constantly monitored signal without us having to know the noise spectral characteristics.

6. Discussion

The results of the previous experiment revealed the hidden resonance frequencies and their decays under unknown nonstationary signal conditions. The CSA of the decay portion has a lot of information related to a structure's impulse response. However, for structural diagnostics, a single-number index is much more useful for a structural health monitoring system. In this section, we discuss the quantitative estimation of structural damage by using the CSA results of the decay portions. From the resonance decays of the CSA, we calculated the reverberant time $T_{R_{60}}$ of each resonance frequency for each condition. To calculate the reverberant time, significant resonance frequencies were selected from the CSA of condition 1 in Figure 7. The resonance frequencies of the other conditions were determined from the similar peaks of condition 1. Moreover, the reverberant time of each resonance frequency was calculated from the CSAs of the simulated observed signals in Figure 7. All reverberant times across conditions were normalized by the fundamental frequency's reverberant time of condition 1. Using the reverberant times and resonance frequencies, the damping ratio ζ can be expressed as

$$\zeta = \frac{6.9}{2\pi f_0 T_{R_{60}}}, \qquad (4)$$

where f_0 is the resonance frequency and $T_{R_{60}}$ is the reverberant time of the resonance frequency [9, 10].

Figure 8 shows damping ratios estimated for each condition under a simulated observed signal. The upper panel shows the damping ratio across significant resonance frequencies. The damping ratio increases in going from condition 1 to condition 4 (as the structural degradation increases). This means that the average of the significant resonance frequency damping ratio could estimate structural degradation. The bottom panel shows a close-up of each resonance frequency. From these panels, it is clear that fine resonance information is difficult to estimate structural degradation because the resonance frequencies do not make simple transitions. In the second panel from the left, the resonance frequency rises and falls in going from condition 1 to condition 4. The maximum damping ratio $\zeta_{\max}$ is also difficult to use as an indicator of structural degradation. $\zeta_{\max}$ is the fundamental frequency of the left panel, and the damping ratio of the fundamental frequencies have almost the same score because the fundamental frequency is robust to structural degradation. These results show that $\zeta_{\max}$ is not suited as an index of structural diagnostics. In contrast, the damping ratios of the other resonant frequencies show that the damping ratio increases as the structural degradation increases. As previously indicated, although the average damping ratio can be used to estimate structural diagnostics, the minimum damping ratio of each condition $\zeta_{\min}$ would also be suited to estimating structural degradation and is a simple way to diagnose structures.

7. Summary

We discussed how to estimate hidden resonance decays using CSAs under nonstationary noise conditions. The first experiment used two scale models to measure the impulse responses and to simulate degraded conditions. The CSAs of the impulse responses revealed clear resonant frequencies and decays. In addition, the peak of the fundamental frequency did not decay, even in very short frame lengths. This means that it is useful to estimate the decay portion of the observed signals because the observed signals do not necessarily have sufficiently long transient decay portions. The random burst noise analysis involved CSAs of different portions. The CSAs show that decay portions only reveal resonant decays. In contrast, the mean CSA of the decay portions shows clear resonant frequencies and their decays. Thus, the mean CSA is a better way to reveal these factors. Finally, we conducted a simulated observed signal experiment. The results showed that the resonance frequency and decay were very similar to the impulse response's CSAs. Moreover, the experiments accurately simulated the decay of the fundamental frequency. From these results, we can conclude that performing CSA on the decay portion is a good way to estimate the resonant frequencies and their decays in the presence of nonstationary noise without using spectral information about the external noise. We also discussed how to estimate structural damage by calculating the damping ratio from the CSA. We confirmed that the average damping ratio increases as the structural degradation increases. This indicates that structural changes can be monitored using a simple score derived from the CSAs. However, the CSAs have a lot of information on the structural conditions related to impulse responses. This means that CSAs might have many uses in structural diagnostics. Moreover, this method would be helpful for structural diagnostics using constantly monitored signals.

Acknowledgments

This work was supported by a Grant from the Japan Society for the Promotion of Science, KAKENHI (24760477). The authors would like to thank Yoshimutsu Hirata who helped in measuring the impulse responses with the scale model.

References

[1] P. J. Halliday and K. Grosh, "Maximum likelihood estimation of structural wave components from noisy data," *Journal of the Acoustical Society of America*, vol. 111, no. 4, pp. 1709–1717, 2002.

[2] J. G. McDaniel and W. S. Shepard, "Estimation of structural wave numbers from spatially sparse response measurements," *Journal of the Acoustical Society of America*, vol. 108, no. 4, pp. 1674–1682, 2000.

[3] Y. Hirata, "A method for monitoring invisible changes in a structure using its non-stationary vibration," *Journal of Sound and Vibration*, vol. 270, no. 4-5, pp. 1041–1044, 2004.

[4] Y. Hirata, M. Tohyama, M. Matsumoto, and S. Gotoh, "Health monitoring for a structure using its nonstationary vibration," *Advances in Acoustics and Vibration*, vol. 2010, Article ID 696512, 5 pages, 2010.

[5] Y. Takahashi, T. Taniguchi, and M. Tohyama, "Structural condition monitoring by cumulative harmonic analysis of random vibration," *Advances in Acoustics and Vibration*, vol. 2008, Article ID 261758, 8 pages, 2008.

[6] Y. Takahashi, S. Goto, and M. Tohyama, "Modal experiment on structural condition monitoring using cumulative harmonic analysis," *Advances in Acoustics and Vibration*, vol. 2012, Article ID 327135, 6 pages, 2012.

[7] J. M. Berman and L. R. Fincham, "The application of digital techniques to the measurement of loudspeakers," *Journal of the Audio Engineering Society*, vol. 25, no. 6, pp. 370–384, 1977.

[8] Y. Takahashi, M. Tohyama, and Y. Yamasaki, "Cumulative spectral analysis for transient decaying signals in a transmission system including a feedback loop," *Journal of the Audio Engineering Society*, vol. 54, no. 7-8, pp. 620–629, 2006.

[9] E. E. Ungar and E. M. Kerwin Jr., "Loss factors of viscoelastic systems in terms of energy concepts," *Journal of the Acoustical Society of America*, vol. 34, no. 7, pp. 954–957, 1962.

[10] R. F. Barron, *Industrial Noise Control and Acoustics*, Marcel Dekker, New York, NY, USA, 2003.

2

Passivity-Based Control of Rotational and Translational Timoshenko Arms

Minoru Sasaki,[1] Takeshi Ueda,[1] Yshihiro Inoue,[2] and Wayne J. Book[3]

[1] *Department of Human and Information Systems Engineering, Gifu University, 1-1 Yanagido, Gifu 501-1193, Japan*
[2] *Department of Mechanical and Systems Engineering, Gifu University, 1-1 Yanagido, Gifu 501-1193, Japan*
[3] *School of Mechanical Engineering, Georgia Institute of Technology Atlanta, J. E. Love Jr. Manufacturing Building, 771 Ferst Drive, GA 30332-0405, USA*

Correspondence should be addressed to Minoru Sasaki, sasaki@gifu-u.ac.jp

Academic Editor: Abul Azad

It is shown that the alternate passivity-based control schemes can be designed which explicitly exploit the passivity properties of the Timoshenko model. This approach has the advantage over the conventional methods in the respect that it allows one to deal directly with the system's partial differential equations without resorting to approximations. Numerical results for the tracking control of a translational and rotational flexible Timoshenko arm are presented and compared. They verify that the proposed control schemes are effective at controlling flexible dynamical systems.

1. Introduction

In recent years, there has been a great deal of interest in the modeling and control of flexible arms [1–11]. This interest has been motivated by the prospect of fast, light, robot whose links, due to material characteristics, will flex under heavy loads. As a first step towards designing controllers for such robots, researchers have begun studying controllers for simple flexible links. These links, in most cases modeled as Euler-Bernoulli beams because of the small deflections involved, are often analyzed through an eigen-function series expansion of the solution to beam equation. The infinite term expansion is typically truncated after a finite number of terms and the controller designed using one of the various available linear control design techniques. Most practical controllers are designed based on models that are truncations of infinite-order transfer function models. There are several reasons for using the reduced-order models. First, these finite-order models agree very well with experimental results, assuming that an adequate number of modes have been retained. Second, there are no infinite-order transfer function models that coincide with experimental results past the first few modes. Finally, the use of a finite-order model greatly simplifies the controller design procedure. Most existing control design methods for flexible system, such as the Independent Modal Space Control, low-authority high-authority control design, frequency-weighted LQG, and positivity require modal truncation. One disadvantage of this kind of approach is that a controller, based on a truncated model, cannot guarantee stability of the closed loop system. This disadvantage arises because the model does not include the higher frequency terms. A second disadvantage is that the criteria for designing the controller for such a model are usually arbitrary and the design becomes more difficult as the truncated model order is increased.

In this paper, it is shown that the alternate passivity-based control schemes can be designed, which explicitly exploit the passivity properties of the Timoshenko model [12, 13]. From standard nonlinear control theory, the passivity theorem tells us that any strictly passive controller with finite gain will stabilize the system in an L_2 sense. This allows the design of very simple controllers for the systems. This approach has the advantage over the conventional methods in the respect that it allows one to deal directly with the system's partial differential equations without resorting to approximations. Moreover, it is potentially applicable to the control of nonlinear flexible systems. Numerical results for the tracking control of a translational and rotational flexible Timoshenko arm are presented and compared. They verify that the proposed control schemes are effective at controlling flexible dynamical systems.

2. Equation of Motions and Boundary Conditions

2.1. Translational Arm. Figure 1 shows a sketch of a translational Timoshenko arm. The equation of motions of the arm and mobile stage are

$$\rho A\left(\frac{\partial^2 y}{\partial t^2}+\frac{\partial^2 W_b}{\partial t^2}\right)-\kappa GA\left(\frac{\partial^2 y}{\partial x^2}-\frac{\partial \phi}{\partial x}\right)=0,$$
$$\rho I\frac{\partial^2 \phi}{\partial t^2}-EI\frac{\partial^2 \phi}{\partial x^2}-\kappa GA\left(\frac{\partial y}{\partial x}-\phi\right)=0, \quad (1)$$
$$M_p\frac{\partial^2 W_b}{\partial t^2}=-\kappa GA\left(\frac{\partial y(0,t)}{\partial x}-\phi(0,t)\right)+u.$$

with boundary conditions:

$$m\frac{\partial^2}{\partial t^2}(y(L,t)+W_b)+\kappa GA\left(\frac{\partial y(L,t)}{\partial x}-\phi(L,t)\right)=0,$$
$$I_c\frac{\partial^2}{\partial t^2}\phi(L,t)+EI\frac{\partial \phi(L,t)}{\partial x}=0, \quad (2)$$
$$y(0,t)=0,$$
$$\phi(0,t)=0.$$

Applying the Laplace transformation method with respect to t to equations (1) and (2) gives

$$\rho As^2(Y+W_b)-\kappa GA\left(\frac{\partial^2 Y}{\partial x^2}-\frac{\partial \Theta}{\partial x}\right)=0, \quad (3)$$

$$\rho Is^2\Phi-EI\frac{\partial^2 \Phi}{\partial x^2}-\kappa GA\left(\frac{\partial Y}{\partial x}-\Phi\right)=0, \quad (4)$$

$$M_ps^2W_b=-\kappa GA\left(\frac{\partial Y(0,t)}{\partial x}-\Phi(0,t)\right)+U, \quad (5)$$

$$ms^2(Y(l,s)+W_b)+\kappa GA\left(\frac{\partial y(L,s)}{\partial x}-\Phi(L,s)\right)=0, \quad (6)$$

$$I_cs^2\Phi(L,s)+EI\frac{\partial \Phi(L,s)}{\partial x}=0, \quad (7)$$

$$Y(0,s)=0, \quad (8)$$

$$\Phi(0,s)=0. \quad (9)$$

2.2. Rotational Arm. The rotational Timoshenko arm depicted in Figure 2 of total length L, area moment of inertia I, cross-sectional area A, density ρ, Young's modulus E, shear modulus G, and shear coefficient κ, is attached at one end to a payload of mass m and inertia I_c and on the other end a hub of inertia I_h, which in turn is connected to an actuator

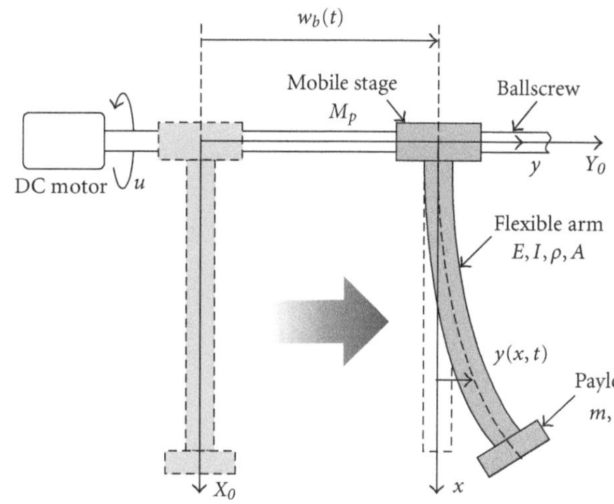

FIGURE 1: Translational Timoshenko arm.

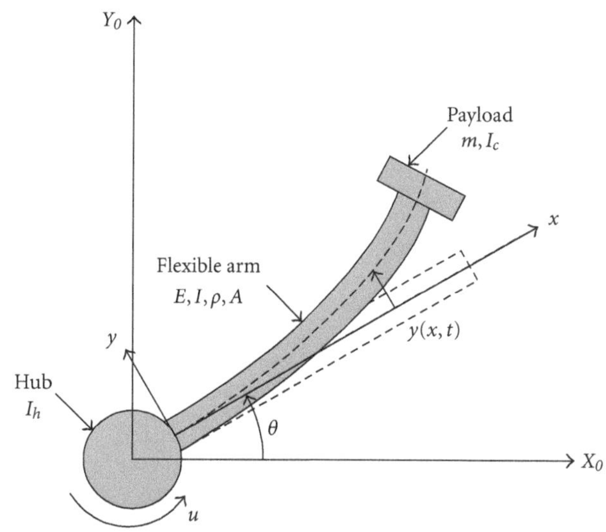

FIGURE 2: Rotational Timoshenko arm.

that supplies a torque u. We obtain the following equations of motion with the boundary conditions.

$$\rho A\left(\frac{\partial^2 y}{\partial t^2}+x\frac{\partial^2 \vartheta}{\partial t^2}\right)-\kappa GA\left(\frac{\partial^2 y}{\partial x^2}-\frac{\partial \phi}{\partial x}\right)=0, \quad (10)$$

$$\rho I\left(\frac{\partial^2 \phi}{\partial t^2}+\frac{\partial^2 \vartheta}{\partial t^2}\right)-EI\frac{\partial^2 \phi}{\partial x^2}-\kappa GA\left(\frac{\partial y}{\partial x}-\phi\right)=0, \quad (11)$$

$$I_h\frac{\partial^2 \vartheta}{\partial t^2}=EI\frac{\partial \phi(t,0)}{\partial x}+u, \quad (12)$$

$$m\frac{\partial^2}{\partial t^2}(y(L,t)+L\vartheta)=-\kappa GA\left(\frac{\partial y(L,t)}{\partial x}-\phi(L,t)\right), \quad (13)$$

$$I_c\frac{\partial^2}{\partial t^2}(\phi(L,t)+\vartheta)=-EI\frac{\partial \phi(L,t)}{\partial x}, \quad (14)$$

$$y(0,t) = 0, \tag{15}$$

$$\phi(0,t) = 0. \tag{16}$$

Taking the Laplace transform of (10)–(16), we get

$$\rho A s^2 (Y + x\Theta) - \kappa GA\left(\frac{\partial^2 Y}{\partial x^2} - \frac{\partial \Theta}{\partial x}\right) = 0, \tag{17}$$

$$\rho I s^2 (\Phi + \Theta) - EI\frac{\partial^2 \Phi}{\partial x^2} - \kappa GA\left(\frac{\partial Y}{\partial x} - \Phi\right) = 0, \tag{18}$$

$$I_h s^2 \Theta = EI\frac{\partial \Phi(s,0)}{\partial x} + U, \tag{19}$$

$$ms^2(Y(l,s) + L\Theta) = -\kappa GA\left(\frac{\partial y(L,s)}{\partial x} - \Phi(L,s)\right), \tag{20}$$

$$I_c s^2 (\Phi(L,s) + \Theta) = -EI\frac{\partial \Phi(L,s)}{\partial x}, \tag{21}$$

$$Y(0,s) = 0, \tag{22}$$

$$\Phi(0,t) = 0\,. \tag{23}$$

3. Stabilizing Feedback Controls

3.1. Translational Arm. The objective is to derive explicit expressions for feedback control u for controlling the arm. The approach taken here is to consider the total energy of the arm at any time t given by

$$\begin{aligned} V = {} & \frac{1}{2}\int_0^L \left[\rho A\left\{\frac{\partial y}{\partial t} + \frac{\partial w_b}{\partial t}\right\}^2 + \rho I\left\{\frac{\partial \phi}{\partial t}\right\}^2\right] dx \\ & + \frac{1}{2}M_p\left\{\frac{\partial w_b}{\partial t}\right\}^2 \\ & + \frac{1}{2}\int_0^L \left[EI\left\{\frac{\partial \phi}{\partial x}\right\}^2 + \kappa GA\left\{\phi - \frac{\partial y}{\partial x}\right\}^2\right] dx, \end{aligned} \tag{24}$$

that is positive definite with respect to $y = 0$, $dy/dt = 0$.

The Time derivative of (24) is given by

$$\begin{aligned} \frac{\partial V}{\partial t} = {} & \int_0^L \left[\rho A\left\{\frac{\partial y}{\partial t} + \frac{\partial w_b}{\partial t}\right\}\left\{\frac{\partial^2 y}{\partial t^2} + \frac{\partial^2 w_b}{\partial t^2}\right\}\right. \\ & \left. + \rho I\left\{\frac{\partial \phi}{\partial t}\right\}\left\{\frac{\partial^2 \phi}{\partial t^2}\right\}\right] dx \\ & + M_p\left\{\frac{\partial w_b}{\partial t}\right\}\left\{\frac{\partial^2 w_b}{\partial t^2}\right\} \\ & + \int_0^L \left[EI\left\{\frac{\partial \phi}{\partial x}\right\}\left\{\frac{\partial^2 \phi}{\partial x \partial t}\right\}\right. \\ & \left. + \kappa GA\left\{\phi - \frac{\partial y}{\partial x}\right\}\left\{\frac{\partial \phi}{\partial t} - \frac{\partial^2 y}{\partial x \partial t}\right\}\right] dx. \end{aligned} \tag{25}$$

By substituting (1) reduces to

$$\begin{aligned} \frac{\partial V}{\partial t} = {} & \int_0^L \left[\left\{\frac{\partial y}{\partial t} + \frac{\partial w_b}{\partial t}\right\}\left\{\kappa GA\left(\frac{\partial^2 y}{\partial x^2} - \frac{\partial \phi}{\partial x}\right)\right\}\right. \\ & \left. + \left\{\frac{\partial \phi}{\partial t}\right\}\left\{EI\frac{\partial^2 \phi}{\partial x^2} - \kappa GA\left(\phi - \frac{\partial y}{\partial x}\right)\right\}\right] dx \\ & + \frac{\partial w_b}{\partial t}\left\{\kappa GA\left(\phi(0,t) - \frac{\partial y(0,t)}{\partial x}\right) + u\right\} \\ & + \int_0^L \left[EI\left\{\frac{\partial \phi}{\partial x}\right\}\left\{\frac{\partial^2 \phi}{\partial x \partial t}\right\}\right. \\ & \left. + \kappa GA\left\{\phi(x,t) - \frac{\partial y}{\partial x}\right\}\left\{\frac{\partial \phi}{\partial t} - \frac{\partial^2 y}{\partial x \partial t}\right\}\right] dx. \end{aligned} \tag{26}$$

We wish to derive stabilizing feedback controls u such that the resulting feedback-controlled arm is dissipative and stable

$$\frac{\partial V}{\partial x} = \frac{\partial w_b}{\partial t}u. \tag{27}$$

Since our goal is to design a stabilizing control input, the most natural choice is to make the time derivative of the Lyapunov function negative in such a way. The following stabilizing feedback control is chosen:

$$u = -\gamma\frac{\partial w_b}{\partial t} \quad (\gamma > 0), \tag{28}$$

which results in

$$\frac{\partial V}{\partial t} = -\gamma\left(\frac{\partial w_b}{\partial t}\right)^2 \le 0. \tag{29}$$

This means that a control input should be chosen with as large as possible and should generate a base velocity that ensures asymptotic stability of the tip bending motion of the arm.

3.2. Rotational Arm. The problem is to find feedback control u such that the arm's total energy given by

$$\begin{aligned} V = {} & \frac{1}{2}\int_0^L \left[\rho A\left\{\frac{\partial y}{\partial t} + x\frac{\partial \theta}{\partial t}\right\}^2 + \rho I\left\{\frac{\partial \phi}{\partial t} + \frac{\partial \theta}{\partial t}\right\}^2\right] dx \\ & + \frac{1}{2}I_h\left\{\frac{\partial \theta}{\partial t}\right\}^2 \\ & + \frac{1}{2}\int_0^L \left[EI\left\{\frac{\partial \phi}{\partial x}\right\}^2 + \kappa GA\left\{\phi - \frac{\partial y}{\partial x}\right\}^2\right] dx. \end{aligned} \tag{30}$$

The time rate-of-change of V is given by

$$\begin{aligned}\frac{\partial V}{\partial t} &= \int_0^L \left[\rho A\left\{\frac{\partial y}{\partial t}+x\frac{\partial \theta}{\partial t}\right\}\left\{\frac{\partial^2 y}{\partial t^2}+x\frac{\partial^2 \theta}{\partial t^2}\right\} + \rho I\left\{\frac{\partial \phi}{\partial t}+\frac{\partial \theta}{\partial t}\right\}\left\{\frac{\partial^2 \phi}{\partial x^2}+\frac{\partial^2 \theta}{\partial t^2}\right\}\right]dx \\ &\quad + I_h\left\{\frac{\partial \theta}{\partial t}\right\}\left\{\frac{\partial^2 \theta}{\partial t^2}\right\} \\ &\quad + \int_0^L \left[EI\left\{\frac{\partial \phi}{\partial x}\right\}\left\{\frac{\partial^2 \phi}{\partial x \partial t}\right\} + \kappa GA\left\{\phi - \frac{\partial y}{\partial x}\right\}\left\{\frac{\partial \phi}{\partial t} - \frac{\partial^2 y}{\partial x \partial t}\right\}\right]dx.\end{aligned} \tag{31}$$

By substituting (10)-(11) reduces to

$$\begin{aligned}\frac{\partial V}{\partial t} &= \int_0^L \left[\left\{\frac{\partial y}{\partial t}+x\frac{\partial \theta}{\partial t}\right\}\left\{\kappa GA\left(\frac{\partial^2 y}{\partial x^2}-\frac{\partial \phi}{\partial x}\right)\right\} + \left\{\frac{\partial \phi}{\partial t}+\frac{\partial \theta}{\partial x}\right\}\left\{EI\frac{\partial^2 \phi}{\partial x^2} - \kappa GA\left(\phi - \frac{\partial y}{\partial x}\right)\right\}\right]dx \\ &\quad + \frac{\partial \theta}{\partial t}\left\{EI\frac{\partial \phi(0,t)}{\partial x}+u\right\} \\ &\quad + \int_0^L \left[EI\left\{\frac{\partial \phi}{\partial x}\right\}\left\{\frac{\partial^2 \phi}{\partial x \partial t}\right\} + \kappa GA\left\{\phi(x,t) - \frac{\partial y}{\partial x}\right\}\left\{\frac{\partial \phi}{\partial t} - \frac{\partial^2 y}{\partial x \partial t}\right\}\right]dx.\end{aligned} \tag{32}$$

The equation (32) is then integrated by parts and boundary conditions, (12)–(16) are substituted into the resulting equation. The result is

$$\frac{\partial V}{\partial t} = \frac{\partial \theta}{\partial t}u. \tag{33}$$

Taking into consideration of the fact that there is a feedback control law which gives nonpositive $\partial V/\partial t$ is

$$u = -\gamma\frac{\partial \theta}{\partial t} \quad (\gamma > 0). \tag{34}$$

Substituting (34) into (33) leads to

$$\frac{\partial V}{\partial t} = -\gamma\left(\frac{\partial \theta}{\partial t}\right)^2 \le 0, \tag{35}$$

which is negative semidefinite. This implies that the closed-loop system is energy dissipative and, hence, stable.

4. Passivity of Timoshenko Arm

Taking the time derivative of the total energy, and then integrating from 0 to t, we get the following restatement of energy balance principle:

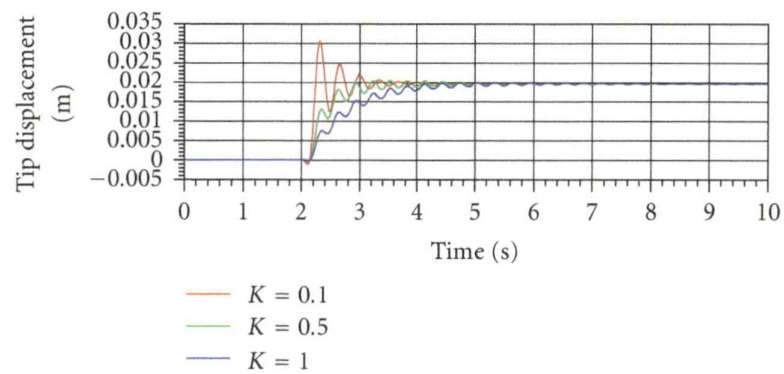

FIGURE 3: Step response of the translational Euler arm.

(1) translational arm

$$\int_0^t \frac{\partial w_b}{\partial \tau}u\,d\tau = \int_0^t \frac{\partial V}{\partial \tau}d\tau = V(t) - V(0) \ge -V(0) = -{y_0}^2, \tag{36}$$

(2) rotational arm

$$\int_0^t \frac{\partial \theta}{\partial \tau}u\,d\tau = \int_0^t \frac{\partial V}{\partial \tau}d\tau = V(t) - V(0) \ge -V(0) = -{y_0}'^2, \tag{37}$$

where the left-hand side is the supplied energy and right-hand side is the energy at time t minus the initial energy. Passivity of the operator u relative to the function V follows from the equation above and positivity of the total energy.

5. Simulation Results

Numerical inversion of Laplace transform is used to obtain the results in the time domain. The computation of the inverse Laplace transform is based on the paper of Hosono [14]. Numerical solutions are obtained by the Bromwich integrating the system corresponding to (A.1) and so forth. In the computer simulation study, we consider a typical arm whose parameters are given in the Table 1. Figures 3–6 show the step response of the arm-tip displacement with feedback gains variations. Figures 3 and 5 show the results of the Euler arm. Figures 4 and 6 show the results of the Timoshenko arm. There are few differences with each other in the case of this system parameters. It can be seen that the arm-tip displacement toward the desired position. With increasing the feedback gains, the settling times are longer and the residual vibrations are more slightly damped. In this case, shear deformations are small and negligible, so that the shear angle is almost zero and the rotary inertia is small and negligible, the so-called Euler-Bernoulli beam equations result. If the beam is long and skinny (aspect ratio, say greater than 10), the Euler-Bernoulli assumptions are appropriate.

6. Conclusion

In this paper, the passivity-based control of rotational and translational Timoshenko arms is studied. This approach has

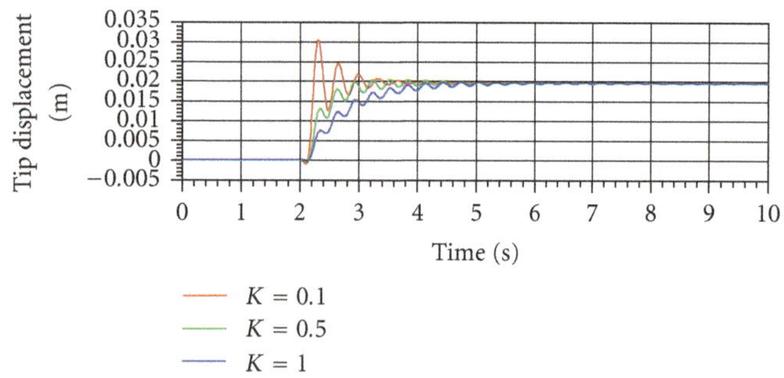

Figure 4: Step response of the translational Timoshenko arm.

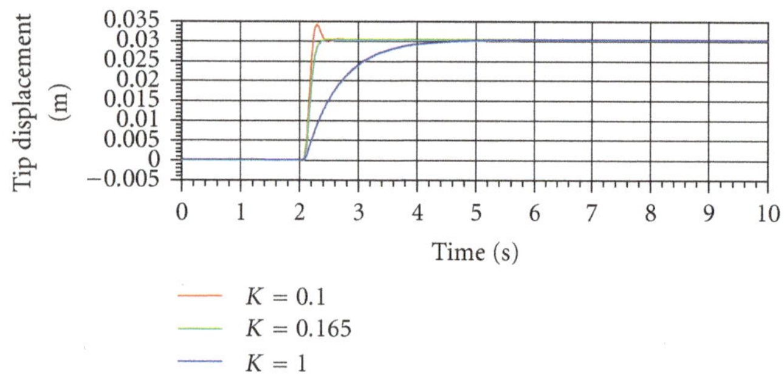

Figure 5: Step response of the rotational Euler arm.

the advantage over the conventional methods in the respect that it allows one to deal directly with the system's partial differential equations without resorting to approximations. Numerical results for the control of a translational and rotational flexible Timoshenko arm are presented and compared in graphical plots. They verify that the proposed control schemes are effective at controlling flexible dynamical systems and vibration control of mechanical systems.

Appendix

In the case of the translational Timoshenko arm, the resulting solution to (3), (4), and (5) is given by

$$Y(x,s) = \frac{\Delta k_1}{\Delta} e^{\lambda_1 x} + \frac{\Delta k_2}{\Delta} e^{\lambda_2 x} - \frac{\Delta k_3}{\Delta} e^{-\lambda_1 x} - \frac{\Delta k_4}{\Delta} e^{-\lambda_2 x} - W_b, \tag{A.1}$$

where

$$a_{11} = \left\{ ms^2 + \frac{\rho A s^2}{\lambda_1} \right\} e^{\lambda_1 L},$$

$$a_{12} = \left\{ ms^2 + \frac{\rho A s^2}{\lambda_2} \right\} e^{\lambda_2 L},$$

$$a_{13} = \left\{ ms^2 - \frac{\rho A s^2}{\lambda_1} \right\} e^{-\lambda_1 L},$$

$$a_{14} = \left\{ ms^2 - \frac{\rho A s^2}{\lambda_2} \right\} e^{-\lambda_2 L},$$

Table 1: System parameters.

Parameter	Nomenclature	Values	Unit
Length of the arm	L	1.2	m
Arm cross-sectional area	A	2.5×10^{-4}	m^2
Volume mass density	ρ	2.7667	Kg/m^3
Young's modulus	E	6.8944×10^{10}	Pa
Internal structural damping	C	0	Kg/sec
Payload moment of inertia	I_c	5×10^{-4}	$Kg \cdot m^2$
Payload mass	m	0.2	Kg
Shear correction factor	κ	5/6	–
Shear modulus	G	2.6517×10^{10}	Pa
Slider mass	M_p	0.5	Kg
Hub moment of inertia	I_h	0.5	$Kg \cdot m^2$

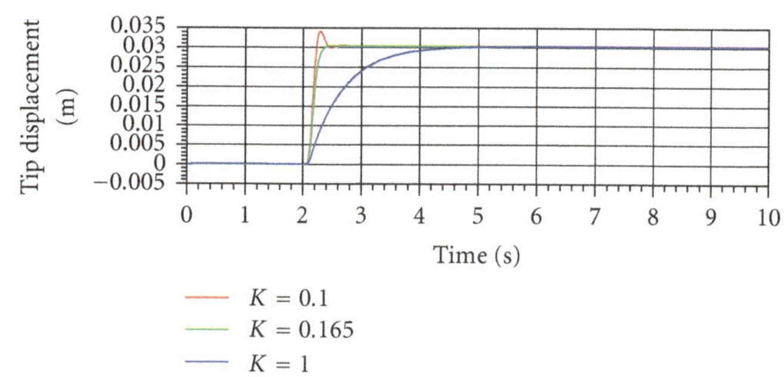

Figure 6: Step response of the rotational Timoshenko arm.

$$a_{21} = \left(\lambda_1 - \frac{\rho s^2}{\kappa G \lambda_1} \right) \left(I_c s^2 + EI(1+Cs)\lambda_1 \right) e^{\lambda_1 L},$$

$$a_{22} = \left(\lambda_2 - \frac{\rho s^2}{\kappa G \lambda_2} \right) \left(I_c s^2 + EI(1+Cs)\lambda_2 \right) e^{\lambda_2 L},$$

$$a_{23} = -\left(\lambda_1 - \frac{\rho s^2}{\kappa G \lambda_1} \right) \left(I_c s^2 - EI(1+Cs)\lambda_1 \right) e^{-\lambda_1 L},$$

$$a_{24} = -\left(\lambda_2 - \frac{\rho s^2}{\kappa G \lambda_2} \right) \left(I_c s^2 - EI(1+Cs)\lambda_2 \right) e^{-\lambda_2 L},$$

$$a_{31} = \left(\lambda_1 - \frac{\rho s^2}{\kappa G \lambda_1} \right),$$

$$a_{32} = \left(\lambda_2 - \frac{\rho s^2}{\kappa G \lambda_2} \right),$$

$$a_{33} = -\left(\lambda_1 - \frac{\rho s^2}{\kappa G \lambda_1} \right),$$

$$a_{34} = -\left(\lambda_2 - \frac{\rho s^2}{\kappa G \lambda_2} \right),$$

$$a_{41} = \left(M_p s^2 - \frac{\rho A s^2}{\lambda_1}\right),$$
$$a_{42} = \left(M_p s^2 - \frac{\rho A s^2}{\lambda_2}\right),$$
$$a_{43} = \left(M_p s^2 + \frac{\rho A s^2}{\lambda_1}\right),$$
$$a_{44} = \left(M_p s^2 + \frac{\rho A s^2}{\lambda_2}\right), \tag{A.2}$$

$$\Delta = \begin{bmatrix} a_{11} & a_{12} & a_{13} & a_{14} \\ a_{21} & a_{22} & a_{23} & a_{24} \\ a_{31} & a_{32} & a_{33} & a_{34} \\ a_{41} & a_{42} & a_{43} & a_{44} \end{bmatrix},$$
$$\Delta k_1 = \begin{bmatrix} 0 & a_{12} & a_{13} & a_{14} \\ 0 & a_{22} & a_{23} & a_{24} \\ 0 & a_{32} & a_{33} & a_{34} \\ U & a_{42} & a_{43} & a_{44} \end{bmatrix},$$
$$\Delta k_2 = \begin{bmatrix} a_{11} & 0 & a_{13} & a_{14} \\ a_{21} & 0 & a_{23} & a_{24} \\ a_{31} & 0 & a_{33} & a_{34} \\ a_{41} & U & a_{43} & a_{44} \end{bmatrix}, \tag{A.3}$$
$$\Delta k_3 = \begin{bmatrix} a_{11} & a_{12} & 0 & a_{14} \\ a_{21} & a_{22} & 0 & a_{24} \\ a_{31} & a_{32} & 0 & a_{34} \\ a_{41} & a_{42} & U & a_{44} \end{bmatrix},$$
$$\Delta k_4 = \begin{bmatrix} a_{11} & a_{12} & a_{13} & 0 \\ a_{21} & a_{22} & a_{23} & 0 \\ a_{31} & a_{32} & a_{33} & 0 \\ a_{41} & a_{42} & a_{43} & U \end{bmatrix},$$

$$\lambda = \pm\left[\frac{1}{2}\left\{\left(\frac{1}{E(1+Cs)} + \frac{1}{\kappa G}\right)\rho s^2 \pm \left(\frac{\rho^2 s^4}{E^2(1+Cs)^2} - \frac{2\rho^2 s^4}{E(1+Cs)\kappa G} + \frac{\rho^2 s^4}{\kappa^2 G^2} - \frac{4\rho A s^2}{EI(1+Cs)}\right)^{1/2}\right\}\right]^{1/2}, \quad \lambda = \pm\lambda_1, \ \pm\lambda_2. \tag{A.4}$$

In the case of the rotational Timoshenko arm, the resulting solution to (17), (18), and (19) can be obtained (omitted details for want of space).

References

[1] K. Yuan and C. M. Hu, "Nonlinear modeling and partial linearizing control of a slewing timoshenko-beam," *Journal of Dynamic Systems, Measurement and Control, Transactions of the ASME*, vol. 118, no. 1, pp. 75–82, 1996.

[2] M. Tadi, "Comparison of two finite-element schemes for feedback control of a Timoshenko beam," in *Proceedings of the ASME Dynamic Systems and Control Division (DSC '97)*, vol. 61, pp. 587–596, November 1997.

[3] M. W. D. White and G. R. Heppler, "Vibration of a rotating timoshenko beam," *Journal of Vibration and Acoustics, Transactions of the ASME*, vol. 118, no. 4, pp. 606–613, 1996.

[4] M. J. Balas, "Direct velocity feedback control of large space structures," *Journal of Guidance, Control, and Dynamics*, vol. 2, no. 3, pp. 252–253, 1979.

[5] Z. H. Luo, "Direct strain feedback control of flexible robot arms: new theoretical and experimental results," *IEEE Transactions on Automatic Control*, vol. 38, no. 11, pp. 1610–1622, 1993.

[6] B. Paden, D. Chen, R. Ledesma, and E. Bayo, "Exponentially stable tracking control for multijoint flexible-link manipulators," *Journal of Dynamic Systems, Measurement and Control, Transactions of the ASME*, vol. 115, no. 1, pp. 53–59, 1993.

[7] M. Karkoub and K. Tamma, "Modelling and μ-synthesis control of flexible manipulators," *Computers and Structures*, vol. 79, no. 5, pp. 543–551, 2001.

[8] S. K. Dwivedy and P. Eberhard, "Dynamic analysis of flexible manipulators, a literature review," *Mechanism and Machine Theory*, vol. 41, no. 7, pp. 749–777, 2006.

[9] G. G. Rigatos, "Model-based and model-free control of flexible-link robots: a comparison between representative methods," *Applied Mathematical Modelling*, vol. 33, no. 10, pp. 3906–3925, 2009.

[10] C. Di Castri and A. Messina, "Vibration analysis of a planar multilink manipulator using a reduced-order matrix formulation," *Mechanics Research Communications*, vol. 38, no. 3, pp. 231–238, 2011.

[11] J. J. Shifman, "Lyapunov functions and the control of the Euler-Bernoulli beam," *International Journal of Control*, vol. 57, no. 4, pp. 971–990, 1993.

[12] J. LaSalle and S. Lefschetz, *Stability by Lyapunov's Direct Method with Applications*, Academic Press, New York, NY, USA, 1961.

[13] R. Ortega, A. Loria, P. J. Nicklasson, and H. Sira-Ramirez, *Passivity-Based Control of Euler-Lagrange Systems*, Springer, London, UK, 1998.

[14] T. Hosono, "Numerical inversion of Laplace transform and some application to wave optics," *Radio Science*, vol. 16, no. 6, pp. 1015–1019, 1980.

3

Analysis of Static Instability of an Asymmetric, Rotating Sand-Wich Beam

P. R. Dash,[1] K. Ray,[2] S. K. Sarangi,[1] and P. K. Pradhan[1]

[1] *Mechanical Engineering Department, VSSUT Burla, Orissa 768018, India*
[2] *Mechanical Engineering Department, IIT Kharagpur, Kharagpur 721302, India*

Correspondence should be addressed to P. K. Pradhan, prasant2001uce@gmail.com

Academic Editor: Joseph CS Lai

The static stability of an asymmetric, rotating sandwich beam subjected to an axial pulsating load has been investigated for pinned-pinned and fixed-free boundary conditions. The equations of motion and associated boundary conditions have been obtained by using the Hamilton's energy principle. Then, these equations of motion and the associated boundary conditions have been nondimensionalised. A set of Hill's equations are obtained from the nondimensional equations of motion by the application of the general Galerkin method. The static buckling loads have been obtained from Hill's equations. The influences of geometric parameters and rotation parameters on the nondimensional static buckling loads have been investigated.

1. Introduction

The dynamic behavior of rotating beams is of great practical interest in the design of steam and gas turbine blades and helicopter blades. Sandwich constructions with high strength facings and a light weight core have been very popular in aerospace applications. Typical sandwich members used varied from structural panels in aircraft to the helicopter rotor blades. In general, high modulus and light weight characteristics of the sandwich construction normally have great advantages of high movability, power saving, and high strength in robotics applications. Extensive publications have been available concerning the design and analysis of sandwich structures.

Bhat [1] studied the natural frequencies and mode shapes of a rotating uniform cantilever beam with tip mass for different values of rotational speed, hub radius, and setting angle using beam characteristic orthogonal polynomials in the Rayleigh-Ritz method. Liu and Yeh [2] presented the natural frequencies of a nonuniform rotating beam with a restrained base for various values of restraint parameters and rotational speed. Unger and Brull [3] made an analytical investigation to determine the simple and combination resonance regions of shafts mounted in a long and short bearings subjected to pulsating torque applied at the ends and found that the most common and dangerous parametric instabilities arose as a result of combination resonance. Kammer and Schlack [4] obtained the boundaries between the stable and unstable regions of an Euler beam rotating with an angular velocity which has a small periodic component. Sri Namachchivaya [5] derived explicit stability conditions for a rotating shaft under parametric excitation comprising a combination of harmonic terms and stationary stochastic processes using the Routh-Hurwitz criterion and found that the addition of nonwhite noise excitation has a stabilizing effect on harmonically excited shafts. The vibration behavior of a rotating beam, oriented perpendicular to the axis of spin, was investigated by Bauer and Eidel [6]. It was observed that the speed of rotation has a very pronounced influence of the rotating beam and an increase in the speed of rotation may increase or decrease the natural frequencies depending on the boundary conditions. Abbas and Thomas [7, 8] studied the effect of rotational speed and root flexibilities on the first-order simple resonance zones of a rotating Timoshenko's beam by using the finite element method. Bauchau and Hong [9] utilized the same method to analyze the effect of viscous damping on the response and stability of parametrically excited beams undergoing large deflections and rotations. Dynamic stability of an ordinary rotating beam with various boundary conditions was studied by Kar and Sujata [10]. The same authors [11] also studied the stability of a rotating, pretwisted and preconed cantilever beam. Parametric instability of a rotating pretwisted beam subjected to sinusoidal compressive axial loads was addressed by Tan et al. [12].

As it has been pointed out by many investigators the shear deformation of the core plays an important role in the flexural and dynamic behavior of a sandwich beam; therefore, the flexural rigidity in the core and shear deformation of the facings were neglected in many analysis. The parametric instability of a cantilever beam with magnetic field and periodic axial load has been studied by Pratiher and Dwivedy [13]. The results obtained from perturbation analysis have been verified by solving the temporal equation of motion using fourth-order Runge-Kutta method. Dwivedy et al. [14] have investigated the parametric instability regions for simple supported, clamp-guided, clamp-free riveted, and clamped-free end condition by modified Hsu's method and found that zones of instability are affected by the shear parameter, core-loss factor and the ratio of core thickness to skin thickness. Zheng et al. [15] investigated the instability of a moving load system with the load approximated as a single-axle mass-spring-damper system. The author showed that instability occurs for a lower mass as compression axial force increases. Metrikine and Verichev [16] introduced two degrees of freedom load and approximated the rail as a Timoshenko beam. They have observed that under these approximations, larger mass of the load lowers the velocity at which instability is observed. The same authors [17] have considered the effect of periodic variations of the foundation stiffness on instability. The parametric instability of viscoelastically supported asymmetric sandwich beam have been studied by Ghosh et al. [18] and found that the viscoelastic constraints improve the stability of the system. The dynamic stability of a rotating sandwich beam using the finite element method has been studied by Lien and Chen [19, 20]. The parametric instability of a rotating asymmetric sandwich beam with viscoelastic core and subjected to a pulsating axial load has been studied by Dash et al. [21].

The present work deals with the static stability of a rotating asymmetric sandwich beam subjected to axial pulsating load, since it has not been studied till now. The equations of motion for transverse vibrations of the beam are obtained using Hamilton's principle. The general Galerkin method is used to reduce the nondimensional equations of motion to a set of coupled Hill's equations with complex coefficients. The static buckling loads are obtained from Hill's equations [22]. The effect of rotation parameters and geometric parameters on the nondimensional static buckling loads is investigated for the pinned-pinned and fixed-free boundary conditions.

2. Formulation of the Problem

2.1. System Configuration. A viscoelastic sandwich beam of length l and width B, set off at a distance b from the axis of rotation (z'-axis) and oriented along the x-axis, perpendicular to the axis of rotation is shown in Figure 1. The beam rotates about the vertical z'-axis at a constant angular velocity Ω_0 and is capable of oscillating in the x-z plane. It is asymmetric with respect to the x-y plane as layer 1 is not same as layer 3 both geometrically and materially. Note that some authors might call this a "rotating column" configuration. However, it is also customary in the literature to call it a "rotating beam". This is followed throughout the paper.

The top layer of the beam is made of an elastic material of thickness $2h_1$ and Young's modulus E_1, and the bottom layer is made of an elastic material of thickness $2h_3$ and Young's modulus E_3. The core is made of a linearly viscoelastic material with a shear modulus $G_2^* = G_2(1 + j\eta)$, where G_2 is the in-phase shear modulus, η is the core loss factor and $j = \sqrt{-1}$.

A pulsating axial load $P(t) = P_0 + P_1\cos(\omega t)$ is applied at the end $x' = b + l$ of the beam. P_0 and P_1 are, respectively, the static and dynamic load amplitudes, ω is the frequency of the dynamic load, and t is time.

2.2. Derivation. The following assumptions are made for deriving the equations of motion.

(1) The beam deflection $w(x,t)$, parallel to z-axis is small and is the same at all points of a given cross-section.

(2) The layers are perfectly bonded so that displacements are continuous across the interfaces. The elastic face layers obey the Euler-Bernoulli beam theory.

(3) The allowance for rotary inertia is neglected while calculating the kinetic energy of the system.

(4) Shear deformation of the facings are neglected.

(5) Damping in the viscoelastic core is predominantly due to shear. Bending and extensional effects in the core are neglected.

(6) The Kerwin's assumption [23] is used.

According to the above assumption, $E_1A_1U_{1,x} + E_3A_3U_{3,x} = 0$

For the system

$$T = \frac{1}{2}m\int_0^l w_t^2\,dx,$$

$$\begin{aligned} V &= \frac{1}{2}\int_0^l E_1A_1U_{1,x}^2\,dx + \frac{1}{2}\int_0^l E_3A_3U_{3,x}^2\,dx \\ &\quad + \frac{1}{2}E_1\int_0^l I_1\left(\frac{\partial^2 w}{\partial x^2}\right)^2 dx + \frac{1}{2}E_3\int_0^l I_3\left(\frac{\partial^2 w}{\partial x^2}\right)^2 dx \\ &\quad + \frac{1}{2}G_2^*\int_0^l A_2v_2^2\,dx \\ &\quad + \frac{1}{2}m\Omega_0^2\int_0^l\left\{(b+x)\int_0^x\left(\frac{\partial w}{\partial x}\right)^2 dx\right\}dx, \end{aligned} \tag{1}$$

$$W = \frac{1}{2}\int_0^l P(t)w_x^2\,dx,$$

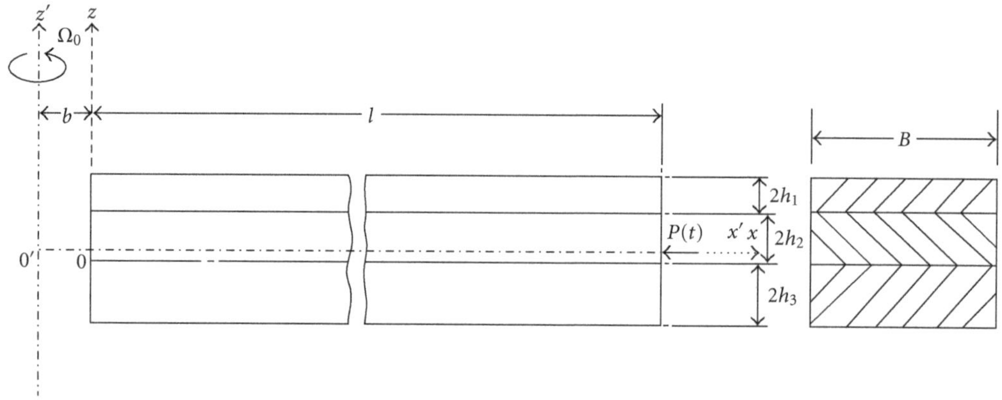

FIGURE 1: System configuration.

where $v_2 = ((U_1 - U_3 - cw_x)/2h_2)$, which is derived according to Kerwin's assumption, where T is the kinetic energy of the system, V is the strain energy of the system and W is the work done on the system by the external force.

Using Hamilton's principle,

$$\delta \int_{t_1}^{t_2} (T - V + W)dt = 0. \tag{2}$$

The equations of motion and the associated boundary conditions are obtained as follows:

$$\begin{aligned}
&\ddot{w} + (E_1 I_1 + E_3 I_3) \\
&\left[\frac{1}{m} + \frac{3\Omega_0^2}{2(E_1A_1 + E_3A_3)}\left\{f - (x+b)^2\right\}\right]w'''' \\
&-3\frac{E_1I_1 + E_3I_3}{E_1A_1 + E_3A_3}\Omega_0^2(x+b)w''' \\
&+\left[-3\frac{E_1I_1 + E_3I_3}{2(E_1A_1 + E_3A_3)}\Omega_0^2 - \frac{1}{2}\Omega_0^2\left\{f - (x+b)^2\right\}\right. \\
&\left.\times\frac{G_2^* A_2 c^2}{m(2h_2)^2} + \frac{P(t)}{m}\right]w'' + \frac{1}{2}\Omega_0^2(x+b)w' \\
&+\frac{G_2^* A_2 c(1+\alpha)}{m(2h_2)^2}u_1' = 0 \\
&u_1'' - \frac{G_2^* A_2(1+\alpha)}{(2h_2)^2(E_1A_1 + \alpha^2 E_3A_3)}\left\{(1+\alpha)u_1 - cw'\right\} = 0.
\end{aligned} \tag{3}$$

The boundary conditions at $x = 0$ and $x = \ell$ are

$$\begin{aligned}
&w'' = 0 \quad \text{or } w' = 0, \\
&(E_1I_1 + E_3I_3)\frac{\partial}{\partial x}\left[\left\{1 + \frac{3m\Omega_0^2}{2(E_1A_1 + E_3A_3)}\left(f - (x+b)^2\right)\right\}w''\right] \\
&-\left[\frac{m\Omega_0^2}{2}\left\{f - (x+b)^2\right\} + \frac{G_2^* A_2 c^2}{(2h_2)^2} + P(t)\right]w' \\
&+\frac{G_2^* A_2 c(1+\alpha)}{(2h_2)^2}u_1 = 0,
\end{aligned} \tag{4}$$

or,

$$\begin{aligned}
&w = 0 \\
&u_1' = 0 \quad \text{or } u_1 = 0.
\end{aligned} \tag{5}$$

Introducing the nondimensional variables $\overline{x} = x/l$, $\overline{w} = w/l$, $\overline{u}_1 = u_1/l$, and $\overline{t} = t/t_0$, where $t_0 = \sqrt{ml^4/(E_1I_1 + E_3I_3)}$ and simplifying, the following nondimensional equations of motion are obtained:

$$\begin{aligned}
&\ddot{\overline{w}} + \left[1 + \frac{\lambda_0^2\left(1 + E_{31}h_{31}^3\right)}{l_{h_1}^2(1 + E_{31}h_{31})}\left\{\frac{f}{l_2} - \left(\overline{x} + \overline{b}\right)^2\right\}\right]\overline{w}'''' \\
&-\frac{2\lambda_0^2\left(1 + E_{31}h_{31}^3\right)}{l_{h1}^2(1 + E_{31}l_{31})}\left(\overline{x} + \overline{b}\right)\overline{w}''' \\
&+\left[\frac{\lambda_0^2\left(1 + E_{31}h_{31}^3\right)}{l_{h1}^2(1 + E_{31}l_{31})} - \lambda_0^2\left\{\frac{f}{l^2} - \left(\overline{x} + \overline{b}\right)^2\right\}\right. \\
&\left.-3g^*\left(1 + \frac{h_{12} + h_{32}}{2}\right)^2 + \overline{P}(\overline{t})\right]\overline{w}'' \\
&+\lambda_0^2\left(\overline{x} + \overline{b}\right)\overline{w}' + \frac{3}{2}g^* l_{h1}h_{12}\left(1 + \frac{h_{12} + h_{32}}{2}\right) \\
&\times(1 + \alpha)\overline{u}_1' = 0, \\
&\overline{u}_1'' - \frac{g^*}{4}h_{12}^2\frac{1 + E_{31}h_{31}^3}{1 + \alpha^2 E_{31}h_{31}}(1 + \alpha) \\
&\times\left[(1 + \alpha)\overline{u}_1 - \left(2\frac{(1 + ((h_{12} + h_{32})/2))}{(l_{h1}h_{12})}\right)\overline{w}'\right] = 0.
\end{aligned} \tag{6}$$

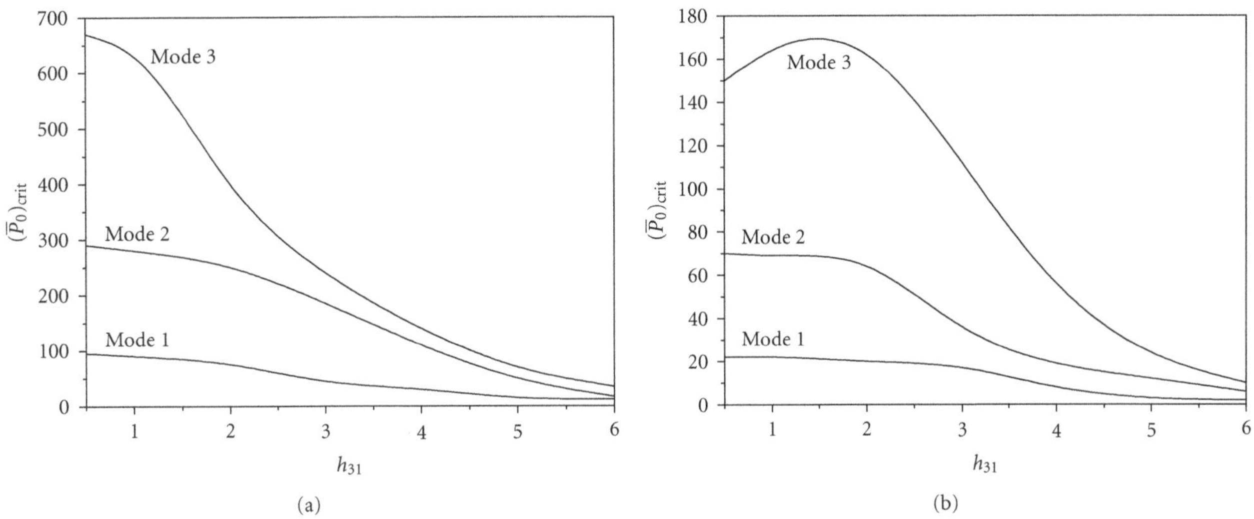

FIGURE 2: (a) Variation static buckling loads with h_{31} (pinned-pinned). (b) Variation static buckling loads with h_{31} (clamped-free).

The associated nondimensional boundary conditions at $\overline{x} = 0$ and $\overline{x} = 1$ are

$$\overline{w}'' = 0 \quad \text{or} \quad \overline{w}' = 0,$$

$$\begin{aligned}
&\frac{\partial}{\partial x}\left[\left(1+\lambda_1{}^2\left\{\frac{f}{l^2}-\left(\overline{x}+\overline{b}\right)^2\right\}\right)\overline{w}''\right] \\
&\quad -\left[\lambda_0^2\left\{\frac{f}{l^2}-\left(\overline{x}+\overline{b}\right)^2\right\}\overline{w}'\right. \\
&\quad\quad \left. +3g^*\left(1+\frac{h_{12}+h_{32}}{2}\right)^2+\overline{P}(\overline{t})\right]\overline{w}' \\
&\quad +\frac{3}{2}g^* l_{h1}h_{12}(1+\alpha)\left(1+\frac{h_{12}+h_{32}}{2}\right)\overline{u}_1 = 0,
\end{aligned} \tag{7}$$

or

$$\begin{aligned}
w &= 0, \\
\overline{u}_1' = 0 \quad &\text{or} \quad \overline{u}_1 = 0
\end{aligned} \tag{8}$$

The various parameters. are defined as

$$\begin{aligned}
E_{31} &= \frac{E_3}{E_1}, \\
h_{31} &= \frac{h_3}{h_1}, \\
l_{h1} &= \frac{l}{h_1}, \\
h_{12} &= \frac{h_1}{h_2}, \\
g^* &= g\ (1+j\eta) \\
&= \frac{G_2^* h_{21} l_{h1}^2}{E_1\left(1+E_{31}h_{31}^3\right)},
\end{aligned} \tag{9}$$

where g being the shear parameter

$$\begin{aligned}
\lambda_0 &= \sqrt{\left(\frac{m\Omega_0^2 l^4}{2(E_1I_1+E_3I_3)}\right)}, \\
\lambda_1 &= \sqrt{\left(\frac{3m\Omega_0^2 l^2}{2(E_1A_1+E_3A_3)}\right)}.
\end{aligned} \tag{10}$$

These are rotation parameters

$$\begin{aligned}
\overline{P}(\overline{t}) &= \overline{P}_0+\overline{P}_1\cos(\overline{\omega}\overline{t}) \\
&= \frac{(P(t)l^2)}{(E_1I_1+E_3I_3)}.
\end{aligned} \tag{11}$$

This is the nondimensional load.

2.3. Approximate Solution. Approximate solution to the nondimensional equations of motion are assumed as

$$\begin{aligned}
\overline{w}(\overline{x},\overline{t}) &= \sum_{i=1}^{i=N} f_i(\overline{t})w_i(\overline{x}), \\
\overline{u}_1(\overline{x},\overline{t}) &= \sum_{k=N+1}^{k=2N} f_k(\overline{t})u_{1k}(\overline{x}),
\end{aligned} \tag{12}$$

where f_r ($r = 1,2,\ldots,2N$) are the generalized coordinates and w_i and u_{1k} are the coordinate functions satisfying as many boundary conditions as possible [24]. For the pinned-pinned case, the shape functions chosen are

$$\begin{aligned}
w_i(\overline{x}) &= \sin(i\pi\overline{x}), \\
u_{1k}(\overline{x}) &= \cos\left(\overline{k}\pi\overline{x}\right), \\
\overline{k} &= k-N.
\end{aligned} \tag{13}$$

For $i = 1,2,\ldots,N$, and $k = N+1, N+2,\ldots,2N$.

For the clamped-free case, the approximating functions are

$$w_i(\bar{x}) = (i+2)(i+3)\bar{x}^{i+1} - 2i(i+3)\bar{x}^{i+2} + i(i+1)\bar{x}^{i+2},$$
$$u_{1k}(\bar{x}) = \bar{x}^{\bar{k}} - \left[\frac{\bar{k}}{\bar{k}+1}\right]\bar{x}^{\bar{k}+1}. \tag{14}$$

Substitution of the series solutions in the nondimensional equations of motion and subsequent application of the general Galerkin method [24] leads to the following matrix equations of motion:

$$[m]\{\ddot{f}_j\} + [k_{11}]\{f_j\} + [k_{12}]\{f_l\} = \{0\}, \tag{15}$$

$$[k_{22}]\{f_l\} + [k_{21}]\{f_j\} = \{0\}, \tag{16}$$

where $j = 1, 2, \ldots, N$ and $l = N+1, \ldots, 2N$. The various matrix elements are given by

$$m_{ij} = \int_0^1 w_i w_j d\bar{x},$$

$$k_{11ij} = \int_0^1 \left[1 + \lambda_1\left\{\frac{f}{l^2} - \left(\bar{x}+\bar{b}\right)^2\right\}\right] w_i'' w_j'' d\bar{x}$$
$$+ \lambda_0^2 \int_0^1 \left\{\frac{f}{l^2} - \left(\bar{x}+\bar{b}\right)^2\right\} w_i' w_j' d\bar{x}$$
$$+ \left\{3g^*\left(1 + \frac{h_{12}+h_{32}}{2}\right)^2 - \overline{P}(\bar{t})\right\}\left(\int_0^1 w_i' w_j' d\bar{x}\right),$$

$$k_{12jl} = -\left(\frac{3}{2}\right) g^* l_{h1} h_{12}(1+\alpha)\left(1 + \frac{h_{12}+h_{32}}{2}\right) \times \left(\int_0^1 u_{1l} w_j' d\bar{x}\right),$$

$$k_{21li} = -\left(\frac{3}{2}\right) g^* l_{h1} h_{12}(1+\alpha)\left(1 + \frac{h_{12}+h_{32}}{2}\right) \times \left(\int_0^1 u_{1l} w_i' d\bar{x}\right),$$

$$k_{22kl} = 3l_{h1}^2 \frac{(1+\alpha^2 E_{31} h_{31})}{\left(1 + E_{31} h_{31}^3\right)}\left(\int_0^1 u_{1k}' u_{1l}' d\bar{x}\right) + \left(\frac{3}{4}\right) g^* l_{h1}^2 h_{12}^2 (1+\alpha)^2 \left(\int_0^1 u_{1k} u_{1l} d\bar{x}\right). \tag{17}$$

From (16)

$$\{f_l\} = -[k_{22}]^{-1}[k_{21}]\{f_j\}, \tag{18}$$

where

$$[k_{21}] = [k_{12}]^T. \tag{19}$$

Substitution of above in (15) and subsequent simplification leads to

$$[m]\{\ddot{f}_j\} + \left[[k] - \overline{P}_0[H]\right]\{f_j\} - \overline{P}_1 \cos(\bar{\omega}\bar{t})[H]\{f_j\} = \{0\}, \tag{20}$$

where

$$[k] = \left[\bar{k}\right] - [k_{12}][k_{22}]^{-1}[k_{12}]^T,$$
$$\bar{k}_{ij} = \int_0^1 \left[1 + \lambda_1\left\{\frac{f}{l^2} - \left(\bar{x}+\bar{b}\right)^2\right\}\right] w_i'' w_j'' d\bar{x}$$
$$+ \lambda_0^2 \int_0^1 \left\{\frac{f}{l^2} - \left(\bar{x}+\bar{b}\right)^2\right\} \times w_i' w_j' d\bar{x} \tag{21}$$
$$+ 3g^*\left(1 + \frac{h_{12}+h_{32}}{2}\right)^2 \int_0^1 w_i' w_j' d\bar{x},$$
$$H_{ij} = \int_0^1 w_i' w_j' d\bar{x}.$$

2.4. Static Buckling Loads. Substitution of $\overline{P}_1 = 0$ and $\{\ddot{f}_j\} = \{0\}$ in (20) leads to the eigen value problem $[k]^{-1}[H]\{f_j\} = 1/P_0\{f_j\}$. The static buckling loads $(\overline{P}_0)_{\text{crit}}$ for the first few modes are obtained as the real parts of the reciprocals of the eigen values of $[k]^{-1}[H]$.

3. Numerical Results and Discussion

Numerical results were obtained for various values of the non-dimensional geometric parameters h_{31}, h_{12}, l_{h1}, $\bar{b}$, the modulus ratio G_2/E_1 and the rotation parameter λ_1. The following values of parameters are used unless otherwise stated: $\eta = 0.05$, $h_{31} = h_{12} = 1.0$, $l_{h1} = 40$, $b = 0.05$, $G_2/E_1 = 0.001$, $g = 0.1$, $\lambda_0 = 0.05$, and $\lambda_1 = 0.1$.

It should be noted that Figures 2(a), 3(a), 4(a), 5(a), 6(a), and 7(a) are for the pinned-pinned case and Figures 2(b), 3(b), 4(b), 5(b), 6(b), and 7(b) are for the clamped-free configuration. Figures from 2(a), 2(b)–7(a), and 7(b) show the variation of the first three nondimensional static buckling loads, $(\overline{P}_0)_{\text{crit}}$ with $h_{31}, h_{12}, l_{h1}, \bar{b}, G_2/E_1$, and λ_1, respectively. The variation of these loads with g and λ_0 are similar to those for G_2/E_1 and l_{h1}, respectively, and are not shown.

With an increase in h_{31} (Figure 2(a)), the buckling loads are seen to decrease monotonically for the pinned-pinned case. For the clamped-free case (Figure 2(b)), these first increase and then decrease. The rate of decrease is maximum for the highest mode for moderate h_{31} values.

As h_{12} (Figures 3(a) and 3(b)) is increased, $(\overline{P}_0)_{\text{crit}}$ for the three modes shows a decreasing trend for low values of h_{12}. It remains fairly constant for moderate values of h_{12} and increases for higher values.

It is observed that $(\overline{P}_0)_{\text{crit}}$ increases nonlinearly with l_{h1} for three modes (Figures 4(a) and 4(b)).

The critical loads are almost constant over the entire range of $\bar{b}$ (Figures 5(a) and 5(b)).

For increasing the G_2/E_1, Figures 6(a) and 6(b), the $(\overline{P}_0)_{\text{crit}}$ values are initially constant and rise appreciably for higher values of G_2/E_1, the rate of increase being higher for the higher modes.

It is seen that $(\overline{P}_0)_{\text{crit}}$ is almost constant over a wide range of λ_1 (Figures 7(a) and 7(b)) and decrease slightly for the values of λ_1 greater than 0.1.

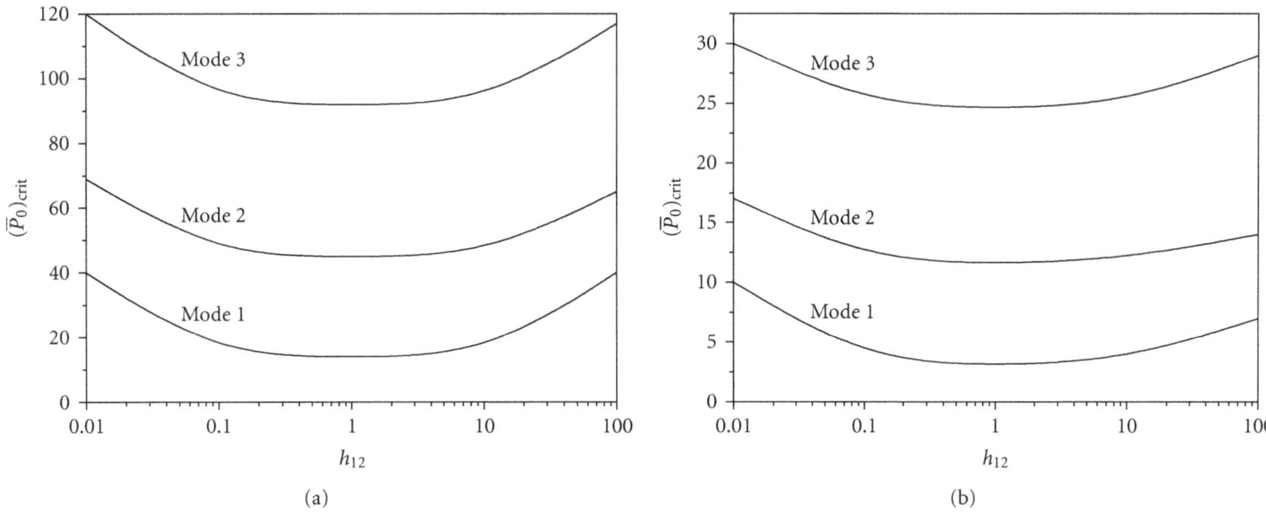

FIGURE 3: (a) Variation static buckling loads with h_{12} (pinned-pinned). (b) Variation static buckling loads with h_{12} (clamped-free).

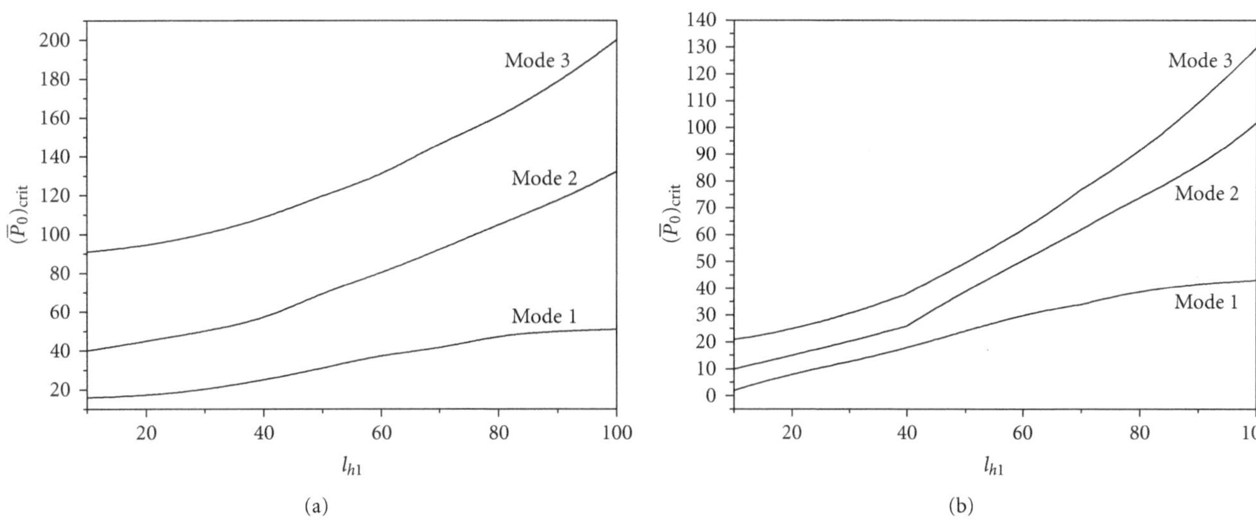

FIGURE 4: (a) Variation static buckling loads with l_{h1} (pinned-pinned). (b) Variation static buckling loads with l_{h1} (clamped-free).

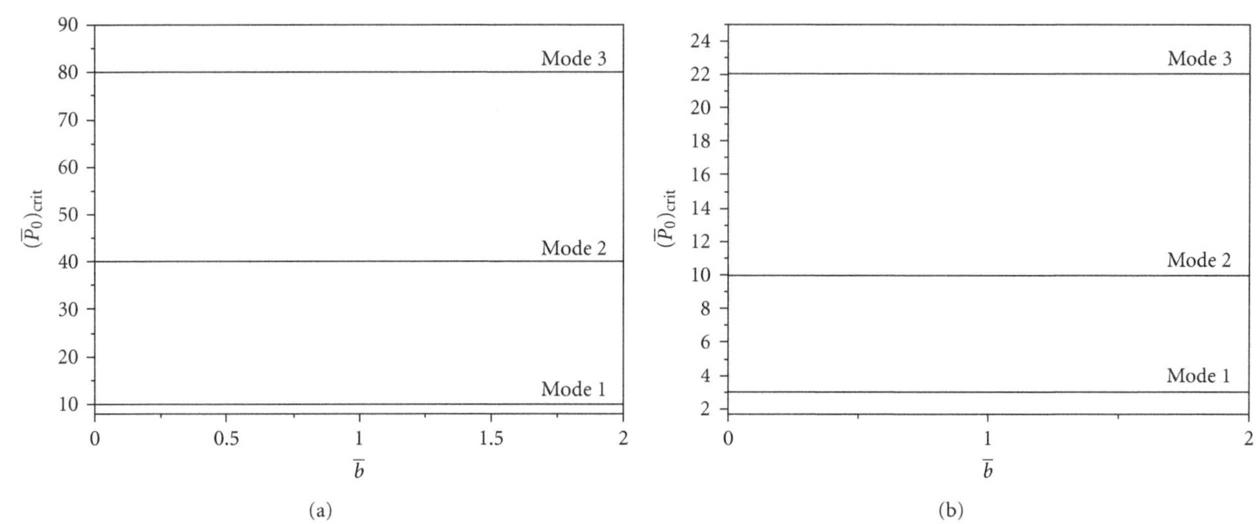

FIGURE 5: (a) Variation static buckling loads with $\overline{b}$ (pinned-pinned). (b) Variation static buckling loads with $\overline{b}$ (clamped-free).

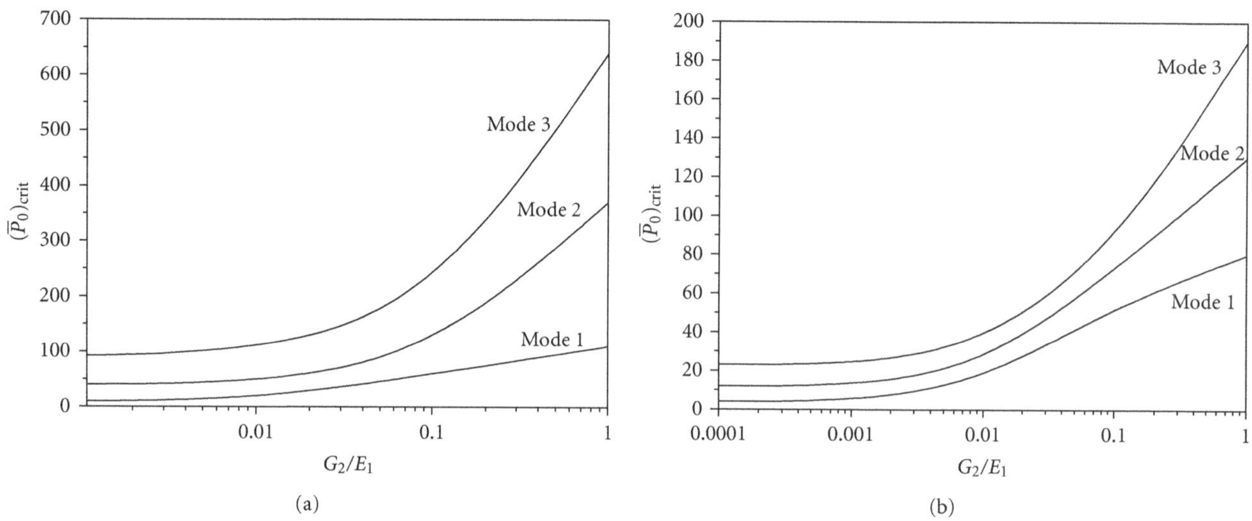

FIGURE 6: (a) Variation static buckling loads with G_2/E_1 (pinned-pinned). (b) Variation static buckling loads with G_2/E_1 (clamped-free).

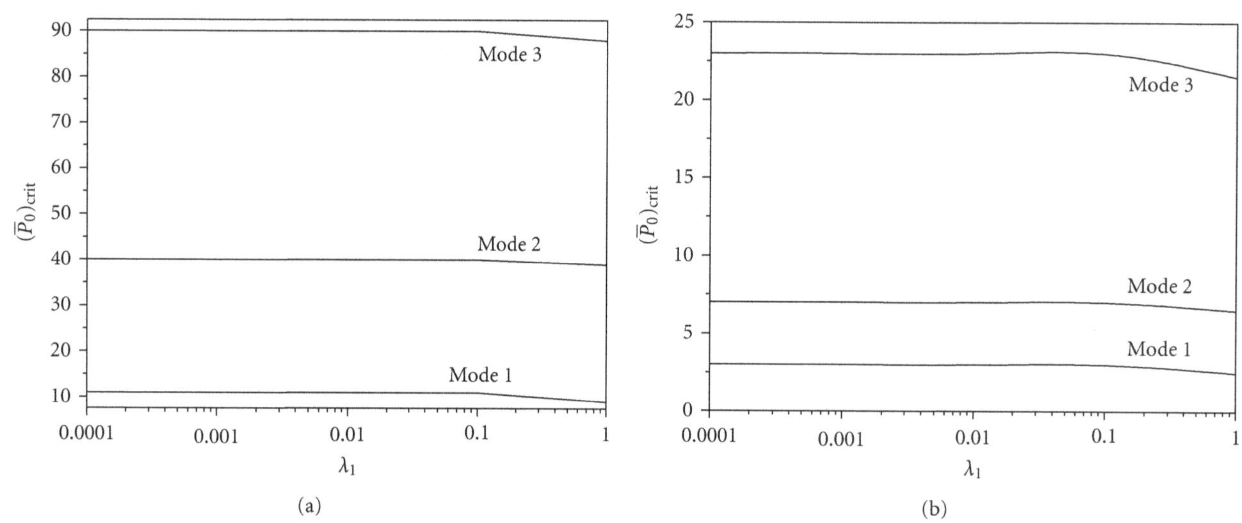

FIGURE 7: (a) Variation static buckling loads with λ_1 (pinned-pinned). (b) Variation static buckling loads with λ_1 (clamped-free).

4. Conclusions

In this paper, the static stability of an asymmetric three-layered rotating sandwich beam with viscoelastic core has been considered. The beam is subjected to an axial pulsating load. Both pinned-pinned and clamped-free configurations have been considered. An increase in h_{31} is seen to have a detrimental effect on the nondimensional static buckling loads. Hence, a symmetric beam is seen to have better resistance against static buckling. Over a wide range of values, h_{12} is seen to have little effect upon the buckling loads. However, for small values, it has a detrimental effect, and for large values, it improves the buckling loads. $l_{h1}, G_2/E_1, \lambda_0$, and g improve static stability, especially for large values of the parameters. The buckling loads are seen to be almost independent of $\overline{b}$ and λ_1.

Nomenclature

$A_i (i = 1, 2, 3)$: Areas of cross section of a three layered beam, $i = 1, 2$, and 3 for top, middle and bottom layer, respectively,

B: Width of beam

b: Distance of nearer end of the beam from the axis of rotation

$\overline{b}$: b/l

c: $h_1 + 2h_2 + h_3$

$E_i (i = 1, 2, 3)$: Young's modulli, $i = 1, 2, 3$ for top, middle and bottom layer, respectively,

$\ddot{f}_j$: $\partial^2 f_j / \partial \overline{t}^2$

G_2: In-phase shear modulus of the viscoelastic core

G_2^*:	$G_2(1 + j\eta)$, complex shear modulus of core
g^*:	$g(1 + j\eta)$, complex shear parameter
g:	Shear parameter
$2h_i (i = 1, 2, 3)$:	Thickness of the ith layer, $i = 1, 2$ and 3 for top, middle and bottom layer, respectively
h_{12}:	h_1/h_2
h_{31}:	h_3/h_1
$I_i (i = 1, 2, 3)$:	Second moments of area of cross section about a relevant axis, $i = 1, 2$ and 3 for top, middle, and bottom layer
j:	$\sqrt{-1}$
l:	Beam length
l_{h1}:	l/h_1
m:	Mass/unit length of beam
$\overline{P}_1$:	Nondimensional amplitude for the dynamic loading
$(\overline{P}_0)_{\text{crit}}$:	Nondimensional static critical buckling load
t:	Time
$\overline{t}$:	Nondimensional time
$u_1(x, t), U_1(x, t)$:	Axial displacement at the middle of the top layer of beam
$u_3(x, t), U_3(x, t)$:	Axial displacement at the middle of the bottom layer of beam
$w(x, t)$:	Transverse deflection of beam
w':	$\partial w/\partial x$
w'':	$\partial^2 w/\partial x^2$
$\overline{w}$:	w/l
$\ddot{\overline{w}}$:	$\partial^2 \overline{w}/\partial \overline{t}^2$
$\overline{w}''$:	$\partial^2 \overline{w}/\partial \overline{x}^2$
t_0:	$\sqrt{ml^4/(E_1 I_1 + E_3 I_3)}$
u_1':	$\partial u_1/\partial x$
$\overline{u_1}''$:	$\partial^2 \overline{u}_1/\partial \overline{x}^2$
f:	$(l + b)^2$, for beam free at $x = l$, $= (l^3/3) + b^2 + bl$ for all other cases
η:	Core-loss factor
λ_0:	Rotation parameter
λ_1:	Rotation parameter
(Φ):	A null matrix
$\overline{\omega}$:	ωt_0
ω:	Frequency of forcing function
$\overline{\omega}$:	Nondimensional forcing frequency
Ω_0:	Speeds of rotation
$\overline{\Omega}_0$:	Nondimensional rotation parameters
α:	$E_1 A_1/E_3 A_3$.

References

[1] R. B. Bhat, "Transverse vibrations of a rotating uniform cantilever beam with tip mass as predicted by using beam characteristic orthogonal polynomials in the Rayleigh-Ritz method," *Journal of Sound and Vibration*, vol. 105, no. 2, pp. 199–210, 1986.

[2] W. H. Liu and F. H. Yeh, "Vibrations of non-uniform rotating beams," *Journal of Sound and Vibration*, vol. 119, no. 2, pp. 379–384, 1987.

[3] A. Unger and M. A. Brull, "Parametric instability of a rotating shaft due to pulsating torque," *Journal of Applied Mechanics, Transactions ASME*, vol. 48, no. 4, pp. 948–958, 1981.

[4] D. C. Kammer and A. L. Schlack, "Dynamic response of a radial beam with nonconstant angular velocity," *Journal of Vibration, Acoustics, Stress, and Reliability in Design*, vol. 109, no. 2, pp. 138–143, 1987.

[5] N. Sri Namachchivaya, "Mean square stability of a rotating shaft under combined harmonic and stochastic excitations," *Journal of Sound and Vibration*, vol. 133, no. 2, pp. 323–336, 1989.

[6] H. F. Bauer and W. Eidel, "Vibration of a rotating uniform beam, Part-II: orientation perpendicular to the axis of rotation," *Journal of Sound and Vibration*, vol. 122, no. 2, pp. 357–375, 1988.

[7] B. A. H. Abbas, "Vibrations of Timoshenko beams with elastically restrained ends," *Journal of Sound and Vibration*, vol. 97, no. 4, pp. 541–548, 1984.

[8] B. A. H. Abbas and J. Thomas, "Dynamic stability of Timoshenko beams resting on an elastic foundation," *Journal of Sound and Vibration*, vol. 108, pp. 25–32, 1986.

[9] O. A. Bauchau and C. H. Hong, "Nonlinear response and stability analysis of beams using finite elements in time," *AIAA Journal*, vol. 26, no. 9, pp. 1135–1142, 1988.

[10] R. C. Kar and T. Sujata, "Dynamic stability of a rotating beam with various boundary conditions," *Computers and Structures*, vol. 40, no. 3, pp. 753–773, 1991.

[11] R. C. Kar and T. Sujata, "Dynamic stability of a rotating, pretwisted and preconed cantilever beam including coriolis effects," *Computers and Structures*, vol. 42, no. 5, pp. 741–750, 1992.

[12] T. H. Tan, H. P. Lee, and G. S. B. Leng, "Parametric instability of spinning pretwisted beams subjected to sinusoidal compressive axial loads," *Computers and Structures*, vol. 66, no. 6, pp. 745–764, 1998.

[13] B. Pratiher and S. K. Dwivedy, "Parametric instability of a cantilever beam with magnetic field and periodic axial load," *Journal of Sound and Vibration*, vol. 305, no. 4-5, pp. 904–917, 2007.

[14] S. K. Dwivedy, K. C. Sahu, and S. Babu, "Parametric instability regions of three-layered soft-cored sandwich beam using higher-order theory," *Journal of Sound and Vibration*, vol. 304, no. 1-2, pp. 326–344, 2007.

[15] D. Y. Zheng, F. T. K. Au, and Y. K. Cheung, "Vibration of vehicle on compressed rail on viscoelastic foundation," *Journal of Engineering Mechanics*, vol. 126, no. 11, pp. 1141–1147, 2000.

[16] A. V. Metrikine and S. N. Verichev, "Instability of vibrations of a moving two-mass oscillator on a flexibly supported Timoshenko beam," *Archive of Applied Mechanics*, vol. 71, no. 9, pp. 613–624, 2001.

[17] A. V. Metrikine and S. N. Verichev, "Instability of vibrations of a mass that moves uniformly along a beam on a periodically inhomogeneous foundation," *Journal of Sound and Vibration*, vol. 260, no. 5, pp. 901–925, 2003.

[18] R. Ghosh, S. Dharmavaram, K. Ray, and P. Dash, "Dynamic stability of a viscoelastically supported sandwich beam," *Structural Engineering and Mechanics*, vol. 19, no. 5, pp. 503–517, 2005.

[19] C. Y. Lin and L. W. Chen, "Dynamic stability of rotating composite beams with a viscoelastic core," *Composite Structures*, vol. 58, no. 2, pp. 185–194, 2002.

[20] C. Y. Lin and L. W. Chen, "Dynamic stability of a rotating beam with a constrained damping layer," *Journal of Sound and Vibration*, vol. 267, no. 2, pp. 209–225, 2003.

[21] P. R. Dash, B. B. Maharathy, R. Mallick, B. B. Pani, and K. Ray, "Parametric instability of an asymmetric rotating sandwich beam," *Journal of Aerospace Sciences and Technologies*, vol. 60, no. 4, pp. 292–309, 2008.

[22] H. Saito and K. Otomi, "Parametric response of viscoelastically supported beams," *Journal of Sound and Vibration*, vol. 63, no. 2, pp. 169–178, 1979.

[23] E. M. Kerwin Jr., "Damping of flexural waves by a constrained viscoelastic layer," *The Journal of the Acoustical society of America*, vol. 31, pp. 952–962, 1959.

[24] H. Leipholz, *Stability Theory*, John Wiley & Sons, Chichester, UK, 2nd edition, 1987.

4

A 3 DOF Model for an Electromagnetic Air Mount

HyungTae Kim,[1] CheolHo Kim,[1] SungBok Kang,[1] KangWon Lee,[1] JaeHo Baek,[2] and HyunHee Han[1]

[1] *Manufacturing System Division, KITECH, 35-3 Hongcheon, Ipjang, Chungnam, Cheonan, Republic of Korea*
[2] *RMS Technology, Techpia Building, 774 Dujeonng, Seobuk,Chungnam, Cheonan, Republic of Korea*

Correspondence should be addressed to HyungTae Kim, htkim@kitech.re.kr

Academic Editor: Luc Gaudiller

A 4 × 4 matrix model with three degrees of freedom is proposed as a means for controlling microvibrations and applied to an electromagnetic isolator. The model was derived from an assumption based on small- and low-frequency vibrations. The coordinates of the 3 DOF was composed of the 4 variables, representing a vertical position, pitch, roll, and a proof term. The coordinates were calculated from the 4 position sensors in the isolator and formulated into a 4 × 4 matrix, which possesses inversive full rank. The electro-magnetic isolator was built for a simulated machine in semiconductor manufacturing and consisted of a heavy surface plate, sensors, amps, a controller, and air springs with electromagnets. The electromagnets are installed in a pneumatic chamber of the individual air spring. The performance of the 3 DOF model was experimented and compared with that of a 1 DOF model in an impact test. The settling time in the result was reduced to 25%.

1. Introduction

Vibration is one of the important factors in semiconductor manufacturing, affecting the processes related to laser optics. Vibration is unpredictable, irregular, and generated by various sources. Studies have been conducted to isolate vibrational effects from the environment. Air spring is one of the solutions and has been used widely in industrial field. Sensitive machines, such as lithography, are placed on air springs in order to isolate vibration. An air spring accumulates air pressure in a chamber and floats a heavy load on a surface plate. This kind of isolator is usually called an air mount. An air mount supports tons of heavy weight under a manufacturing machine. The number of the air springs is determined by the weight which the air mount will support. The weight can be several tons including a surface plate and a manufacturing machine. The physical combination of the heavy load and the springs usually causes resonance at low-frequency vibrations. This causes long oscillation due to vibrational disturbances, producing problems in the manufacturing process. When a high-speed machine immediately stops at an instance, several seconds are required to regulate the vibration in the resonance frequency.

Active controls, such as pneumatic, electromagnetic (EM), and magnetorheological (MR) methods, are coupled with the air springs in an air mount. A pneumatic control is a commonly employed method for active control. Air pressure in the air chambers is controlled by adjusting air valves [1–3]. Although other active controls are applied in the air mount, the pneumatic control is essential because EM and MR cannot generate power as high as pneumatics. The pneumatics is in charge of large vibrations, the MR manages small vibrations, and EM controls micro-scale vibration. MR fluid can be used to reduce the oscillation in an air mount [4]. When vibration occurs in the mount, a damping force will increase and reduce oscillation due to the viscosity of the MR fluid. Piezoelectric (PZT) actuators can be applied to the air mount [5]. The PZT actuator is coupled with an air spring through a serial connection. EM actuators are another method to reduce the resonance [6]. The vibration in an air spring is varied by the EM force.

An air mount usually has multiple actuators, vibration occurs in a six DOF space, and a multi-DOF model is reported in many cases [7–9]. A common method isolating an air mount is controlling the air flow. Time delay control [10], predictive control [11], and conventional PID

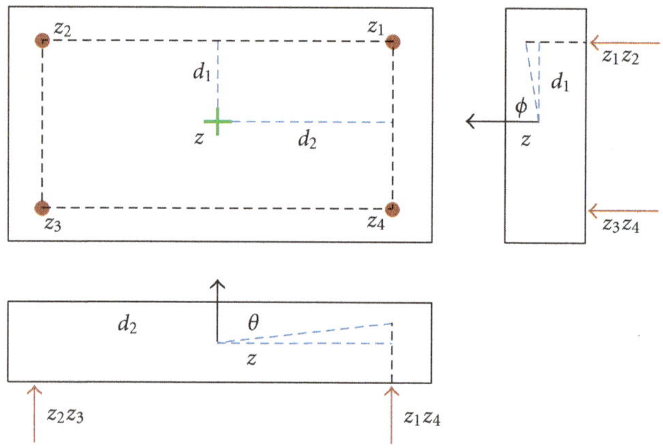

FIGURE 1: Variables in an air mount.

control [12] adjust the pneumatic status in order to isolate vibration.

Although EM may be applied as a means for controlling vibrations [13–16], the load capacity is not enough to sustain tons of mass; thus, it cannot be used in the manufacturing system. Microvibration can be managed by combining EM and pneumatic forces. Most of this air mount regulates vibration energy by vertical motion, so a rigid surface plate shows vertical, pitch, and roll motions. Planar and rotational motion is negligible for cases in which moving mass is excluded. A 3 DOF model of the air mount with these vertical, pitch, and roll motions; these can be considered under the assumption that microvibration occurs. However, a 3 DOF model in an air mount has 3 coordinate variables and four isolating devices. Thus, difficulty arises when finding the inverse matrix for the control algorithm. This study proposes a 3 DOF model for microvibration in typical isolators using 4 isolating devices. The model includes a control algorithm and contains an inversible matrix in the coordinate conversion. The proposed model was applied to an air mount which provided 4 EM air springs. The experiment was carried out for a single DOF and 3 DOF model.

2. Full-Rank 3 DOF Model

An air mount supported four air springs under a surface plate. The surface plate was assumed to be a rigid body exhibiting no deformation. Vertical, pitch, and roll motions of a surface plate can be derived from the motion of the four air springs. Noncontact sensors measured the displacements of individual air springs. The surface plate was heavy and large, making the vertical displacements of the air springs relatively small. Planar and rotational motion was not considered excluding moving mass. Motion of the air mount can be described by the variables in Figure 1. Vertical displacement of individual actuators is z_n and that of center in the surface plate is z. The displacement of z can be calculated by averaging the vertical displacements of all of the actuators, as shown in

$$z = \frac{z_1 + z_2 + z_3 + z_4}{4}. \quad (1)$$

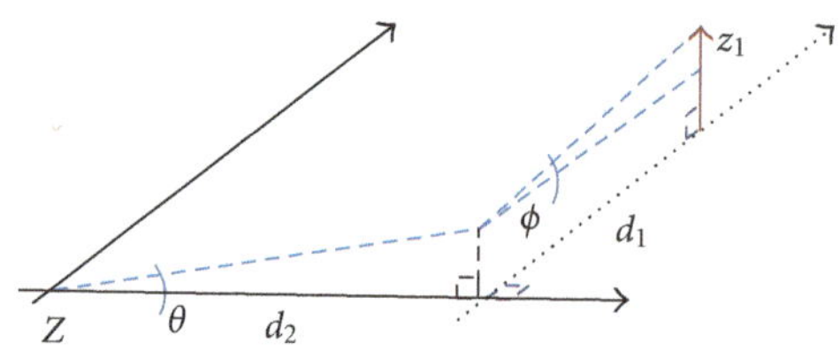

FIGURE 2: Coordinate conversion in an air mount.

The planar distance between the center and the actuators is designated as d_1 and d_2. The roll and pitch angles are ϕ and θ, respectively. ϕ and θ can be derived from the z_n, as shown in Figure 2. Equation (2) is the relative distance between z and z_1:

$$z_1 - z = d_1 \sin\phi + d_2 \sin\theta. \quad (2)$$

The motions of other points can be written similarly, as given by the following equations:

$$\begin{aligned} z_2 - z &= d_1 \sin\phi - d_2 \sin\theta, \\ z_3 - z &= -d_1 \sin\phi - d_2 \sin\theta, \\ z_4 - z &= -d_1 \sin\phi + d_2 \sin\theta. \end{aligned} \quad (3)$$

The equation can be arranged for ϕ and θ without z:

$$\begin{aligned} \sin\phi &= \frac{1}{4d_1}(z_1 + z_2 - z_3 - z_4), \\ \sin\theta &= \frac{1}{4d_2}(z_1 - z_2 - z_3 + z_4), \end{aligned} \quad (4)$$

ϕ and θ can be assumed as infinitesimal because d_1 and d_2 are relatively larger than z_n. The d_1 and d_2 are 1000–3000 mm for the manufacturing machines, and z_n is under hundreds of micron in common state and 3 mm in exceptional state. The angle due to the motion is approximately under 0.2 degree. If infinitesimal motion is assumed, $\sin\phi$ and $\sin\theta$ are approximated as ϕ and θ:

$$(\sin\phi, \sin\theta) \approx (\phi, \theta). \quad (5)$$

An additional coordinate, the proof term, ε, can be defined as follows:

$$\varepsilon = \frac{1}{4}(z_1 - z_2 + z_3 - z_4). \quad (6)$$

The ε is the difference of diagonal pairs in the mount and should theoretically be zero. But the value of ε can be a small constant because of sensor noise and geometrical error. Therefore, ε can be used as an index for modelling errors. From (2) through (6), a full-rank matrix form can be derived as follows:

$$\begin{pmatrix} z \\ \phi \\ \theta \\ \varepsilon \end{pmatrix} = \frac{1}{4} \begin{pmatrix} 1 & 1 & 1 & 1 \\ \frac{1}{d_1} & \frac{1}{d_1} & \frac{-1}{d_1} & \frac{-1}{d_1} \\ \frac{1}{d_2} & \frac{-1}{d_2} & \frac{-1}{d_2} & \frac{1}{d_2} \\ 1 & -1 & 1 & -1 \end{pmatrix} \begin{pmatrix} z_1 \\ z_2 \\ z_3 \\ z_4 \end{pmatrix}. \quad (7)$$

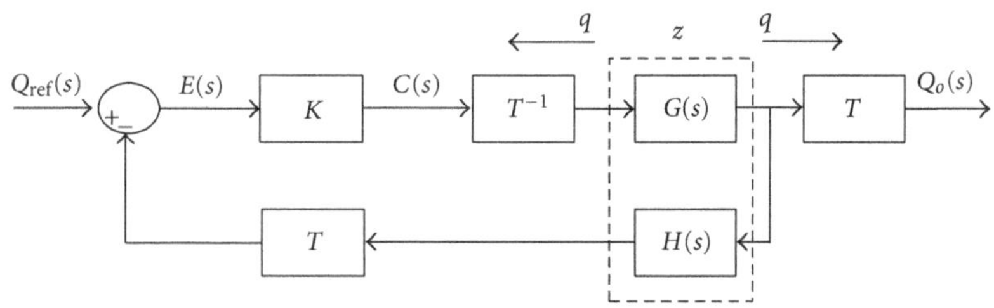

$$C(s) = KE(s) = (K_p + K_d s)E(s)$$

: PD control

$$\frac{E(s)}{Q_{\text{ref}}(s)} = \frac{1}{1 + KT^{-1}G(s)H(s)T}$$

Figure 3: Control loop of the 3 DOF model for an air mount.

This matrix converts the actuator coordinate to that of the surface plate. d_1 and d_2 are positive, and the matrix is always inversible. The matrix can be symbolized as T for coordinate conversion:

$$Q = TZ. \tag{8}$$

T^{-1} is calculated as follows:

$$T^{-1} = \frac{1}{4}\begin{pmatrix} 1 & d_1 & d_2 & 1 \\ 1 & d_1 & -d_2 & -1 \\ 1 & -d_1 & -d_2 & 1 \\ 1 & -d_1 & d_2 & -1 \end{pmatrix}. \tag{9}$$

Differentiation of the coordinate can be obtained from

$$\dot{z}_1 - \dot{z} = d_1\dot{\phi}\cos\phi + d_2\dot{\theta}\cos\theta. \tag{10}$$

Differentiation is applied to (3)–(6) under the condition that $\cos\phi \approx 1$, $\cos\theta \approx 1$ for infinitesimal motion. Then, $\dot{T}$ can be defined as follows:

$$\dot{Q} = \dot{T}\dot{Z}, \quad \dot{T} = T. \tag{11}$$

Double differentiation can be formulated with

$$\ddot{z}_1 - \ddot{z} = d_1\ddot{\phi}\cos\phi - \dot{\phi}^2\sin\phi + d_2\ddot{\theta}\cos\theta - \dot{\theta}^2\sin\theta. \tag{12}$$

The following equation can be derived from infinitesimal motion, $\ddot{\phi} \gg \dot{\phi}^2\phi$, and $\ddot{\theta} \gg \dot{\theta}^2\theta$. Thus, the following relation can be obtained:

$$\ddot{Q} = \ddot{T}\ddot{Z}, \quad \ddot{T} = T. \tag{13}$$

3. Control Method

An EM air spring is composed of EM parts, the pneumatics and attach assemblies. The pneumatics, which possesses damping and elastic characteristics, supports the heavy load. The characteristics are linear and in the microvibration range. The EM component generates the control force F, which reduces vibration. A disturbance force is exerted from the environment:

$$m\ddot{z}_n + c\dot{z}_n + kz_n = F + D. \tag{14}$$

The EM force is governed by the electric current and the size of the gap between a coil and a magnet. If the gap is large enough to be placed in the linear range of the current under an assumption of microvibration, then the force is function of the current, $F(I)$. The electric current is formed by voltage level V which is determined from control input and sensor outputs in a controller:

$$L\dot{I} + RI = V. \tag{15}$$

The V is determined by the error between the reference level and the measured level:

$$V = V(e). \tag{16}$$

The P control is applied to the pneumatics using the feedback from mechanical levelers. A PD feedback control method is applied to the EM components with a noncontact displacement sensor. The 1 DOF model is defined by individual actuators that control the reference positions. The motion of the surface plate is defined as $G(s)$, feedback is $H(s)$, error is $E(s)$, and the controller is K. When vibration occurs, a sensor in each actuator measures the vibration from a reference level and a controller determines the output of the actuator. The loop is formed for each sensor and actuator. The control loop is easy and simple to build and effective for uniaxial vertical vibration. However, the 1 DOF model is weak at accounting for multiaxial vibration because each actuator controls its own vibration without assisting other actuators.

The 3 DOF model attempts to control the vibration within a view of a surface plate. The control output of each actuator is determined in consideration of other sensors, so it can be better at managing multiaxial vibration. The 3 DOF model's control loop can be constructed with the coordinate conversion. The air mount system has an actuator coordinate, so the conversion matrix to 3 DOF motion is necessary to form the feedback loop, as shown in Figure 3. The transfer function of the system can be summarized as follows:

$$\frac{E(s)}{Q_{\text{ref}}(s)} = \frac{1}{1 + KT^{-1}G(s)H(s)T}. \tag{17}$$

Figure 4: An air mount with EM actuators.

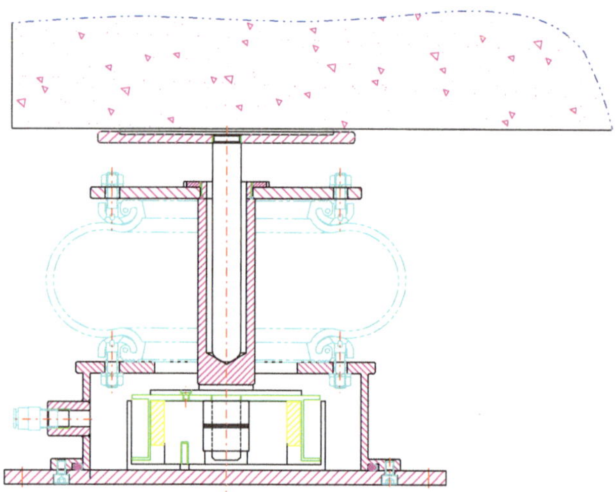

Figure 5: Cross section of an EM air spring.

1 DOF model forms a PD feedback loop and controls the vibration in individual measuring point. An EM spring is activated using its corresponding sensor, but the action is not coupled with other sensors. So, the 1 DOF model is actually SISO. However, the 3 DOF model is coupled with other sensors and generates multiple DOF motion, so the 3 DOF model is MIMO.

4. Experiment

Figure 4 shows the air mount that was constructed for the experiment. The weight of the surface plate was 3.0 ton, and its size was 2000 mm × 2000 mm. Sensors and actuators were installed on each corner of the surface plate. Displacement due to vibration in common operating conditions of air mount is within hundreds of microns and ± 3 mm in exceptional situation. The range of infinitesimal vibration was within $\pm 50\,\mu$m. The internal structure of an EM air spring is shown in Figure 5. A magnet and a coil were installed in the air chamber. The magnet was fixed on the base, and a coil was attached under an upper plate. The upper plate was floated by compressed air, allowing relative motion between the magnet and the coil.

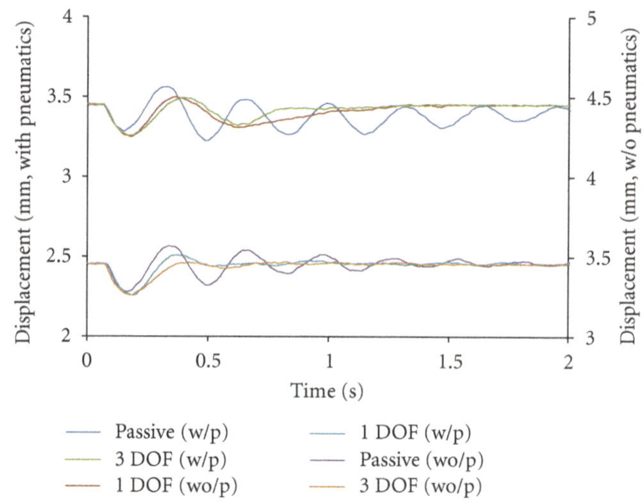

Figure 6: Motion of z-axis in an air mount.

The impact point was a corner in the air mount, giving a 3 DOF vibration. A PC-based embedded system (NI PXI-8108) was used as a controller. The vibration was measured by gap sensors (Santec LS-500) with the accuracy of 3 um. An A/D board (NI PXI-6250) received the analog output of sensors. The control program was built by C++ and control period was 10 ms. The controller output was transferred to a 4-channel linear AMP via a D/A board (NI PXI-6733). The maximum power of the linear AMP was 250 W for each channel. The sensor output was also measured with an oscilloscope (Tek 2024B).

The test was conducted using the pneumatic control and not using the pneumatic control. EM force was not sufficiently large enough to recover the heavy weight, and air leakage was unavoidable in the air mount—thus, actual situations require a pneumatic control. However, if operating time is short, then the leak is negligible. The effect of the EM control can be easily monitored in the case in which a pneumatic control is not used. The PD gain for the 1 DOF model was 1.9 and 0.7. The PD gain for the 3 DOF model was 1.6/0.2, 1.3/0.2, 1.3/0.2, and 15.0/3.0 for $(z, \phi, \theta, \varepsilon)$. An impact was acted on a corner of the surface plate to generate 3 DOF vibration and check the control performance.

The ε was monitored with a passive control using an impact applied at the center of the surface plate. Four triaxial accelerometers (PCB 3713D1FD3G) were used to measure acceleration. A high-speed A/D converter measured the signal of acceleration; a TMS-C6713-based DSP was used (DSignT C6713).

5. Results

The control response of the 3 DOF model is shown in Figures 6–8. The horizontal axis represents time, and the vertical axis is displacement in Figure 6, roll in Figure 7, and pitch in Figure 8. The response of 6 experimental conditions

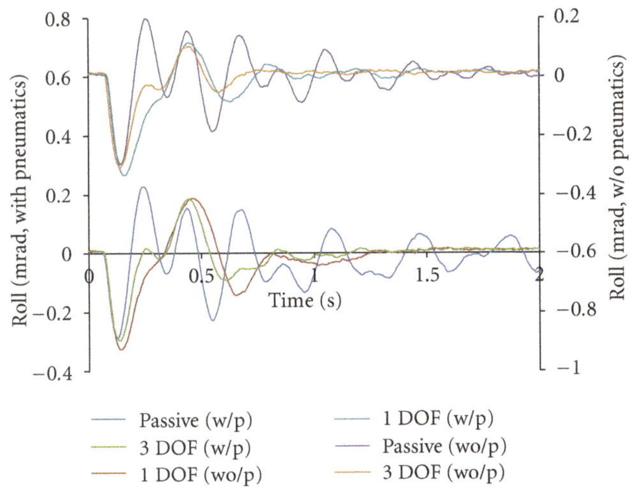

FIGURE 7: Motion of ϕ-axis in an air mount.

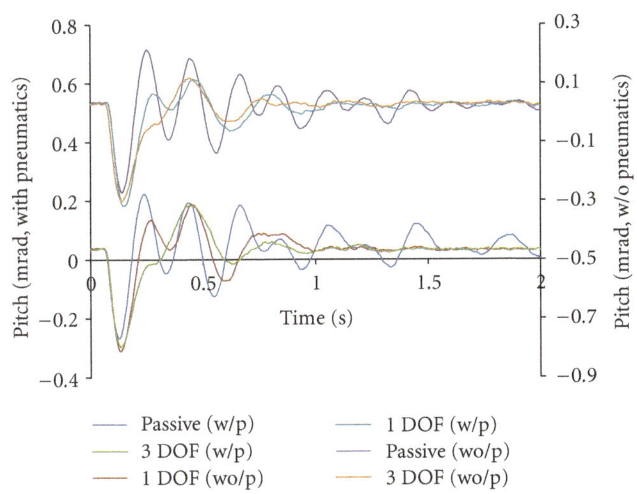

FIGURE 8: Motion of θ-axis in an air mount.

is plotted in each figure. The conditions are passive control, 1 DOF EM control, and 3 DOF EM control with and without pneumatic control. Passive control signifies the state in which the pneumatic control was employed without an EM control. The settling time without the pneumatic control is faster than that with a pneumatic control. The pneumatic control allows air to leak in the mount and has a longer settling time to compensate the leakage. However, small amounts of air escape from the air mount, thus the mount can be used in short periods of time without a pneumatic control. Passive control takes the most amount of time to reach steady state. Roll motion is similar to pitch motion because the top shape of the surface plate is square. The settling time of 1 and 3 DOF with a pneumatic control was 0.6 sec. The settling time of 1 and 3 DOF without the pneumatic control was 1.2 and 0.9 sec, respectively. The response of the 3 DOF model was 25% faster than that of the 1 DOF model. Oscillation in the cases with an EM control is regulated faster than in those without an EM control. Exponential decay is shown in the plot due to slower response of a pneumatic control. An

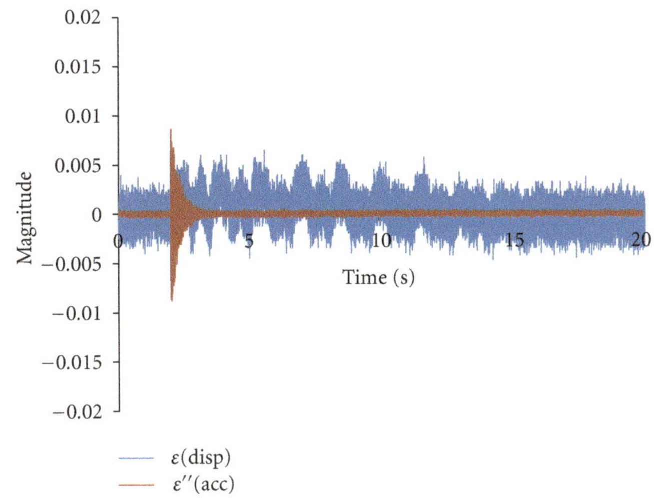

FIGURE 9: Acceleration response of ε.

EM control was effective in the oscillation, and a pneumatic control is useful to reduce steady-state errors.

The response of ε is shown in Figure 9. The horizontal axis represents time, and the vertical axis is the magnitude of displacement and acceleration. Both of the responses fluctuated after impact and converged to zero in steady state. The level of the responses was small in value and was approximated in the noise level. A transient acceleration response occurred at the moment of impact. The transient response was 428 Hz and was considered to be a solid vibration of the surface plate due to the impact. Therefore, the transient response was not related to the motion of the surface plate and the proposed model is correct. The magnitude of ε was in noise level in the experiments.

6. Conclusion

This study proposed a 3 DOF control method for an air mount that reduces microvibration and resonance in semiconductor manufacturing. The air mount was composed of a heavy weight and active air springs. The heavy weight was supported by 4 active air springs. The active air spring consisted of pneumatic and EM components. The EM components generated an EM force that reduced oscillation. The 3 DOF model was derived from the geometric conditions under an assumption of microvibration. The logic contained a 4 by 4 matrix which was full rank and inversible. The control signal of the EM components was determined by the 3 DOF model from a gap sensor. The 3 DOF model was experimentally applied to an air mount. The proposed model effectively controlled microvibration.

References

[1] D. K. Han and P. H. Chang, "A robust two-time-scale control design for a pneumatic vibration isolator," in *the 46th IEEE Conference on Decision and Control (CDC '07)*, pp. 1666–1672, December 2007.

[2] T. Kato, K. Kawashima, T. Funaki, H. L. Gunter, and T. Kagawa, "Active control of a pneumatic vibration isolation table using a newly developed precise and high response pressure regulator," in *the 16th IEEE International Conference on Control Applications (CCA '07)*, pp. 497–502, October 2007.

[3] T. Mizuno, M. Murashita, M. Takasaki, and Y. Ishino, "Pneumatic three-axis vibration isolation system using negative stiffness," in *the 44th IEEE Conference on Decision and Control, and the European Control Conference (CDC-ECC '05)*, pp. 8254–8259, December 2005.

[4] M. S. Seong, S. B. Choi, and C. H Kim, "Modeling and analysis of MR damper for integrated isolation mount," in *Proceeding of the KSNVE Spring Conference*, pp. 190–191, 2009.

[5] N. B. Roozen, P. P. H. Philips, D. Biloen, P. Limpens, and H. H. Tuithof, "Active vibration isolation applied to a magnetic resonance imaging (MRI) system," in *International Congress on Sound and Vibration*, 2005.

[6] H. T. Kim, K. W. Lee, C. H. Kim, G. S. Lee, and S. W. Son, "An electro-magneto-pneumatic spring for vibration control in semiconductor manufacturing," in *IEEE International Conference on Mechatronics (ICM '09)*, April 2009.

[7] M. Emdadul Hoque, T. Mizuno, Y. Ishino, and M. Takasaki, "A six-axis hybrid vibration isolation system using active zero-power control supported by passive weight support mechanism," *Journal of Sound and Vibration*, vol. 329, no. 17, pp. 3417–3430, 2010.

[8] F. Kerbera, S. Hurlebausb, B. M. Beadlec, and U. Stobener, "Control concepts for an active vibration isolation system," *Mechanical Systems and Signal Processing*, vol. 21, pp. 3042–3059, 2007.

[9] B. M. Beadle, S. Hurlebaus, L. Gaul, and U. Stobener, "Active control stratgies for vibration isolation," in *IUTAM Symposium on Vibration Control of Nonlinear Mechanisms and Structures*, pp. 91–100, 2005.

[10] P. H. Cang, D. K. Han, Y. H. Shin, and K. J. Kim, "Effective suppression of pneumatic vibration isolators byusing input-output linearization and time delay control," *Journal of Sound and Vibration*, vol. 329, pp. 1636–1652, 2010.

[11] H. Z. Fei, G. T. Zheng, and Z. G. Liu, "An investigation into active vibration isolation based on predictive control. Part I: energy source control," *Journal of Sound and Vibration*, vol. 296, no. 1-2, pp. 195–208, 2006.

[12] T. Kato, K. Kawashima, T. Funaki, K. Tadano, and T. Kagawa, "A new, high precision, quick response pressure regulator for active control of pneumatic vibration isolation tables," *Precision Engineering*, vol. 34, no. 1, pp. 43–48, 2010.

[13] D. I. Jones and R. G. Owen, "A magnetically levitated anti-vibration mount," *IEEE Transactions on Magnetics*, vol. 20, no. 5, pp. 1687–1689, 1984.

[14] M. E. Hoque, M. Takasaki, Y. Ishino, and T. Mizuno, "Development of a three-axis active vibration isolator using zero-power control," *IEEE/ASME Transactions on Mechatronics*, vol. 11, no. 4, pp. 462–469, 2006.

[15] T. Mizuno, M. Takasaki, D. Kishita, and K. Hirakawa, "Vibration isolation system combining zero-power magnetic suspension with springs," *Control Engineering Practice*, vol. 15, no. 2, pp. 187–196, 2007.

[16] M. Unsal, C. Niezrecki, and C. Crane, "Two semi-active approaches for vibration isolation: Piezoelectric friction damper and magnetorheological damper," in *Proceedings of IEEE International Conference on Mechatronics (ICM '04)*, pp. 60–65, June 2004.

5

Classes of Strongly Stabilizing Bandpass Controllers for Flexible Structures

Alberto Cavallo, Giuseppe De Maria, Ciro Natale, and Salvatore Pirozzi

Dipartimento di Ingegneria dell'Informazione, Seconda Università degli Studi di Napoli, Via Roma 29, 81031 Aversa, Italy

Correspondence should be addressed to Alberto Cavallo, alberto.cavallo@unina2.it

Academic Editor: Emil Manoach

This paper proposes different design strategies of robust controllers for high-order plants. The design is tailored on the structure of the equations resulting from modeling flexible structures by using modal coordinates. Moreover, the control laws have some characteristics which make them specially suited for active vibration reduction, such as strong stabilization property and bandpass frequency shape. The approach is also targeted the case of more sensors than actuators, which is very frequent in practical applications. Indeed, actuators are often rather heavy and bulky, while small and light sensors may be placed more freely. In such cases, sensors can be usefully placed in the locations where the primary force fields act on the structure, so as to provide the controller with a direct information on the disturbance effects in terms of structural vibrations. Eventually, this approach may lead to uncolocated control strategies. The design problem is here solved by resorting to a Linear Matrix Inequality technique, which allows also to select the performance weights based on different design requirements, for example, a suitable bandpass frequency shape. Experimental results are presented for a vibration reduction problem of a stiffened aeronautical panel controlled by piezoelectric actuators.

1. Introduction

Reduction of vibrations in flexible systems has been considered for years a challenging problem, due to some specific aspects. First, as the model of the structure is described by partial differential equations that can be treated as an infinite set of ordinary differential equations, any finite-order controller must rely on a limited (truncated) description of the plant. Thus, neglected dynamics actions must be taken into account by robustness properties of the controller, otherwise, spillover effects [1] may lead to control problems, even to the instability of the closed-loop system, in the most unfavorable situation. A second problem is that, in order to base the design technique for the controller on a meaningful model of the structure, many structural modes must be included in the model, especially for structures with high modal density, so that control problems for flexible structures are intrinsically large-scale problems. Only in the simpler case of low modal density nonoptimal control strategies (e.g., sliding manifold-based control [2] may be used). Finally, stability of the controller (in order to reduce sensitivity to disturbances) and limited bandwidth, for example, to avoid spillover, due to interaction of unmodelled dynamics with control action, must be enforced. From a practical point of view, vibrations are often measured by using accelerometers, which are prone to offsets which can even saturate actuators if not suitably filtered out in the feedback loop. Thus, the need for a filtering action both at high and at low frequency arises. Hence, a limited-bandwidth strongly stabilizing controller is needed. Further discussions on this topic can be found in [3, 4]. Furthermore, in [5] a broader list of references on the topic is presented and discussed.

In this paper the selection of a stable stabilizing controller with bandpass frequency shape is discussed. First, optimal controllers are discussed, with analytical expression in closed form that is useful when high-order models may result into increased computational burden. The main limitation of this approach is that it can deal only with square plants, that is, as many inputs as outputs must be present. Next, a further technique, able to deal with rectangular plants (i.e., with more outputs than inputs), is proposed. The motivation is provided by the requirements of many applications, where the rejection of a disturbance acting on the structure is desired in locations where the installation of actuator/sensor couples is prohibited. Therefore, the possibility to exploit the vibration measurement in those locations appears the only viable option. The second control strategy proposed in this paper is specifically designed to usefully exploit the knowledge of the disturbance entry point on the structure to place additional sensors able to provide the controller with the necessary information on the vibration status caused by the disturbance. The resulting controller thus has more inputs than outputs, but it is still guaranteed to be strongly stabilizing and with bandpass frequency response. Moreover, selection of free control parameters in order to improve effectiveness of the control, by emphasizing the action of the most controllable and observable modes, is discussed. To achieve such an objective, the control design procedure, still based on an $\mathcal{H}_2$ approach, adopts an LMI (linear matrix inequality) technique to compute the control parameters to allow the user to select the design weights based on various design requirements or by taking into account modal controllability/observability indices. The paper reports experimental results of broadband vibration reduction for a stiffened aeronautical panel controlled with piezoelectric actuators.

A final consideration is in order in this section. It is clear that the placement of the actuators and sensors is of paramount importance in the whole response of the flexible structure to external disturbances. Thus, an efficient control system should in general address also the problem of the optimal allocation of actuators and sensors, for example, based on maximizing vibrational energy reduction. However, in practical situations the position of sensors and actuators is often constrained by different considerations (geometry, weight limitations, safety, and cost); thus the control designer has very limited degrees of freedom in deciding sensors and actuators position and even number. For this reason, in this paper the location of sensors and actuators is assumed to be given.

2. The Model

In this paper a flexible structure with m control inputs and $m + n_d$ measured (control) outputs is considered. The structure is assumed to be infinite-dimensional, linear, and time invariant. Since in general $n_d \geq 0$, the system is not necessarily square, that is, it may have more outputs than inputs. This assumption is quite reasonable in flexible systems control applications, where actuators are heavy and expensive, while sensors (accelerometers) are light, small, and relatively cheap.

As extensively discussed in [5], the mathematical model of a flexible structure can be written as

$$\begin{aligned} \dot{x} &= Ax + \overline{B}_u \overline{u}, \\ \overline{y} &= \overline{C}_y x, \end{aligned} \tag{1}$$

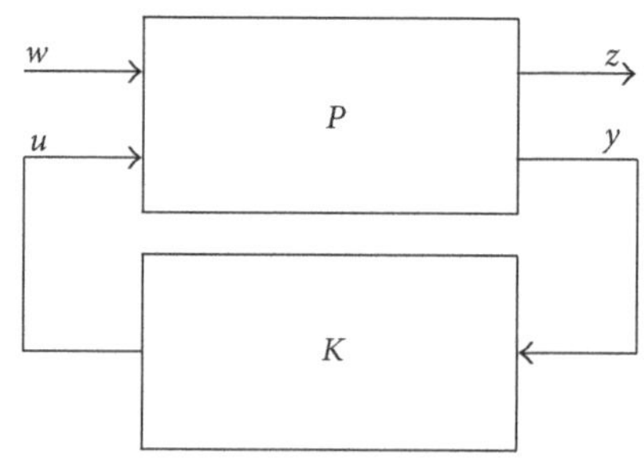

FIGURE 1: Standard control problem scheme.

where $x = (x_1^T \; x_2^T)^T \in \mathbb{R}^{2n}$ is the state vector, $\overline{u} \in \mathbb{R}^m$ the control input, and $\overline{y} \in \mathbb{R}^{m+n_d}$ the measured (control) output. Obviously, since we are approximating an infinite-order system, the number of retained modes n is generally very high (with systems of order of hundreds). By resorting to the "modal coordinates" the matrices of the state space model assume the form

$$A = \begin{pmatrix} 0 & I \\ -\Omega & -\Lambda \end{pmatrix}, \tag{2}$$

$$\overline{B}_u = \begin{pmatrix} 0 \\ \overline{B}_{u_2} \end{pmatrix}, \tag{3}$$

$$\overline{C}_y = \begin{pmatrix} 0 & \overline{B}_{d_2}^T & \overline{B}_{u_2}^T \end{pmatrix}, \tag{4}$$

where

$$\Omega = \operatorname{diag}(\omega_1^2, \ldots, \omega_n^2), \qquad \Lambda = \operatorname{diag}(2\zeta_1\omega_1, \ldots, 2\zeta_n\omega_n), \tag{5}$$

with ω_i and $\zeta_i > 0$ being the natural frequency and damping coefficient of the ith mode ($i = 1, \ldots, n$). In this case x_1 are the modal displacements and x_2 the modal velocities. Moreover, assuming that the velocity measurement is available, the output matrix $\overline{C}_y$ has the form (4). By following the gray-box approach proposed by Cavallo et al. [5], the model above can be accurately identified through a measurement campaign carried out on the structure. The gray-box approach assumes that the structure of the system to identify is exactly the one in (2)–(4), and this assumption is of paramount importance in the design of the controller, as it will be illustrated in Section 3.

3. The Control Strategy

In order to design a stabilizing controller for the flexible system it is convenient to resort to the standard TITO (two-input-two-output) control problem framework depicted in Figure 1. In this figure, P is the system to control, with measured output y, control input u, generalized disturbance w, and performance output z. Augmenting the flexible system with suitable fictitious "disturbance" and "performance" matrices, the following generalized plant P is obtained:

$$\dot{x} = Ax + B_w w + B_u u, \tag{6}$$

$$z = C_z x + D_{12} u, \tag{7}$$

$$y = C_y x + D_{21} w, \tag{8}$$

where $w \in \mathbb{R}^{m+n_d+n}$ is a fictitious disturbance, $z \in \mathbb{R}^{m+n}$ a fictitious performance output, and

$$u = R_1^{1/2}\overline{u} \in \mathbb{R}^m, \qquad y = R_2^{-1/2}\overline{y} \in \mathbb{R}^{m+n_d}, \tag{9}$$

$$B_u = \overline{B}_u R_1^{-1/2} = \begin{pmatrix} 0 \\ B_{u_2} \end{pmatrix}, \tag{10}$$

$$C_y = R_2^{-1/2}\overline{C}_y = \begin{pmatrix} 0 & B_{d_2}^T & B_{u_2}^T \end{pmatrix}, \tag{11}$$

with R_1 and R_2 being the weighting matrices to be selected. The matrices B_w and C_z are $2n \times (m+n_d+n)$ and $(m+n)\times 2n$ real matrices, respectively, with the following structure:

$$B_w = \begin{pmatrix} 0_{n\times n} & 0_{n\times(n_d+m)} \\ W^{1/2} & 0_{n\times(n_d+m)} \end{pmatrix}, \qquad C_z = \begin{pmatrix} 0_{n\times n} & Q^{1/2} \\ 0_{m\times n} & 0_{m\times n} \end{pmatrix}, \tag{12}$$

where the matrices W and Q are to be selected. Finally, for the sake of simplicity, the matrices D_{12} and D_{21} are chosen as to satisfy the following conditions:

$$D_{12}^T C_z = 0, \qquad B_w D_{21}^T = 0. \tag{13}$$

Although not essential, conditions (13) are standard in $\mathcal{H}_2$ or $\mathcal{H}_\infty$ control [6]. It is possible to remove them by slightly increasing the complexity of the control strategy. A possible selection is

$$D_{12} = \begin{pmatrix} 0_{n\times m} \\ D_{12_2} \end{pmatrix}, \qquad D_{21} = \begin{pmatrix} 0_{n_d\times n} & D_{21_1} & 0_{n_d\times m} \\ 0_{m\times n} & 0_{m\times n_d} & D_{21_2} \end{pmatrix}, \tag{14}$$

where

$$R_1 = D_{12}^T D_{12} = D_{12_2}^T D_{12_2} > 0, \tag{15}$$

$$R_2 = \begin{pmatrix} D_{21_1} D_{21_1}^T & 0_{n_d\times m} \\ 0_{m\times n_d} & D_{21_2} D_{21_2}^T \end{pmatrix} = \begin{pmatrix} R_{2_1} & 0 \\ 0 & R_{2_2} \end{pmatrix} > 0. \tag{16}$$

Finally, let T_{zw} denote the lower LFT (linear fractional transformation) $T_{zw} = \mathcal{F}_l(P, K)$ [6].

Before computing the controller, a discussion on the required properties of the controller is in order. As stated in the introduction, stable bandpass controllers are preferred options in the control of flexible systems, due to their ability to filter out accelerometer biases. The following lemma is useful for characterizing the state space representation of a general class of bandpass systems.

Lemma 1. *Let $H(s)$ be a linear time invariant (LTI) $2n$-(McMillan) degree system with minimal state space representation*

$$H(s) = \left[\begin{array}{c|c} A & B \\ \hline C & 0 \end{array}\right], \tag{17}$$

where

$$A = \begin{pmatrix} 0 & I \\ A_1 & A_2 \end{pmatrix}, \qquad B = \begin{pmatrix} 0 \\ B_2 \end{pmatrix}, \qquad C = \begin{pmatrix} 0 & C_2 \end{pmatrix}, \tag{18}$$

where all the matrices are partitioned assuming the state to be split into two n-dimensional substates and the square $n \times n$ matrix A_1 is invertible. Then, the system $H(s)$ is bandpass, in the sense that

$$\lim_{s\to 0} H(s) = \lim_{s\to\infty} H(s) = 0. \tag{19}$$

Proof. Although the lemma is an extension of Theorem 1 in Cavallo et al. [3], it can be easily proved observing that

$$\lim_{s\to 0}(sI - A)^{-1} = \begin{pmatrix} A_1^{-1}A_2 & A_1^{-1} \\ -I & 0 \end{pmatrix}. \tag{20}$$

Thus, direct computation shows that the static gain is zero. The high-frequency behavior is trivial. □

The above lemma has a key role in our context. Indeed, it guarantees that a controller with a bandpass frequency shape, as required in the introduction, can be imposed by satisfying the rather mild conditions (18). Thus, the lemma is useful to generate different families of bandpass controllers.

The last issue to address in order to design a useful controller for flexible structures is stability of the controller itself. It is well known that inserting unstable elements into the control loop increases sensitivity to disturbances and calls for high-bandwidth controller, thus also increasing the effect of measurement noise [7]. Moreover, in the case of flexible structures the high-bandwidth requirement conflicts with the need to avoid spillover effects.

A first possible family of controllers is obtained in the case of square plants, that is, when $n_d = 0$. Indeed, in this case it is possible to compute the explicit solution of a stable stabilizing controller satisfying an $\mathcal{H}_\infty$ index. Preliminarily, the following lemma is needed.

Lemma 2. *Consider the system (6)–(8), assume, $n_d = 0$, and let X_∞ and Y_∞ be $(2n \times 2n)$ positive semidefinite matrices satisfying*

$$A^T X_\infty + X_\infty A + X_\infty\left(\gamma^{-2}B_w B_w^T - B_u B_u^T\right)X_\infty + C_z^T C_z = 0,$$
$$A Y_\infty + Y_\infty A^T + Y_\infty\left(\gamma^{-2}C_z^T C_z - C_y^T C_y\right)Y_\infty + B_w B_w^T = 0, \tag{21}$$

for a given $\gamma > 0$. Then, a necessary condition for the solution of the AREs (21) is to be of the form

$$X_\infty = \beta^2 Y_\infty^{-1}, \tag{22}$$

where $\beta < \gamma$ is a real scalar, that is, X_∞ and Y_∞ are block diagonal.

Although rather technical, Lemma 2 is useful to characterize the closed-form solution to an $\mathcal{H}_\infty$ control problem, as stated by the following theorem.

Theorem 3. *Consider the system (6)–(8), with matrices given by (2), (10), and (11), assume, $n_d = 0$, let the pair (A, B_u) be stabilizable and the pair (A, C_y) detectable, and let*

$$B_w = \begin{pmatrix} 0 & 0 \\ B_{w_2} & 0 \end{pmatrix}, \qquad C_y = \begin{pmatrix} 0 & B_{w_2}^T \\ 0 & 0 \end{pmatrix}, \tag{23}$$

where

$$B_{w_2} = \left(\alpha\Lambda + B_{u_2}B_{u_2}^T\right)^{1/2} \tag{24}$$

and $\alpha > 0$ is a scalar parameter.

Then, the following controller is strongly stabilizing and has m zeros at the origin

$$K_\infty(s, \gamma) = \left[\begin{array}{c|c} A_\infty & B_u \\ \hline -\dfrac{\gamma^2}{\gamma^2 - 2}B_u^T & 0 \end{array}\right], \tag{25}$$

where

$$A_\infty = \begin{pmatrix} 0 & I \\ -\Omega & -\Lambda_\infty \end{pmatrix}, \tag{26}$$

and

$$\Lambda_\infty = \left(1 - \frac{2}{\gamma^2}\right)\Lambda + 2\frac{\gamma^2 - 1}{\gamma^2 - 2}\sqrt{\frac{\gamma^2 - 1}{\gamma^2}}B_{u_2}B_{u_2}^T. \tag{27}$$

Moreover, if

$$\alpha = 2\sqrt{1 - \gamma^{-2}}, \tag{28}$$

the LFT T_{zw} satisfies

$$\|T_{zw}\|_\infty < \gamma. \tag{29}$$

Finally, inf $\gamma = \sqrt{2}$ *and the controller*

$$K_{\text{opt}} = B_{u_2}^T\left(B_{u_2}B_{u_2}^T\right)^\dagger B_{u_2}, \tag{30}$$

where $(\cdot)^\dagger$ denotes the pseudoinverse, achieves the optimal value

$$\min_{K \in \mathcal{K}}\|T_{zw}\|_\infty = \sqrt{2}, \tag{31}$$

where $\mathcal{K}$ is the set of stable, stabilizing controllers with m zeros at the origin $s = 0$.

The proofs of Lemma 2 and Theorem 3 can be found in Cavallo et al. [5]. They are based on the classic solution to the $\mathcal{H}_\infty$ problem involving two coupled Riccati equations [6], while the lower bound for γ is obtained by resorting to the theory of singularly perturbed systems. It is worth noticing that the above theorem yields as a byproduct the closed-form solution of the $\mathcal{H}_2$ problem, obtained by letting $\gamma \to \infty$.

Corollary 4. *With the same hypotheses as those in Theorem 3, the controller*

$$K_2(s) = \left[\begin{array}{c|c} A - 2B_uB_u^T & B_u \\ \hline -B_u^T & 0 \end{array}\right] \tag{32}$$

minimizes the norm $\|T_{zw}\|_2$.

Proof. It is sufficient to let $\gamma \to \infty$ in the controller in Theorem 3. □

Finally, by resorting to the equations for the parameterization of all the stabilizing controllers [8], it is possible to give a general expression for the whole family of $\mathcal{H}_\infty$ optimal controllers and $\mathcal{H}_2$ suboptimal controllers as follows.

Let

$$Q(s) = C_q\left(sI - A_q\right)^{-1}B_q + D_q \tag{33}$$

be a stable transfer matrix with $\|Q\|_\infty < \gamma$. Then, the family of controllers such that $\|T_{zw}\|_\infty < \gamma$ is

$$K_{\infty,Q}(s) = \left[\begin{array}{c|c} A_{\infty,Q} & B_{\infty,Q} \\ \hline C_{\infty,Q} & D_q \end{array}\right], \tag{34}$$

where

$$\begin{aligned} A_{\infty,Q} &= \begin{pmatrix} A_\infty - \dfrac{\gamma^2 - 1}{\gamma^2 - 2}B_uD_qB_u^T & \dfrac{\gamma^2 - 1}{\gamma^2 - 2}B_uC_q \\ -B_qB_u^T & A_q \end{pmatrix}, \\ B_{\infty,Q} &= \begin{pmatrix} \dfrac{\gamma^2 - 1}{\gamma^2 - 2}B_u\left(D_q + \dfrac{\gamma}{\sqrt{\gamma^2 - 1}}I\right) \\ B_q \end{pmatrix}, \\ C_{\infty,Q} &= \left(-\left(D_q + \dfrac{\gamma}{\sqrt{\gamma^2 - 1}}I\right)B_u^T \;\; B_q\right). \end{aligned} \tag{35}$$

Analogously, an (simpler) expression for the $\mathcal{H}_2$ suboptimal controller may be obtained as follows:

$$K_{2,Q}(s) = \left[\begin{array}{cc|c} A - 2B_uB_u^T & B_uC_q & B_u \\ -B_qB_u^T & A_q & B_q \\ \hline -B_u^T & C_q & 0 \end{array}\right], \tag{36}$$

where, it has been, obviously, assumed that $D_q = 0$.

In the general case of $n_d \neq 0$ the above formulation does not hold any longer, and the design of the controller has to be redefined. The main reason is that the Riccati equations to be solved for the $\mathcal{H}_\infty$ problem are strongly coupled. The choice of a stabilizing controller with the structure in Lemma 1 can be carried out by exploiting the following theorem.

Theorem 5. *Any system of the form (6)–(8) with matrices given by (2), (10), and (11) is closed-loop stabilized by a controller with state space representation*

$$K(s) = \left[\begin{array}{c|c} A_c & B_c \\ \hline C_c & 0 \end{array}\right], \tag{37}$$

where

$$\begin{aligned} B_c &= \begin{pmatrix} 0_{m\times n} & 0_{m\times n} \\ Y_2B_{d_2}R_{2_1}^{-1} & Y_2B_{u_2}R_{2_2}^{-1} \end{pmatrix}, \\ C_c &= -R_1^{-1}\left(0_{m\times n} \;\; B_{u_2}^TX_2\right), \\ A_c &= A + B_uC_c - B_cC_y, \end{aligned} \tag{38}$$

where X_2 and Y_2 are $n \times n$ diagonal matrices with nonnegative entries.

Proof. Let

$$Q = 2X_2\Lambda + X_2 B_{u_2} R_1^{-1} B_{u_2}^T X_2, \tag{39}$$

$$W = 2Y_2\Lambda + Y_2\left(B_{d_2} R_{2_1}^{-1} B_{d_2}^T + B_{u_2} R_{2_2}^{-1} B_{u_2}^T\right) Y_2. \tag{40}$$

It is clear that both Q and W are positive semidefinite for any nonnegative matrices X_2 and Y_2. Moreover, by defining the matrices

$$X = \begin{pmatrix} \Omega X_2 & 0 \\ 0 & X_2 \end{pmatrix}, \quad Y = \begin{pmatrix} \Omega^{-1} Y_2 & 0 \\ 0 & Y_2 \end{pmatrix}, \tag{41}$$

it is easy to show that $X = \mathrm{Ric}(H_2)$ and $Y = \mathrm{Ric}(J_2)$, where, for a given Hamiltonian matrix Z, $\mathrm{Ric}(Z)$ denotes the *stabilizing solution* of the Riccati equation associated to the Hamiltonian matrix [6], and the Hamiltonian matrices H_2 and J_2 are

$$H_2 = \begin{pmatrix} A & -B_u R_1^{-1} B_u^T \\ -\widetilde{Q} & -A^T \end{pmatrix}, \tag{42}$$

$$J_2 = \begin{pmatrix} A^T & -C_y^T R_2^{-1} C_y \\ -\widetilde{W} & -A \end{pmatrix}. \tag{43}$$

Finally, the matrices $\widetilde{Q}$ and $\widetilde{W}$ are

$$\widetilde{Q} = \begin{pmatrix} 0 & 0 \\ 0 & Q \end{pmatrix}, \quad \widetilde{W} = \begin{pmatrix} 0 & 0 \\ 0 & W \end{pmatrix}. \tag{44}$$

Thus, for any matrices X_2 and Y_2 the controller (38) solves an $\mathcal{H}_2$ problem and is therefore stabilizing. □

Controller stability is not guaranteed by the above theorem; thus it must be enforced as follows. Preliminarily, define

$$\widetilde{B}_X = B_u R_1^{-1} B_u^T, \quad \widetilde{B}_Y = C_y R_2^{-1} C_y^T. \tag{45}$$

From the above formulation it is easy to deduce the following theorem.

Theorem 6. *Consider the system $K_s(s)$ with minimal state space representation*

$$K_s(s) = \left[\begin{array}{c|c} A_c & B_c \\ \hline C_c & 0 \end{array}\right], \tag{46}$$

where the matrices A_c, B_c, and $C_c$are given in (38). *If the diagonal matrices X_2 and Y_2 are chosen so as to satisfy the LMI*

$$X_2\widetilde{B}_X + \widetilde{B}_X X_2 + Y_2\widetilde{B}_Y + \widetilde{B}_Y Y_2 > 0, \quad X_2 > 0, \quad Y_2 > 0, \tag{47}$$

then the system $K_s(s)$ strongly stabilizes the system (6)–(8) and is bandpass.

Proof. The theorem can be easily proved by considering the Lyapunov function

$$V(x) = x^T \begin{pmatrix} \Omega^{-1} & 0 \\ 0 & I \end{pmatrix} x. \tag{48}$$

The rest of the theorem follows from Theorem 5 and Lemma 1. □

The above theorem shows that a parametrization of stable, stabilizing, and bandpass controller is obtained, with $2n$ free parameters, namely, the diagonal entries of the matrices X_2 and Y_2. Since these matrices define the controller gain, a further constraint in the form of LMI

$$\mathrm{tr}(X_2) + \mathrm{tr}(Y_2) < \delta, \tag{49}$$

where $\mathrm{tr}(X)$ is the trace of the matrix X and δ is an upper bound on the controller gain depending on the problem, can be imposed. Moreover, also the matrices R_1 and R_2 have still to be selected. Expressions (39) and (40) can suggest how to choose the degrees of freedom. According to them, X_2 and Y_2 can weigh modes on which to operate, while the matrices R_1 and R_2 define the authority of the control.

The above considerations can be summarized in a procedure for controller design, namely,

(1) preliminarily, collect the diagonal elements of X_2 and Y_2 in the vector $\xi = (x_{2_{11}}, \ldots, x_{2_{nn}}, y_{2_{11}}, \ldots, y_{2_{nn}})^T$, define a vector of performance weights $c \in \mathbb{R}^n$, whose jth element is the weight of the jth mode,

(2) solve the optimization problem

$$\max \quad \left(c^T \; c^T\right)\xi, \quad \text{subject to} \tag{50}$$

$$X_2(\xi)\widetilde{B}_X + \widetilde{B}_X X_2(\xi) + Y_2(\xi)\widetilde{B}_Y + \widetilde{B}_Y Y_2(\xi) + 2\Lambda > 0, \tag{51}$$

$$X_2(\xi) > 0, \tag{52}$$

$$Y_2(\xi) > 0, \tag{53}$$

$$1_n^T X_2(\xi) 1_n + 1_n^T Y_2(\xi) 1_n < \delta, \tag{54}$$

where $\delta > 0$ and 1_n is the n-dimensional column vector with all elements equal to 1;

(3) compute the controller according to (37)–(38).

In the above procedure, the scalar $\delta > 0$ is chosen to limit the controller gain since (54) can be written as

$$\mathrm{tr}(X_2) + \mathrm{tr}(Y_2) < \delta. \tag{55}$$

Obviously, in the above algorithm c_j, the jth element of c can assume any real value. For instance, choosing $c_1 = 10$,

$c_2 = 1$, and $c_3 = 0.1$ would pose different penalties on different modes, thus focussing the control action more on mode one than on mode two and on mode three. Cavallo et al. [9] showed that the optimal choice of the input scaling matrices is performed based on the controllability Gramian of the open-loop and closed-loop systems. However, it is clear that the efficiency of the control action depends on both the controllability and the observability Gramians. As an example, a well-known strategy for model order reduction [10] consists of the computation of a balanced realization and a reduction based on neglecting the modes associated with the Hankel singular values, that is, square roots of the eigenvalues of the product of the controllability Gramian and the observability Gramian, with the lowest magnitude. This kind of consideration leads to the idea of selecting the modal performance weights in (50) based on the Hankel singular values. Another aspect of the presented control strategy to highlight is the selection of the matrices R_1 and, even more importantly, R_2 in (15) and (16), respectively. The former is responsible for the authority of the control. The latter is responsible for the relative weights of colocated measurements with respect to noncolocated ones. By varying the magnitude of R_{2_1} relative to R_{2_2} the case of noncolocated feedback can be considered with stability guaranteed.

4. Experimental Results

Control procedures described in the previous section have been experimentally tested on a very complex flexible structure. In particular, the selected test article is a fuselage skin panel of a BOEING 717 (depicted in Figure 2). The panel is stiffened by two bulkheads and three orthogonal stiffeners riveted to the rear of the panel itself. In order to reproduce the conditions usually adopted in the structural testing phase, the panel has been suspended by a couple of soft springs to simulate free boundary conditions. In the experiments presented in this section piezoelectric ceramic actuators have been used both for producing the primary disturbance field and for generating the control forces on the structure aimed at counteracting the effects of the primary field.

4.1. Square Case. In detail, the control inputs are produced using three of the four piezoelectric patches bonded on the panel surface. The output acceleration is measured by means of the three colocated accelerometers placed on the rear part of the panel behind the piezoelectric actuators. The structure has a high modal density in the considered frequency range $\Omega = [100, 600]$ Hz and a model with 148 modes has been identified as proposed by Cavallo et al. [11]. The experiments were performed by using a dSPACE rapid prototyping real-time control system with 16-bit A/D channels and 14-bit D/A channels at a sampling frequency of 20 kHz.

A fourth piezo is used only as a disturbance input so as to introduce a disturbance contribution not belonging to the range of the input matrix B_u in (6).

The effectiveness of the control strategy has been evaluated comparing the frequency response functions (FRFs) from disturbance input to acceleration outputs and in terms

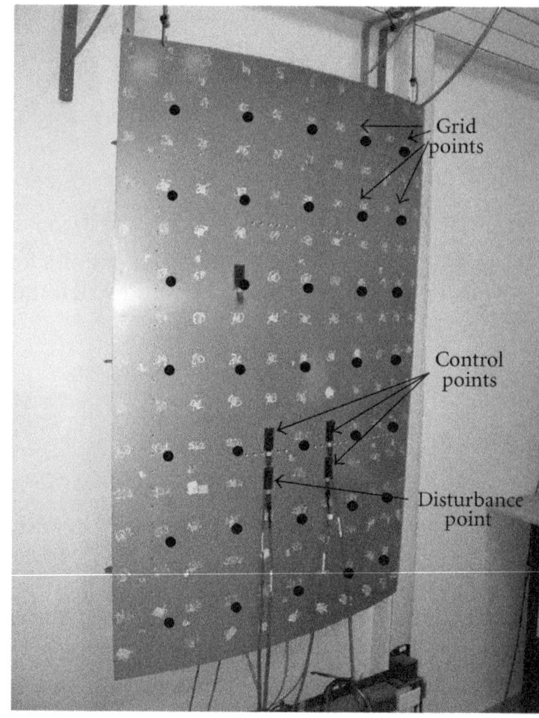

FIGURE 2: Front view of the panel with piezoelectric actuators and grid performance point.

of an energy-based index defined as follows. First, the energy of the measured acceleration in the given frequency range Ω has been estimated for the kth of the five used sensors as

$$J_k = \sqrt{\sum_{h=1}^{M} |\Phi_k(\omega_h)|^2}, \quad k = 1, \ldots, 35, \tag{56}$$

where $\Phi_k(\omega_h)$ is the acceleration spectrum measured at the kth point at the frequency $\omega_h \in \{\omega_1, \ldots, \omega_M\}$, a set of $M = 2500$ linearly spaced frequencies within Ω. Then, the relative energy reduction at the kth point has been quantified as

$$R_{k\%} = \frac{J_{k,\mathrm{ol}} - J_{k,\mathrm{cl}}}{J_{k,\mathrm{ol}}} 100, \tag{57}$$

where the subscripts "ol" and "cl" stand for "open loop" and "closed loop," respectively.

In a first experimental campaign, the optimal $\mathcal{H}_\infty$ and $\mathcal{H}_2$ controllers are tested. One of the known drawbacks in the application of optimal feedback control to vibration reduction problems is that the control action may focus only on the performance objectives, reducing vibrations only at the controlled points, while increasing vibration levels at other points of the structure. In order to check that the controller actually has a beneficial effect on the *whole* structure a matrix of 35 points evenly spaced on the panel has been considered, as shown in Figure 2.

A band-limited white noise in the range [100, 600] Hz has been chosen as primary field, to excite all the system

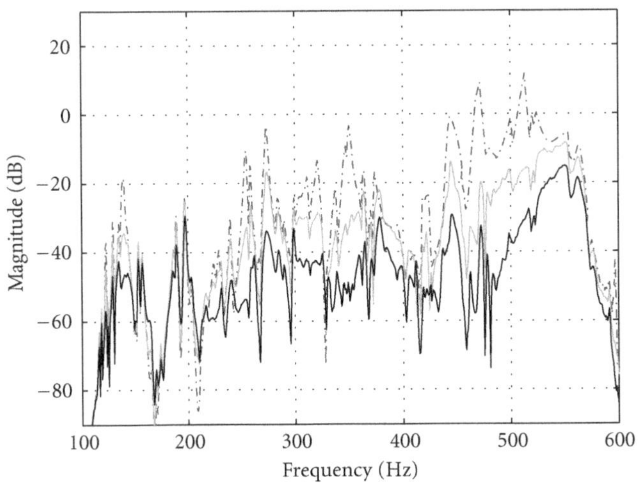

FIGURE 3: Acceleration in a colocated point of the panel: open loop (*dashed grey line*), $\mathcal{H}_2$ (*solid grey line*), and $\mathcal{H}_\infty$ (*solid black line*).

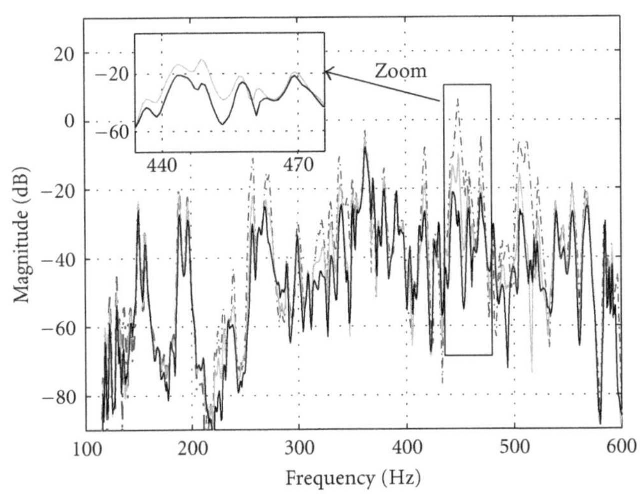

FIGURE 4: Acceleration in a noncolocated point of the panel: open loop (*dashed grey line*), $\mathcal{H}_2$ (*solid grey line*), and $\mathcal{H}_\infty$ (*solid black line*).

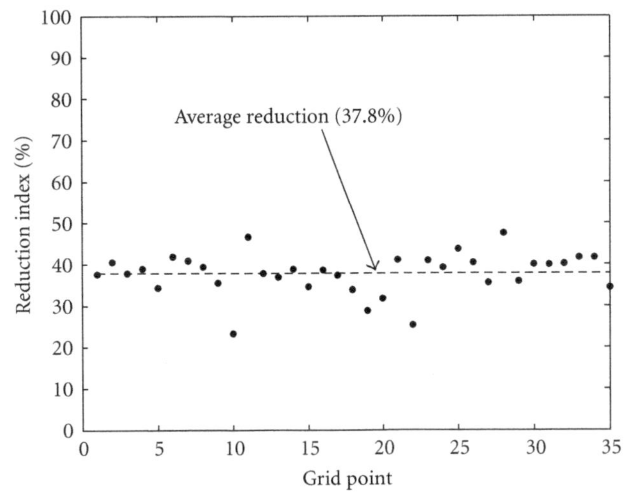

FIGURE 5: Reduction energy index at grid points of the panel with $\mathcal{H}_2$ controller.

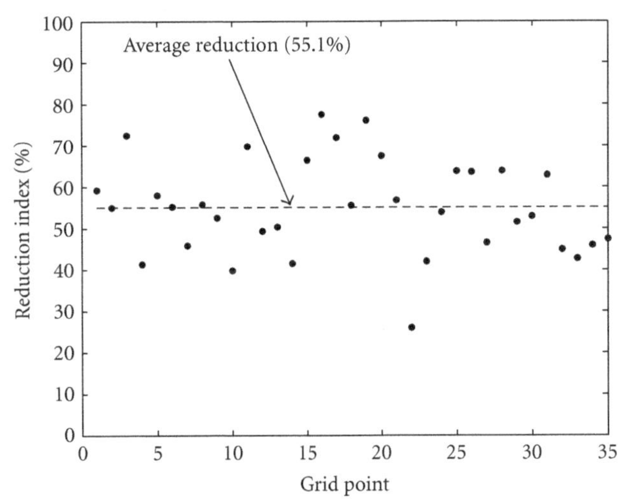

FIGURE 6: Reduction energy index at grid points of the panel with $\mathcal{H}_\infty$ controller.

natural modes in the frequency range Ω. The noise has been applied to the structure on all four input channels so that the disturbance has components both in the range of B_u and off range. Figures 3 and 4 present results for both $\mathcal{H}_2$ and $\mathcal{H}_\infty$ cases. The $\mathcal{H}_\infty$ controller has been designed fixing $\gamma = 1.5$ in (25). The figures show the acceleration spectrum measured in a control point and in a performance test point. Obviously, performance is better at the control points than at the test points, where a vibration reduction is still apparent, anyway.

Figures 5 and 6 report the values of the relative energy reduction index as defined in expression (57) at each grid point for the $\mathcal{H}_2$ and $\mathcal{H}_\infty$ cases, respectively, showing the effectiveness of both controllers on the whole panel. The average vibration reductions are 37.8% with the $\mathcal{H}_2$ controller $\mathcal{H}_2$ and 55.1% with the $\mathcal{H}_\infty$ controller almost uniformly on the whole panel, despite the small number of control actuators, the high modal density, and the broadband nature of the disturbance.

Figures 7 and 8 further clarify the effectiveness of the proposed controllers reporting the values of the J_k energy index in all the grid points with the same color scale in order to compare open loop with closed loop.

Finally, Figure 9 presents the time history of a control input (piezoelectric actuator voltage) in both the $\mathcal{H}_2$ and the $\mathcal{H}_\infty$ cases, showing that the control voltage with the $\mathcal{H}_\infty$ controller is generally higher than the $\mathcal{H}_2$ case (this explains why the $\mathcal{H}_\infty$ controller behaves better). Nevertheless, in both cases the input voltage is far below the saturation limits of the actuators (200 V).

4.2. Rectangular Case. The next campaign of experiments addresses the case of more measurements than controls available. In this case, a fifth accelerometer is used to measure the acceleration in a performance point far from both control

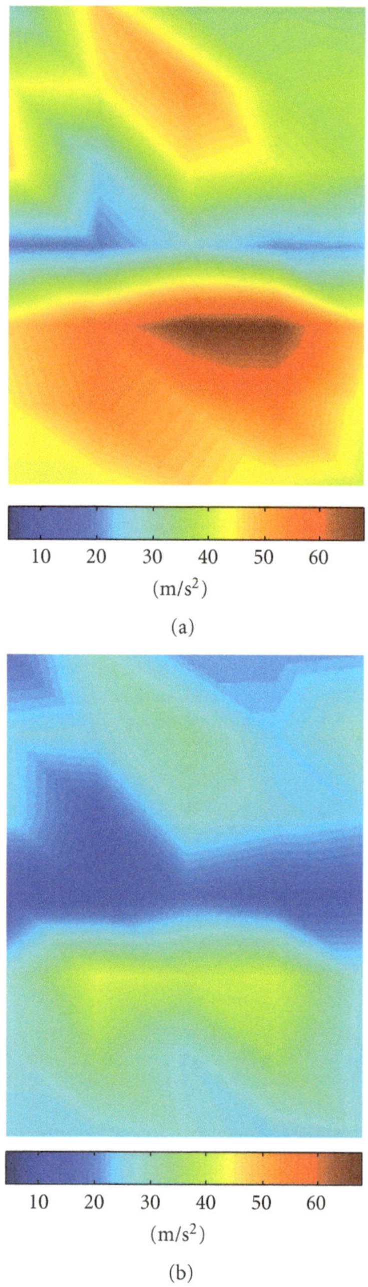

FIGURE 7: Performance of the $\mathcal{H}_2$ controller on the whole panel: open loop (a) and closed loop (b).

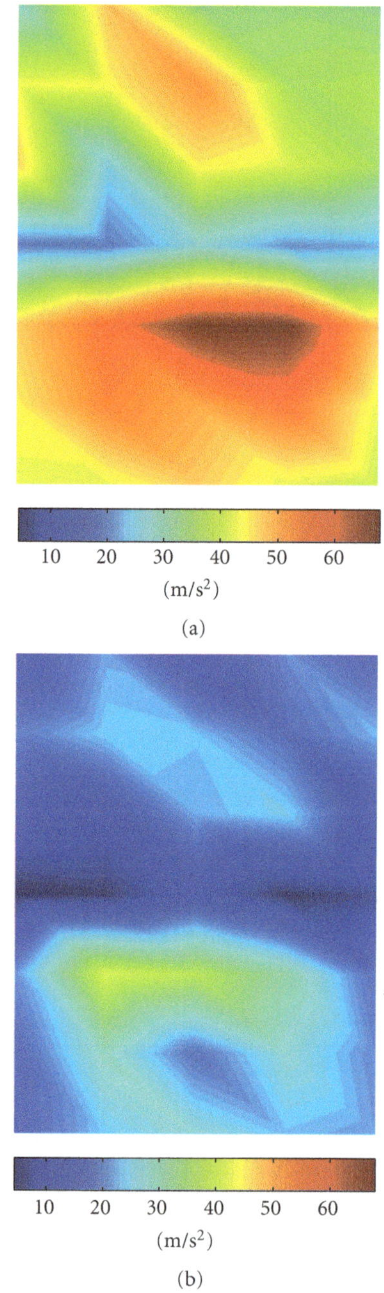

FIGURE 8: Performance of the $\mathcal{H}_\infty$ controller on the whole panel: open loop (a) and closed loop (b).

and disturbance inputs. Moreover, the noise has been applied to the structure *only through the fourth piezo actuator* not used as control input, thus the disturbance is totally out of the range of the input matrix B_u. Although this makes the two sets of experiments not comparable, this choice allows better highlighting of the feature of the controller when used with additional sensors located near the entry point of the primary field. Since the controller computation depends on the choice of modal performance weights and on the number of feedback sensors, the guidelines described at the end of Section 3 have been followed. As a consequence, three controllers have been computed and compared:

A: colocated controller with modal performance weights based on the Henkel singular values,

B: rectangular controller (with an additional sensor) with modal performance weights based on the Henkel singular values,

C: noncolocated controller with modal performance weights based on the Henkel singular values.

All these controllers use the same three piezoelectric actuators as control inputs and the four accelerometers as measured outputs and are designed by selecting the modal

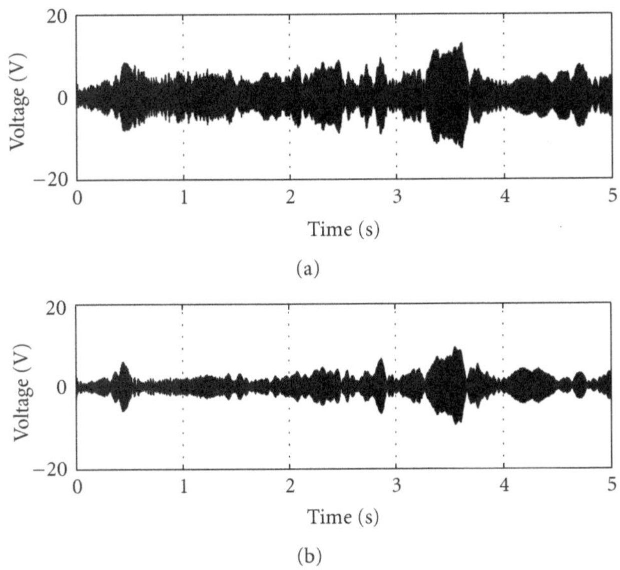

FIGURE 9: Control signals in the $\mathcal{H}_\infty$ case (a) and $\mathcal{H}_2$ case (b).

performance weights in (50) according to the Henkel singular values. However, controller A practically uses, as output accelerations, only the measurements provided by the three colocated sensors, since weighting matrices are selected as $R_1 = I$, $R_{2_1} = 10^8$, and $R_{2_2} = I$. The large value of R_{2_1} makes the information of the fourth sensor meaningless for the control action. Controller B takes into account the problem of vibration reduction when information on the location of the application point of the disturbance force is available. Of course, in such a case, the optimal solution would be to locate an actuator/sensor pair in this entry point. However, assume that mounting the actuator in such a location is not possible for some reason, while placement of the sensor is still allowed. Under this assumption, controller B uses the information provided by the fourth sensor by selecting the weighting matrices $R_1 = I$, $R_{2_1} = 1$, and $R_{2_2} = I$, where a lower value for R_{2_1} has been set. The controller C has been considered to assess the behaviour of the active vibration control system when the application at hand does not allow to place actuators and sensors in colocated pairs. In particular, it practically uses only the signal of the fourth sensor placed in the disturbance entry point, by selecting the weighting matrices $R_1 = 10^{-4}I$, $R_{2_1} = 10^{-4}$, and $R_{2_2} = 10^8 I$, where a large value for R_{2_2} has been set. The rectangular noncolocated controller C has no information on what happens to the structure in the control points.

The performance obtained with the controllers above are evaluated both in terms of vibration reduction quantified according to the index in (57) and by comparing the FRFs from disturbance input point to a control point, the disturbance point, and the performance point. In the first experiment controller A, used as reference, has been compared to controller B. Table 1 shows that the reduction energy index for controller B is significantly higher in the disturbance and performance points. Moreover, the effects of controller B, compared with controller A and open loop, on

TABLE 1: Values of the reduction energy index $R_{k\%}$ in (57) obtained in the frequency range [100, 600] Hz for disturbance and performance points.

	Disturbance pt.	Performance pt.
$R_{k\%}$ controller A	30.74	30.58
$R_{k\%}$ controller B	39.16	39.50

TABLE 2: Control energy reduction index $R_{e\%}$ for controllers B and C.

	Controller B	Controller C
Channel 1	11.18	56.35
Channel 2	12.83	64.59
Channel 3	10.14	75.57

the entire frequency range Ω are shown in Figure 10, where over 20 dB of vibration reduction peaks are visible and no spillover appears.

The controllers have been also compared in terms of required control energy according to the index defined below. First, the energy E_s of a generic discrete-time signal $s(k)$ has been evaluated as

$$E_s = \sum_{k=1}^{N} s^2(k), \qquad (58)$$

where N is a fixed number of samples equal for all experiments. Taking into account that the primary field has been randomly generated, the energy $E^r_{u,Y}$ of the rth control signal in the experiment Y (Y denotes the control label as listed above) has been normalized with respect to the actual energy of the disturbance $E_{d,Y}$ in the same experiment, that is,

$$\widehat{E}^r_{u,Y} = \frac{E^r_{u,Y}}{E_{d,Y}}, \quad r = 1,2,3, \quad Y = A, B, C. \qquad (59)$$

The control energy required by controllers B and C has been then compared to the one required by controller A, by considering the control energy reduction index

$$R_{e\%} = \frac{\widehat{E}^r_{u,A} - \widehat{E}^r_{u,X}}{\widehat{E}^r_{u,A}} 100, \quad r = 1,2,3, \quad X = B, C. \qquad (60)$$

The results are summarized in Table 2, which shows that controller B requires less control energy than controller A despite its higher performance. It is apparent that the information from the additional sensor improves the control action.

In a subsequent experiment controller C has been tested. Although in the design phase the weighting matrices have been selected positive definite according to (15) and (16), in the real experiment, only the fourth sensor placed in the disturbance point has been physically used, in order to demonstrate the possibility to use the proposed controller as a true non-colocated one. The performance in terms of reduction energy index is reported in Table 3. The reduction energy index values are lower than in the previous cases, but the same applies to the required control energy indicated

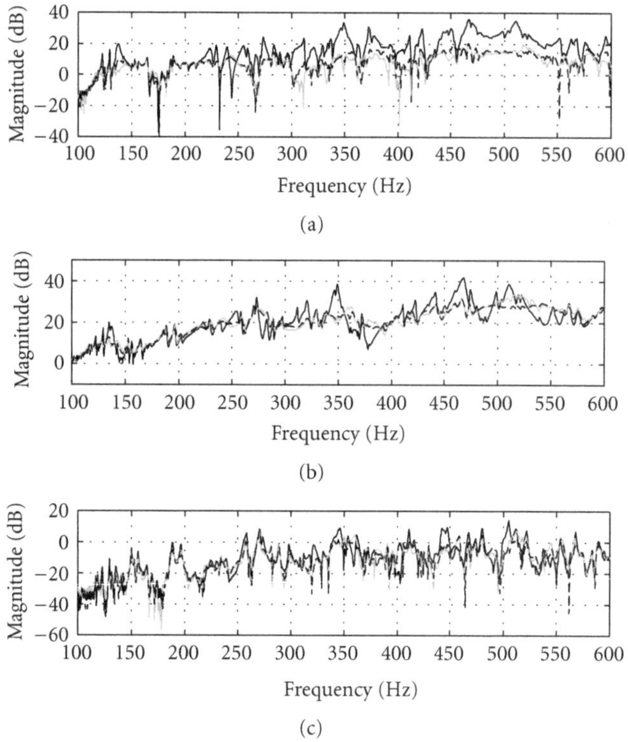

Figure 10: FRFs from disturbance input to acceleration output in a control point (a), in the disturbance point (b), and in the performance point (c): open loop (*solid black lines*), and controller A (*solid grey lines*), controller B (*dashed black lines*).

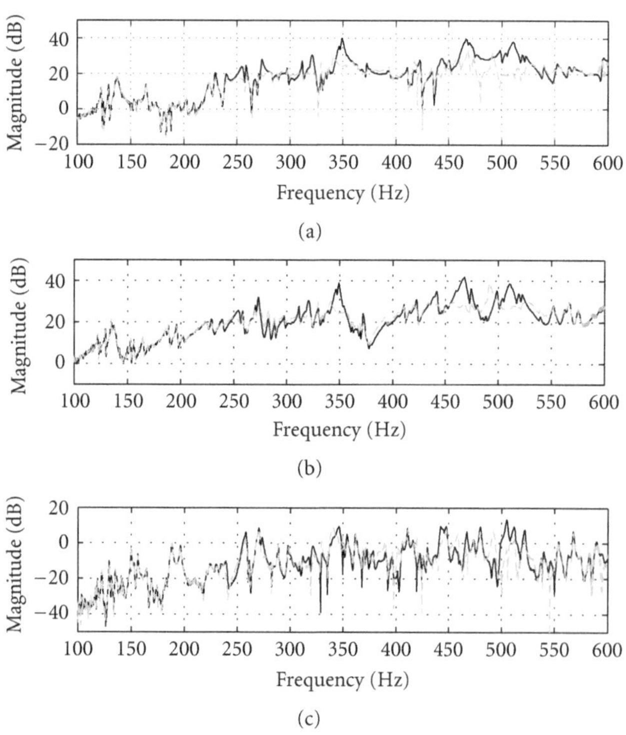

Figure 11: FRFs from disturbance input to acceleration output in a control point (a), in the disturbance point (b), and in the performance point (c): open loop (*solid black lines*) and noncolocated controller C (*dashed grey lines*).

Table 3: Values of the reduction energy index $R_{k\%}$ in (57) obtained for noncolocated controller in the frequency range $[100, 600]$ Hz for disturbance and performance points.

	Disturbance pt.	Performance pt.
$R_{k\%}$ controller C	27.44	18.25

in Table 2. Interestingly enough, even in this actually noncolocated control no spillover phenomenon occurs, as evident from the FRFs reported in Figure 11, still obtaining over 10 dB of vibration reduction peaks.

5. Conclusions

The paper reported a procedure for designing closed-form optimal ($\mathcal{H}_\infty$ and $\mathcal{H}_2$) controllers for large-scale square systems, together with an extension to nonsquare plants to tackle the vibration reduction problem of flexible structure with the strong requirement of stable and bandpass controller. The original control law is here extended to the case of more sensors than actuators. The more complex design is tackled by solving an LMI problem, which allows also to select the performance weights based on controllability/observability indices. The experimental results obtained for a vibration reduction problem of a stiffened aeronautical panel controlled by piezoelectric actuators confirm the applicability of the strategy in actual systems and its advantages with respect to a controller assuming only colocated actuator/sensor pairs.

References

[1] M. Balas, "Trends in large space structure control theory: fodest hopes, widest dreams," *IEEE Transactions on Automatic Control*, vol. 27, pp. 522–535, 1982.

[2] A. Cavallo, G. De Maria, and R. Setola, "Sliding manifold approach for vibration reduction of flexible systems," *Automatica*, vol. 35, no. 10, pp. 1689–1696, 1999.

[3] A. Cavallo, G. De Maria, C. Natale, and S. Pirozzi, "Robust control of flexible structures with stable bandpass controllers," *Automatica*, vol. 44, no. 5, pp. 1251–1260, 2008.

[4] A. Cavallo, G. De Maria, C. Natale, and S. Pirozzi, "$\mathcal{H}\infty$ strongly stabilizing bandpass controllers for flexible systems," in *Proceedings of the IEEE Conference on Decision and Control*, pp. 6543–6548, 2006.

[5] A. Cavallo, G. De Maria, C. Natale, and S. Pirozzi, *Active Control of Flexible Structures: From Modelling to Implementation*, Springer, London, UK, 2010.

[6] K. Zhou and J.C. Doyle, *Essentials of Robust Control*, Prentice Hall, New Jersey, NJ, USA, 1995.

[7] J. Maciejowski, *Multivariable Feedback Design*, Addison-Wesley, Wokingham, UK, 1989.

[8] D. C. Youla, J. J. Bongiorno, and C. N. Lu, "Single-loop feedback-stabilization of linear multivariable dynamical plants," *Automatica*, vol. 10, no. 2, pp. 159–173, 1974.

[9] A. Cavallo, G. De Maria, C. Natale, and S. Pirozzi, "On scaling matrices optimal selection for H_2 strongly stabilizing bandpass controllers," in *Proceedings of the IEEE Conference on Decision and Control*, pp. 4749–4754, New Orleans, La, USA, 2007.

[10] B. C. Moore, "Principal components analysis in linear systems: controllability, observability and model reduction," *IEEE Transactions on Automatic Control*, vol. AC-26, no. 1, pp. 17–32, 1981.

[11] A. Cavallo, G. De Maria, C. Natale, and S. Pirozzi, "Gray-box identification of continuous-time models of flexible structures," *IEEE Transactions on Control Systems Technology*, vol. 15, no. 5, pp. 967–981, 2007.

6

Nonlinear Model of Pseudoelastic Shape Memory Alloy Damper Considering Residual Martensite Strain Effect

Y. M. Parulekar and G. R. Reddy

Reactor Safety Division, Bhabha Atomic Research Centre, Mumbai 400085, India

Correspondence should be addressed to Y. M. Parulekar, yogitap@barc.gov.in

Academic Editor: Mohammad Tawfik

Recently, there has been increasing interest in using superelastic shape memory alloys for applications in seismic resistant-design. Shape memory alloys (SMAs) have a unique property by which they can recover their original shape after experiencing large strains up to 8% either by heating (shape memory effect) or removing stress (pseudoelastic effect). Many simplified shape memory alloy models are suggested in the past literature for capturing the pseudoelastic response of SMAs in passive vibration control of structures. Most of these models do not consider the cyclic effects of SMA's and resulting residual martensite deformation. Therefore, a suitable constitutive model of shape memory alloy damper which represents the nonlinear hysterical dynamic system appropriately is essential. In this paper a multilinear hysteretic model incorporating residual martensite strain effect of pseudoelastic shape memory alloy damper is developed and experimentally validated using SMA wire, based damper device. A sensitivity analysis is done using the proposed model along with three other simplified SMA models. The models are implemented on a steel frame representing an SDOF system and the comparison of seismic response of structure with all the models is made in the numerical study.

1. Introduction

Shape memory alloys (Ni-Ti alloys) are unique alloys that have the ability to undergo large deformations but can return to their undeformed shape by heating (shape memory effect) or through removal of the stress (super elastic effect). Recently, there has been increasing trend in using super elastic shape memory alloy devices based on wires for applications in structural vibration control.

The significant properties of SMAs like Pseudoelasticity, large ductility, excellent corrosion, and fatigue resistance make these alloys attractive material for structural vibration control. Graessar and Cozzarelli [1] in 1991 first proposed the use of Nitinol (SMA) as damping materials. They studied the effect of loading frequency and history on the energy dissipation characteristics of SMA wires. They also proposed a one-dimensional constitutive model for pseudoelastic behavior of the model. The study on the use of SMAs for passive structural damping is presented in Thomson et al. [2] in 1995, where three different quasistatic models of hysterisis were reviewed and compared with experimental investigation of a cantilever beam restrained by two SMA wires. Later, Clark et al. [3] and Dolce et al. [4] in 2000 demonstrated the feasibility of the concept of SMA wire device conducting large number of experiments. It is observed from the tests conducted that there exist the residual martensite strain effects in the hysterisis loop of the SMA device. This effect increases the hysterisis energy dissipated and thus further helps in the control of seismic response. Dolce et al. [5] also studied the implementation of various configurations of SMA-based special dampers by shaking table tests on RC structures. Researchers also have done representative works in thermomechanical modeling like Tanaka [6], Liang and Rogers [7], Boyd and Lagoudas [8]. Based on their work, Khan et al. [9] developed simplified hysterical model of device based on SMA tubes. Motahari and Ghassemieh [10] proposed a simple multilinear model and proved it to be accurate. However, all these models do not take into account the residual martensite strain accumulation effects due to cyclic loading during earthquake. Experimental studies by

Lexcellent and Bourbon [11] and Miyazaki et al. [12] have shown that SMAs change in their mechanical cyclic behavior when subjected to cyclic loading and there is 20% reduction in transformation stress. They also observed that there is accumulation of residual strain for austenitic Ni-Ti wires with number of cycles under repeated cyclic loads which attains a value of 0.4–0.5% at room temperature after about 50 cycles. In the past, most of the works simply neglected this residual strain in the model of SMA damper subjected to earthquake loads. Thus, in the present paper, the model proposed in [10] is modified to include the effect of residual martensite strains. Work presented by Andrawes and DesRosches [13] shows that the response of SMA model is more sensitive to cyclic effects in case of earthquakes with long duration or large intensities. Parulekar et al. [14] modified an existing thermomechanical [10] to incorporate the residual martensite deformation observed in austenitic SMA wires and thus take into account the cyclic effects of SMAs. However, in order to accurately represent the dynamic experimental hysterical loop of pseudoelastic SMA wire device, the thermomechanical model is further modified in this paper considering the decrease of apparent modulus of elasticity of the austenite while unloading. The computationally efficient physically based proposed model quantifies the effect of pseudo-elasticity on wide range of system parameters like SMA operating temperature, hysterisis, structural stiffness, hardening, softening, displacement due to phase transformations, residual martensite strains, and reduction of modulus of austenite during unloading. Although many other simplified models are available in the literature, this model gives accurate prediction of the experimental response in view of modeling the residual strains. The modified multilinear model of SMA device also predicts the experiment precisely. Comparison of different simplified SMA models is made with the proposed model using an application of the wire device on a steel frame representing a single degree of freedom system and subjected to earthquake time histories with increasing level of excitations.

2. Pseudoelastic Hysterisis

Pseudo elastic effect is the property by which the shape memory alloy recovers its initial shape when external load is removed. It occurs in stress induced austenitic to martensitic phase transformation of shape memory alloys. Figure 1(a) shows the experimental stress strain behavior of an austenitic SMA wire stressed uniaxially. On loading (o-a-b-c), the wire, when the stress reaches the transformation level σ_{ms}, transformation from austenite to martensite will be induced and this transformation will continue until all the austenite has been transformed to martensite at almost constant stress level (if strain rate is very low). Upon release of stress during unloading, (c-d) martensite unloads elastically down to σ_{as}, where it will transform back to austenite (d-e), once again at almost constant stress level (if strain rate is very low). When this transformation is completed, there is final elastic unloading (e-o) of austenite phase. This is called pseudoelastic (superelastic) effect as there is no permanent deformation though the behaviour is nonlinear. However, researchers [12] have observed experimentally that subjecting the austenitic Ni-Ti shape memory alloys to cycles show that some residual martensite is retained in superelastic Ni-Ti shape memory alloys subjected to repeated cyclic loading. This residual martensite progressively increases while increasing the number of cycles and tends to reach a value of 0.4%–0.5% in terms of residual strain after about 50 cycles. This is due to the local stress created in particular at the grain boundaries which locally induces martensite and retains it. Moreover, Liu and Xiang [15] measured the apparent modulus of austenite while loading and unloading and found that the measurement decreased continuously with increasing magnitude of deformation. The slope of e′o′ during unloading is 80% of the slope during loading. In Figure 1, oo′ is the residual martensite which is retained at the end of 50 cycles.

3. SMA Damper Device

Energy dissipation device was designed making use of the energy dissipation properties of Ni-Ti wires. The wires are austenitic wires which give the energy dissipating as well as the recentering property. The device as shown in Figure 2(a) consists of two concentric pipes which will move mutually. Three studs are attached to inner pipe at angles 120° apart at two locations. Six studs are attached to outer pipe in the centre equidistant (100 mm) from the studs attached to the inner pipe. Super-elastic SMA wires having length 100 mm are attached between these two studs. However, to get a good stroke length of the damper, the length of the wires can be increased. In order to get a good grip on the wires and prevent slipping of the wires during tensile tests, they were fixed to the studs using collets. Thus, the damper consists of 6 wires each of 1.2 mm diameter connected to studs. The testing of the damper for tension compression sinusoidal loading was done at room temperature of 35°C. First, the tests were carried out on wires (Figure 1) and then the damper device was tested. It is observed from the figure that the wires can be used effectively for energy dissipation upto a strain of 8%. During testing of the damper one end of the damper is fixed and other end is connected to an actuator. When the damper is loaded, at a time three of the wires are in tension and three are slack. The wires, which will be in tension dissipate earthquake energy and less energy will be transmitted to the structural system. Thus two independent groups of wire loops will act as energy dissipating group. Testing has been carried out on SMA Damper with 3 numbers of 1.2 mm diameter wires in tension and 3 wires becoming slack at a time. Tests have been carried out by increasing the loading rate from quasistatic conditions (0.01 Hz) to the loading rate of predominant earthquake frequencies (0.1 Hz–5 Hz). Combined graphs showing one cycle for 3 Hz loading rate with increasing amplitude are shown in Figure 2(b). It is observed that with change in amplitude for same frequency, the stress at austenite start transformation (σ_{as}) remains the same while the unloading paths are different at different amplitude of loadings.

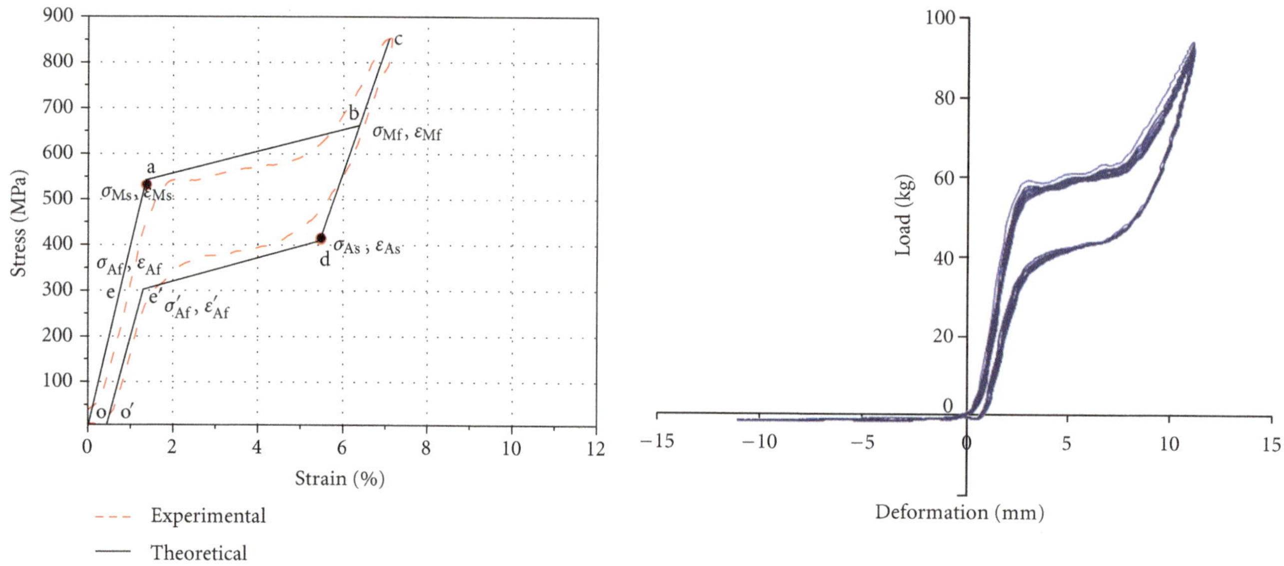

(a) (b)

Figure 1: (a) Pseudoelastic Hysterisis of SMA wire. (b) Force displacement curve for cyclic tensile tests on 1.2 mm dia SMA wire.

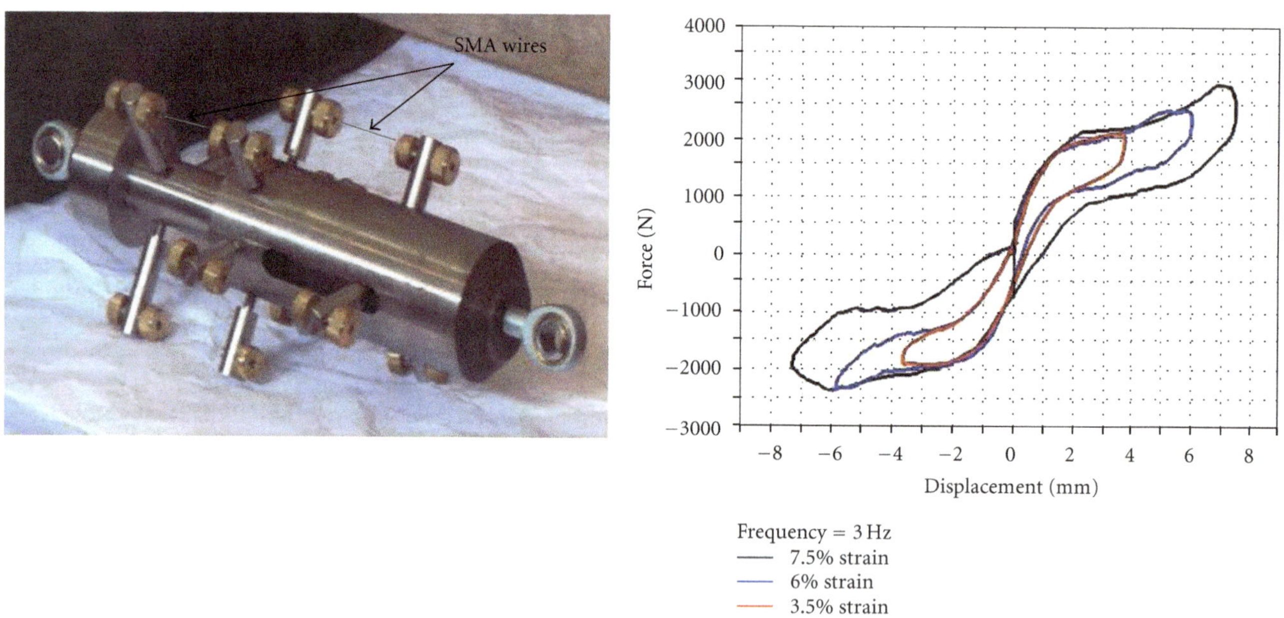

(a) (b)

Figure 2: Shape memory alloy damper device. (a) Schematic. (b) Force displacement relation.

4. Thermomechanical Model of SMA Damper

Thermomechanical model based on the original work of Boyd and Lagoudas [8] which is modified [10] to give multi-linear stress train relationship is used in the present study. This model is simple and has the ability to capture the cyclic loading effects on SMAs in addition to sub looping behavior resulting from incomplete phase transformation cycles. The uniaxial themo-mechanical behavior of SMAs can be described by constitutive relation

$$\sigma = E(\xi)\varepsilon - \alpha(\xi)E(\xi)(T - T_s) - E(\xi)\xi\varepsilon_l, \quad (1)$$

where σ, ε, T, T_s are the stress, strain, temperature, and start temperature of transformation for a mixture of austenite and martensite material. E and α are the effective Young's Modulus and effective coefficient of thermal expansion attributable to transformation. ε_l denotes the maximum residual strain which is material constant and ξ denotes the volume fraction of the martensite phase. Different researchers have proposed different thermomechanical models in which the evolution equation for martensite fraction ξ has exponential and polynomial relation [7, 8]. The model used in the present study is proposed by Motahari and Ghassemieh [10] in which the evolution equation enforces linear relationship between stress and strain. This model is modified to take into account residual martensite strain of 0.45% in the stress strain relation path when the pseudo-elastic wire is unloaded.

Considering isothermal process ($T = T_s$) for super-elastic behaviour and using (1)

$$\sigma = E(\xi)\varepsilon - E(\xi)\xi\varepsilon_l. \tag{2}$$

Now considering E_A and E_M Effective Young's moduli of austenite and martensite state, respectively, the subsequent equation is good approximation for polycrystalline SMAs:

$$E(\xi) = E_A + (E_M - E_A)\xi, \tag{3}$$

where function $\xi = f(\varepsilon)$ will be such that

$$\sigma = E(\xi)\varepsilon - E(\xi)\xi\varepsilon_l = a\varepsilon + b, \tag{4}$$

where a and b can be obtained enforcing the model to pass though finish critical stresses\and strains.

The critical stresses (σ_{Ms}, σ_{Mf}, σ_{As}, and σ_{Af}) and strains (ε_{Ms}, ε_{Mf}, ε_{As}, and ε_{Af}) are obtained using the stress strain relationship obtained from wire tests. Sinusoidal cyclic tests were performed on 0.4 mm diameter and 1.2 mm diameter superelastic wires for varying displacement amplitudes from 5 mm to 11 mm (4% strain to 9% strain). The force displacement relationship for 1.2 mm diameter wire is shown in Figure 1(b). It is observed from the figure that the wires can be used effectively for energy dissipation upto a strain of 8%. With strains greater than 8%, the hysterisis loops narrow and translate upwards, while branches of the curve relevant to the phase transformations harden, thus yielding an increase in stress levels. Similar behaviour was observed by Clark et al. [3]. Tests were also carried out with increasing number of cycles to obtain the fatigue characteristics and it is observed that 1.2 mm diameter wire is capable of taking 1200 cycles of 7 mm amplitude (6% strain) without decrease in the load taken by the wire. Thus, the wires can withstand considerable amount of fatigue loading within a strain range of 2–10%.

Using Figure 1(a), the transformation critical stresses are obtained as $\sigma_{Ms} = 551$ MPa, $\sigma_{Mf} = 667$ MPa, $\sigma_{As} = 402$ MPa and $\sigma_{Af} = 288.1$ MPa whereas the critical strains are obtained as $\varepsilon_{Ms} = 0.013$, $\varepsilon_{Mf} = 0.064$, $\varepsilon_{As} = 0.054$, and $\varepsilon_{Af} = 0.0068$. The Young's moduli $E_A = 42308$ MPa, $E_M = 28571$ MPa and the total phase strain value, $\varepsilon_l = 0.04$, is also obtained from the uniaxial tests in wires. The maximum value of residual matrensite strain, $\varepsilon_{ir,max}$, is obtained from tests on wires as 0.0045. Modulus of elasticity of Austenite Ni-Ti was measured for unloading and reloading by Liu and Xiang and it was observed that the modulus of elasticity for unloading is 15–20% lesser than that for loading [15]. This occurs due to contribution of a second deformation mechanism operating during the apparent elastic deformation of a specimen. This second deformation mechanism may be either the stress-induced martensitic transformation or martensite reorientation, depending on the starting structure. Thus, considering the modulus of elasticity of unloading as 35055, it is observed that the values of σ'_{Af} and ε'_{Af} are obtained as 303.44 MPa and 0.0132, respectively.

The transformation temperatures for the SMA considered in present study are $M_f = -80°C$, $M_s = -60°C$, $A_s = -25°C$, and $A_f = -8°C$. The tests on SMA wires have been carried out at room temperature hence $T_s = 35°C$. Using the phase diagram (Figure 3(a)) and the pseudoelastic multilinear loading path (Figure 3(b)), the transformation constants C_A and C_M are obtained as 6.7 Mpa and 5.8 MPa, respectively. These values are in agreement with those obtained by Boyd and Lagoudas [8] for pseudo-elastic NiTi alloys. The stress strain relations on the paths in Figure 3(b) are as follows.

Elastic fully austenitic (loading), Path O-A and E-O:

$$\sigma = E_A\varepsilon. \tag{5}$$

Forward Transformation (loading), Path A-B:

$$\sigma = \sigma_{Ms} + \frac{\sigma_{Mf} - \sigma_{Ms}}{\varepsilon_{Mf} - \varepsilon_{Ms}}(\varepsilon - \varepsilon_{Ms}). \tag{6}$$

Elastic fully martensite (unloading), Path B-C and C-D:

$$\sigma = \sigma_{Mf} + E_M(\varepsilon - \varepsilon_{Mf}) \tag{7}$$

Reverse Transformation (unloading), Path D-E′:

$$\sigma = \sigma_{As} + \frac{\sigma'_{Af} - \sigma_{As}}{\varepsilon'_{Af} - \varepsilon_{As}}(\varepsilon - \varepsilon_{As}). \tag{8}$$

Unloading in elastic Austenitic with residual martensite strain for $\varepsilon < \varepsilon'_{Af}$ (i.e., 0.0132), Path E′O″:

$$\sigma = \frac{\sigma'_{Af}}{\varepsilon'_{Af} - \varepsilon_{ir}}(\varepsilon - \varepsilon_{ir}), \tag{9}$$

σ'_{Af} and ε'_{Af} are the stress and the strain corresponding to austenite finish states taking into account the residual strain, ε_{ir}. The maximum value of this strain is limited to $\varepsilon_{ir\,max}$ (i.e., 0.45%) for complete martensite transformation. Lexcellent and Bourbon [11] stated that for cycles with incomplete martensite transformations residual stress will depend on the residual martensite fraction Z_{ir}. The maximum residual strain is proportional to maximum residual martensite fraction by the relation

$$\varepsilon_{ir\,max} = \varepsilon_l Z_{ir}, \tag{10}$$

where $\varepsilon_l = 0.04$ is the total phase strain. Thus for the incomplete phase transformation strain say γ the residual martensite strain will be given by relation

$$\varepsilon_{ir} = \frac{\varepsilon_{ir\,max}}{\varepsilon_l}\gamma. \tag{11}$$

If unloading occurs before the completion of forward transformation or reloading begins before the completion

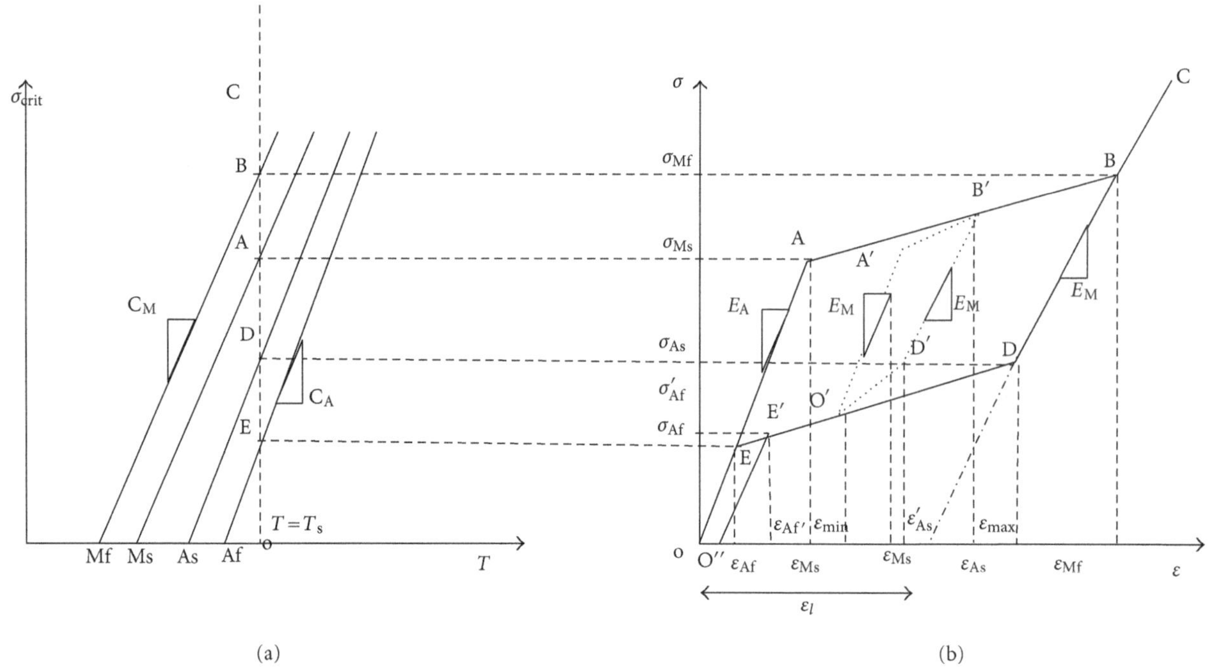

FIGURE 3: (a) Phase diagram for pseudoelasticity. (b) Pseudoelasticity loop and subloop.

of reverse transformation, then the elastic stiffness would be different from both austenite and martensite phases. If L_m and L_a are the compliances of mixture constituents and they have linear relationship with strain [1]:

$$L_m = xL_M + (1 - x)L_A, \tag{12}$$

in which x is $(\varepsilon_{max} - \varepsilon_{Ms})/(\varepsilon_{Mf} - \varepsilon_{Ms})$ for unloading and x is $(\varepsilon_{min} - \varepsilon_{Af})/(\varepsilon_{As} - \varepsilon_{Af})$ for reloading.

Where ε_{max} and ε_{min} are maximum and minimum strains before unloading or reloading as shown in Figure 3(b), the elastic stiffness would become:

$$E_m = \frac{E_M E_A}{x(E_A - E_M) + E_M}. \tag{13}$$

The new values for critical austenite strain will be obtained as:

$$\varepsilon'_{As} = \varepsilon_{max} + \frac{\sigma_{As} - \sigma_{max}}{E_m}. \tag{14}$$

Similarly, new value of critical martensite strain, ε'_{Ms} is obtained by substituting ε_{max},σ_{max} and σ_{As} by ε_{min}, σ_{min} and σ_{Ms}, respectively. Where σ_{max} and σ_{min} are stresses corresponding to strains ε_{max} and ε_{min}, the stress strain relations on the subloop paths shown in Figure 3(b) are as follows:

Path O′A′:

$$\sigma = \sigma_{min} + E_m(\varepsilon - \varepsilon_{min}), \tag{15}$$

Path A′B′

$$\sigma = \sigma_{Ms} + \frac{\sigma_{Mf} - \sigma_{Ms}}{\varepsilon_{Mf} - \varepsilon'_{Ms}}(\varepsilon - \varepsilon'_{Ms}), \tag{16}$$

Path B′D′

$$\sigma = \sigma_{max} + E_m(\varepsilon - \varepsilon_{max}), \tag{17}$$

Path D′O′

$$\sigma = \sigma_{As} + \frac{\sigma_{Af} - \sigma_{As}}{\varepsilon_{Af} - \varepsilon'_{As}}(\varepsilon - \varepsilon'_{As}). \tag{18}$$

For the complete stress strain relationship of the SMA damper device subjected to cyclic loading, it is essential to define the stress strain path when the damper will undergo compression load. During compressive load, another set of three wires go in tension according to the design of the damper. Thus, when the load becomes negative during unloading, that is, Path O″F (see Figure 4), stress-strain relation ship, will be equivalent to that of unloading in elastic austenitic with reduced modulus of Elasticity, equivalent to 80% of initial modulus E_A and with residual martensite strain (Path E′O″).

Path O″F

$$\sigma = \frac{\sigma'_{Af}}{\varepsilon'_{Af} - \varepsilon_{ir}}(\varepsilon - \varepsilon_{ir}), \tag{19}$$

where σ'_{Af} and ε'_{Af} are the stress and the strain corresponding to austenite finish states taking into account the residual strain.

During further unloading Path FG

$$\sigma = E_A\varepsilon, \tag{20}$$

where E_A is the initial elastic modulus of austenite. For further unloading (Path GH) stress strain characteristics will be similar to Path AB.

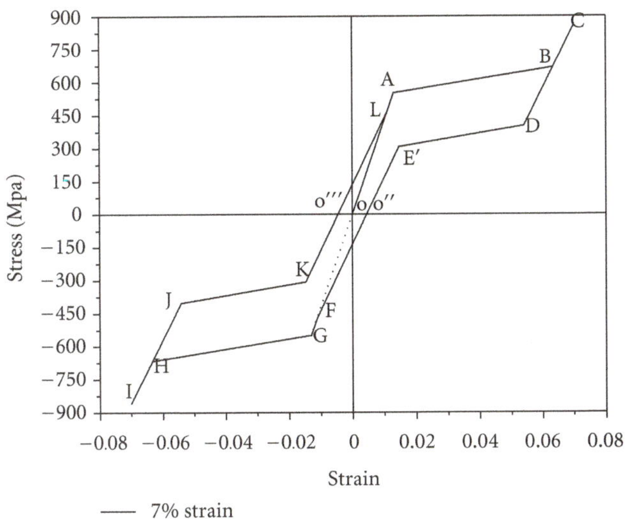

FIGURE 4: Stress strain relationship of damper.

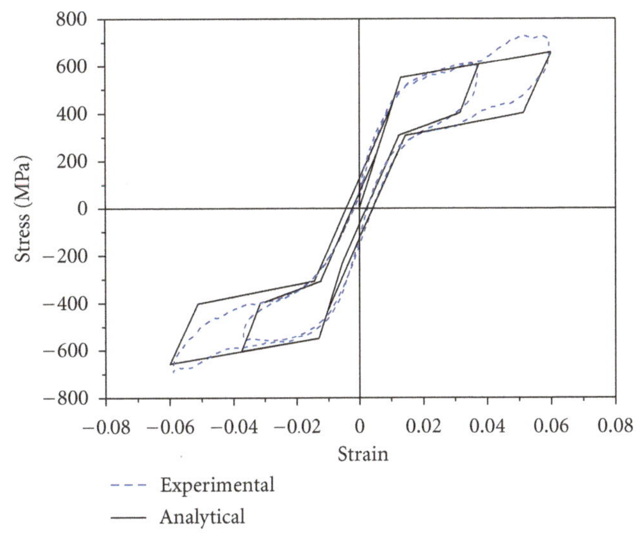

FIGURE 5: Comparison of experimental and analytical stress strain relation ship.

Using (5)–(20) and the critical stress and strains obtained from the wires characteristics, the load deflection characteristics of the damper device is obtained. This theoretical load deflection characteristic is compared with that obtained by experiments. Figure 5 shows the comparison for 3.75% and 6% strain. The hysterisis loop of the experiments and theory are in good agreement. The analytical model shows the residual deformation in the hysterisis loop when the load is less than austenite finish state. This model is an improvement of the models proposed in literature [9, 10] where the force displacement characteristics shows same linear path for loading and unloading when the load is less than Austenite finish state. The force displacement characteristics of the analytical model thus well predict the energy dissipated in SMA wires compared to the previous models mentioned in literature. The cyclic stress strain relationship for varying peak strain (2–7%) using the analytical model is shown in Figure 6. The shaft friction between the two cylinders of the damper device is negligible due to lubrication between the two surfaces using oil lubricants. It is observed that the theoretical load deflection characteristic is higher than experimental in some parts while it is lower in some other parts. However considering the overall area of energy dissipation the theoretical and experimental results are in good agreement.

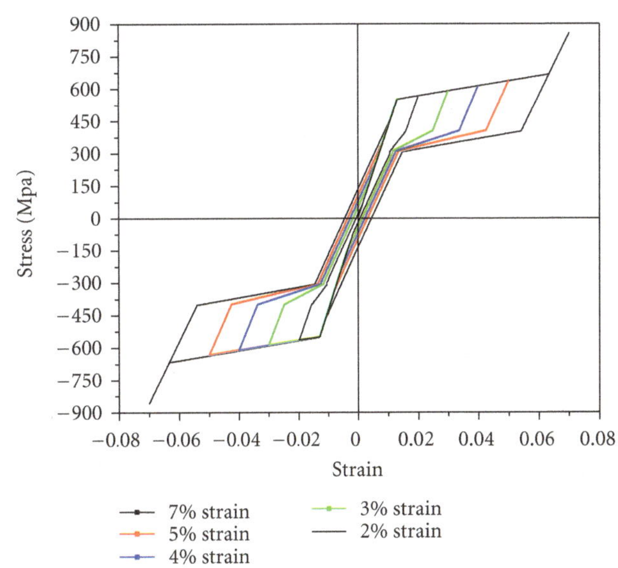

FIGURE 6: Cyclic stress strain relationship for varying peak strain.

5. SMA Analytical Models

In this paper three SMA models from literature are investigated and response of a steel framed structure with these models is compared with that of the proposed SMA model. The first model considered is a simplified model which is experimental based as given by Thomson et al. [2] in which the hysteretic properties are predefined and are independent of thermomechanical properties of SMA. The second model considered is the multilinear thermomechanical model given by Boyd and Lagoudas [8] and modified by Motahari and Ghassemieh [10]. This model has an ability to capture the cyclic loading effects on SMAs in addition to the sub looping behaviour resulting from incomplete phase transformations. The third model considered is mentioned in Parulekar et al. [14]. This model is the modified version of the model brought out by Motahari and Ghassemieh which further takes into account the effect of residual martensite strain accumulated during cyclic loading.

6. Experiment and Analysis of Steel Structure

The six-storey steel structural model shown in Figure 7(a) tested on shake table is considered for the analysis. Mass of 6.0 tons representing the mass of a tank is placed on the 4th storey in form of plates and the total mass of the structure is 6.9 tons.

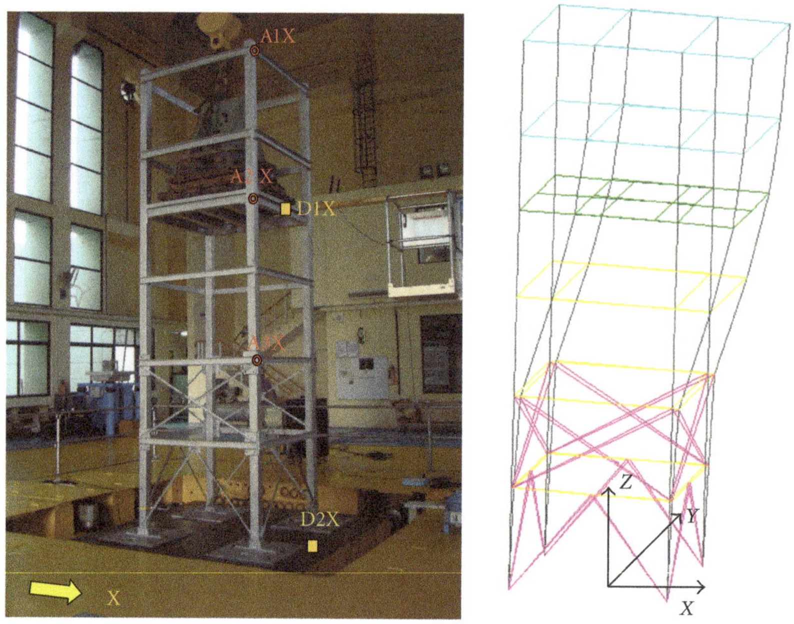

(a) (b)

FIGURE 7: Model of the steel structure. (a) Test model. (b) FE model showing mode shape (Frequency = 3.1 Hz).

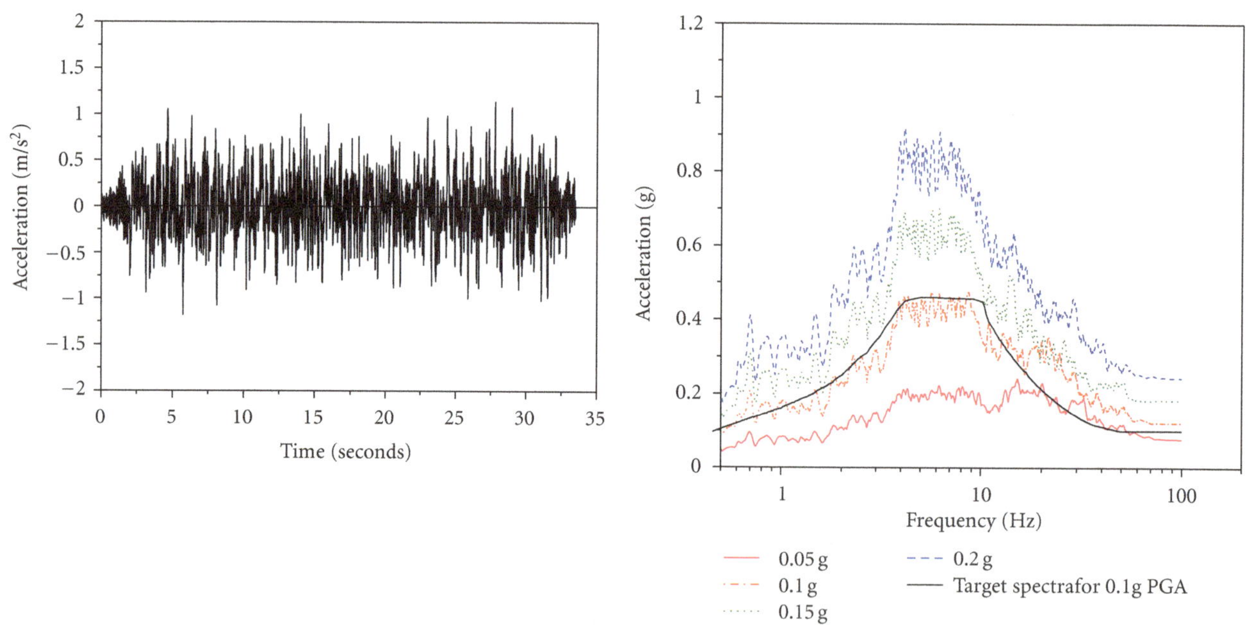

(a) (b)

FIGURE 8: (a) Spectrum compatible time history. (b) Response spectrum for 2% damping.

Sine sweep tests are carried out on the structure. Transfer function is obtained and the resonance frequency is 3.05 Hz and the damping is obtained as 2%. The model is then subjected to spectrum compatible time histories shown in Figure 8(a) with peak acceleration increasing from 0.05 g to 0.2 g. The response spectra for the test time histories and target response spectra are shown in Figure 8(b). The spectra are plotted for damping of 2% which is obtained from the experiments. Time history analysis is carried out for the steel structure for the spectrum compatible time history of 0.05 g to 0.2 g PGA with increment of 0.05 g. The analysis results are compared with the test results. The first fundamental frequency of the structure is 3.1 Hz with 96% mass participation in the first mode (Figure 7(b)). Thus, the structure can be idealized as a single degree of freedom system. The comparison of test and analysis displacement time histories at

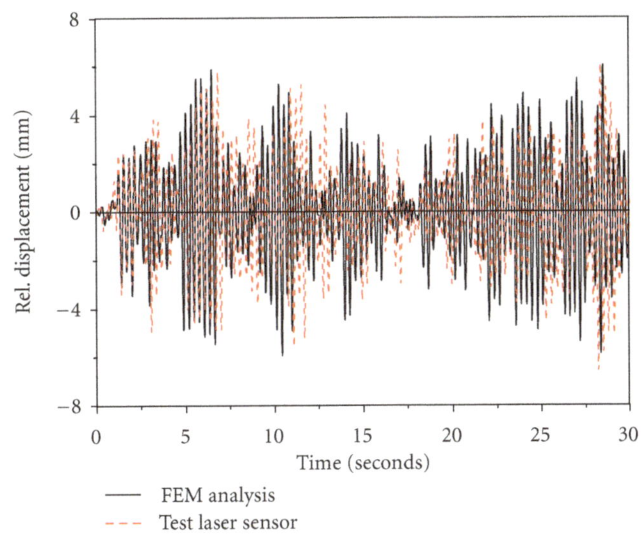

FIGURE 9: Comparison of test and analysis displacement time histories at mass level for 0.1 g PGA.

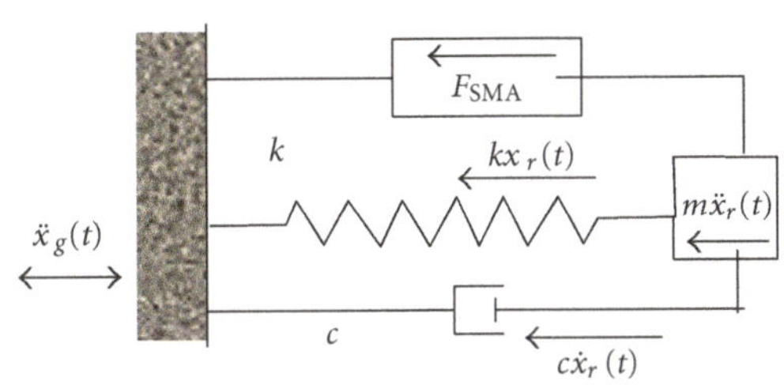

FIGURE 10: SDOF system with SMA damper.

the mass level of the frame is shown in Figure 9 for 0.1 g PGA. It is observed from the figure that the analysis and test response are in good agreement with each other. The maximum difference in peak response in test and analysis is about 10%. The displacements and accelerations obtained at various floors of the structure in linear analysis at 0.1 g peak base acceleration are obtained from the analysis of single degree of freedom system by multiplying the displacement and acceleration obtained at the mass with the spatial representation vector $\{\phi\}$.

7. Nonlinear Analysis of Frame with SMA Models

The SMA damper consisting of 4 wires of 1.2 mm dia acting simultaneously in tension is to be placed on either side of the steel frame. Thus, at a time 8 wires (4 on either side of the frame) will be in tension with its force and displacement required for martensitic transformation as 4950 N and 1.3 mm, respectively. The structure can be idealized as a single degree of freedom system connected with SMA damper as shown in Figure 10. The equation of motion for such a case is written as:

$$m\ddot{x}(t) + kx(t) + c\dot{x}(t) + F_{SMA}(t) = -m\ddot{x}_g(t), \quad (21)$$

where m is the mass attached to the structure, c is the damping coefficient, k is the stiffness of the structure, $x(t)$, $\dot{x}(t)$, $\ddot{x}(t)$ are the displacement velocity and acceleration of the structure at the lumped mass and $\ddot{x}_g(t)$ is the base acceleration. $F_{SMA}(t)$ is the force exerted on the structure by SMA damper. Stress strain relations of SMA damper can be used to describe the F_{SMA} when the damper has complete and incomplete phase transformations. Nonlinear time history (TH) analysis is performed for the structure implemented with damper considering four different models. Analysis is performed for the spectrum compatible earthquake time history shown in Figure 8 with peak acceleration from 0.05 g to 0.2 g.

8. Results and Discussion

The comparison of the force displacement relationship of the damper considering all the four models for 0.05 g and 0.15 g is shown in Figures 11 and 12, respectively. The force deformation characteristics for thermomechanical model show slight increase in maximum deflection than the simplified model for 0.05 g and 0.15 g peak excitation. This is because the thermomechanical model showed more realistic unloading path than the simplified model. Moreover, there was reduction in level of force during unloading thus the energy dissipated was less than the simplified model. This resulted

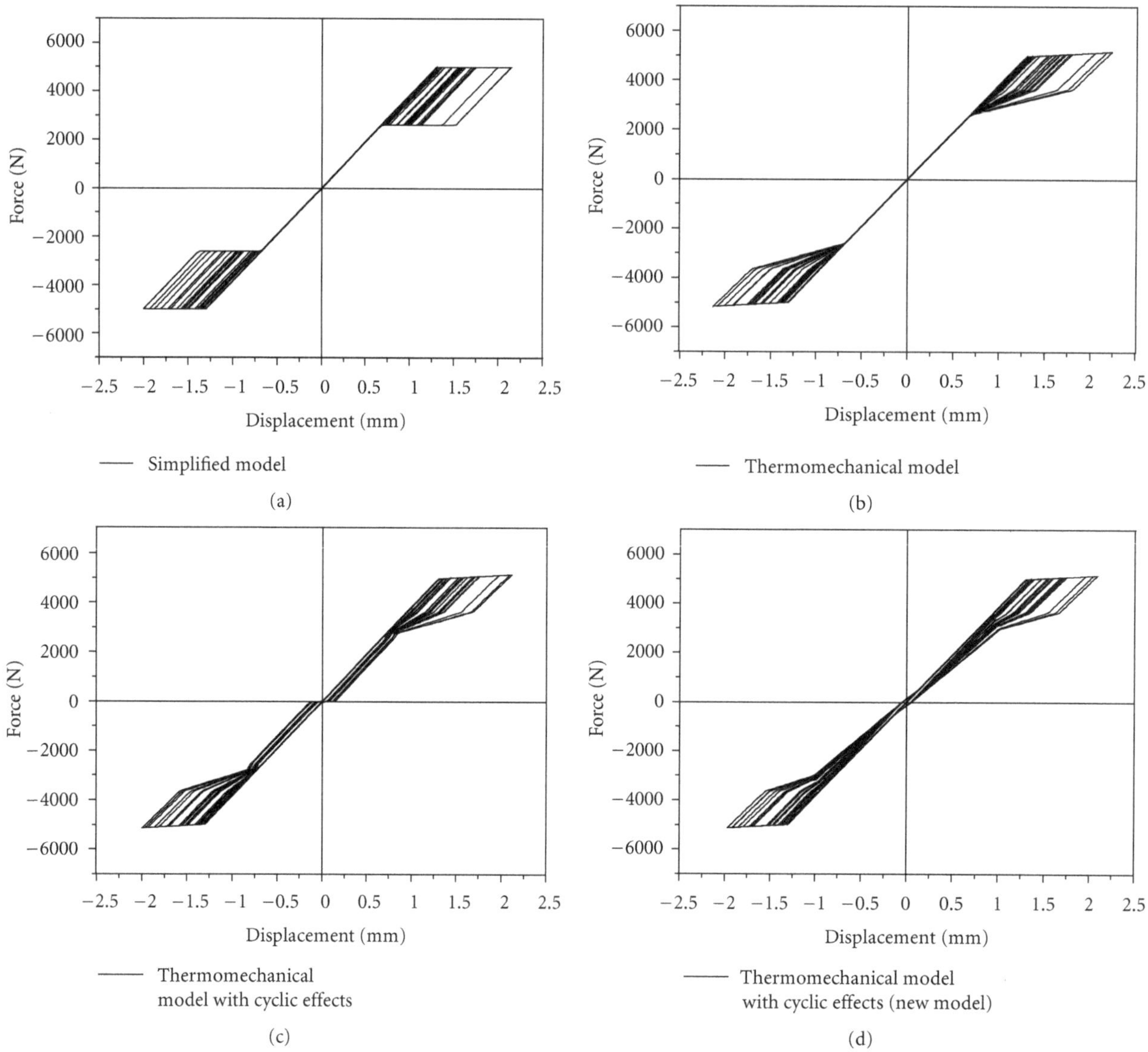

FIGURE 11: Comparison of force displacement characteristics of SMA damper models for 0.05 g peak base excitation.

in slight increase in the maximum displacement. The force deflection characteristics of Thermomechanical model with cyclic effects show slightly lesser maximum displacement than the simplified model and the thermomechanical model. The energy dissipated in thermomechanical model with cyclic effects is higher due to the residual strain accumulation which reduces the maximum displacement. The reduction in the response of the structure with SMA damper for increasing base excitation considering the different SMA models is shown in Figure 13.

Figure 13 shows that the response reduction of thermomechanical model with cyclic effects is higher than that of simplified model and the thermomechanical model. The displacement response of simplified model is quite close to the thermomechanical model and average difference in displacement response between the two models is about 4%. It is also observed from the figure that the response reduction of the structure with damper decreases with increase in peak base excitation level. The damping due to hysterisis deformation of SMA is proportional to ratio of hysterisis energy to strain energy. Due to increase in strain energy of structure with increase in peak base excitation the damping decreases with excitation level and thus the response reduction decreases. It is also observed that the newly proposed model with residual martensite strain is more effective in capturing the energy dissipation of the SMA damper device. For higher excitation, if the damper undergoes nonlinear displacement greater than 6.4 mm (i.e., higher than complete martensite transformation strain, ε_{Af} of 6.4%), the wires undergo martensitic hardening of the SMA. As the PGA further increases, the elastic response with modulus E_M of pure martensite follows. Figure 12 shows that at 0.15 g peak base excitation, the peak response displacement of structure with damper is 6.8 mm and thus the wires undergo maximum strain of 6.8%. It is observed experimentally that the maximum strain taken by SMA wires is about 10% and hence the damper designed

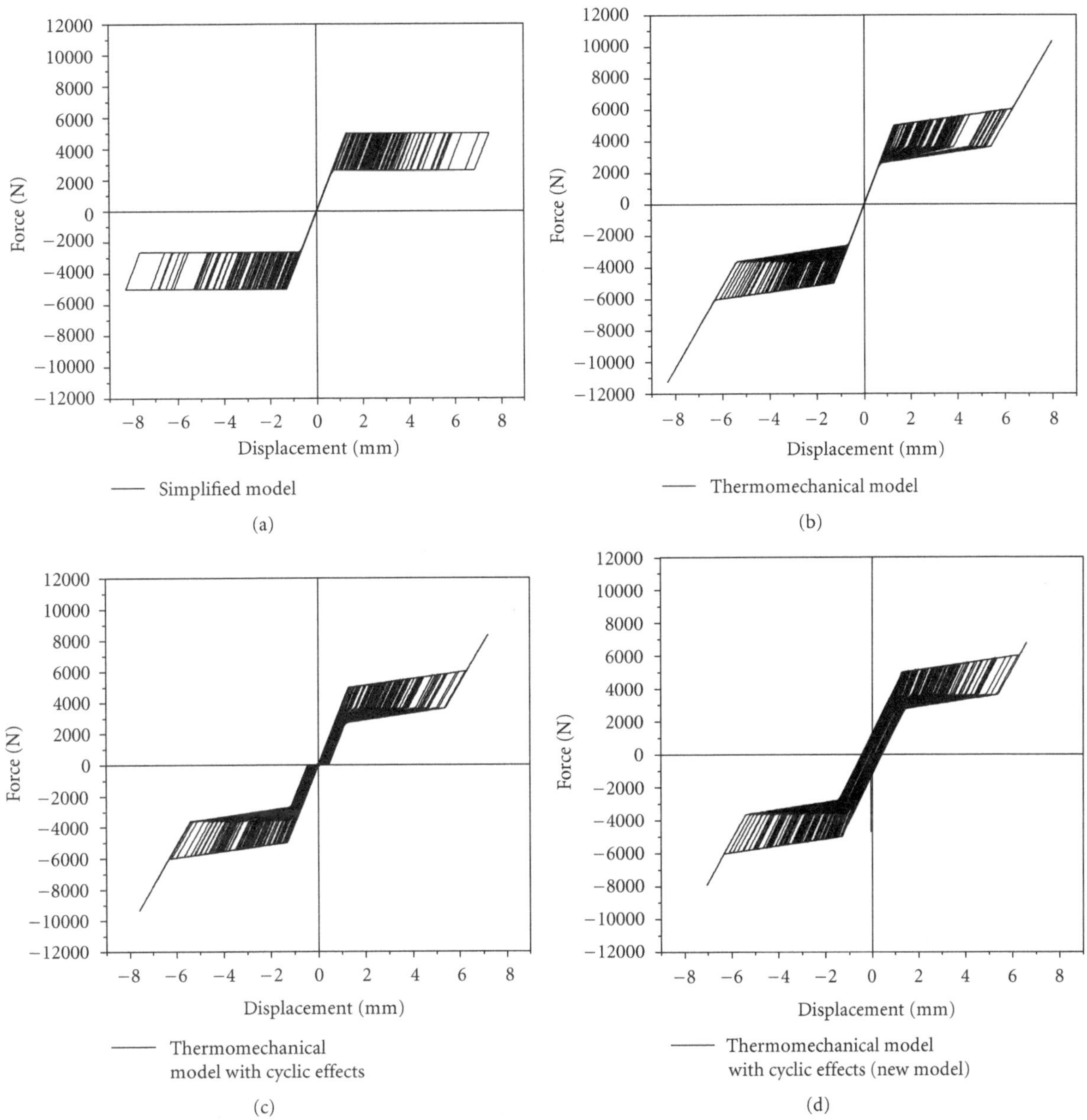

Figure 12: Comparison of force displacement characteristics of SMA damper models for 0.15 g peak base excitation.

can be used for peak base excitation of 0.2 g. However, for higher excitation damper should be designed for better stroke by increasing the length of the wires used. It is also observed from Figure 13 that the reduction in displacement response for 0.05 g, 0.1 g, 0.15 g, and 0.2 g is 41%, 43%, 31%, and 22%, respectively, for the new SMA model considering cyclic effects. For higher excitation (greater than 0.15 g), the damper undergoes martensite hardening and the effectiveness of damper decreases. Moreover, for higher excitation the increase is strain energy effect also decreases the damping given by the SMA damper. Hence, it is essential that the number of wires and the length of the wires should be increased to maintain the effectiveness of damper subjected to higher excitation.

9. Conclusions

The mechanical behavior of superelastic shape memory alloys suits the optimal requirements of a seismic control device. A thermomechanical model of pseudoelastic shape memory alloy damper device is developed. This model includes the cyclic effects of SMAs considering accumulation of martensite strain. This damper device is implemented on a steel-framed structure and the structural response using three simplified models from literature is compared with that obtained using proposed model. It is observed that the force deflection characteristics of the thermomechanical model with cyclic effects show slightly lesser maximum displacement than the simplified model and the thermomechanical

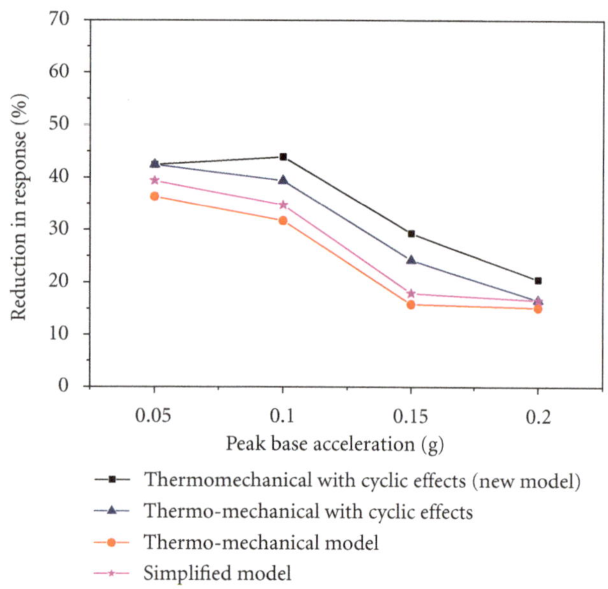

FIGURE 13: Reduction in response for different SMA damper models with peak base excitation.

model. It is observed that the proposed model is more effective in capturing the real characteristics of SMA than the other models. For higher excitation, the effectiveness of SMA damper decreases due to martensitic hardening of SMAs for higher strains. It is observed that the average reduction in response of the steel structure with SMA damper device is about 35%.

References

[1] E. J. Graesser and F. A. Cozzarelli, "Shape-memory alloys as new materials for aseismic isolation," *Journal of Engineering Mechanics*, vol. 117, no. 11, pp. 2590–2608, 1991.

[2] P. Thomson, G. J. Balas, and P. H. Leo, "The use of shape memory alloys for passive structural damping," *Smart Materials and Structures*, vol. 4, no. 1, pp. 36–41, 1995.

[3] P. W. Clark, I. D. Aiken, J. M. Kelly, M. Higashino, and R. C. Krumme, "Experimental and analytical studies of shape-memory alloy dampers for structural control," in *Proceedings of SPIE—The International Society for Optical Engineering*, vol. 2445, pp. 241–251, 1995.

[4] M. Dolce, D. Cardone, and R. Marnetto, "Implementation and testing of passive control devices based on shape memory alloys," *Earthquake Engineering and Structural Dynamics*, vol. 29, no. 7, pp. 945–968, 2000.

[5] M. Dolce, D. Cardone, F. Ponzo, and C. Valente, "Shaking table tests on reinforced concrete frames without and with passive control systems," *Earthquake Engineering and Structural Dynamics*, vol. 34, no. 14, pp. 1687–1717, 2005.

[6] K. Tanaka, "Thermo-mechanical sketch of shape memory effect: one dimensional tensile behaviour," *Res Mechanica*, vol. 18, no. 3, pp. 251–263, 1986.

[7] C. Liang and C. A. Rogers, "A multi-dimensional constitutive model for shape memory alloys," *Journal of Engineering Mathematics*, vol. 26, no. 3, pp. 429–443, 1992.

[8] J. G. Boyd and D. C. Lagoudas, "A thermodynamical constitutive model for shape memory materials. Part I. The monolithic shape memory alloy," *International Journal of Plasticity*, vol. 12, no. 6, pp. 805–842, 1996.

[9] M. M. Khan, D. C. Lagoudas, J. J. Mayes, and B. K. Henderson, "Pseudoelastic SMA spring elements for passive vibration isolation: part I—modeling," *Journal of Intelligent Material Systems and Structures*, vol. 15, no. 6, pp. 415–441, 2004.

[10] S. A. Motahari and M. Ghassemieh, "Multilinear one-dimensional shape memory material model for use in structural engineering applications," *Engineering Structures*, vol. 29, no. 6, pp. 904–913, 2007.

[11] C. Lexcellent and G. Bourbon, "Thermodynamical model of cyclic behaviour of Ti-Ni and Cu-Zn-Al shape memory alloys under isothermal undulated tensile tests," *Mechanics of Materials*, vol. 24, no. 1, pp. 59–73, 1996.

[12] S. Miyazaki, T. Imai, Y. Igo, and K. Otsuka, "Effect of cyclic deformation on the pseudoelastic characteristics of Ti-Ni alloys," *Metallurgical Transactions A*, vol. 17, no. 1, pp. 115–120, 1986.

[13] B. Andrawes and R. DesRoches, "Sensitivity of seismic applications to different shape memory alloy models," *Journal of Engineering Mechanics*, vol. 134, no. 2, pp. 173–183, 2008.

[14] Y. M. Parulekar, G. R. Reddy, K. K. Vaze et al., "Seismic response attenuation using of structures using shape memory alloy dampers," *Journal of Structural Control and Health Monitoring*, 2011.

[15] Y. Liu and H. Xiang, "Apparent modulus of elasticity of near-equiatomic NiTi," *Journal of Alloys and Compounds*, vol. 270, no. 1-2, pp. 154–159, 1998.

7

Effects of Structural Parameters on the Dynamics of a Beam Structure with a Beam-Type Vibration Absorber

Mothanna Y. Abd,[1] Azma Putra,[2] Nawal A. A. Jalil,[1] and Jamaludin Noorzaei[3]

[1] *Department of Mechanical and Manufacturing Engineering, Faculty of Engineering, Universiti Putra Malaysia, 43400 Serdang, Selangor, Malaysia*
[2] *Department of Structure and Materials, Faculty of Mechanical Engineering, Universiti Teknikal Malaysia Melaka, Hang Tuah Jaya, 76100 Durian Tunggal, Melaka, Malaysia*
[3] *Department of Civil Engineering, Faculty of Engineering, Universiti Putra Malaysia, 43400 Serdang, Selangor, Malaysia*

Correspondence should be addressed to Nawal A. A. Jalil, nawal@eng.upm.edu.my

Academic Editor: Abul Azad

A beam-type absorber has been known as one of the dynamic vibration absorbers used to suppress excessive vibration of an engineering structure. This paper studies an absorbing beam which is attached through a visco-elastic layer on a primary beam structure. Solutions of the dynamic response are presented at the midspan of the primary and absorbing beams in simply supported edges subjected to a stationary harmonic load. The effect of structural parameters, namely, rigidity ratio, mass ratio, and damping of the layer and the structure as well as the layer stiffness on the response is investigated to reduce the vibration amplitude at the fundamental frequency of the original single primary beam. It is found that this can considerably reduce the amplitude at the corresponding troublesome frequency, but compromised situation should be noted by controlling the structural parameters. The model is also validated with measured data with reasonable agreement.

1. Introduction

A beam-type absorber is one of the techniques to reduce undesirable vibration of many vibrating systems, such as a synchronous machine, mounting structure for a sensitive instrument, and other continuous structure in engineering. The absorber system usually consists of a beam attached to the host structure using an elastic element. The natural frequency of the absorber is then tuned to be the same as the troublesome operating frequency of the host structure to create counter force, which in return reduces the vibration of the structure. As beams are important structures in civil or mechanical engineering, several works have also been established to investigate the performance of the absorbing beam which is attached also to a beam structure.

Among the earliest studies of the double-beam system is one proposed by Yamaguchi [1], which investigated the effectiveness of the dynamic vibration absorber consisting of double-cantilever visco-elastic beam connected by spring and viscous damper. The auxiliary beam is attached to the center of the main beam excited at its end by a sinusoidal force. It is found that the amplitude at resonances of the main beam is sensitive to the stiffness and mass of the absorbing beam. The damping ratio was formulated as a function of mass and layer stiffness of the absorber. Vu et al. [2] studied the distributed vibration absorber under stationary distributed force. A closed form was developed by utilizing change of variables and modal analysis to decouple and solve differential equations. Oniszczuk [3] studied the free vibrations of two identical parallel simply supported beams continuously joined by an elastic layer. The eigen frequencies and mode shapes of vibration of the double-beam system were found using the classical assumed mode summation.

Another theoretical study for finding the optimum design of beam-type absorber was presented by Aida et al. [4]. A uniform absorbing beam with the same boundary

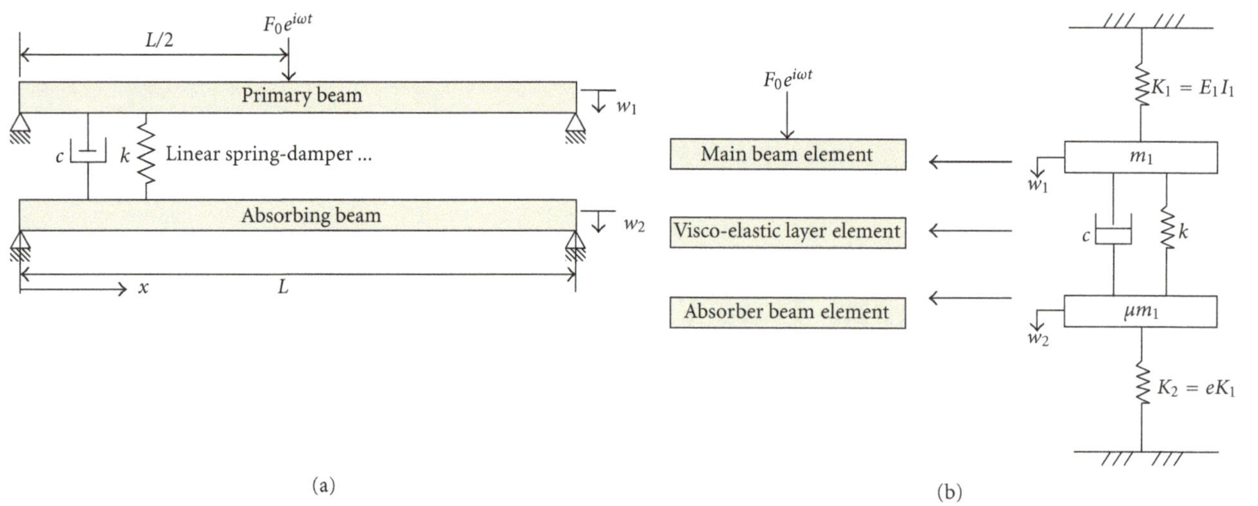

FIGURE 1: (a) Schematic diagram of a visco-elastically connected simply supported double-beam system and (b) the two-degree-of-freedom system of distributed lumped parameter model.

condition is connected to the main beam by spring and damper components. In the study, an optimum tuning method to reduce the main beam vibration was proposed. Another structural analysis and optimum design of a dynamic absorbing beam with free edges was studied by Chen and Lin [5]. The effect of mass ratio and the stiffness layer on the vibration response was discussed.

The effect of different forcing types on the double-beam system has also been discussed by several authors. Zhang et al. [6] studied vibration characteristics of the double-beam system under axial compressive load. The studies were limited for two identical simply supported double-beam system and the effect of the axial load on the beams vibration amplitude was reported. Abu-Hilal [7] studied the effect of a moving constant load on the dynamic response of the beams. This was done for different values of speed parameter, damping ratio, and stiffness parameter.

Several works focus on the development of mathematical model to provide solutions of the vibration response. De Rosa and Lippiello [8] studied free vibration of double-identical beam system using the Differential Quadrature Method (DQM). Sadek et al. [9] presented a computational method for solving optimal control of transverse vibration of a two-parallel-beam system based on legendry wavelets approach. It is found here that the reduction of the beam vibration depends on the spring location on the beam.

This paper investigates the effect of structural parameters, namely, the rigidity ratio, mass ratio, damping loss factor, the stiffness, as well as the damping ratio of the layer on the dynamic response of simply supported double-beam system to provide a thorough analysis. The discussion is limited on controlling the fundamental mode of a single-beam structure using a dynamic absorbing beam. The following section discusses the derivation of the mathematical model.

2. Mathematical Modeling

The schematic diagram of a beam connected with an absorbing beam having the same length L and simply supported is shown in Figure 1(a). Here distributed lumped system of two degrees of freedom is assumed, where the visco-elastic element between the beams consists of parallel distributed springs and dampers as shown in Figure 1(b).

The equation of motion of the dynamic model can therefore be written as

$$E_1 I_1 \frac{\partial^4 w_1}{\partial x^4} + m_1 \ddot{w}_1 + c(\dot{w}_1 - \dot{w}_2) + k(w_1 - w_2) = Pe^{i\omega t}, \quad (1)$$

$$E_2 I_2 \frac{\partial^4 w_2}{\partial x^4} + m_2 \ddot{w}_2 + c(\dot{w}_2 - \dot{w}_1) + k(w_2 - w_1) = 0, \quad (2)$$

where $P = F_0 \delta(x - L/2)$ is the external point force with frequency ω at the midspan of the beam, F_0 is the force magnitude, $K = EI$is the flexural rigidity, $c$is the damping constant, $k$is the stiffness constant of the viscous layer, and $m$is the mass of the beam. The subscripts 1 and 2 refer to the main beam and the absorber beam, respectively. The damping of the beam can be introduced by replacing the flexural rigidity in (1) and (2) by $K(1 + i\eta)$, where η is the damping loss factor. The vertical displacement of the main beam w_1 and the absorber w_2 can be expressed as a series expansion in terms of mode shape function $X_n(x)$ for the nth mode.The amplitudes in generalized time coordinates q_{1n} are given by

$$w_1(x, t) = \sum_{n=1}^{\infty} X_n(x)\, q_{1n}(t), \quad (3)$$

$$w_2(x, t) = \sum_{n=1}^{\infty} X_n(x)\, q_{2n}(t), \quad (4)$$

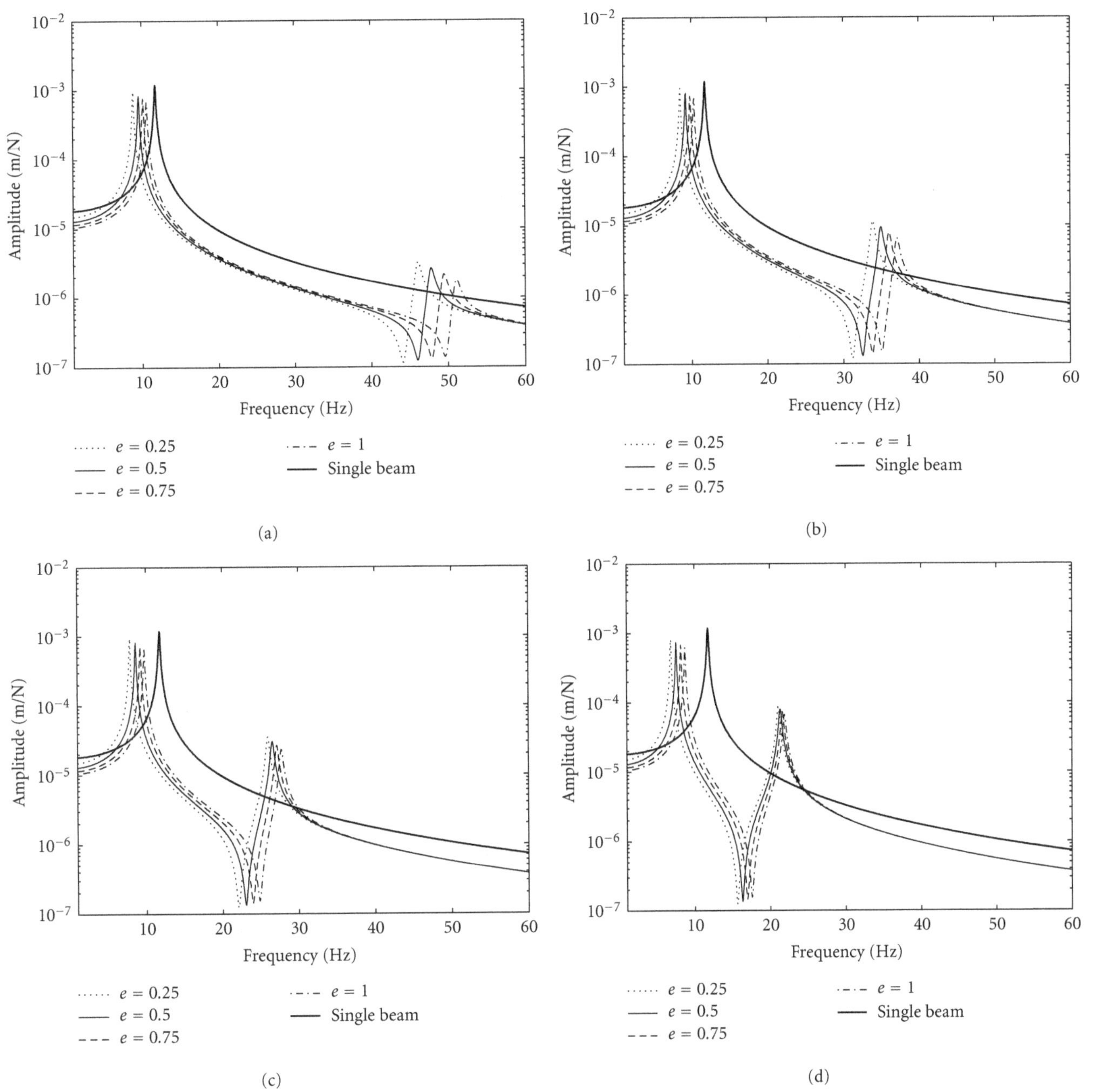

FIGURE 2: Effect of the rigidity ratio e and the mass ratio μ on the vibration amplitude at the midspan location of the primary beam: (a) $\mu = 0.1$, (b) $\mu = 0.2$, (c) $\mu = 0.4$, and (d) $\mu = 0.8$.

where x is the position on the beam at which the load is applied. For the simply supported boundary condition, the nth mode shape function can be written as

$$X_n(x) = \sin(\sigma_n x), \tag{5}$$

where σ_n is the eigenvalue of the mode shape which can be expressed as:

$$\sigma_n = \frac{n\pi}{L}. \tag{6}$$

The orthogonality conditions can be used for simplifying the equations of motion which is represented in the following form [1]:

$$\int_0^L X_n(x) X_m(x) dx = 0 , \quad n \neq m. \tag{7}$$

Equations (3) and (4) can be substituted to (1) and (2). Multiplying both sides of the equations by mode shape function X_m then integrating through the beam length and applying the orthogonality condition as in (7) yield

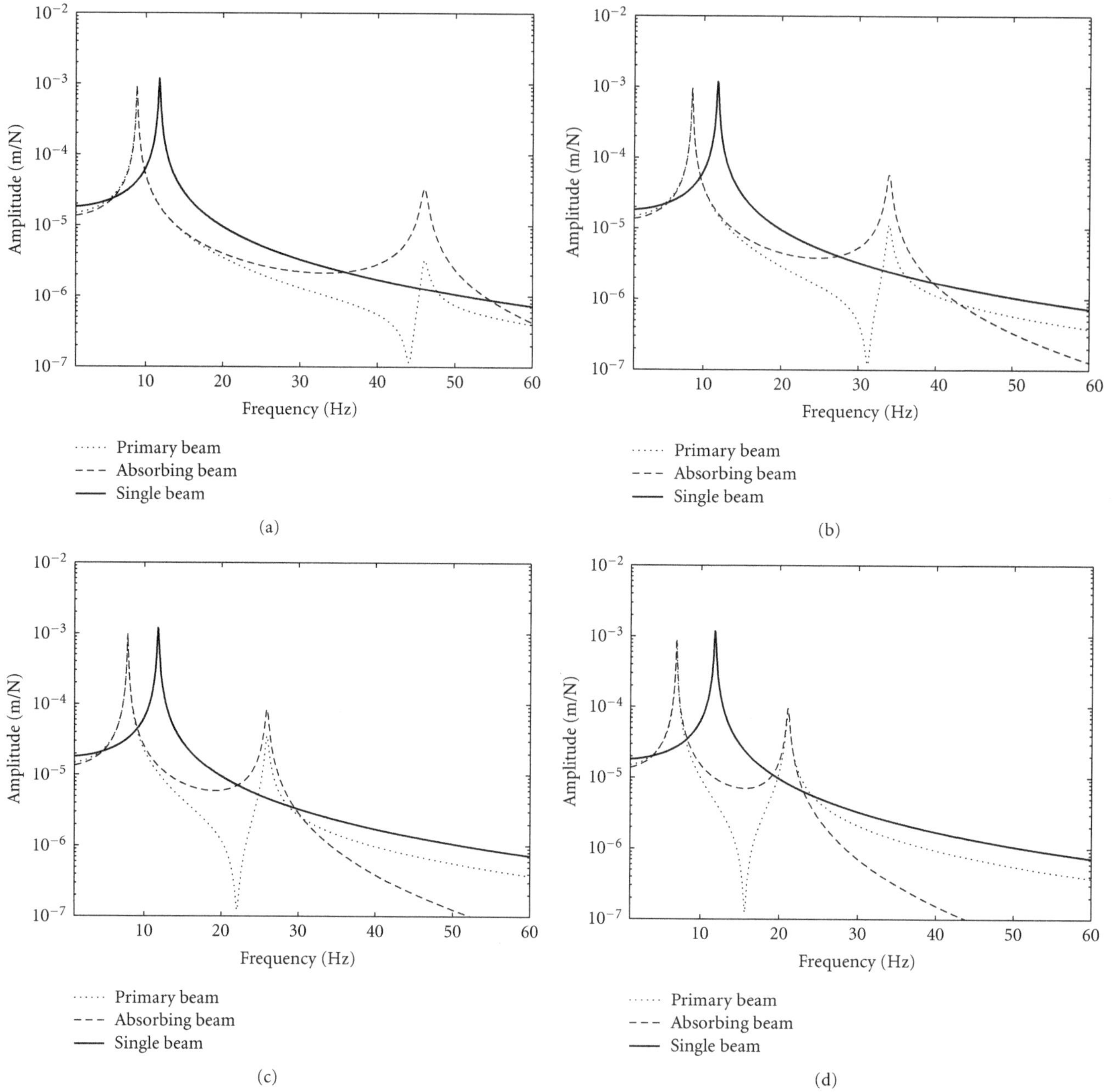

FIGURE 3: The vibration amplitude at the midspan location of the double-beam system ($e = 0.25$): (a) $\mu = 0.1$, (b) $\mu = 0.2$, (c) $\mu = 0.4$, and (d) $\mu = 0.8$.

a matrix form for the equations of motion expressed as (see Appendix)

$$\begin{bmatrix} m_1 & 0 \\ 0 & \mu m_1 \end{bmatrix}\begin{bmatrix} \ddot{q}_1 \\ \ddot{q}_2 \end{bmatrix} + \begin{bmatrix} c & -c \\ -c & c \end{bmatrix}\begin{bmatrix} \dot{q}_1 \\ \dot{q}_2 \end{bmatrix} + \begin{bmatrix} K_1\sigma_n^4 + k & -k \\ -k & eK_1\sigma_n^4 + k \end{bmatrix}\begin{bmatrix} q_1 \\ q_2 \end{bmatrix} = \begin{bmatrix} F_0 \\ 0 \end{bmatrix}\frac{2}{L}\sin\left(\frac{n\pi}{2}\right)e^{i\omega t}, \quad (8)$$

where $\mu = m_2/m_1$ is the mass ratio and $e = E_2I_2/E_1I_1$ is the rigidity ratio. The damping of the layer between the main beam and the absorbing beam can be approached by using the concept of "mixed damping ratio" in discrete dynamic vibration absorber [10]. The damping ratio of the visco-elastic layer can be defined as

$$\xi = \frac{c}{2m_2\omega_n} = \frac{c}{2\mu m_1\omega_n}, \quad (9)$$

where ω_n is the original natural frequency of the primary beam (without the absorber attached) which is given by

$$\omega_n = \sqrt{\frac{E_1I_1\,\sigma_n^4}{m_1}}. \quad (10)$$

For a stationary harmonic load, it is therefore necessary to analyze the system performance in the frequency domain

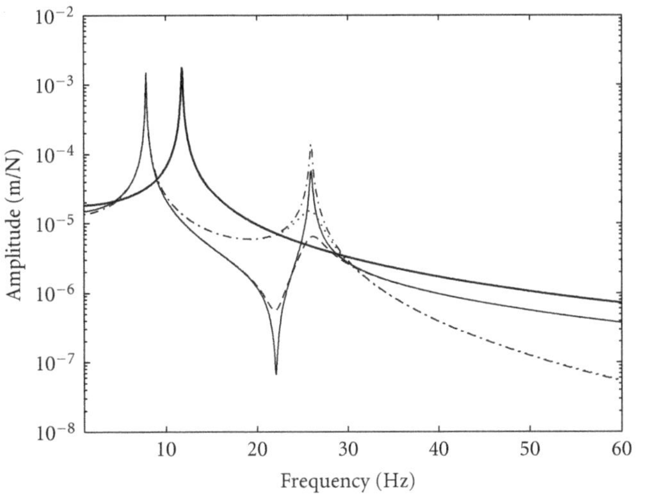

Figure 4: Effect of visco-elastic layer damping on the vibration amplitude at the midspan location of the double-beam system ($e = 0.25$, $\mu = 0.4$, $\eta_1 = \eta_2 = 0.01$).

to clearly identify the distinct responses in the resonant frequencies. By defining the amplitude q in terms of the complex exponential notation it gives

$$\begin{bmatrix} q_1 \\ q_2 \end{bmatrix} = \mathrm{Re}\left\{ \begin{bmatrix} Q_1 \\ Q_2 \end{bmatrix} e^{i\omega t} \right\}, \quad (11)$$

where Q is the complex amplitude of q.

Substituting (11) into (8) yields

$$\begin{bmatrix} K_1\sigma_n^4 + k + i\omega c - \omega^2 m_1 & -i\omega c - k \\ -i\omega c - k & eK_1\sigma_n^4 + k + i\omega c - \omega^2 \mu m_1 \end{bmatrix} \begin{bmatrix} Q_1 \\ Q_2 \end{bmatrix} = \begin{bmatrix} F_0 \\ 0 \end{bmatrix} \frac{2}{L} \sin\left(\frac{n\pi}{2}\right), \quad (12)$$

where the solutions for the complex amplitude of the frequency response function in terms of the receptance, that is, Q/F_0 can then be obtained for each mode n which are given by

$$\frac{Q_1}{F_0} = \frac{2(eK_1\sigma_n^4 + k + i\omega c - \omega^2 \mu m_1)\sin(n\pi/2)}{\left[(K_1\sigma_n^4 + k + i\omega c - \omega^2 m_1)(eK_1\sigma_n^4 + k + i\omega c - \omega^2 \mu m_1) - (i\omega c + k)^2\right]L}, \quad (13)$$

$$\frac{Q_2}{F_0} = \frac{2(i\omega c + k)\sin(n\pi/2)}{\left[(K_1\sigma_n^4 + k + i\omega c - \omega^2 m_1)(eK_1\sigma_n^4 + k + i\omega c - \omega^2 \mu m_1) - (i\omega c + k)^2\right]L}. \quad (14)$$

As observed from (13) and (14), the double-beam structure is expected to have two natural frequencies for each mode of vibration, as the system is modeled using two-degree-of-freedom system. It can also be seen in (13), where for the lumped parameter system with $K_1 = 0$ and for an undamped case where $c = 0$, the magnitude of the numerator is proportional to $k - \omega^2 m_2$. The main beam amplitude can therefore be suppressed to zero by tuning the layer stiffness and the mass of the absorbing beam to be equal to the forcing frequency, that is, $\omega = \sqrt{k/m_2}$. The subsequent sections discuss the effect of structural parameters on the response in (13) and (14).

3. Effects of Rigidity and Mass Ratio

In this investigation, the rigidity ratio and mass ratio are varied to observe their effects on the main beam response. This is calculated for a steel beam (Young's modulus $E = 2.1 \times 10^{11}$ Pa, density $\rho = 7800\,\mathrm{kg/m^3}$) having length 2 m, width 0.065 m, and thickness 0.02 m. The stiffness of the layer is assumed 100 kN/m. The calculation here is done for the first mode of vibration ($n = 1$) for each beam. Figure 2 shows the vibration amplitude for the first two resonances of the primary beam plotted in logarithmic scale. Here the damping ratio of the visco-elastic layer and the structural damping loss factor of the beams are assumed very small, that is, $\zeta = \eta = 0.01$.

It can be seen that the double-beam system successfully suppresses the amplitude at the fundamental frequency of the single beam (at 12 Hz) as a result of countering some of the energy force from the main beam. However, the system now behaves as a two-degree-of-freedom system which, in consequence, creates new resonances. Thus, a significant level of vibration amplitude appears at lower frequency around 8–10 Hz and much lower amplitude level for the second resonance around 25–30 Hz. Increasing the elasticity ratio gives insignificant effect to reduce the amplitude at the first resonance. Instead, this shifts both resonant frequencies to higher frequency, which causes the first resonance to approach that of the single beam; a situation which should be avoided especially if forcing frequency from the primary beam is not stable, for example, a nonsynchronous machine having certain range of operating frequency. Meanwhile, increasing the mass ratio of the double-beam is shown to decrease the frequency of both first and second resonances, although it has less effect on the former. This hence reduces the frequency gap of the two resonances. However, one can observe that the amplitude of the primary beam is reduced with the increasing mass ratio (note the dips just before the second resonance which approaches the fundamental frequency of the single primary beam). It can

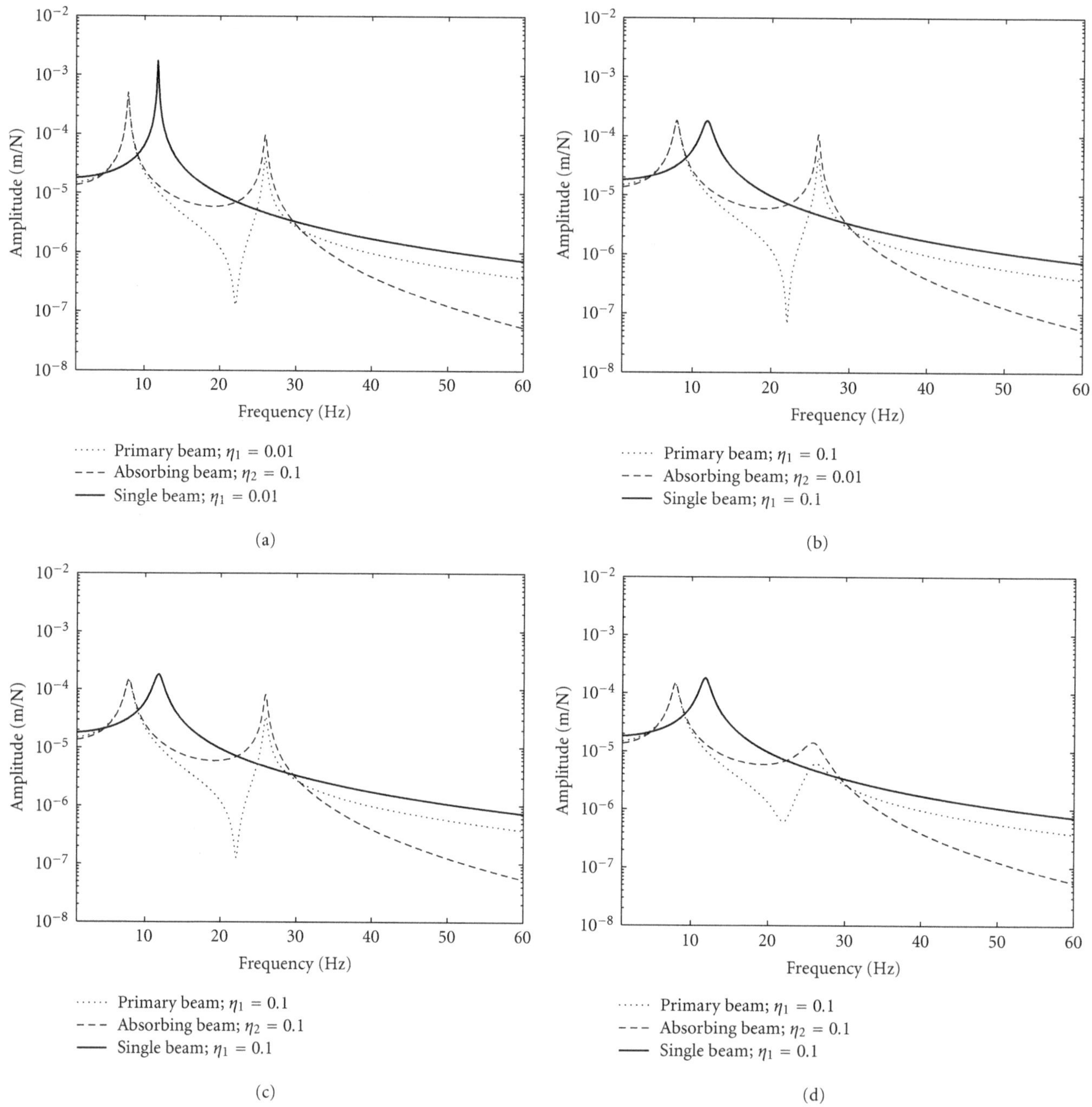

FIGURE 5: Effect of structural damping of the beam on the vibration amplitude at the midspan location of the double-beam system ($e = 0.25$, $\mu = 0.4$): (a)–(c) $\zeta = 0.01$ and (d) $\zeta = 0.1$.

also be observed that the amplitude at the second resonance increases as the mass ratio is increased.

Figure 3 presents the amplitudes of the primary beam and the absorbing beam. Both amplitudes at the first resonance can be seen not affected by the mass ratio. However, the response of the primary beam at the second resonance is considerably lower than that of the absorbing beam for low mass ratio, but both amplitudes become comparable as the mass ratio increases. As also seen in Figure 2, the compromised situation is that the vibration amplitude of the primary beam becomes lower as the mass ratio increases. The amplitude of the absorber at the second resonance can also be observed to have negligible effect due to the change of mass ratio, and it steeply decreases above this resonant frequency.

4. Effect of Damping and Layer Stiffness

Previous results in Figures 2 and 3 assume very small damping of the visco-elastic layer as well as the structural damping loss factor of the beam. Figure 4 plots the effect of the layer damping, which yields reduction of vibration amplitude only at the second resonance of both the primary

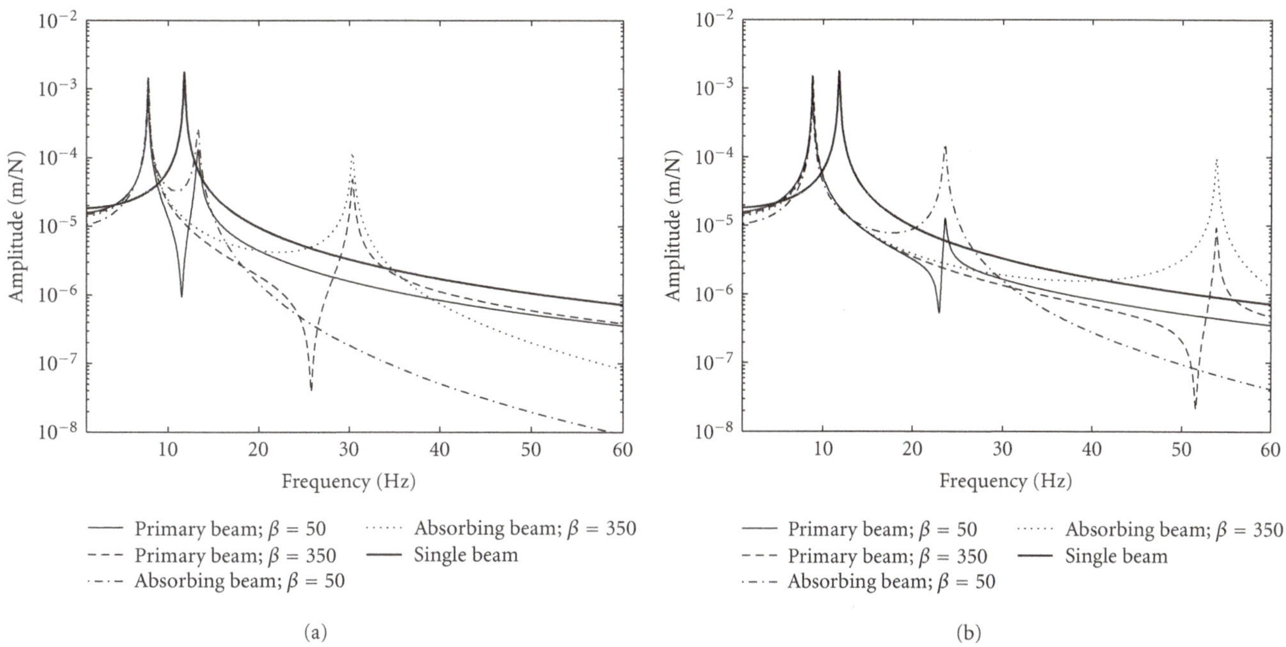

FIGURE 6: Effect of visco-elastic stiffness on the vibration amplitude at the midspan location of the double-beam system ($e = 0.25$): (a) $\mu = 0.1$ and (b) $\mu = 0.4$.

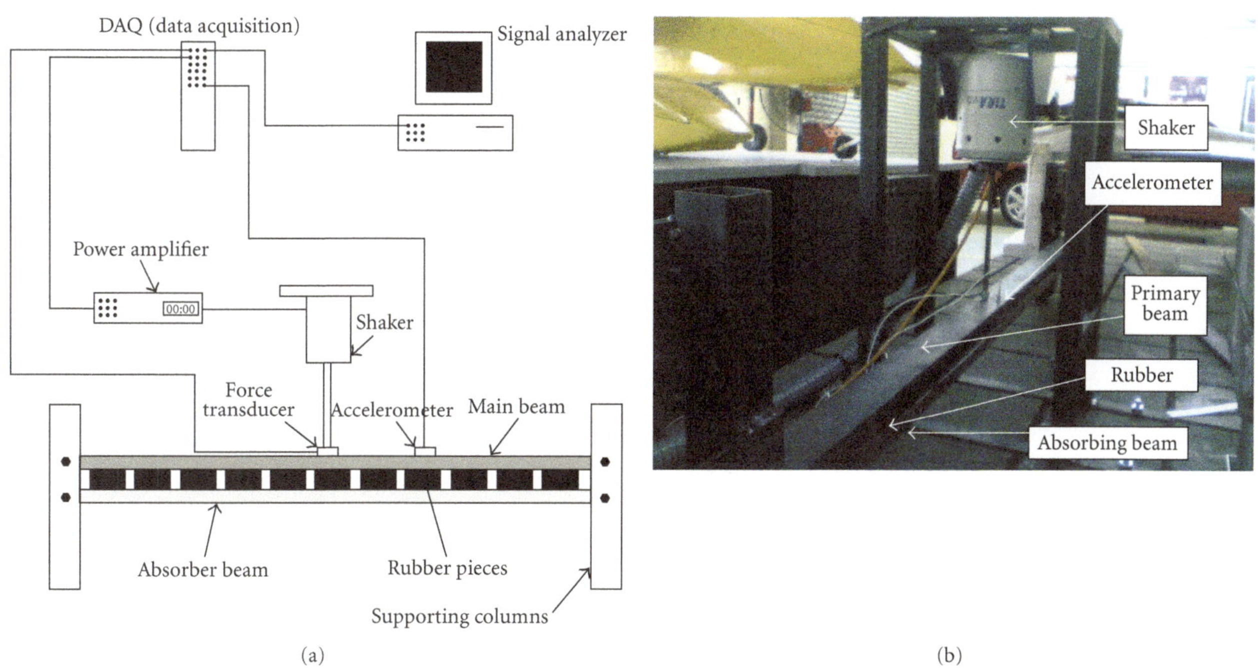

FIGURE 7: (a) Diagram of the experimental setup and (b) laboratory test.

beam and the absorbing beam as the damping is increased. At this frequency, the primary beam and the absorber move out of phase for which the role of the visco-elastic layer damping is important to absorb the vibration energy. At the first resonance, as the two beams move in-phase, the layer damping can therefore be seen to have negligible effect on the vibration amplitude.

Figure 5 shows the effect of the structural damping. It can be seen in Figure 5(a) that increasing the damping of the absorbing beam does not give significant effect to reduce the vibration amplitude. Increasing only the damping of the primary beam reduces only the amplitude at the first resonance, but not at the second resonance as shown in Figure 5(b). The same applies if the damping of the absorbing beam is

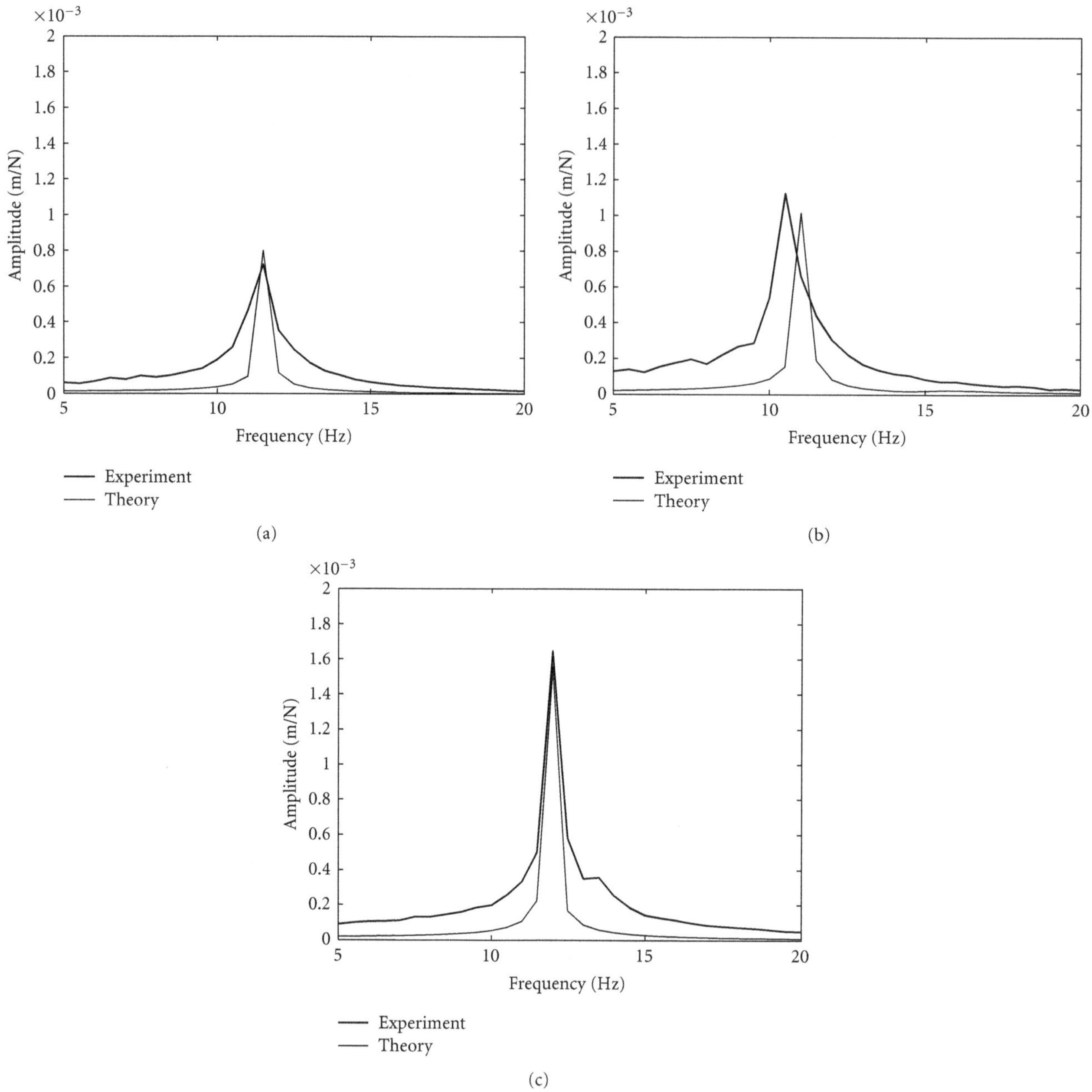

Figure 8: Comparison of experimental and theoretical results of the vibration amplitude of the double-beam system for the case of (a) ST-ST, (b) ST-WD, and (c) ST-AL.

also increased (see Figure 5(c)). Figure 5(d) shows that the amplitude of the second resonance is mainly controlled by the damping of the visco-elastic layer.

Figure 6 plots the results for the effect of stiffness of the visco-elastic layer. Here the stiffness is presented using a non-dimensional parameter as a function of the rigidity and length of the main beam, that is, $\beta = kL^4/K_1$. By reducing the stiffness of the layer, the second resonance shifts to low frequency, while no effect is given to the first resonance. Again same for the case of the layer damping, the in-phase motion of both beams at the first resonance will not be affected by the hardness (stiffness) of the layer. As the mass ratio increases, this frequency shift becomes greater, which also reduces the frequency gap of the two resonances.

5. Experimental Validation

Figure 7 shows the schematic diagram and the laboratory setup of the experiment. As in the simulation, the primary beam is placed at the upper side of the absorbing beam. However, it was difficult to realize a simply supported boundary condition. Both beams were therefore clamped to the supporting columns. As the measurement point is at

TABLE 1: Physical parameters of the materials used in the experiment.

Material	Cross section	Mass per unit length (kg/m)	Young's modulus (GPa)
Steel	Solid	10.27	210
Wood	Solid	1.12	10.3
Aluminum	Hollow	0.77	69

the midspan of the primary beam, the edge condition has negligible effect on the vibration amplitude.

The primary beam was excited by broadband pseudo-random signal at location close to the midspan of the primary beam using TIRA electromagnetic shaker type TV50018. The input force was measured by a Dytran force gauge type 1051V1. A stud was tightly bolted on the transducer surface and then glued to the beam surface using an epoxy glue. A 200 mm long stinger was used to connect the force gauge and the shaker to minimize the effect of moments transmitted from the shaker. A Dytran accelerometer type 3225F1 was attached exactly at the mid point to measure the vibration amplitude.

The experimental test was conducted with the primary beam made of steel having thickness 20 mm, width 60 mm, and length 2 m. Four measurement cases were made with different configuration and material of the absorber beam as follows:

(a) steel single beam without absorber (ST),

(b) steel main beam with steel absorber beam; double-identical beams (ST-ST),

(c) steel main beam with wood absorber beam (ST-WD), and

(d) steel main beam with aluminum absorber beam (ST-AL).

A rubber material was used as the viscoelastic layer between the beams having stiffness of 131 kN/m.

The physical parameters of the beam materials are listed in Table 1.

Figure 8 presents the experimental results for the double-beam structure. It can be seen that it demonstrates good agreement with the theory especially around the resonance, although for each case, broader frequency response from the measured results can also be observed, which slightly overestimated the model.

6. Conclusions

The effect of structural parameters on the dynamic response of a beam structure attached with a beam vibration absorber through a visco-elastic layer under a stationary harmonic load has been studied. The amplitude of the original primary single beam at the fundamental frequency can be considerably reduced. It is found that increasing the rigidity ratio shifts the resulting resonances of the double-system to higher frequency but gives small effect on reducing the vibration amplitude of the resonances. Reducing the mass ratio reduces considerable level of the second resonance of the primary beam as well as widens the gap between the resonant frequencies. However, this increases the first resonance to approach the troublesome resonant frequency of the original single beam but further reduces the vibration of the primary beam, a compromised situation which should be taken into account in the absorber design. The same phenomena apply for the absorbing beam, but it does not affect its vibration amplitude. Adding more damping to the layer has been shown to reduce only the second resonant amplitude. The amplitude at the first resonance can be reduced by increasing the damping of the primary beam. Meanwhile, increasing the layer elasticity (reducing stiffness) reduces the second resonance. The theoretical results have been validated by experimental data with a reasonable agreement.

Appendix

Substituting (3) and (4) into (1) and (2) gives

$$\begin{aligned} & E_1 I_1 \frac{\partial^4 X_n(x)}{\partial x^4} q_{1n}(t) + m_1 X_n(x) \ddot{q}_{1n}(t) \\ & \quad + c X_n(x)\left[\dot{q}_{1n}(t) - \dot{q}_{2n}(t)\right] + k X_n(x)\left[q_{1n}(t) - q_{2n}(t)\right] \\ & = P e^{i\omega t}, \end{aligned} \tag{A.1}$$

$$\begin{aligned} & E_2 I_2 \frac{\partial^4 X_n(x)}{\partial x^4} q_{2n}(t) + m_2 X_n(x) \ddot{q}_{2n}(t) \\ & \quad + c X_n(x)\left[\dot{q}_{2n}(t) - \dot{q}_{1n}(t)\right] + k X_n(x)\left[q_{1n}(t) - q_{2n}(t)\right] = 0. \end{aligned} \tag{A.2}$$

Substituting (5) into (A.1) and (A.2) with $P = F_0 \delta(x - L/2)$ yields

$$\begin{aligned} & E_1 I_1 \sigma_n^4 X_n(x)\, q_{1n}(t) + m_1 X_n(x) \ddot{q}_{1n}(t) \\ & \quad + c X_n(x)\left[\dot{q}_{1n}(t) - \dot{q}_{2n}(t)\right] + k X_n(x)\left[q_{1n}(t) - q_{2n}(t)\right] \\ & = F_0\, \delta\left(x - \frac{L}{2}\right) e^{i\omega t}, \end{aligned} \tag{A.3}$$

$$\begin{aligned} & E_2 I_2 \sigma_n^4 X_n(x)\, q_{2n}(t) + m_2 X_n(x) \ddot{q}_{2n}(t) \\ & \quad + c X_n(x)\left[\dot{q}_{1n}(t) - \dot{q}_{2n}(t)\right] + k X_n(x)\left[q_{1n}(t) - q_{2n}(t)\right] = 0. \end{aligned} \tag{A.4}$$

Equations (A.3) and (A.4) can be simplified to

$$\begin{aligned}&[m_1\ddot{q}_{1n}(t)+c\dot{q}_{1n}(t)-c\dot{q}_{2n}(t)\\&\quad+(E_1I_1\sigma_n^4+k)q_{1n}(t)-\ kq_{2n}(t)]X_n(x)\\&=F_0\ \delta\left(x\ -\frac{L}{2}\right)\ e^{i\omega t},\end{aligned}\tag{A.5}$$

$$\begin{aligned}&[m_2\ddot{q}_{2n}(t)+\ c\dot{q}_{2n}(t)-c\dot{q}_{1n}(t)\\&\quad+(E_2I_2\sigma_n^4+k)q_{2n}(t)-kq_{1n}(t)]X_n(x)=0.\end{aligned}\tag{A.6}$$

Multiplying both sides of (A.5) and (A.6) by mode shape function X_m

$$\begin{aligned}&[m_1\ddot{q}_{1n}(t)+c\dot{q}_{1n}(t)-c\dot{q}_{2n}(t)\\&\quad+(E_1I_1\sigma_n^4+k)q_{1n}(t)-\ kq_{2n}(t)]X_n(x)X_m(x)\\&=F_0\delta\left(x\ -\frac{L}{2}\right)X_m(x)\ e^{i\omega t},\end{aligned}\tag{A.7}$$

$$\begin{aligned}&[m_2\ddot{q}_{2n}(t)+\ c\dot{q}_{2n}(t)-c\dot{q}_{1n}(t)\\&\quad+(E_2I_2\sigma_n^4+k)q_{2n}(t)-kq_{1n}(t)]X_n(x)X_m(x)=0\end{aligned}\tag{A.8}$$

Integrating (A.7) and (A.8) through the beam length

$$\begin{aligned}&[m_1\ddot{q}_{1n}(t)+c\dot{q}_{1n}(t)-c\dot{q}_{2n}(t)\\&\quad+(E_1I_1\sigma_n^4+k)q_{1n}(t)-\ kq_{2n}(t)]\int_0^L X_n(x)X_m(x)\\&=F_0e^{i\omega t}\int_0^L X_m(x)\delta\left(x\ -\frac{L}{2}\right)=F_0e^{i\omega t}\sin\left(\frac{n\pi}{2}\right)\end{aligned}\tag{A.9}$$

$$\begin{aligned}&[m_2\ddot{q}_{2n}(t)+c\dot{q}_{2n}(t)-c\dot{q}_{1n}(t)\\&\quad+(E_2I_2\sigma_n^4+k)q_{2n}(t)-kq_{1n}(t)]\int_0^L X_n(x)X_m(x)=0.\end{aligned}\tag{A.10}$$

Applying the orthogonality conditions

$$\int_0^L X_n(x)X_m(x)dx=0\ ,\qquad n\neq m,\tag{A.11}$$

and for $n = m$, (A-11) gives

$$\int_0^L X_n^2(x)dx=\int_0^L\sin^2(\sigma_nx)\mathrm{d}x=\left[\frac{x}{2}-\frac{\sin(2\sigma_nx)}{4\sigma_n}\right]_2^L=\frac{L}{2}.\tag{A.12}$$

Equations (A-9) and (A-10) can therefore be expressed in a matrix form in terms of mass ratio ($\mu = m_2/m_1$) and rigidity ratio ($e = E_1I_1/E_2I_2$),

$$\begin{aligned}&\begin{bmatrix}m_1&0\\0&\mu m_1\end{bmatrix}\begin{bmatrix}\ddot{q}_1\\\ddot{q}_2\end{bmatrix}+\begin{bmatrix}c&-c\\-c&c\end{bmatrix}\begin{bmatrix}\dot{q}_1\\\dot{q}_2\end{bmatrix}\\&+\begin{bmatrix}K_1\sigma_n^4+k&-k\\-k&eK_1\sigma_n^4+k\end{bmatrix}\begin{bmatrix}q_1\\q_2\end{bmatrix}\\&=\begin{bmatrix}F_0\\0\end{bmatrix}\frac{2}{L}\sin\left(\frac{n\pi}{2}\right)e^{i\omega t}\end{aligned}\tag{A.13}$$

Nomenclature

E: Modulus of elasticity of the beam (N/m^2)
I: Area moment of inertia of the beam (m^4)
w: Vertical displacement of the beam (m)
x: Position coordinate (m)
t: Time (s)
m: Mass per unit length (kg/m)
k: Layer stiffness (N/m^2)
β: Non-dimensional layer stiffness parameter
ω: Radial frequency (rad/s)
ω_n: Natural frequency at the nth mode (rad/s)
μ: Mass ratio of absorbing beam to primary beam
ζ: Damping ratio of layer
e: Rigidity ratio of absorbing beam to primary beam
F_0: Magnitude of the external load (N)
σ: Eigenvalue of the mode shape function
X: Mode shape function
q: Generalized time function of amplitude (m)
Q: Complex displacement amplitude (m)
δ: Dirac delta function
L: Length of beam (m)
N: Newton, unit of force.

References

[1] H. Yamaguchi, "Vibrations of a beam with an absorber consisting of a viscoelastic beam and a spring-viscous damper," *Journal of Sound and Vibration*, vol. 103, no. 3, pp. 417–425, 1985.
[2] H. V. Vu, A. M. Ordóñez, and B. H. Karnopp, "Vibration of a double-beam system," *Journal of Sound and Vibration*, vol. 229, no. 4, pp. 807–822, 2000.
[3] Z. Oniszczuk, "Free transverse vibrations of elastically connected simply supported double-beam complex system," *Journal of Sound and Vibration*, vol. 232, no. 2, pp. 387–403, 2000.
[4] T. Aida, S. Toda, N. Ogawa, and Y. Imada, "Vibration control of beams by beam-type dynamic vibration absorbers," *Journal of Engineering Mechanics*, vol. 118, no. 2, pp. 248–258, 1992.
[5] Y. H. Chen and C. Y. Lin, "Structural analysis and optimal design of a dynamic absorbing beam," *Journal of Sound and Vibration*, vol. 212, no. 5, pp. 759–769, 1998.
[6] Y. Q. Zhang, Y. Lu, S. L. Wang, and X. Liu, "Vibration and buckling of a double-beam system under compressive axial loading," *Journal of Sound and Vibration*, vol. 318, no. 1-2, pp. 341–352, 2008.
[7] M. Abu-Hilal, "Dynamic response of a double Euler-Bernoulli beam due to a moving constant load," *Journal of Sound and Vibration*, vol. 297, no. 3–5, pp. 477–491, 2006.
[8] M. A. De Rosa and M. Lippiello, "Non-classical boundary conditions and DQM for double-beams," *Mechanics Research Communications*, vol. 34, no. 7-8, pp. 538–544, 2007.
[9] I. Sadek, T. Abualrub, and M. Abukhaled, "A computational method for solving optimal control of a system of parallel beams using Legendre wavelets," *Mathematical and Computer Modelling*, vol. 45, no. 9-10, pp. 1253–1264, 2007.
[10] D. J. Inman, *Engineering Vibrations*, Pearson Prentice Hall, 2008.

8

Cantilever Beam Vibration Suppression Using Modal Energy Redistribution

Shahin S. Nudehi

Mechanical Engineering Department, Valparaiso University, Valparaiso, IN 46383, USA

Correspondence should be addressed to Shahin S. Nudehi, shahin.nudehi@valpo.edu

Academic Editor: Luc Gaudiller

Modal energy redistribution is utilized to suppress vibrations of a cantilever beam. The energy redistribution between the modes of the beam is achieved by switching on/off an end force which causes varying stiffness of the beam. The control methodology to suppress all the modes relies on continuously funneling the energy from the higher to the lower modes with the end force and dissipating the energy associated with the fundamental mode. We present an analytical framework for control design exploiting the modal energy redistribution and verify the results through simulations and experiments.

1. Introduction

There have been extensive studies on active vibration of flexible structures in the literatures due to its many applications such as space and terrestrial structures (e.g., [1–3]). In many of these studies, the control scheme for these systems is based on a finite number of modes, and the resulting closed-loop system is prone to instability due to spillover [4, 5]. There are different ways to avoid spillover in flexile structures. One approach is to have a collocated structure guarantee closed-loop system stability despite model truncation [6–8]. Another approach to avoid spillover is redistribution of energy in vibrating structures from mode to mode and, in space, from one region of the structure to another [9, 10]. In the latter approach, stiffness variations of the structure and the difference between modes in two stiffness states are exploited to gain control authority and suppress vibrations. Some examples of stiffness variations in vibration suppression are [11–13] where the stiffness variation and information of the modes are used to reduce energy of the system. This research also uses stiffness variation to redistribute the modal energy and suppress vibration. However, in our study, all the modal states of the structure except the first (fundamental) one are not available or reliable to be used in a feedback closed loop. As a result, displacement and velocity of the structure at different places are used to turn on and off an end force to redistribute the energy of the beam such that higher modes' energy funnel to the lower modes or get suppressed. The energy associated with the first mode is dissipated by employing a feedback closed-loop system. We show that in this scheme that energy of the beam decreases if the switching of the end force (i.e., turning on/off the end force) does not add significant energy to the beam or removes its energy.

This paper is organized as follows. In Section 2, we show a proof of the concept of why modal energy gets redistributed using an end force and how the energy of the beam is reduced. In Section 3, we consider a cantilever beam with an end force and derive its mathematical model. Additionally, we look at the modal energy of the beam and introduce a switching strategy which minimally affects the energy of the beam. Sections 4 and 5 provide simulation and experimental results, respectively. Concluding remarks and future research directions are given in Section 6.

2. Background and Notation

2.1. Proof of Concept. A cantilever beam with an end force is shown in Figure 1. The goal is to change the mode shapes of the beam using the end force and exploit these changes in beam vibration suppression.

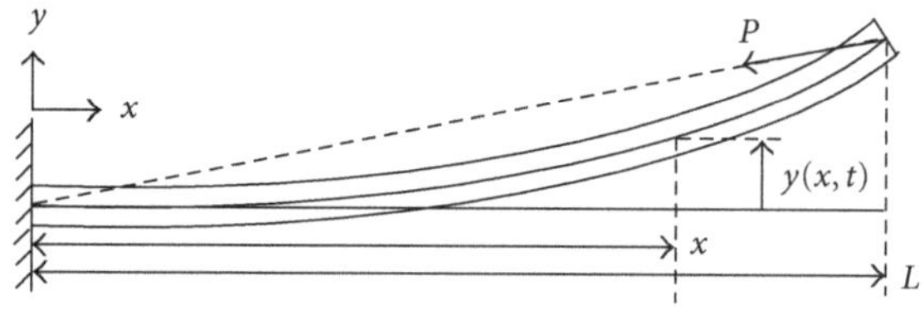

FIGURE 1: A flexible cantilever beam with an end force.

Consider an idealized static problem wherein the end force is instantaneously switched between two values P_0 and P_1, and the fundamental modal components (first mode) are repeatedly removed from the system after each switch. We will show that such a strategy removes all the energy from the beam including higher modes. In addition, this strategy only requires that one only be able to suppress the fundamental mode corresponding to P_0 and P_1 rather than all the modes of the beam. For the calculations, we denote the attendant mode corresponding to each of mode shapes for P_0 and P_1 as $\phi_j(x)$ and $\psi_j(x)$, respectively. To get an idea how the modes shapes compare with different end forces, in Figure 2 the first four mode shapes of a cantilever beam for $P = 0$ N and $P_1 = 40$ N are shown. In this figure, the beam length, width, and thickness are 1 m, 0.05 m, and 0.003 m, respectively, with aluminum properties. These dimensions and property are used in the simulation and the experimental section of this paper.

If one starts with a beam deflection $y_0(x,t)$ and end-load of P_0, then we can write

$$y_0(x,t) = \sum_{j=1}^{N} \delta_j(t)\phi_j(x), \qquad (1)$$

where N denotes the modal truncation level, and $\delta_j(t)$ is the jth modal amplitude component. Assuming that one can remove the first mode, the resulting shape is given by

$$y_1(x,t) = y_0(x,t) - \delta_1(t)\phi_1(x) = \sum_{j=2}^{N} \delta_j(t)\phi_j(x). \qquad (2)$$

At this point, the end load is switched to P_1, and the shape is now conveniently expressed as

$$y_1(x,t) = \sum_{j=1}^{N} \beta_j(t)\psi_j(x). \qquad (3)$$

It is assumed that the first mode is again removed while $P = P_1$, resulting in the shape

$$y_2(x,t) = \sum_{j=2}^{N} \beta_j(t)\psi_j(x). \qquad (4)$$

The end load is then switched back to P_0 where the shape can then be expressed by

$$y_2(x,t) = \sum_{j=1}^{N} \gamma_j(t)\phi_j(x). \qquad (5)$$

This completes one cycle of the process, and one is interested in how the new modal coefficients, the $\gamma_j(t)$, are related to the originals, the $\delta_j(t)$. This is conveniently described by a linear mapping

$$\Gamma = M\Delta, \qquad (6)$$

where Γ and Δ are the vectors of modal coefficients

$$\Gamma = (\gamma_1, \gamma_2, \ldots, \gamma_n)^T, \quad \Delta = (\delta_1, \delta_2, \ldots, \delta_n)^T, \qquad (7)$$

and M, the mapping matrix, can be developed by a sequence of calculations that use modal projections for each level of the end force as follows:

$$\gamma_i(t) = \langle y_2(x,t), \phi_i(x)\rangle = \left\langle \sum_{j=2}^{N} \beta_j(t)\psi_j(x), \phi_i(x) \right\rangle = \sum_{j=2}^{N} \beta_j(t) \left\langle \psi_j(x), \phi_i(x) \right\rangle. \qquad (8)$$

($\langle f, g\rangle$ is the inner product of functions f and g denoted by $\int_0^L fg\,dx$), and similarly,

$$\beta_j(t) = \sum_{k=2}^{N} \delta_k(t) \left\langle \psi_j(x), \phi_k(x) \right\rangle. \qquad (9)$$

Now, substituting for $\beta_j(t)$ from (9) in (8) results in

$$\gamma_i(t) = \sum_{j=2}^{N}\sum_{k=2}^{N} \left\langle \phi_i(x), \psi_j(x) \right\rangle \left\langle \psi_j(x), \phi_k(x) \right\rangle \delta_k, \quad i = 1, \ldots, N. \qquad (10)$$

Comparing this equation with (6) reveals the structure of the mapping matrix M. It should be noticed that the convergence of this process depends on the $N \times N$ linear operator M, which can be constructed as follows: the first column contains all zeros since the first modal coefficient was zeroed out (note that this implies that M will always have at least one zero eigenvalue). The remaining columns are filled in by the coefficient $\langle \phi_i(x), \psi_j(x)\rangle\langle \psi_j(x), \phi_k(x)\rangle$, $i = 1, 2, \ldots, N$, $j, k = 2, 3, \ldots, N$. If all eigenvalues of M lie inside the unit circle, the process will converge, implying that all modes under consideration die out under repeated cycling and removal of the first relevant mode. In fact, rate of convergence (or divergence) is dictated by these eigenvalues. Figure 3 depicts a schematic of the concept of the modal energy redistribution using the end force.

The above process for the cantilever beam whose mode shapes are shown in Figure 2 is carried out for the first four modes, and the resulting eigenvalues for the mapping matrix M are

$$\lambda_i = \{0, 0.51, 0.87, 0.92\}. \qquad (11)$$

We can see that the rate of convergence for the higher modes is much smaller as compared to the lower modes. In

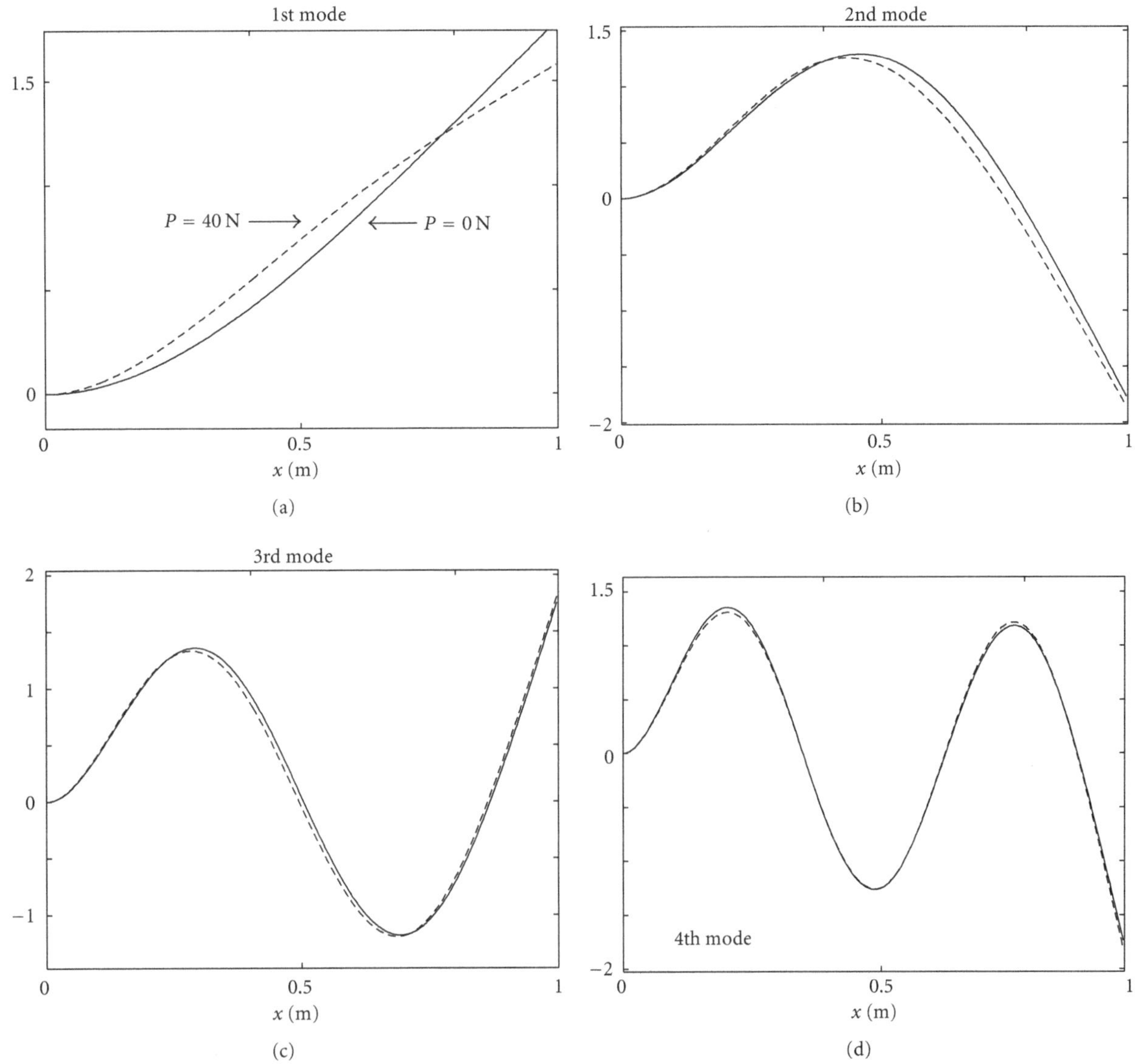

FIGURE 2: Cantilever beam mode shapes for $P = 0$ N and $P = 40$ N.

other words, the absolute value of the higher modes is close to unity. Note that in this calculation it is assumed that the end force does not add or remove energy to or from the beam in the time interval that is switched on which; this is very unlikely to happen. In the next section, we study a switched system that is comprised of two subsystems (corresponding to the case where the end force is on or off) to model the effect of switching on energy of each modes.

3. Dynamic Analysis

3.1. Mathematical Modeling. Consider the cantilever beam of length L and uniform cross-sectional area A, shown in Figure 1. Let P be the force acting at the free end of the beam such that the line of action always passes through the fixed end of the beam. Under assumption of Euler-Bernoulli theory and small deflections, the equation of motion of the beam in Figure 1 can be written as follows:

$$EIy'''' + Py'' + \rho A\ddot{y} = 0, \tag{12}$$

where E, I, and ρ are the Young's modulus of elasticity, area moment of inertia, and density of the beam, respectively, and y' and $\dot{y}$ denote the partial derivatives of $y(x,t)$ with respect to x and t, respectively. Equation (12) is identical to the equation of a beam with a follower end force [14] and a beam with an axial end force whose line of action remains parallel to the undeformed axis of the beam. However, this configuration is easier and more practical to implement as compared to a follower force. The boundary conditions of the beam in Figure 1 are, however, different from beams with

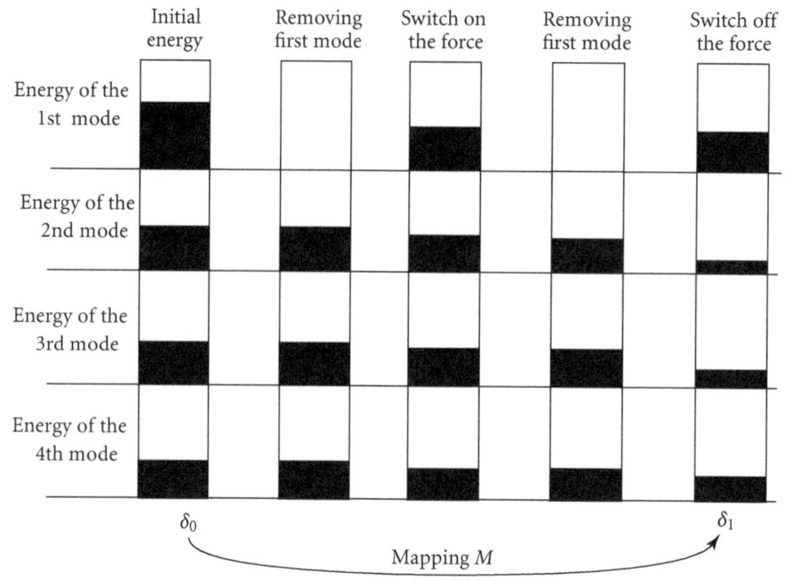

FIGURE 3: Concept of the control strategy to redistribute and remove the beam's energy.

follower and axial end forces. The geometric boundary conditions, which are related to zero deflection and zero slope at the fixed end, are

$$y(0,t) = 0, \qquad y'(0,t) = 0. \tag{13}$$

The natural boundary conditions corresponding to zero moment and nonzero shear force at the free end are given by the relations [15]

$$y''(L,t) = 0, \qquad EIy'''(L,t) + P\left\{y'(L,t) - \frac{1}{L}y(L,t)\right\} = 0. \tag{14}$$

In order to obtain an approximate solution to (12) subject to the boundary conditions in (13) and (14), we multiply (12) with a weight function $w(x)$ and integrate it over the length of the beam

$$\int_0^L EIw(x)y''''(x,t)dx + \int_0^L Pw(x)y''(x,t)dx + \int_0^L \rho A w(x)\ddot{y}(x,t)dx = 0. \tag{15}$$

We simply assume that $w(x)$ is continuous, twice differentiable and satisfies the two geometric boundary conditions in (13), namely,

$$w(0) = 0, \qquad w'(0) = 0. \tag{16}$$

In order to distribute the derivatives equally between $y(x,t)$ and $w(x)$, we integrate the first integral in (15) twice by parts, and the second integral once by parts, to get

$$(EIwy''' - EIw'y'' + Pwy')\big|_0^L + \int_0^L (EIw''y'' - Pw'y' + \rho A w\ddot{y})dx = 0. \tag{17}$$

Equation (17) is called the weak form [16] of (12). Using the natural boundary conditions of y in (14) and the geometric boundary conditions of w in (16), we get

$$P\frac{w(L)y(L,t)}{L} + \int_0^L (EIw''y'' - Pw'y' + \rho A w\ddot{y})dx = 0. \tag{18}$$

We now use Rayleigh-Ritz approximation [16] to express $y(x,t)$ as a linear combination of N suitable functions that satisfy the geometric boundary conditions in (13). Specifically, we use the first N normalized mode shapes of the clamped-free cantilever beam, as follows:

$$y_N(x,t) = \sum_{i=1}^{N} \delta_i(t)\phi_i(x). \tag{19}$$

In the above equation, $\phi_i(x)$, $i = 1,2,\ldots,N$, are the assumed modes, N is the number of desired modes, and $\delta_i(t)$, $i = 1,2,\ldots,N$, are the corresponding modal displacements. Now, by choosing $y(x,t) \approx y_N(x,t)$ and $w(x) = \phi_j(x)$, $j = 1,2,\ldots,N$, consistent with (16), we get N differential equations from (18), as follows:

$$EI\sum_{i=1}^{N}\left(\int_0^L \phi_i''\phi_j''dx\right)\delta_i - P\sum_{i=1}^{N}\left(\int_0^L \phi_i'\phi_j'dx - \frac{\phi_i(L)\phi_j(L)}{L}\right)\delta_i + \rho A\sum_{i=1}^{N}\left(\int_0^L \phi_i\phi_j dx\right)\ddot{\delta}_i = 0, \quad j = 1,2,\ldots,N. \tag{20}$$

Using the orthogonality property of the assumed mode shapes, the above N equations can be written as follows:

$$\rho A\ddot{\delta} + (K - PC)\delta = 0, \tag{21}$$

where $\delta = (\delta_1, \delta_2, \ldots, \delta_N)^T$, $K \in R^{N\times N}$ is a diagonal positive definite matrix with elements K_{ii}, $C \in R^{N\times N}$ is a positive definite symmetric matrix [17, 18] with elements C_{ij}, and P is assumed to be positive in the direction shown in Figure 1. The elements K_{ii} and C_{ij} have the following expressions:

$$K_{ii} = EI\int_0^L [\phi_i''(x)]^2 dx,$$
$$C_{ij} = \int_0^L \phi_i'(x)\phi_j'(x)dx - \frac{\phi_i(L)\phi_j(L)}{L}. \tag{22}$$

Equation (21) shows that the end force in Figure 1 is conservative in nature. This can now be claimed [19] from the symmetric nature of the matrix $(K - PC)$. Additionally, when $P = 0$, the eigenvalues of $(K - PC)$ are the same as the eigenvalues of K, which are all positive. As P is increased, all the eigenvalues start moving towards the origin before crossing the imaginary axis and becoming negative. This can be claimed from the physics of the buckling problem and has also been verified using numerical simulations. In the derivation of (21), structural damping was assumed to be absent. If structural damping is present, we can use

$$\rho A\ddot{\delta} + D\dot{\delta} + (K - PC)\delta = 0, \tag{23}$$

where $D \in R^{N\times N}$ is a diagonal positive definite matrix of modal damping.

3.2. Switching Strategy and Stability Proof. The goal of this section is to introduce a switching strategy for the end force P to funnel energy from the higher modes of vibrations to the first mode with minimal contribution to the total energy of the system. The following shows (23) modified for a beam subjected to the end load P with a control law to remove the first mode of vibration:

$$\rho A\ddot{\delta} + D\dot{\delta} + (K - P(t)C)\delta + H_1\left(\delta_1, \dot{\delta}_1\right) = 0. \tag{24}$$

Here, $P(t)$ is zero ($P_0 = 0$) when the end force is switched off and is equal to ($P_1 < P_{\text{buckling}}$) when the force is switched on, and H_1 is the diagonal control law to remove the oscillations of the fundamental frequency. By expanding this equation and multiplying the first row by $\dot{\delta}_1$, the second row by $\dot{\delta}_2$, and so on, we have

$$\frac{dE_1}{dt} = -\dot{\delta}_1 D_{11}\dot{\delta}_1 - \dot{\delta}_1 h_1\left(\dot{\delta}_1, \delta_1\right) + P\dot{w}_1\left(\dot{\delta}_1, \delta\right),$$
$$\frac{dE_2}{dt} = -\dot{\delta}_2 D_{22}\dot{\delta}_2 + P\dot{w}_2\left(\dot{\delta}_2, \delta\right),$$
$$\vdots \tag{25}$$
$$\frac{dE_n}{dt} = -\dot{\delta}_n D_{nn}\dot{\delta}_n + P\dot{w}_n\left(\dot{\delta}_n, \delta\right),$$

where $E_i = (1/2)(\rho A\dot{\delta}_i^2 + K_{ii}\delta_i^2)$ is the modal energy corresponds to the ith mode, $\dot{w}_i(\dot{\delta}_i, \delta) = \dot{\delta}_i[C_i]\delta$, D_{ii} is the ith element of the $[D]$ matrix, and $[C_i]$ refers to the ith row of the $[C]$ matrix.

Summing up all the rows of (25) results in

$$\frac{d}{dt}(E_1 + E_2 + \cdots + E_n)$$
$$= -\sum_{i=1}^{n}\dot{\delta}_i D_{ii}\dot{\delta}_i + P\sum_{i=1}^{n}\dot{w}_i\left(\dot{\delta}_i, \delta\right) - \dot{\delta}_1 h_1\left(\dot{\delta}_1, \delta_1\right). \tag{26}$$

In (26), the term $-\sum_{i=1}^{n}\dot{\delta}_i D_{ii}\dot{\delta}_i$ is negative. In addition, the control law for the fundamental mode is assumed to be designed properly and also removes energy of the first mode, that is, $-\dot{\delta}_1 h_1(\dot{\delta}_1, \delta_1) < 0$. As a result, as long as the end force is turned on, whenever the value of $\dot{W}_P = \sum_{i=1}^{n}\dot{w}_i(\dot{\delta}_i, \delta) = 0$, the total energy of the system will be reducing, and the end-load does not affect the total energy of the beam. In fact, one could also reduce the energy of the beam further by turning the end force on when it does negative mechanical work [17], that is, $\dot{W}_P < 0$. The work of the end force (per unit time), $\dot{W}_p$, consists of two terms [20]:

$$\dot{W}_P = P\int_0^L \frac{\partial\dot{y}(x,t)}{\partial x}\frac{\partial y(x,t)}{\partial x}dx - \frac{P\dot{y}(L,t)y(L,t)}{L}. \tag{27}$$

The first integral term represents the work done (per unit time) by the axial component of the end force P against the change in the length of the beam due to elastic deformation [20], and the second component is the work done by the lateral component of the force P at the tip of the beam due to lateral motion of the beam. It should be noted that due to small deformation of the beam, the work done by the lateral component of the force P is a magnitude of order smaller than the axial component. As a result,

$$\dot{W}_P \cong P\int_0^L \frac{\partial\dot{y}(x,t)}{\partial x}\frac{\partial y(x,t)}{\partial x}dx. \tag{28}$$

From (28), one can calculate the work done by the end force P at every instant using information of the displacement and velocity of every location along the beam. In practice, (28) can be discretized to consider a finite number of location displacements and velocities to calculate the work done by P as seen in the following:

$$\dot{W}_P \approx P\sum_{i=0}^{i=r-1}\frac{[\dot{y}(x_{i+1},t) - \dot{y}(x_i,t)][y(x_{i+1},t) - y(x_i,t)]}{x_{i+1} - x_i}. \tag{29}$$

Here, r is the number of locations used to calculate the integral, x_i is the position at location $i = 0\cdots r-1$, and $y(x_i,t)$ and $\dot{y}(x_i,t)$ are the displacement and velocity of the beam at location x_i, respectively. Equation (29) shows that the displacement and velocity information of the beam at different locations can be used to estimate the work done by the end force P without information of the modal state estimations. In other words, one can use the end force to redistribute or remove the modal energy without knowledge of the modes or number of modes involved in the beam's oscillations.

TABLE 1

Material	Aluminum
Young's modulus	70 GPa
Mass density	2730 kg/m^3
Dimensions	$1.00 \times 0.05 \times 0.003$ m

4. Simulation Results

In this section, the efficacy of the idea of redistributing the modal energy with an end force is investigated using simulation results. We assumed the material and geometric properties of the beam to be as shown in Table 1.

For a four-mode approximation of beam dynamics, the K and C matrices of our mathematical model in (22) were computed as

$$K = \begin{pmatrix} 97.3 & 0 & 0 & 0 \\ 0 & 3824.4 & 0 & 0 \\ 0 & 0 & 29976.5 & 0 \\ 0 & 0 & 0 & 115111.1 \end{pmatrix},$$
$$C = \begin{pmatrix} 0.65 & -3.38 & -0.06 & -2.59 \\ -3.38 & 28.42 & -18.35 & 9.58 \\ -0.06 & -18.35 & 73.30 & -31.65 \\ -2.59 & 9.58 & -31.65 & 138.90 \end{pmatrix}. \quad (30)$$

The square roots of the diagonal entries of the $K/\rho A$ matrix are the natural frequencies of the beam and are equal to 15.42, 96.63, 270.6, and 530.2 rad/s, respectively.

In the first simulation, to better illustrate energy redistribution between the modal states with an end force, we assume the modal damping to be absent, that is, the diagonal entries of the D matrix in (23) are zero. Additionally, the control law to remove the first mode oscillations is also considered to be $h_1(\dot{\delta}_1, \delta_1) = 0$. We used the exact expression for the $\dot{W}_P$ to turn on/off the end force to avoid changing the total energy of the beam (i.e., $\dot{W}_P = 0$) after the switchings. We also assume that P_1 in (24) is selected to be 40 N which is less than the critical buckling load (77 N) of the beam. The simulation results are shown in Figure 4 for the following initial conditions in metric units:

$$\delta(0) = [0, -0.5, 0, 0.1]^T,$$
$$\dot{\delta}(0) = [0, -1.2, 0, 3.5]^T. \quad (31)$$

This simulation shows clearly that the end force causes modal energy redistribution. The oscillations of the beam consist of the first and third modes, although the initial conditions for these two modes are chosen as zero. Clearly, the end force causes energy associated with the second and fourth modes to be redistributed among all the modes of the beam with nonzero end force. It should be noted that the energy transferred by the fourth mode to the other modes is considerably smaller compared to the second mode. This was expected from the values of the mapping matrix from (11).

In the second simulation, the end force P is used for two purposes: to redistribute the modal energy between the modes and to remove the modal energy by doing negative

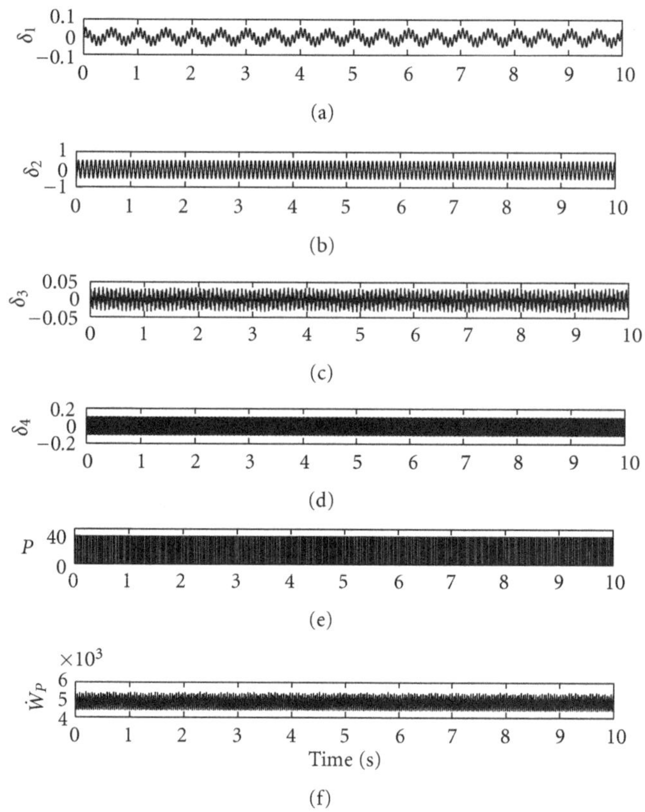

FIGURE 4: Modal redistribution of the beam with the switching end force.

work, that is, $\dot{W}_P \le 0$. This is to address the issue of a small percentage of the higher modes energy that gets funneled to the lower modes. We assume that modal states of the beam are not observable except for the first mode and information of the displacement and velocity is used to determine the work of the end force using (29). The control law to remove the first mode oscillation is chosen as $h_1(\dot{\delta}_1, \delta_1) = 2\dot{\delta}_1$, and the modal damping is assumed to be absent as well. To calculate $\dot{W}_P$ from (29), velocities and displacement at the base, the midpoint, and the end point locations of the beam are used

$$\dot{W}_P \approx P\left(\frac{[\dot{y}(L,t) - \dot{y}(L/2,t)][y(L,t) - y(L/2,t)]}{L/2} + \frac{[\dot{y}(L/2,t) - \dot{y}(0,t)][y(L/2,t) - y(0,t)]}{L/2}\right), \quad (32)$$

where $\dot{y}(0,t) = y(0,t) = 0$. In Figure 5, the simulation results are plotted for arbitrary initial conditions in SI units

$$\delta(0) = [0.3, -0.5, -0.1, 0.1]^T,$$
$$\dot{\delta}(0) = [-0.3, -1.2, 0.2, 3.5]^T. \quad (33)$$

This figure shows the modal amplitudes and energy of the beam rapidly decaying to zero, as expected, but the control

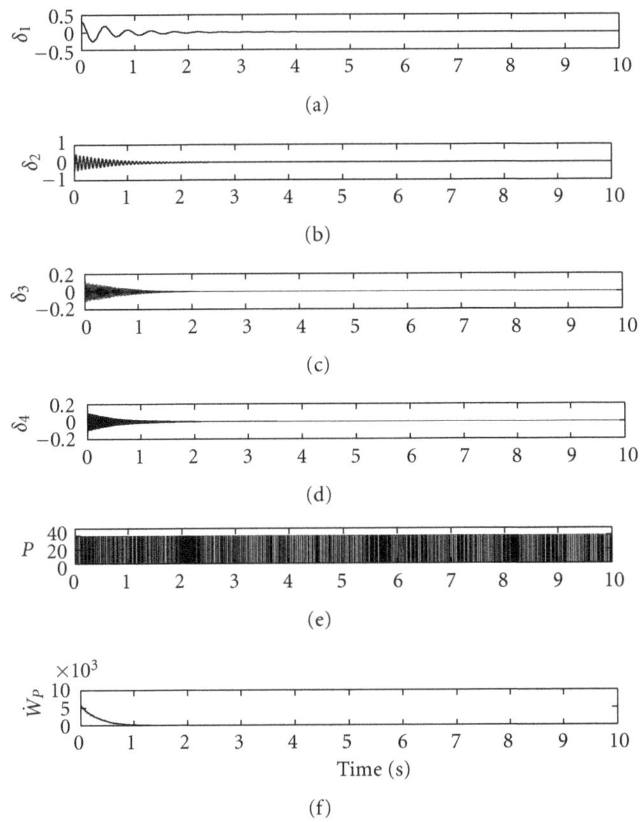

FIGURE 5: Modal redistribution and vibration suppression of the beam with the switching end force and the control law.

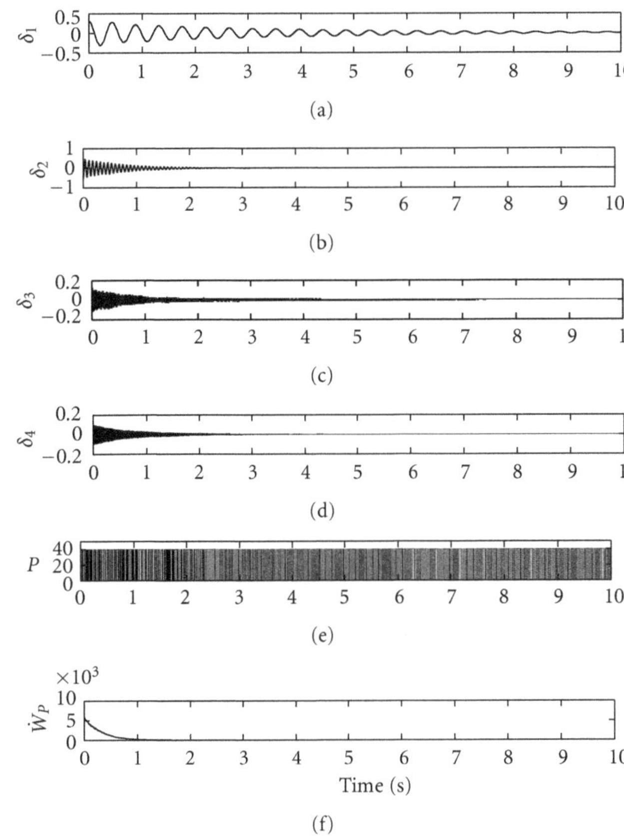

FIGURE 6: Modal redistribution and vibration suppression of the beam with just the switching end force.

input continues switching. This can be attributed to small numerical errors causing frequent change in sign of $\dot{W}_P$. This problem can be easily rectified during implementation. It should be noted that the approximation of $\dot{W}_P$ in (33) causes at some tiny time intervals that the end force adds energy to the beam instead of redistributing or removing it. This can be resolved by adding more terms (i.e., displacement and velocities measurements) to approximate $\dot{W}_P$ in (29). In this result, the beam's energy is dissipated by the control law $h_1(\dot{\delta}_1, \delta_1) = 2\dot{\delta}_1$ and the negative work of the end force P.

To see better the relative influence of the end force negative work and the control law in the modal attenuation, another simulation is performed in which the control law is set to zero, $h_1(\dot{\delta}_1, \delta_1) = 0$, and the end force P is used for both modal distributions and modal attenuation. The simulation results are shown in Figure 6 for the same initial conditions used in (33).

These results show that the end force P is capable of removing all the modal energy since $\dot{W}_P \leq 0$ at the times the end force is on. In this specific problem, the switching end force can be used solely to remove all the modes without needing the control law $h_1(\dot{\delta}_1, \delta_1)$. However, in practice, adding the control law makes the attenuation faster as seen in Figures 5 and 6 and (26). More importantly, it could remove any addition in the modal energy due to end force switching time not being exact, which may happen due to the approximation in (32) or a slow response time of the actuator used for turning on/off the end force. It should be noted that the bandwidth of the actuator used to switch the end force should be larger than the highest frequency in which $\dot{W}_P$ sign changes. If the actuator does not have the requisite bandwidth, there is no guarantee that the end force removes energy of the beam.

5. Experimental Verification

We conducted an experiment with a cantilever beam to prove the principal of redistributing and removing modal energies with an end force. In our experimental setup, the end force is applied with a Kevlar cable, wrapped around a pulley at the front face of the beam and then tied to a high-bandwidth servo DC motor [21] shaft as shown in Figure 7.

The motor was driven by a National Instruments NI 9505, brushed DC drive module [22] to pull or release the cable which causes the end force to turn on/off. Two miniature single-axis accelerometers from PCB Electronics [23] are mounted on the midpoint and end point of the beam to derive the displacement and velocity of the beam at these two locations. In the experiment, we did not implement the control law (i.e., $h_1(\dot{\delta}_1, \delta_1) = 0$) and only used the end force to redistribute/attenuate the modal energy. (One can easily implement the control's law using a piezoelectric sensor and actuator attached to the beam; see, for instance, [24]).

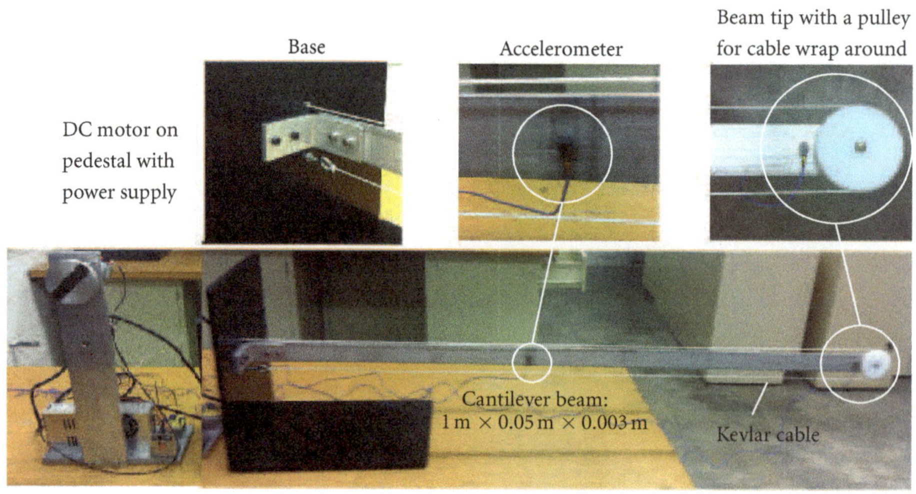

FIGURE 7: Experimental setup.

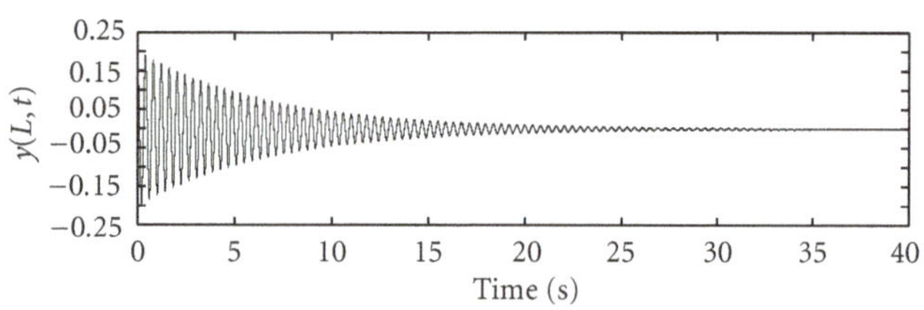

FIGURE 8: Experimental results: free vibration.

We programmed the switching law and the integration processers to achieve displacements and velocities in the LabVIEW FPGA [22] environment and downloaded it to our CompactRIO 9022 processing unit (not shown). The flexible beam and material used in the experiment are the same as the ones used in the simulations presented in Section 4.

The first experiment pertains to free vibration of the beam in the first mode. The time history for the displacement of the beam's tip is shown in Figure 8 in metric units. In the absence of the switching end force, the beam vibration is attenuated slowly by the structural damping. The time required for the amplitude of vibration to decay by a factor of 40 approximately is 25 seconds.

The second experiment pertains to vibration suppression of the beam with the switching end force. The time history for the displacement and velocity of the beam's tip and the end force in the experiment are shown in Figure 9 in metric units. In these results, the initial condition is achieved by deflecting the beam arbitrarily. These results indicate that the control strategy attenuates all the modes of vibration which corroborates our earlier simulation results. It should be noted that the structural damping also contributes to the removal of energy from the beam especially at higher frequencies. In fact, the modal damping of the structure increases whenever the end force is switched on due to an increase in axial stress of the beam [15] which further helps vibration suppression of the beam.

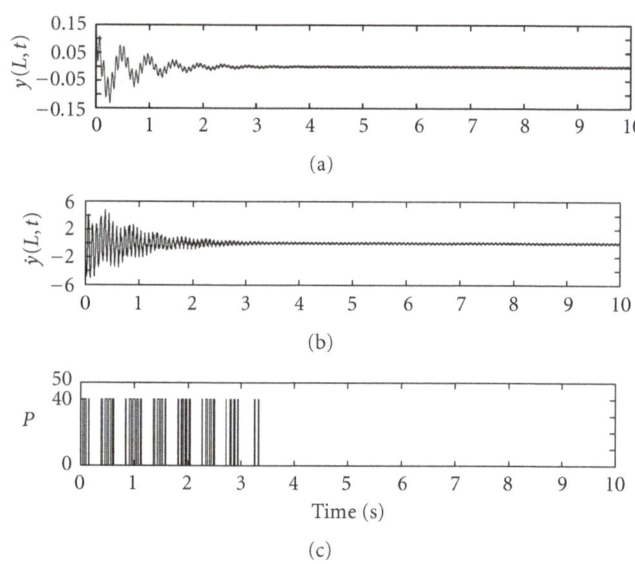

FIGURE 9: Experimental results: vibration suppression using the end force.

6. Conclusions

Modal redistribution and vibration suppression of a cantilever beam with an end force is mathematically studied and experimentally verified. We showed that it is possible to redistribute modal energy of the beam by changing its stiffness using an end force. In addition, the end force was used to suppress the vibrations of the beam since mathematical and simulation results showed that the rate of energy distribution from higher modes to the lower modes of the beam is small. As a result, the end force was also used to remove the energy of the beam. The end force was turned on whenever it either removed or did not change the modal energy of the beam; otherwise, it was turned off. For this purpose, we used lateral displacements and velocities of the beam at different locations to construct and estimate the rate of the mechanical work done by the end force without information of the modal amplitudes. Our goal for future work is to employ modal energy redistribution for two- and three-dimensional flexible structures in which the stiffness changes with cables at multiple locations along the structure.

Acknowledgments

The author would like to thank Professor Ranjan Mukherjee of Michigan State University for his useful comments and suggestions during the course of this research. In addition, the author gratefully acknowledges the support provided by Valparaiso University's Creative Work and Research Committee that made this research possible.

References

[1] D. C. Hyland, J. L. Junkins, and R. W. Longman, "Active control technology for large space structures," *Journal of Guidance, Control, and Dynamics*, vol. 16, no. 5, pp. 801–821, 1993.

[2] G. S. Nurre, R. S. Ryan, H. N. Scofield, and J. L. Sims, "Dynamics and control of large space structures," *Journal of Guidance, Control, and Dynamics*, vol. 7, no. 5, pp. 514–526, 1984.

[3] M. J. Balas, "Trends in large space structure control theory: fondest hopes, wildest dreams," *IEEE Transactions on Automatic Control*, vol. 27, no. 3, pp. 522–535, 1982.

[4] M. J. Balas, "Active control of flexible systems," *Journal of Optimization Theory and Applications*, vol. 25, no. 3, pp. 415–436, 1978.

[5] M. J. Balas, "Feedback control of flexible systems," *IEEE Transactions on Automatic Control*, vol. 23, no. 4, pp. 673–679, 1978.

[6] J. J. Dosch, D. J. Inman, and E. Garcia, "Self-sensing piezoelectric actuator for collocated control," *Journal of Intelligent Material Systems and Structures*, vol. 3, no. 1, pp. 166–185, 1992.

[7] D. Halim and S. O. R. Moheimani, "Spatial resonant control of flexible structures—application to a piezoelectric laminate beam," *IEEE Transactions on Control Systems Technology*, vol. 9, no. 1, pp. 37–53, 2001.

[8] E. H. Anderson, N. W. Hagood, and J. M. Goodliffe, "Modelling of piezoelectric actuator dynamics for active structural control," in *in Proceeding of the AIAA/ASME/ASCE/AHS Structures, Structural Dynamics, Materials Conference*, pp. 2241–2255, Dallas, Tex, USA, April 1992.

[9] A. Diaz and R. Mukherjee, "Modal disparity enhancement through optimal insertion of nonstructural masses," *Structural and Multidisciplinary Optimization*, vol. 31, no. 1, pp. 1–7, 2006.

[10] A. R. Diaz and R. Mukherjee, "A topology optimization problem in control of structures using modal disparity," *Journal of Mechanical Design*, vol. 128, no. 3, pp. 536–541, 2006.

[11] J. Issa, R. Mukherjee, A. R. Diaz, and S. W. Shaw, "Modal disparity and its experimental verification," *Journal of Sound and Vibration*, vol. 311, no. 3–5, pp. 1465–1475, 2008.

[12] A. Kurdila, W. Clark, W. Wang, and D. McDaniel, "Stability of a class of piezoelectric state-switching methods," in *Adaptive Structures and Material Systems—2000*, pp. 509–517, American Society of Mechanical Engineers, New York, NY, USA, 2000.

[13] A. Ramaratnam and N. Jalili, "A switched stiffness approach for structural vibration control: theory and real-time implementation," *Journal of Sound and Vibration*, vol. 291, no. 1-2, pp. 258–274, 2006.

[14] V. Bolotin and G. Herrmann, *Nonconservative Problems of the Theory of Elastic Stability*, vol. 1991, Pergamon Press, London, UK, 1963.

[15] S. R. M. Nudehi, *Active Vibration Suppression of Flexible Structures Using Switching Control [Ph.D. thesis]*, Michigan State University, East Lansing, Mich, USA, 2005.

[16] J. Reddy, *An Introduction to the Finite Element Method*, vol. 2, McGraw-Hill, New York, NY, USA, 1993.

[17] S. Nudehi, R. Mukherjee, and S. W. Shaw, "Active vibration control of a flexible beam using a buckling-type end force," *Journal of Dynamic Systems, Measurement and Control*, vol. 128, no. 2, pp. 278–286, 2006.

[18] H. P. Lee, "Divergence and flutter of a cantilever rod with an intermediate spring support," *International Journal of Solids and Structures*, vol. 32, no. 10, pp. 1371–1382, 1995.

[19] M. A. Langthjem and Y. Sugiyama, "Optimum design of cantilevered columns under the combined action of conservative and nonconservative loads. Part I: the undamped case," *Computers and Structures*, vol. 74, no. 4, pp. 385–398, 2000.

[20] S. Rao, *Vibration of Continuous Systems*, Wiley, New York, NY, USA, 2007.

[21] http://faulhaber group.com/.

[22] http://www.ni.com/.

[23] http://www.pcb.com/.

[24] S. S. Nudehi and R. Mukherjee, "Enhancing controllability and observability in underactuated and undersensed systems through switching: application to vibration control," *Journal of Dynamic Systems, Measurement, and Control*, vol. 126, no. 4, pp. 790–799, 2004.

Gearbox Fault Features Extraction Using Vibration Measurements and Novel Adaptive Filtering Scheme

Ghalib R. Ibrahim[1] and A. Albarbar[2]

[1] *Mechanical Engineering Department, College of Engineering, University of Anbar, Anbar, Iraq*
[2] *Advanced Industrial Diagnostic Centre, School of Engineering, Manchester Metropolitan University, Manchester M1 5GD, UK*

Correspondence should be addressed to A. Albarbar, a.albarbar@mmu.ac.uk

Academic Editor: Akira Ikuta

Vibration signals measured from a gearbox are complex multicomponent signals, generated by tooth meshing, gear shaft rotation, gearbox resonance vibration signatures, and a substantial amount of noise. This paper presents a novel scheme for extracting gearbox fault features using adaptive filtering techniques for enhancing condition features, meshing frequency sidebands. A modified least mean square (LMS) algorithm is examined and validated using only one accelerometer, instead of using two accelerometers in traditional arrangement, as the main signal and a desired signal is artificially generated from the measured shaft speed and gear meshing frequencies. The proposed scheme is applied to a signal simulated from gearbox frequencies with a numerous values of step size. Findings confirm that 10^{-5} step size invariably produces more accurate results and there has been a substantial improvement in signal clarity (better signal-to-noise ratio), which makes meshing frequency sidebands more discernible. The developed scheme is validated via a number of experiments carried out using two-stage helical gearbox for a healthy pair of gears and a pair suffering from a tooth breakage with severity fault 1 (25% tooth removal) and fault 2 (50% tooth removal) under loads (0%, and 80% of the total load). The experimental results show remarkable improvements and enhance gear condition features. This paper illustrates that the new approach offers a more effective way to detect early faults.

1. Introduction

Fault detection techniques include many parameters which are based on time domain statistical analysis, frequency domain (fast Fourier transform), and time-frequency techniques to detect damage of the gearbox. Liao et al. [1] introduced a self-organizing feature maps (SOFM) network technique as a tool for gearbox CM. Gearbox vibration signals were captured under different operating conditions: normal, tooth cracked, and tooth broken. The results showed that the SOFM network gave a good representation for understanding the distribution of the vibration signal. Moreover, the growth of damage can early be monitored via the path of the image points for the feature data in two-dimensional space. He et al. [2] employed the kernel principal component analysis (KPCA) to extract nonlinear features from the time- and frequency-domains of the vibration signals. To classify the gearbox conditions, they constructed two different KPCA-based subspace structures. A fatigue test of an automobile transmission gearbox was carried out and the data set was analysed using the proposed method. The results illustrated that the KPCA features were better than PCA features in terms of clustering capability, and both the two KPCA-based subspace methods can be effectively applied to gearbox CM. Rafiee and Tse [3] used autocorrelation of continuous wavelet coefficients (CWT) to detect a fault in a gear. A motorcycle gearbox system was used under different operating conditions; normal gearbox (NG), slight-worn (SW), medium-worn (MW), and broken-tooth gear (BT). Their conclusions were (a) autocorrelation CWT is a suitable method for feature extraction from non-stationary signals in gearbox for CM, (b) the proposed technique could be used to monitor other vibration signals and diagnose faults in bearings. Zhou et al. [4] used the ensemble empirical mode decomposition (EEMD) method to diagnose localized damage of the feed-axis gearbox.

Built-in position sensors such as motor encoders and linear scales with high resolution and high precision were employed in their study. Their results were that, (a) the EEMD method was cheaper to use as it was simpler and did not require an extra sensor, (b) the EEMD was efficient and effective in determining the problems in the system. Eftekharnejad and Mba [5] investigated the effectiveness of acoustic emission (AE) in identifying seeded defects in helical gears. The results identified the RMS levels from the acoustic signal are more capable of correctly identifying the seeded defects on helical gears than the vibration signals from an accelerometer mounted on the bearing pedestal. Saravanan et al. [6] presented a method that used ANN and proximal support vector machines (PSVM) to classify the fault in a gearbox. Statistical techniques were used to calculate Morlet wavelet coefficients and were fed as input for training the ANN and PSVM, and testing their relative efficiency in classifying the faults in a bevel gear box. The findings showed that the PSVM had an edge over ANN in classification of parameters. Albarbar et al. [7] presented adaptive filtering techniques to enhance diesel fuel-injector needle impact excitations contained within the air-borne acoustic signals. The results showed that the proposed technique was useful in enhancing other impulsive events within the airborne acoustic signals such as intake and exhaust valves operation and piston slap excitations. Recently, Ibrahim and Albarbar [8] presented a comparison between empirical mode decomposition (EMD) and smoothed pseudo Wigner-Ville distribution (SPWVD) methods based on vibration signature and energy calculation procedure for monitoring gearbox systems. The results showed that the calculation of energy using EMD techniques offers a more effective way to detect early faults than that computations using the SPWVD method, and that the computation of energy using the EMD technique is faster than the calculations done using the SPWVD method.

Adaptive noise cancelling is an effective technique for enhancing signal to noise ratio of corrupted signals. The effectiveness of the signal enhancement depends on the correlation degrees between primary signal and the reference signal. Many adaptive filter parameters search algorithms such as least mean square (LMS) algorithm, and recursive least squares (RLS) algorithm are used to remove a noise from signal [9].

This work presents a novel adaptive filtering scheme to the detection of damage in gears. The developed technique suits cases where a reference signal is not available; it is based on an—operating-condition dependent—artificially synthesized gearbox vibration signal, which can be assumed as the desired signal by which to extract a gear meshing signal from main signal via adaptive noise cancelling.

This paper is organized as follows: Adaptive noise cancelling and least mean squares algorithms are presented in Sections 2 and 3, respectively. The proposed scheme and numerical validation are illustrated in Section 4. Experimental work is presented in Section 5. The developed monitoring scheme is applied for detecting gear-tooth breakage in Section 6. Conclusions are outlined in Section 7.

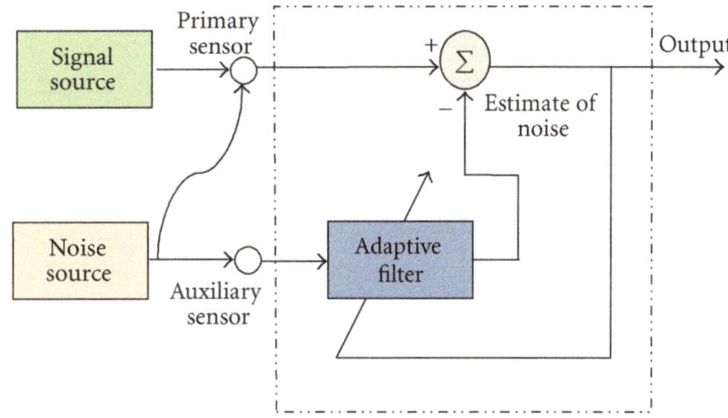

Figure 1: Schematic of adaptive noise cancellation system.

2. Adaptive Filtering and Noise Cancellation

One technique to improve the signal-to-noise ratio is adaptive noise cancelling. This is a technique whereby noise is subtracted from the original signal. Separate from the transducer detecting, the primary signal is a second or auxiliary transducer placed in a position where it is subject to much the same noise as the primary transducer but where the primary signal is very weak. The noise may now be eliminated by a process of subtraction. The problem is, of course, that the noise picked up by the auxiliary transducer will not usually exactly match the noise picked up by the primary transducer with the risk that part of the original signal may be lost. However, the risk of distorting the original signal (or decreasing signal-to-noise ratio) is substantially reduced if the noise signal to be subtracted is treated by a suitable filtering and adaptive process [10].

Figure 1 shows the essential components of an adaptive noise cancellation system: a dual-input, closed-loop, and adaptive feedback system [11].

The desired signal from the primary sensor $s(n)$ is contaminated with noise $N_o(n)$ which is assumed to be uncorrelated with $s(n)$. The combined signal from the primary sensor into the cancellation device is $(s(n)+N_o(n))$. The auxiliary transducer receives noise $N_1(n)$ which is also assumed uncorrelated with the primary signal but correlated—though in an unknown way—with the noise $N_o(n)$. For example, both primary and secondary transducers might be microphones and pick-up airborne noise from a compressor. If the noise travels different distances to the two transducers there will be correlation between the two noise signals picked up by the microphones but the signals will differ in both phase and amplitude, and possibly have random differences due to gusting winds. It is usual to assume $s(n)$, $N_o(n)$, $N_1(n)$, and $y(n)$ are statistically stationary with zero means.

The noise signal $N_1(n)$ is filtered so that it is as close a replica as possible to $N_o(n)$. This filtered signal, $y(n)$, is subtracted from the combined signal to produce the output [12]:

$$e(n) = s(n) + N_o(n) - y(n). \quad (1)$$

In Figure 1 the input from the auxiliary sensor is treated by an adaptive filter, which responds to an error signal that depends on the system output and automatically adjusts its impulse response until the closest fit between $e(n)$ and $s(n)$ is obtained. With noise cancellation the adaptive filter usually uses a least mean squares algorithm (LMS) to minimize the error in the output signal until $e(n) = s(n)$.

3. Least Mean Squares Algorithms

The LMS algorithm is a useful tool used with which to find optimal filter coefficients in real time. Its two great advantages are its simplicity and its almost universal use. The core of the LMS algorithm is the gradient search technique whereby an initial estimate is made of the coefficient vector, which is then iteratively updated so that it changes from its current estimate to a better estimate of the optimal filter coefficients [13].

Because R and P, which appear in the Wiener-Hopf equation, are not known in advance and are time variant, the Wiener-Hopf equation is not usually immediately suitable for real-time applications. If higher filter orders are necessary then the inversion of R can be very time consuming. The LMS algorithm is a means for calculating W_{opt} without having to know either R or P, and because it does not involve matrix inversion it is thus much more practical to use in real time. The LMS seeks to minimise the running average of the squared error signal (e^2) rather than the expectation value $E[e^2]$. The LMS performs iterative calculations to determine the next set of Finite duration Impulse Response (FIR) coefficient, $w(n+1)$, from the values at step n. The speed of convergence of the LMS algorithm is controlled by introducing a factor m which controls the step width of the iteration [14].

4. The Proposed Scheme and Numerical Validation

This section describes how LMS algorithms were applied to a signal simulated from gearbox frequencies. A sinusoidal approximation is used to represent a gear mesh signal (GMS) as [15]:

$$\text{GMS}(N,\theta) = \sin(N \times \theta), \tag{2}$$

where N represents the number of gear teeth, and θ the degree of shaft rotation.

The sidebands around the fundamental gear mesh frequency in the frequency domain are defined as sum and difference of the mesh frequency and gear or pinion speed [16]:

$$\begin{aligned} f_{sp} &= f_m \pm f_p, \\ f_{sg} &= f_m \pm f_g. \end{aligned} \tag{3}$$

With a gearbox operating at shaft frequency 22.53 Hz, the gear meshing frequency at the first stage is 766.1 Hz, and the gear meshing frequency at the second stage is 317.4 Hz, a simulated signal can be assumed to be:

$$\begin{aligned} x(t) = {} & 0.5 * \sin(2 * \pi * 22.53 * t) \\ & + \sin(2 * \pi * 766.1 * t) \\ & + 0.8 * \sin(2 * \pi * 317.4 * t) \\ & + 0.1 * \sin(2 * \pi * 788.63 * t) \\ & + 0.1 * \sin(2 * \pi * 743.57 * t). \end{aligned} \tag{4}$$

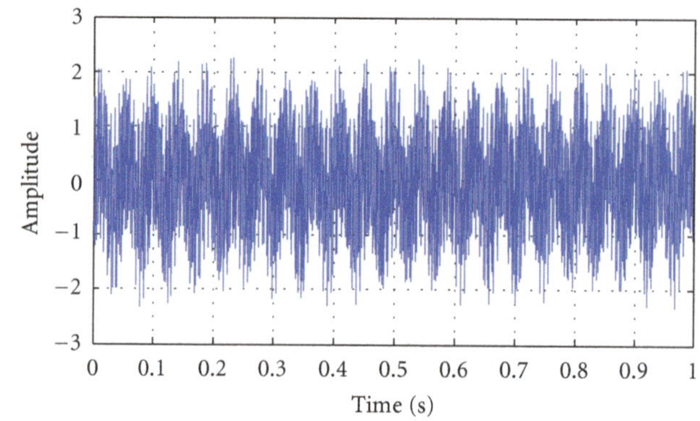

Figure 2: Time domain of simulated signal.

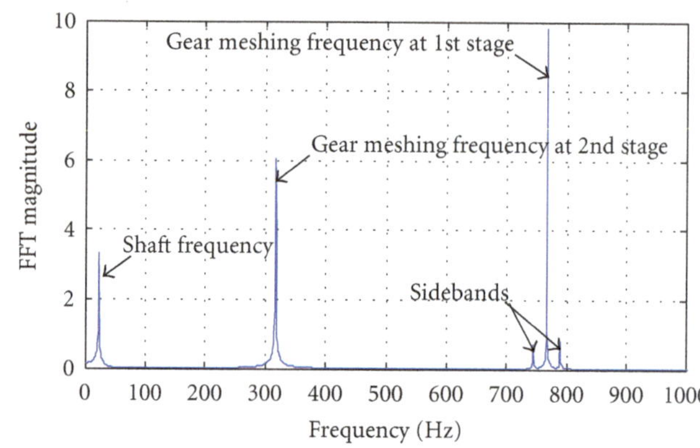

Figure 3: Frequency domain of simulated signal.

Figures 2 and 3 represent time domain and FFT, respectively for signal $x(t)$.

Figure 3 shows sidebands around the 1st-stage gear meshing frequency at frequencies of 743.57 Hz (766.1− 22.53) and 788.63 Hz (766.1 + 22.53). In a real operating gearbox there will be different noises from different sources and the sidebands will be affected. To understand the effect of noise on a sideband let signal $x(t)$ be corrupted by random noise as shown in Figure 4. The FFT of the corrupted signal appears in Figure 5 and it can be clearly seen that the sidebands are no longer discernible.

An LMS algorithm was applied to remove noise from signal and the results can be seen in Figures 6 and 7 which also show the effects of the step size chosen. To examine step size of LMS on signal, two values of step size were chosen

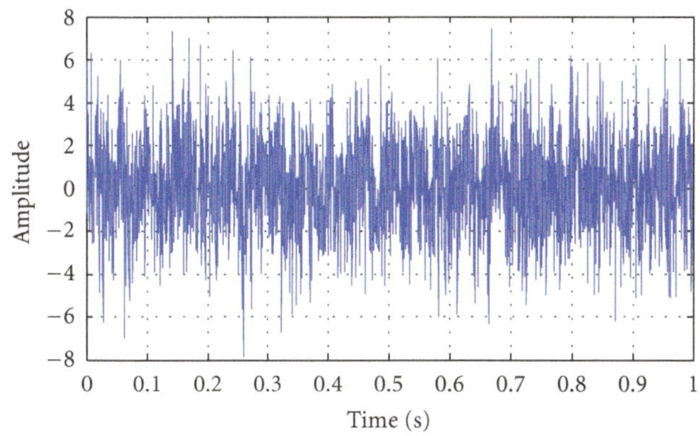

FIGURE 4: Time domain of simulated corrupted signal.

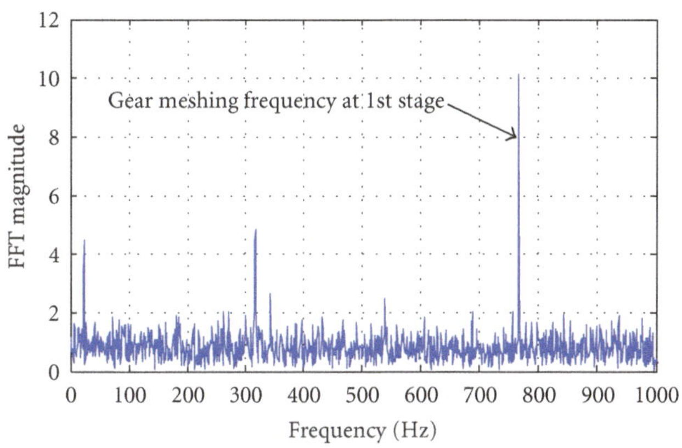

FIGURE 5: Frequency domain for simulated corrupted signal.

for the simulation: 0.0001 and 0.00001. Figure 6 shows the adaptive noise cancelling applied to the simulated corrupted signal to remove noise with step size of 0.0001. It can be seen that there is an improvement in the signal with the sidebands beginning to appear either side of gear meshing frequency. Figure 7 shows the results when the step size was 0.00001. Clearly there has been a substantial improvement in signal clarity (much greater signal-to-noise ratio) and the sidebands are now clearly discernible.

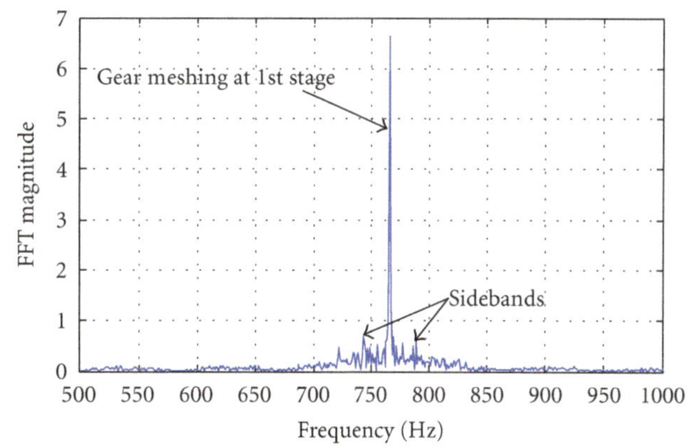

FIGURE 6: Frequency domain of corrupted signal obtained using noise cancellation with LMS step size 0.0001.

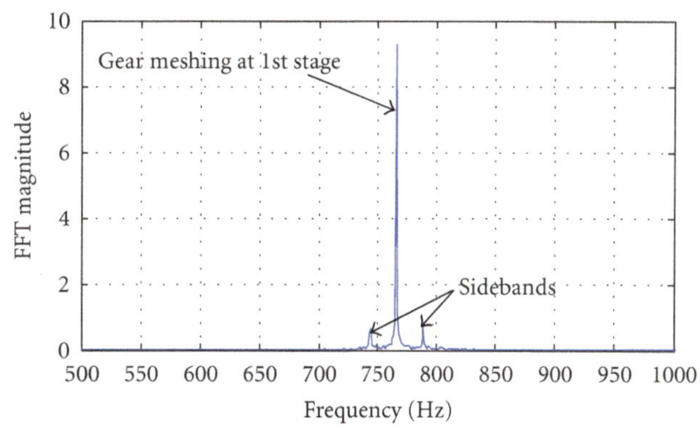

FIGURE 7: Frequency domain of corrupted signal obtained using noise cancellation with LMS step size 0.00001.

5. Experimental Setup

The experimental data were collected using the test rig shown in Figure 8. It comprises a two-stage, 11 Kw, helical gearbox drive by three-phase induction motor and connected to a DC generator and adjacent resistor banks.

Test were carried out using healthy pair of gears and one pair suffering from a tooth breakage with magnitude fault 1 (25% tooth removal), fault 2 (50% tooth removal), and fault 3 (75% tooth removal) of the tooth under varying loads (0%, 20%, 40%, 60%, and 80%). The drive pinion at the first stage had 34 teeth meshing with a 70-tooth wheel. The pinion gear at the second stage had 29 teeth meshing with 52-tooth wheel. The vibration signals were collected using an accelerometer mounted vertically on the gearbox housing. The accelerometers were B&K type 4371 with sensitivity of 10 mV/g and suitable for vibration measurements within a range of 1 Hz to 12 kHz. The sampling frequency was set to 50 kHz. Before the accelerometer signals were fed to the analogue-to-digital converter NI USB 9233 card, they passed through a B&K type 2635 charge amplifier to condition the signal. The charge amplifier converts the accelerometer high impedance, low charge (in the range of Pico-coulomb) signal into low impedance and high voltage (in the range of mV) and the cut off frequency for initialising filter was set to 10 kHz.

6. Gear Fault Features Enhancement

This section reports the application of the LMS algorithm to the detection of gear-tooth breakage. The data sets measured under operating conditions of: no fault, 25% tooth removal, and 50% tooth removal, carries many sinusoidal waveforms of different frequencies. A sinusoidal approximation was assumed as the desired signal and used to extract the gear

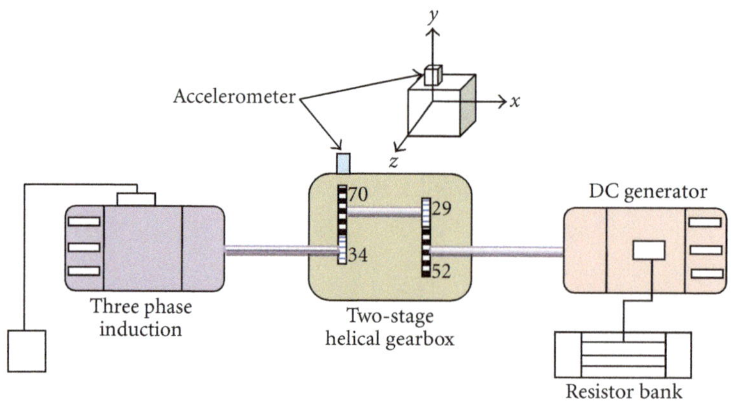

FIGURE 8: Schematic layout of the gearbox test rig.

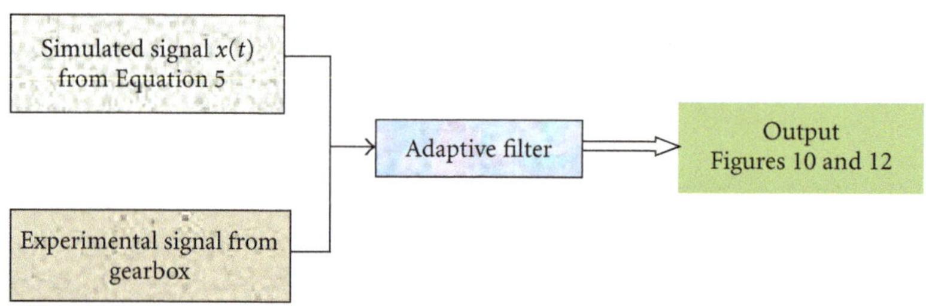

FIGURE 9: The proposed scheme architecture.

meshing signal from main signal. A desired signal can be expressed as:

$$x(t) = \frac{\text{RMS}(\text{main signal})}{3} * \sin(2 * \pi * f_s * t) + \text{RMS}(\text{main signal}) * \sin(2 * \pi * f_{m1} * t) + \frac{\text{RMS}(\text{main signal})}{2} * \sin(2 * \pi * f_{m2} * t) + \frac{\text{RMS}(\text{main signal})}{3} * \sin(2 * \pi * f_{\text{side1}} * t) + \frac{\text{RMS}(\text{main signal})}{3} * \sin(2 * \pi * f_{\text{side2}} * t), \quad (5)$$

where RMS(main signal) is the root mean square of the main signal, f_s is shaft frequency, f_{m1} is gear meshing frequency at 1st stage, f_{m2} is gear meshing frequency at 2nd stage and f_{side1}, f_{side2} are sidebands frequencies.

Vibration signals collected from the gearbox housing under no load and 80% of the total load was considered as reference when using the LMS algorithm. Figure 9 illustrates the proposed scheme architecture for simulated and experimental signals.

Figures 10 and 12 show the vibration signal in frequency domain, under no load and 80% of the total load, after applying the LMS algorithm. Figures 11 and 13 zoom in on the upper sideband. The FFT magnitude is 5.8, 7, and 10.1 for healthy, 25% tooth removal and 50% tooth removal, respectively, under zero load; and it is 33, 40, and 48 for healthy, 25% tooth removal and 50% tooth removal, respectively, under 80% of total load. From these figures, it is clear that there is a relationship between the fault level and the amplitude of sidebands for all cases considered. The amplitude of the sidebands increased as the damage level increased. It is concluded that there is advantage to applying the LMS algorithm for fault detection.

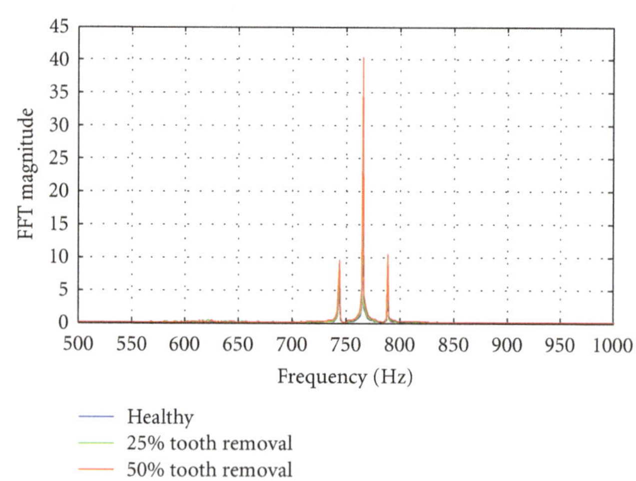

FIGURE 10: Frequency domain of gear vibration obtained after applying LMS algorithm (zero load).

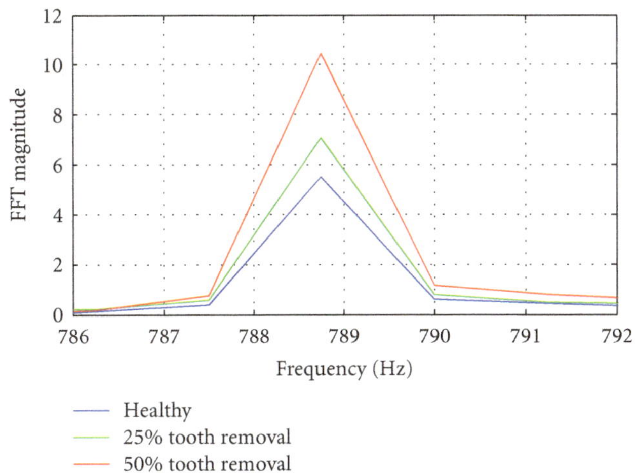

FIGURE 11: Zoom onto sideband (zero load).

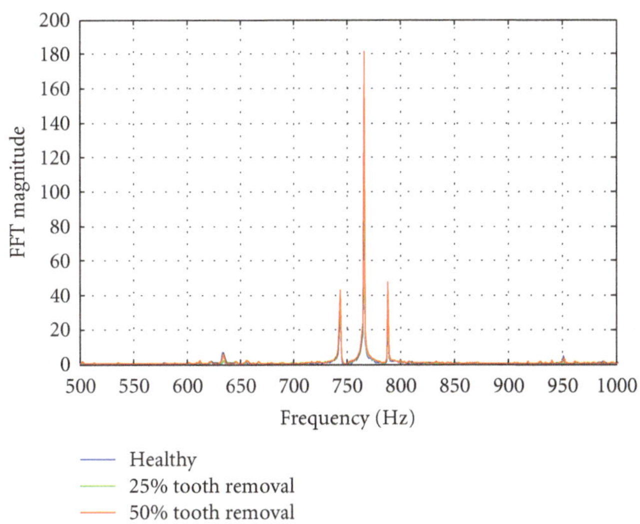

FIGURE 12: Frequency-domain of gear vibration obtained after applying LMS algorithm (80% of total load).

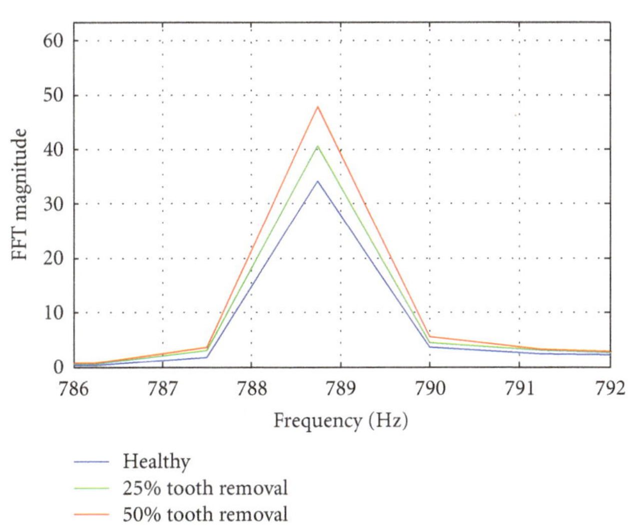

FIGURE 13: Zoom onto sideband (80% of total load).

7. Conclusion

To improve signal-to-noise ratio the LMS algorithm was adopted, modified, and applied to signals which simulated gearbox frequencies and signals collected experimentally from a gearbox. The novelty lies on using only one accelerometer as the main signal and a desired signal was artificially generated from the measured shaft and gear meshing frequencies. The conclusions can be summarized as follows.

(1) A sinusoidal approximation (5) was used as the desired signal to extract the gear meshing signal from the main signal.

(2) The new proposed procedure was effective at revealing faults by producing larger amplitude sidebands as shown Figures 11 and 13.

(3) All previous researchers used at least two accelerometers in adaptive filtering but in this work one accelerometer was used and the results appeared good, which could mean reduced monitoring costs.

References

[1] G. Liao, S. Liu, T. Shi, and G. Zhang, "Gearbox condition monitoring using self-organizing feature maps," *Proceedings of the Institution of Mechanical Engineers, Part C*, vol. 218, no. 1, pp. 119–130, 2004.

[2] Q. He, F. Kong, and R. Yan, "Subspace-based gearbox condition monitoring by kernel principal component analysis," *Mechanical Systems and Signal Processing*, vol. 21, no. 4, pp. 1755–1772, 2007.

[3] J. Rafiee and P. W. Tse, "Use of autocorrelation of wavelet coefficients for fault diagnosis," *Mechanical Systems and Signal Processing*, vol. 23, no. 5, pp. 1554–1572, 2009.

[4] Y. Zhou, T. Tao, X. Mei, G. Jiang, and N. Sun, "Feed-axis gearbox condition monitoring using built-in position sensors and EEMD method," *Robotics and Computer-Integrated Manufacturing*, vol. 27, no. 4, pp. 785–793, 2011.

[5] B. Eftekharnejad and D. Mba, "Seeded fault detection on helical gears with acoustic emission," *Applied Acoustics*, vol. 70, no. 4, pp. 547–555, 2009.

[6] N. Saravanan, V. N. S. K. Siddabattuni, and K. I. Ramachandran, "Fault diagnosis of spur bevel gear box using artificial neural network (ANN), and proximal support vector machine (PSVM)," *Applied Soft Computing Journal*, vol. 10, no. 1, pp. 344–360, 2010.

[7] A. Albarbar, F. Gu, A. D. Ball, and A. Starr, "Acoustic monitoring of engine fuel injection based on adaptive filtering techniques," *Applied Acoustics*, vol. 71, no. 12, pp. 1132–1141, 2010.

[8] G. R. Ibrahim and A. Albarbar, "Comparison between Wigner-Ville distribution- and empirical mode decomposition vibration-based techniques for helical gearbox monitoring," *Proceedings of the Institution of Mechanical Engineers, Part C*, vol. 225, no. 8, pp. 1833–1846, 2011.

[9] X. Zhou, Y. Shao, D. Zhen, F. Gu, and A. Ball, "Gear fault signal detection based on an adaptive fractional Fourier transform filter," *Journal of Physics*, vol. 305, no. 1, Article ID 012022, 2011.

[10] B. Widrow, J. R. Glover, and J. M. McCool, "Adaptive noise cancelling: principles and applications," *Proceedings of the IEEE*, vol. 63, no. 12, pp. 1692–1716, 1975.

[11] S. Haykin, *Adaptive Filter Theory*, 3rd edition, 1995.
[12] M. Abdulmagid, D. Krusienski, S. Pal, and W. Jenkins, "Principles of adaptive noise cancelling," *Annual Research Journal*, vol. 2, 2004.
[13] V. Bajic, *Design and implementation of an adaptive noise canceling system in wavelet transform domain [M.S. thesis]*, University of Akron, 2005.
[14] W. Fohl, J. Matthies, and B. Schwarz, "A FPGA-based adaptive noise cancelling system," in *Proceedings of the 12th International Conference on Digital Audio Effects (DAFx '09)*, Como, Italy, 2009.
[15] C. J. Stander and P. S. Heyns, "Instantaneous angular speed monitoring of gearboxes under non-cyclic stationary load conditions," *Mechanical Systems and Signal Processing*, vol. 19, no. 4, pp. 817–835, 2005.
[16] Barron, *Engineering Condition Monitoring Practice, Methods and Application*, Addison-Wesley/Longman, New York, NY, USA, 1996.

10

Fault Detection of Inline Reciprocating Diesel Engine: A Mass and Gas-Torque Approach

S. H. Gawande,[1] L. G. Navale,[1] M. R. Nandgaonkar,[2] D. S. Butala,[3] and S. Kunamalla[3]

[1] *Department of Mechanical Engineering, M. E. Society's College of Engineering, Pune-411 001, Maharashtra, India*
[2] *Department of Mechanical Engineering, Government College of Engineering, Pune-411 005, Maharashtra, India*
[3] *Research and Development, Greaves Cotton Limited, Pune-411 019, Maharashtra, India*

Correspondence should be addressed to S. H. Gawande, shgawande@yahoo.co.in

Academic Editor: Abul Azad

Early fault detection and diagnosis for medium-speed diesel engines are important to ensure reliable operation throughout the course of their service. This work presents an investigation of the diesel engine combustion-related fault detection capability of crankshaft torsional vibrations. Proposed methodology state the way of early fault detection in the operating six-cylinder diesel engine. The model of six cylinders DI Diesel engine is developed appropriately. As per the earlier work by the same author the torsional vibration amplitudes are used to superimpose the mass and gas torque. Further mass and gas torque analysis is used to detect fault in the operating engine. The DFT of the measured crankshaft's speed, under steady-state operating conditions at constant load shows significant variation of the amplitude of the lowest major harmonic order. This is valid both for uniform operating and faulty conditions and the lowest harmonic orders may be used to correlate its amplitude to the gas pressure torque and mass torque for a given engine. The amplitudes of the lowest harmonic orders (0.5, 1, and 1.5) of the gas pressure torque and mass torque are used to map the fault. A method capable to detect faulty cylinder of operating Kirloskar diesel engine of SL90 Engine-SL8800TA type is developed, based on the phases of the lowest three harmonic orders.

1. Introduction

The great interest of engine manufacturing industry and research centers on the crankshaft vibration [1] and instantaneous speed analysis for the fault or misfire detection in internal combustion engines is due to the positive results obtained using different approaches. A misfire event can be considered as an impulsive excitation on the crankshaft, related to the lack of torque resulting from the missing combustion. This excitation causes a sudden engine speed decrease followed by an oscillation with characteristics amplitude and frequency associated with the torsional behavior of the engine-load system.

A number of methods are devised to detect nonuniformities in the contributions of the cylinders to the total engine output by analyzing the variation of the measured crankshaft's speed may be classified in different groups such as observer-based method, pattern recognition, statistical estimators, order domain methods, and a combination of these methods. In [2], a good overview of the earlier state of the art has been presented. The model developed in [3] has been extensively used with very good results to predict and control torsional vibrations to correlate the amplitudes of the major harmonic orders of the crankshaft's speed variation with the average indicated mean effective pressure of the engine. Balancing the cylinder-wise torque contributions of automotive and other high-speed engines is usually addressed by reconstructing the cylinder-wise net-indicated torques, by direct use of the angular acceleration [4]. The method suggested [5, 6], in which the measured angular speed was directly related to the nonuniform torque contribution to detect misfire. In order to determine the cylinder-wise net-indicated torque, the oscillating torque applied on the flywheel was reconstructed from measurements of the angular speeds of the crankshaft. The oscillating torque can be reconstructed by using a single mass engine model and one angular speed measurement [7] by assuming that the crankshaft is rigid and that the engine is sufficiently decoupled from the transmission and load,. Obtaining an estimate of the prevailing load torque [7, 8] uses the fact that the instantaneous torque from the engine is zero at top-dead center (TDC) and bottom-dead-center (BDC). In [9], misfire was simulated by disconnection of fuel supply of one cylinder of the engine which describes the measurements of vibro-acoustic exhaust locomotive engine signals for misfire phenomenon simulation and some results of their nonlinear analysis. Instead of estimating the engine load torque [10] proposed an observer-based method, which uses the measured engine load torque directly to reconstruct the cylinder-wise net-indicated torques of a six-cylinder engine. In [11], a novel method, based on the polar coordinate system of the instantaneous angular speed (IAS) waveform, has also been introduced: to improve the discrimination features of the faults compared to the FFT-based approach of the IAS signal. In [11], two typical experimental studies were performed on 16- and 20-cylinder engines, with and without faults, and diagnose the results by the proposed polar presentation method. Method for evaluating the torque nonuniformity between the various cylinders of an internal combustion engine due to statistical dispersion in manufacturing or aging in the injection system is proposed in [12], where the problem of the crankshaft torsional vibration effects is presented on the engine speed fluctuations, that make the misfire detection critical, when multiple misfires are present in the same cycle. In [13], a simple algorithm based on an energy model of the engine which requires the measurement of the instantaneous angular speed for misfire detection in reciprocating engines is proposed. By processing the engine dynamics in the angular domain, variations in the working parameters of the engine, such as external load and mean angular speed, are compensated. A dimensionless feature has been abstracted for evaluation of the combustion as well as compression process of each cylinder.

In this work, a method of detecting early fault for medium-speed power plant diesel engines is proposed, in which as in [14, 15] torsional vibration amplitudes are used to superimpose the mass and gas torque. Further mass and gas torque analysis is used to detect fault in the operating engine.

2. Problem Definition

Problem studying in this work is to identify the early faulty cylinder by using the amplitudes of torsional vibration and lowest harmonic orders (0.5, 1, and 1.5) of the inertia masses and DFT of measured engine speed signal based on gas forces. In this paper, the authors focus and present how amplitudes of torsional vibration play an important role in performance diagnosis of internal combustion engine which will be the recent area of research for engine manufacturing industry and research centers.

3. Engine Model

Figure 1 illustrates a complete mass-elastic model of a four-stroke six-cylinder inline diesel engine system. There is a flywheel on the engine shaft to even the rotational speed of the engine shaft. The rectangles in Figure 1 represent masses with link rotating in relation to the shafts. For example, the mass of the piston/crank mechanism of each cylinder is illustrated by rectangles. Dampings are represented by the box/plate symbol and stiffness of shafts is represented by spring symbol. In Figure 1, $J_1, \ldots, J_7$ are the inertia of the each cylinder and pulley, $k_1, \ldots, k_8$ are the stiffness, and $c_1, \ldots, c_6$ are the viscous damping's, respectively.

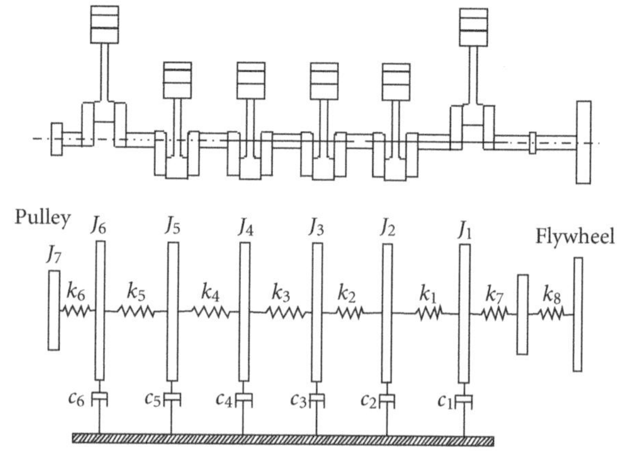

FIGURE 1: - Mass-elastic model of a four-stroke six-cylinder inline diesel engine.

4. Results and Discussion

In this work, as per [1, 6, 14, 15], detailed torsional vibration analysis was carried out and is used to calculate the torsional vibration amplitude for the given operating engine. The information in Table 1 pertains to the four-stroke, six-cylinder inline diesel engine. Tables 2, 3, and 4 show detailed torsional vibration analysis for the engine under consideration.

5. Interpretation and Calculation of Resultant Phase Vector Based on Reciprocating Masses

Here, the engine under consideration is SL90 model manufactured by Kirloskar Oil Engine Pune-03. For the considered six-cylinder engine used as an example, the crankshaft has crank throws spaced 120° apart. The firing sequence selected is l-5-3-6-2-4. Considering the order number is the number of complete cycles occurring during one crankshaft revolution, and then the first-order vectors will be spaced 120° apart in the sequence corresponding to the firing order. second-order vectors rotate at twice crank intervals so each succeeding second-order vector will be spaced at $2 \times 120° = 240°$ apart in their correct sequence and so on.

5.1. Torsional Resonance Calculation by Holzer Tabulation Method. A phase vector diagram as shown in Figure 2 is drawn for torsional resonance of 1387.22 rpm for the given engine to analyze the minor and major critical harmonic orders. It is seen that the third order is the major critical

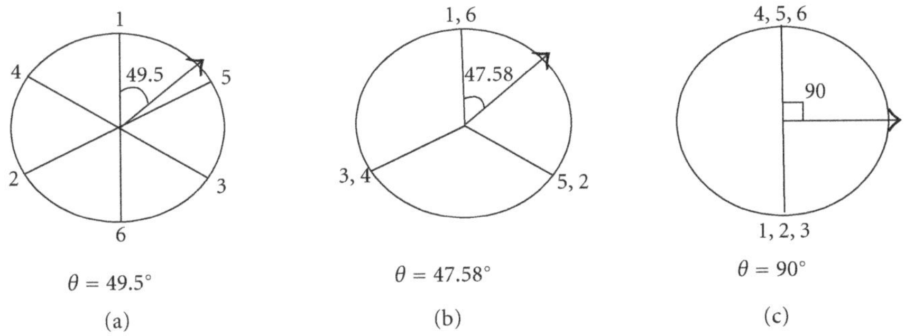

FIGURE 2: Phase vector diagram for 0.5, 1, 1.5 order.

TABLE 1: Engine basic data.

Engine type		Inline, four stroke, DI
Number of cylinder		6
Direction of rotation		Anticlockwise
Firing order		1-5-3-6-2-4
Power output at 1500 rpm		309 H.P
Cylinder diameter		118 mm
Piston stroke		135 mm
Compression ratio		15.5
Engine mass-elastic system		
Mass	Inertia Kg m^2	Stiffness MN m/rad
Pulley	0.094	0.391
Cyl.1	0.075	1.952
Cyl.2	0.075	1.952
Cyl.3	0.075	1.952
Cyl.4	0.075	1.952
Cyl.5	0.075	1.952
Cyl.5	0.075	1.952
Cyl.6	0.075	3.52
Flywheel	1.97	

orders of excitation for calculated torsional resonance. The corresponding critical speeds of engine operation would be;

$$= 1148 \text{ vpm}/3 = 191.330 \text{ rpm}$$
$$= 2379 \text{ vpm}/3 = 396.500 \text{ rpm} \quad (1)$$
$$= 4162 \text{ vpm}/3 = 1387.33 \text{ rpm} \Longleftarrow$$

Since the 3rd order excitation of 3-node vibration mode falls within the operating range of 750 to 2200 rpm, hence a phase vector diagram as shown in Figure 2 is drawn for torsional resonance of 1387.33 rpm or 4162 vpm (Vibration per Minute).

Figures 2(a) and 2(b) shows the phase angle diagram with resultant vectors for the 0.5 order and 1 order with phase angles 60° and 120°. The cylinder number and the amplitude vectors corresponding to it are shown in Figure 2 are drawn by using data in Table 5. It is found that more than one vector has the same angular position, so these have been arithmetically added making the diagram to the right. Then, by taking cosine components and algebraically adding the total cosine (vertical) component for the resultant vector is found, and by taking the sine components the resultant sine component is similarly found. The resultant vector magnitude is then

$$[\text{Resultant Vector}]^2 = [\text{Sine Component}]^2 + [\text{Cosine Component}]^2. \quad (2)$$

It is found that the fourth-order successive vectors are spaced at $4 \times 120 = 480° = (360 + 120°)$ apart and, therefore, have the same pattern as for the first order, similarly for the 7th and 10th orders. Figure 2(a) shows the 0.5, 2.5, 3.5, 5.5, 6.5 orders. The 1.5, 4.5, 7.5 orders as shown in Figure 2(c) all act up and down, as drawn, and have a moderately large resultant. In the case of order numbers which are multiples of half the number of cylinders, that is, 3, 6, 9, and so forth, for the considered six-cylinder engine, all the vectors act in the same direction (not shown in Figure 2), for this reason, they are called "Major Orders."

For 0.5 Order. Phase vector angle, $\theta = \tan^{-1}$ (sum of all vertical components/sum of all horizontal components) $\theta = -49.50°$.

For 1 Order. Phase vector angle, $\theta = \tan^{-1}$ (sum of all vertical components/sum of all horizontal components) $\theta = -47.58°$.

For 1.5 Order. Phase-vector angle $\theta = \tan^{-1}$ (sum of all vertical components/sum of all horizontal components) $\theta = 90°$.

For given engine the resultant phase vector can be calculated as above, and this Resultant phase Vector can be represented (by arrow having magnitude equal to amplitude of torsional vibration) as in Figures 2(a), 2(b), and 2(c).

6. Interpretation and Calculation of Resultant Phase Vector Based on Gas Forces

Extensive experiments were conducted on a four-stroke six cylinder direct injection diesel engine (Kirloskar SL90-SL8800TA). The engine was operated at constant speed and

Table 2: Natural frequency of whole system = 1148 vpm.

	Mass no.	Moment of inertia J	Torque per unit deflection $J\omega^2$	Deflection in plane of mass θ	Torque in plane of mass $J\omega^2\theta$	Total torque $\Sigma J\omega^2\theta$	Shaft stiffness C	Change in deflection $\Delta\theta$
		Kg m^2	MN m	±rad	MN m	MN m	MN m/rad	rad
Pulley	1	0.094	0.1238829	1	0.123882976	0.123882976	0.391	0.31683626
Cyl.1	2	0.075	0.0988428	0.683163744	0.067525817	0.191408793	1.952	0.09805778
Cyl.2	3	0.075	0.0988428	0.585105961	0.057833511	0.249242305	1.952	0.12768561
Cyl.3	4	0.075	0.0988428	0.457420354	0.045212709	0.294455013	1.952	0.15084786
Cyl.4	5	0.075	0.0988428	0.306572499	0.030302484	0.324757498	1.952	0.16637167
Cyl.5	6	0.075	0.0988428	0.14020083	0.013857843	0.33861534	1.952	0.17347097
Cyl.6	7	0.075	0.0988428	−0.03327014	−0.00328851	0.335326826	3.52	0.0952633
Flywheel	8	1.97	2.5962708	−0.12853344	−0.33370764	0.001619182		

F = 1148 vpm (vibration per min.)

Table 3: Natural frequency of whole system = 2379 vpm.

	Mass no.	Moment of inertia J	Torque per unit deflection $J\omega^2$	Deflection in plane of mass θ	Torque in plane of mass $J\omega^2\theta$	Total torque $\Sigma J\omega^2\theta$	Shaft stiffness C	Change in deflection $\Delta\theta$
		Kg m^2	MN m	±rad	MN m	MN m	MN m/rad	rad
Pulley	1	0.094	0.5320062	1	0.532006254	0.532006254	0.391	1.36062981
Cyl.1	2	0.075	0.4244730	−0.36062981	−0.15307764	0.378928611	1.952	0.19412326
Cyl.2	3	0.075	0.4244730	−0.55475307	−0.23547774	0.14345087	1.952	0.07348918
Cyl.3	4	0.075	0.4244730	−0.62824225	−0.26667192	−0.12322105	1.952	−0.0631255
Cyl.4	5	0.075	0.4244730	−0.56511671	−0.23987683	−0.36309787	1.952	−0.1860133
Cyl.5	6	0.075	0.4244730	−0.37910345	−0.16091922	−0.52401708	1.952	−0.2684514
Cyl.6	7	0.075	0.4244730	−0.11065208	−0.04696883	−0.57098591	3.52	−0.1622119
Flywheel	8	1.97	11.149492	0.051559828	0.574865928	0.003880017		

F = 2379 vpm (vibration per min.)

full loads. To simulate a early faulty cylinder, the engine was operated under normal working condition. The pressures were measured in all cylinders by piezoelectric pressure transducers. The mean-indicated pressure (MIP) and the gas pressure torque (GPT) of each cylinder were calculated from the pressure traces. The GPT and the measured speed were subjected to a discrete Fourier transform (DFT) to determine the amplitudes and phases of their harmonic components. Figure 3 shows an actual pressure curve generated by the pressure transducers on all six cylinders when engine was working under normal condition. The results of the application of this technique to given DI Diesel engine are graphically shown in Figures 4 and 5.

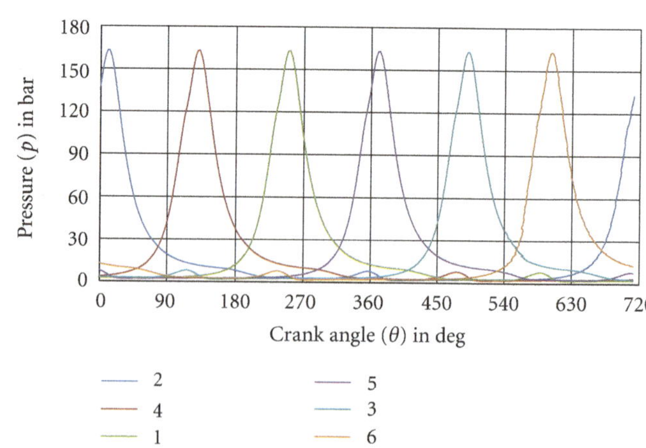

Figure 3: Gas pressure versus crank angle.

It is observed that when the cylinders are uniformly contributing to the total engine torque, the first three harmonic orders (K = 0.5, 1, 1.5) play a significant role in the frequency spectrum of the total gas-pressure torque and, consequently, appear with a very low contribution in the frequency spectrum of the crankshaft's speed [5, 6]. If the frequency spectrum of the crankshaft's speed corresponding to uniform cylinders operation is compared to the spectrum corresponding to a faulty cylinder, one may see that the major difference is produced by the amplitudes of the first three harmonic orders. As far as the cylinders operate uniformly, these amplitudes are maintained under a certain limit. Once a cylinder starts to reduce its contribution, the amplitudes of the first three harmonic orders start increasing. These amplitudes may be used to determine the degree by which

Table 4: Natural frequency of whole system = 4162 vpm.

	Mass no.	Moment of inertia J	Torque per unit deflection $J\omega^2$	Deflection in plane of mass θ	Torque in plane of mass $J\omega^2\theta$	Total torque $\Sigma J\omega^2\theta$	Shaft stiffness C	Change in deflection $\Delta\theta$
		Kg m^2	MN m	±rad	MN m	MN m	MN m/rad	rad
Pulley	1	0.094	2.016259	1	2.016259	2.016259	0.391	0.974415
Cyl.1	2	0.075	0.07293367	−3.164426946	0.018660	2.034920	1.952	0.914119
Cyl.2	3	0.075	0.07293367	−1.8924842	−0.648040	1.386879	1.952	0.623009
Cyl.3	4	0.075	0.07293367	0.639015657	−1.102423	0.2844561	1.952	0.127782
Cyl.4	5	0.075	0.07293367	2.745213832	−1.195620	−0.911163	1.952	−0.40931
Cyl.5	6	0.075	0.07293367	3.024314267	−0.897095	−1.808259	1.952	−0.81235
Cyl.6	7	0.075	13.164734	1.290559567	2.125865	0.01295095	3.52	0.161948
Flywheel	8	1.97	27.805941	−0.147208403	−0.0129509	$0.03691E-6$		
				F = 4162 vpm (vibration per min.)				

Table 5: Summary of the values of torsional vibration amplitude from Holzer Tables 2, 3, and 4.

Firing order	Values of torsional vibration amplitude from computerize Holzer table		
Frequency	1148 vpm	2379 vpm	4162 vpm
Cyl.1	0.683163744	−0.36062981	−3.164426946
Cyl.5	0.14020083	−0.37910345	3.024314267
Cyl.3	0.457420354	−0.62824225	0.639015657
Cyl.6	−0.03327014	−0.11065208	1.290559567
Cyl.2	0.585105961	−0.55475307	−1.8924842
Cyl.4	0.306572499	−0.56511671	2.745213832

a cylinder reduces its contribution to the total gas pressure torque. The identification of the faulty cylinder may be achieved by analyzing the phases of the lowest three harmonic orders.

Figure 4 is drawn by reconstructing the pressure traces of the six cylinders in a sequence corresponding to the firing order (1-5-3-6-2-4) with engine operating under normal conditions. Figure 5 shows the lowest three harmonic orders of the measured speed, respecting the measured amplitudes and phases. It is seen that only for the expansion stroke of cylinder 5, all three harmonic curves have, simultaneously, a negative slope. In the phase-angle diagrams (Figure 5) of these orders, the resultant vector corresponding to the harmonic component of the measured speed is also represented (shown by arrow having magnitude equal to amplitude of vibration) is drawn by using data in Figure 6. One may see that, for each of the three considered orders, the vectors are pointing toward the group of cylinders that produces less work. The cylinder that is identified three times among the less productive cylinders is the faulty as shown in Figure 5.

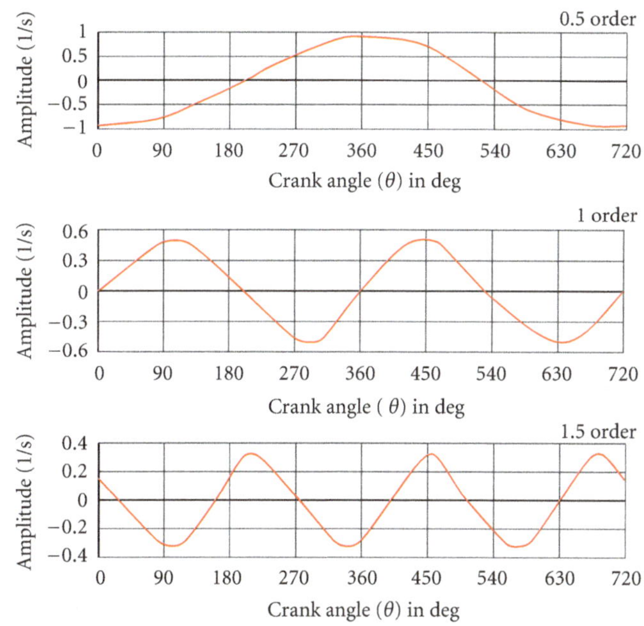

Figure 4: Detection of a faulty cylinder from the phases of the lowest three harmonic orders.

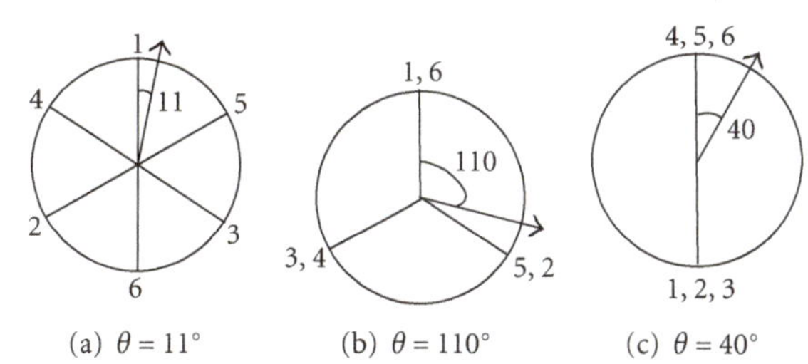

Figure 5: Phase angle Diagram 0.5, 1, 1.5 order.

7. Method for Detecting Fault in Operating Engine

Based on the Figures 2 and 5, the following method is developed.

(1) The phase-angle diagrams, considering the firing order of the engine are drawn for the lowest three harmonic orders placing in the top dead centre (TDC) the cylinder that fires at 0° in the considered cycle.

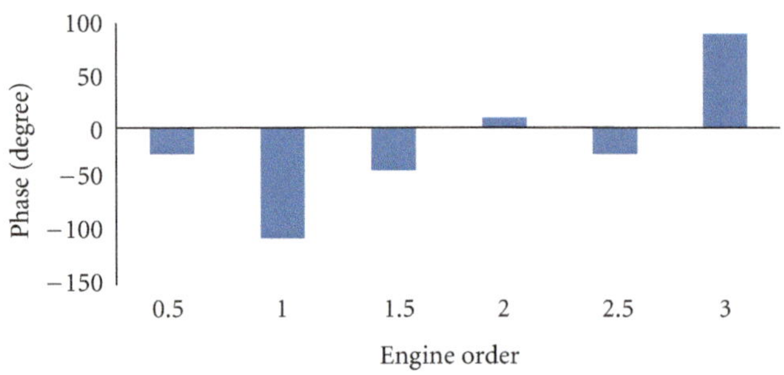

Figure 6: Phase versus engine order.

Table 6: Identification of the faulty cylinder from the position, in the phase-angle diagrams from Figure 2.

K	Cylinders					
	1	5	3	6	2	4
0.5	—	—				
1	—	—		—	—	
1.5		—		—		—
I_i	0	−1	0	0	0	0

Table 7: Identification of the faulty cylinder from the position, in the phase-angle diagrams Figure 5.

K	Cylinders					
	1	5	3	6	2	4
0.5	—	—				
1	—	—		—	—	
1.5		—		—		—
I_i	0	−1	0	0	0	0

(2) On these phase angle diagrams, the corresponding vectors of the measured speed are represented in a system of coordinate axes.

(3) The cylinders toward which the vectors are pointing are the less contributors and receive a "-" mark. If there are cylinders that receive a "-" mark for all three harmonic orders they are clearly identified as less contributors to the engine total output.

(4) This method is able to identify a faulty cylinder at very early stage, that is, before starting the engine manufacturing (Table 6) and as soon as its contribution reduces from rated value to zero with respect to the contribution of the other cylinders (Table 7).

8. Conclusions

Proposed methodology state the way of early fault detection in the operating diesel engine and Tables 6 and 7 shows validation of developed methodology. Here, torsional vibration amplitudes Table 2 are used to detect early fault in the operating diesel engine. In this paper, combination of mass torque and gas torque is used to map the fault. From this work it is found that the amplitudes of the lowest harmonic orders (0.5, 1, and 1.5) of the inertia masses and DFT of measured speed based on gas forces may be effectively used to map the fault. This method is capable to detect early faulty cylinders (which is now cylinder number 5) based on the phases of the lowest three harmonic as soon as its contribution starts to drop from rated value to zero due to its inertia forces as compared to the contribution of the other cylinders. This methodology can be also used for early fault diagnosis in the operating SI engine.

References

[1] E. J. Nestrides, *A Handbook on Torsional Vibration, B.I.C.E.R.A, Research Library*, Cambridge University Press, Cambridge, UK, 1958.

[2] J. Williams, "An overview of misfiring cylinder engine diagnostic techniques based on crankshaft angular velocity measurements," SAE Technical Paper 960039, 1996.

[3] K. S. Chen and T. Chang, "Crankshaft torsional and damping simulation-an update and correlation with test results," SAE Paper 861226, 1986.

[4] M. J. Van Nieuwstadt and I. V. Kolmanovsky, "Detecting and correcting cylinder imbalance in direct injection engines," *ASME Journal of Dynamic Systems, Measurement and Control*, vol. 123, no. 3, pp. 413–424, 2001.

[5] D. Taraza, N. Henein, and W. Bryzik, "Diesel engine diagnostics based on analysis of the crankshaft's speed variations," SAE Paper 982540, 1998.

[6] S. H. Gawande, L. G. Navale, M. R. Nandgaonkar D, S. Butala, and S. Kunamalla, "Detecting power imbalance in multi-cylinder inline diesel engine genst set," *Journal of Electronic Science and Technology*, vol. 8, no. 3, pp. 273–279, 2010.

[7] U. Kiencke and L. Nielson, *Automotive Control Systems, for Engine, Driveline and Vehicle*, Springer, Italy, 2005.

[8] U. Kiencke, "Engine misfire detection," *Control Engineering Practice*, vol. 7, no. 2, pp. 203–208, 1999.

[9] P. Boguś and J. Merkisz, "Misfire detection of locomotive diesel engine by non-linear analysis," *Mechanical Systems and Signal Processing*, vol. 19, no. 4, pp. 881–899, 2005.

[10] Y. W. Kim, G. Rizzoni, and Y. Y. Wang, "Design of an IC engine torque estimator using unknown input observer," *Transactions of ASME Journal of Dynamic Systems Measurement and Control*, vol. 121, no. 3, pp. 487–495, 1999.

[11] P. Charles, J. K. Sinha, F. Gu, L. Lidstone, and A. D. Ball, "Detecting the crankshaft torsional vibration of diesel engines for combustion related diagnosis," *Journal of Sound and Vibration*, vol. 321, no. 3–5, pp. 1171–1185, 2009.

[12] N. Cavina and F. Ponti, "Engine torque nonuniformity evaluation using instantaneous crankshaft speed signal," *ASME Journal of Engineering for Gas Turbines and Power*, vol. 125, no. 4, pp. 1050–1058, 2003.

[13] F. V. Tinaut, A. Melgar, H. Laget, and J. L. Domínguez, "Misfire and compression fault detection through the energy model," *Mechanical Systems and Signal Processing*, vol. 21, no. 3, pp. 1521–1535, 2007.

[14] S. H. Gawande, L. G. Navale, M. R. Nandgaonkar, and D. Butala, "Torsional frequency analysis of multi-cylinder inline diesel engine generator system," in *Proceedings of the International Conference on Mechanical, Industrial, and Manufacturing Technologies (MIMT '10)*, pp. 519–524, ASME, January 2010.

[15] S. H. Gawande, L. G. Navale, M. R. Nandgaonkar, D. Butala, and S. Kunamalla, "Cylinder imbalance detection of six cylinder DI diesel engine using pressure variation," *International Journal of Engineering Science and Technology*, vol. 2, no. 4, pp. 433–441, 2010.

11

Gas Turbine Blade Damper Optimization Methodology

R. K. Giridhar,[1] P. V. Ramaiah,[2] G. Krishnaiah,[2] and S. G. Barad[1]

[1] *Vibration Engineering Group, Gas Turbine Research Establishment, CV Raman Nagar, Bangalore 560093, India*
[2] *Mechanical Engineering Department, Sri Venkateswara University, Tirupati 517502, India*

Correspondence should be addressed to R. K. Giridhar, giridhardrdo@gmail.com

Academic Editor: Benjamin Soenarko

The friction damping concept is widely used to reduce resonance stresses in gas turbines. A friction damper has been designed for high pressure turbine stage of a turbojet engine. The objective of this work is to find out effectiveness of the damper while minimizing resonant stresses for sixth and ninth engine order excitation of first flexure mode. This paper presents a methodology that combines three essential phases of friction damping optimization in turbo-machinery. The first phase is to develop an analytical model of blade damper system. The second phase is experimentation and model tuning necessary for response studies while the third phase is evaluating damper performance. The reduced model of blade is developed corresponding to the mode under investigation incorporating the friction damper then the simulations were carried out to arrive at an optimum design point of the damper. Bench tests were carried out in two phases. Phase-1 deals with characterization of the blade dynamically and the phase-2 deals with finding optimal normal load at which the blade resonating response is minimal for a given excitation. The test results are discussed, and are corroborated with simulated results, are in good agreement.

1. Introduction

The friction damping concept is widely applied in turbomachinery applications, especially at hot end parts, to reduce resonance stresses. A typical application of this in gas turbines. They are popularly called as "friction damper," "cottage-roof damper" or "under platform damper." This damper is loaded by centrifugal force against the underside of the platforms of two adjacent blades. The main design criterion for such devices is to determine the optimum damper configuration or the damper mass or both in order to reduce the dynamic stresses to maximum possible extent. For example, if the damper mass is too small for a given configuration, the friction force will not be large enough to dissipate sufficient energy. On the other hand, if the damper mass is too large, it will get into "stick" condition, thereby limiting the relative motion across the interface and hence the amount of energy dissipation. In both cases, the friction damper will be inefficient, and between these two extremes there exists an optima.

A good review of the friction damping concept in turbomachinery applications is given by Griffin [1]. Theoretical analysis and the optimization of this simple device is difficult because of marked nonlinearity and assumptions about the contact characteristics and damper behavior. Several friction damper models and analysis methods have been proposed in past. The simplest and most commonly used model reported in the literature is a macroslip contact model [2–4]. There are also several micro-slip friction models reported which are more appropriate in case of high normal loads [5–10]. The macroslip model under this high normal load shows damper in lockup condition.

Although significant advances are made in theoretical modeling of friction dampers and analysis methods, turbomachinary manufacturers still rely on previous experience and empirical data rather than computer-based predictions alone for friction damper optimization. This has been mainly due to the oversimplification introduced in the models regarding the basic contact behavior and/or damper geometry and the inability to analyze representative-size models due to excessive computational cost [4].

The latest and most advanced contact models make use of three parameters to characterize the contact behavior, namely, friction coefficient, tangential contact stiffness and normal contact stiffness [11–13]. Szwedowicz et al. [13] investigated numerically and experimentally the performance of a thin-walled damper mounted under the platform of two rotating free-standing high-pressure turbine blades. Characterization of friction contact of nonspherical contact geometries obeying the Coulomb friction law with constant friction coefficient and constant normal load is proposed by [14]. Firrone et al. [15] undertook forced response studies of bladed disc under platform damper considering both static and dynamic displacements.

This paper presents a methodology which combines three essential phases of friction damping optimization in turbomachinery. The first phase is to develop an analytical model of the blade damper system. The second phase is experimentation and model tuning necessary for response studies; while the third phase is evaluating damper performance and arriving at optimal damper design point. Figure 1 gives the flow chart of complete design methodology. In the present work, we are limiting our discussions to static bench test and using this data along with results from analytical studies, studying the damper performance, and then arriving at an optimal damper design.

2. Development of a Mathematical Model

The development of a mathematical model for a blade damper involves following steps:

(1) estimation of eigenvalues and establishment of stress distribution for a specific mode of interest;

(2) development of a friction model;

(3) conducting response studies using reduced model.

2.1. Eigenvalue Estimation and Response Studies. The objective of this phase is to obtain modal characteristics of the blade and stress distribution along the aerofoil for a particular mode of interest along with estimation of nondimensional parameters. Figure 2 presents a finite element 3D model of a turbine blade, with boundary condition in the form of root fixation. The blade material is transversally anisotropic with minimal rigidity axis directed along the blade radius.

Analysis is performed for natural frequencies, distribution of stresses, deformations, and displacements for first six modes. Table 1 gives the various frequencies estimated for various conditions; all the frequencies are normalized with respect to first flexural mode of the blade.

The optimization study is undertaken for the first flexure mode of the blade alone; therefore, we are limiting our discussions to this only. Figures 3 and 4 show the 1st flexural mode of the blade and corresponding stress distribution, respectively. The values corroborate well with the experimental results given in Table 2. The maximum stress is located near the fillet area close to the first lobe of the fir tree.

With the input drawn from above data, the following coefficients are estimated in order to establish the damper design parameters. These are

(1) all the natural frequencies normalized with respect to 1st flexural mode of the blade;

(2) ratio of maximum stresses in the attachment to stresses in aerofoil root, $k_{\sigma\sigma} = 1.86$;

(3) ratio of maximum total displacement to stresses in aerofoil, $k_{u\sigma} = 2.12e^{-3}\ \text{mm}^3/\text{N}$;

(4) ratio of displacement (circumferential direction) at contact location of damping insert to maximum displacement of blade aerofoil, $k_{uu} = 0.07$;

(5) ratio of tip displacement to load amplitude, $k_{uP} = 0.026$.

Further results from the 3D model calculations were used for identifying strain gauge locations for experimental studies.

2.2. Friction Model. The friction interface between the blade root and the damper insert under investigation has a rectangular shape. It is thus appropriate to use a friction interface model that has a rectangular contact surface, such as the model by [5, 9], further the normal load distribution is assumed to be constant over the interface. The model that is used in this paper, shown in Figure 5, is derived from both Menq et al's [5] and the Csaba's one-bar model [9]. The normal load on the damper is assumed as uniform. Further displacement and force is represented in terms of slip length, this feature is derived from Csaba's work [8].

The interface is considered as a rectangular cross-section and modeled as a bar pressed against a rigid surface with a normal load $q(x)$, and the force of P is applied at the end of the bar. The bar has a modulus of elasticity E and area of cross-section A. L, is the length of the bar. The coefficient of friction μ is assumed as constant across the interface and independent of motion of the bar. The normal load across the interface is assumed uniform and constant.

The assumed function of normal load is $q(x) = q_0$, and the total normal load is found by integrating the normal load function over the length of the bar:

$$N = \int_0^L q(x)dx. \tag{1}$$

In the present work, the derivation of displacement function is not presented, but relevant equations as and when required are given.

(1) *Governing Equation.* It is assumed that the bar deforms elastically and the friction is governed by Coulomb's law at each contact point. By applying the load P at the one end of the bar, the onset of slip takes place. Further by increase in amplitude of "P" propagates of slip zone until a gross level slip is achieved. In order to formulate the equation, this bar length is divided into two parts, sliding and not sliding, as shown in Figure 6. The sliding length, that is, the slip zone length is given by "δ" and its maximum value is given by δ_a, corresponding to max force P_a. The friction force is defined by F, and its direction depends on the direction of resultant strain. The displacement at the end of bar is defined as "u."

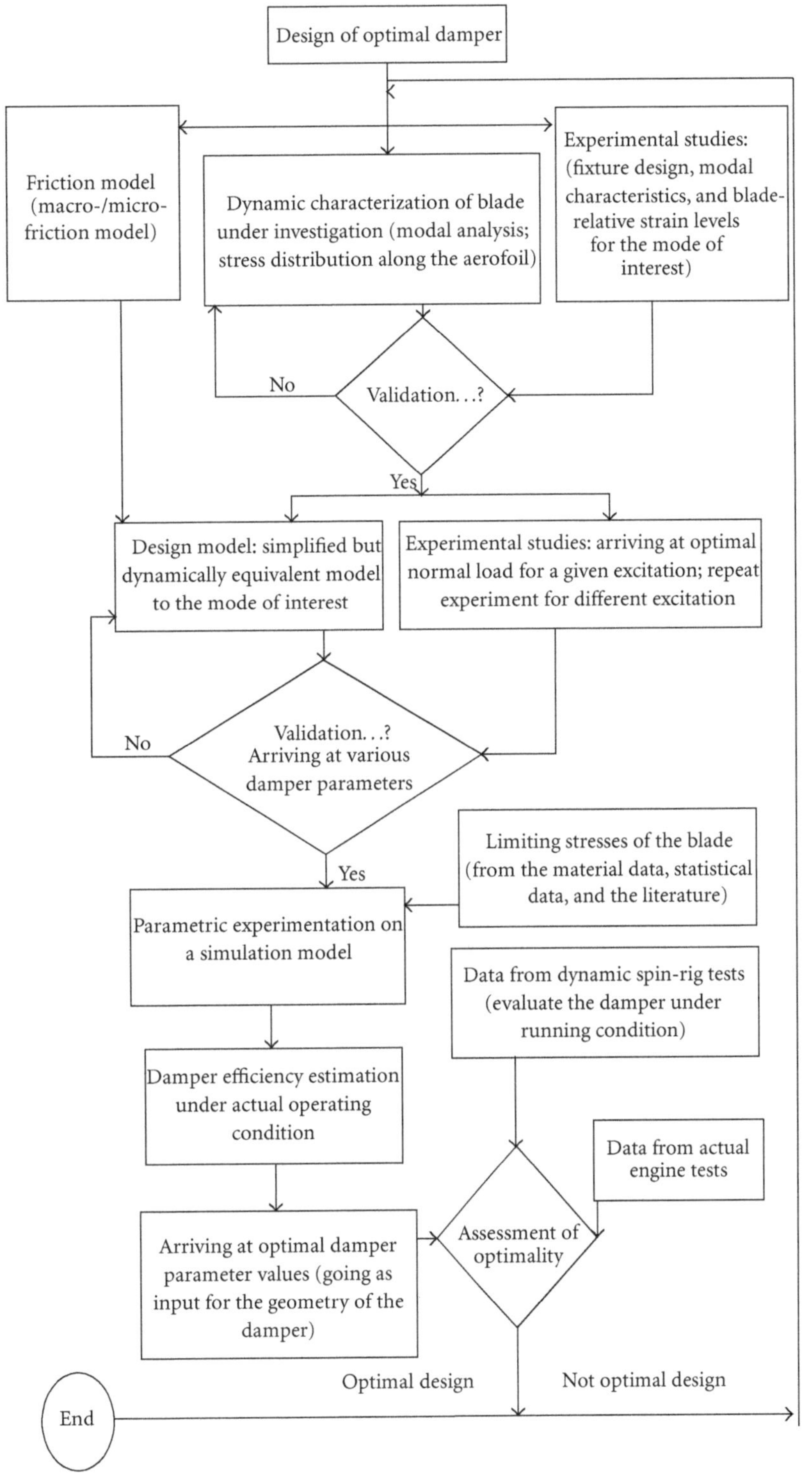

Figure 1: The flow chart indicating the damper design methodology.

The procedure for derivation of displacement function and the force function is followed from [8].

The force is considered as monotonic between its peak values, with amplitude P_a, then the analysis may be divided into three parts with the bar initially at rest [8]:

P is increased from 0 to P_a;

P is decreased from P_a to $-P_a$;

P is increased from $-P_a$ to P_a.

(2) *Formulation of Initial Loading Relations.* The Coulomb's friction law says

$$F = \mu q(x). \quad (2)$$

TABLE 1: Estimated eigenvalues for various conditions normalized with 1st natural frequency at room temperature.

Mode number and frequency (Hz)				Simulated condition/constraints
1	2	3	4	
f1	1.74f1	2.26f1	3.06f1	Nonrotating blade at room temperature
1.008f1	1.755f1	2.27f1	3.08f1	Rotating at 50% RPM at room temperature
1.056f1	1.79f1	2.30f1	3.16f1	Rotating at 100% RPM at room temperature
0.95f1	1.61f1	2.06f1	2.85f1	Rotating at 100% RPM at 800°C average blade temperature

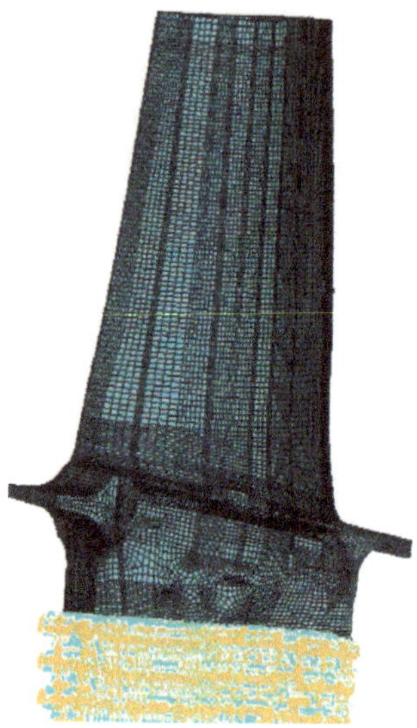

FIGURE 2: Finite element model of the blade.

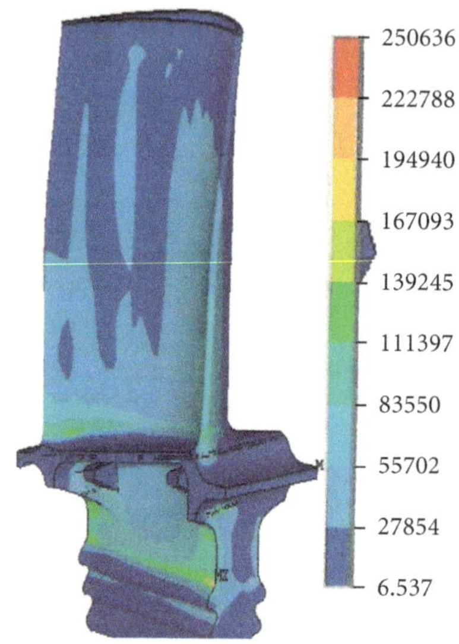

FIGURE 4: Distribution of equivalent blade stresses at the 1st vibration mode.

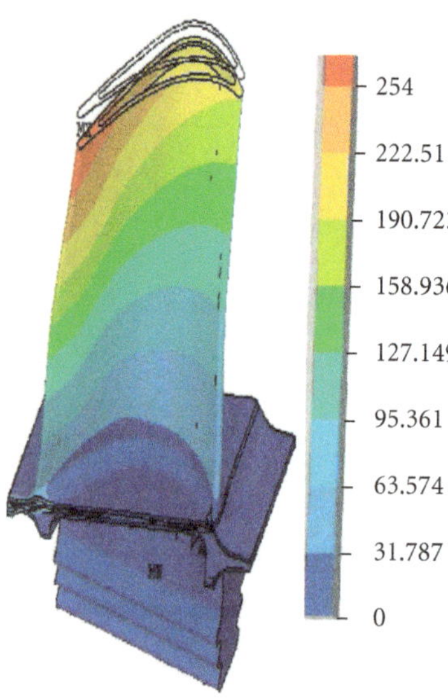

FIGURE 3: Mode shape for the 1st flexure mode of the blade.

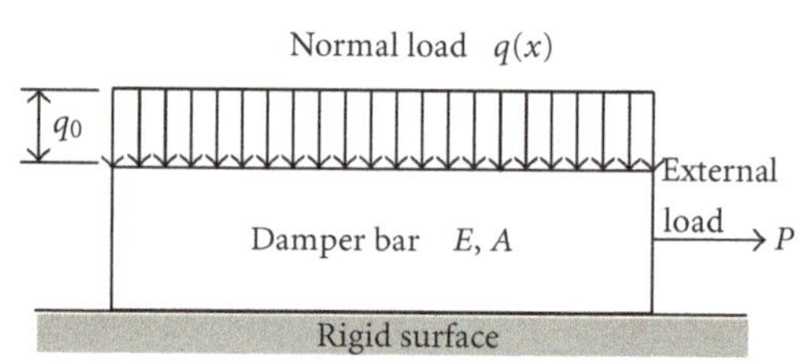

FIGURE 5: Microslip model for friction interface.

TABLE 2: Comparison of estimated eigenvalues with experimentally measured values.

Mode number	Estimated values	Experimental values
1	f1	1.01f1
2	1.74f1	1.76f1
3	2.26f1	2.27f1
4	3.06f1	3.12f1

The various forces acting on a small element of length dx of the sliding zone are given in Figure 7. The direction of "F" depends on whether force "P" is increasing or decreasing with respect to "x."

Equilibrium of forces yields

$$F\,dx + P + \frac{dP}{dx}dx = P. \tag{3}$$

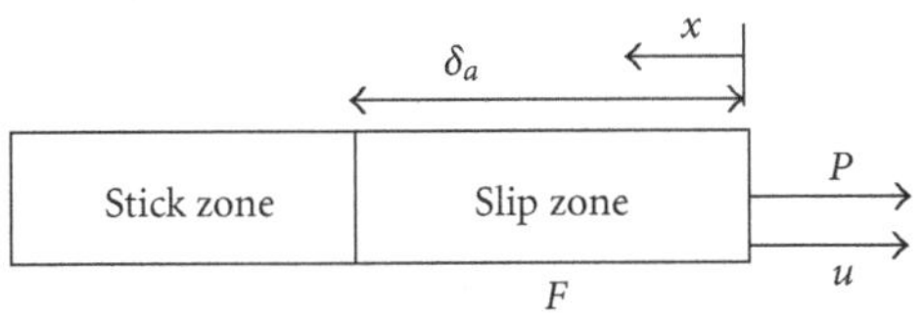

FIGURE 6: Slip and Stick zones, when the bar is subjected to initial loading.

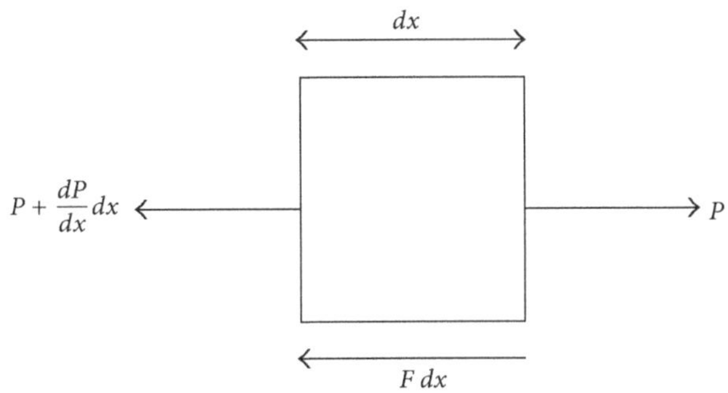

FIGURE 7: Forces acting on a small element.

The strain in bar is

$$\varepsilon = -\frac{du}{dx}. \quad (4)$$

The tensile force in the bar is defined as

$$P = -EA\left.\frac{du}{dx}\right|_{x=0}. \quad (5)$$

Substituting (5) in (3) we get

$$F dx - EA\frac{d^2u}{d^2x} = 0, \quad (6)$$

where du/dx is the strain in the bar. The displacement "u" is found by double integrating over the slip zone length using the following boundary conditions [8]:

$$x = 0, \quad P = -EA\frac{du}{dx}, \qquad x = \delta,\ u = 0. \quad (7)$$

Therefore, the displacement as a function of force is

$$u(x, P) = P\frac{(x-\delta)}{EA} + \mu q_0\frac{(\delta^2 - x^2)}{2EA}. \quad (8)$$

The force at the bar end can be expressed as a function of the slip length. This is done by integrating the friction force over the slip length [8]:

$$P(\delta) = \int_0^{\delta} F_x dx = \mu q_0 \delta. \quad (9)$$

The displacement in terms of slip length is given by substituting (9) in (8):

$$u(\delta, x) = \mu q_0 \delta\frac{(x-\delta)}{EA} + \mu q_0\frac{(\delta^2 - x^2)}{2EA}, \quad u(\delta, x) = \mu q_0\frac{(\delta - x)^2}{2EA}. \quad (10)$$

After having the force and displacement as functions of slip length. Applying "P_{amp}" will give amplitude slip length "δ_a". The force amplitude function is defined by

$$P(\delta_a) = \mu q_0 \delta_a. \quad (11)$$

And the displacement function is given as

$$u(\delta_a) = \frac{\mu q_0 \delta_a^2}{2EA}. \quad (12)$$

(3) *Load Decreasing Relations.* The section of the bar that was slipping is stretched after an initial loading of $P = P_{\text{amp}}$. This is given as broken line in Figure 8. The bar is divided into three zones shown in Figure 8, as given in [8]. Zone "Z," where tension is changing in to compression will extend, as the force is decreasing from P_{amp} to $-P_{\text{amp}}$. The length of the compression zone length is denoted as δ_d. Zone Z will increase, and zone Y will decrease as P is decreasing, this will continue till $P = -P_{\text{amp}}$. Zone Y is then eliminated, and δ_d equals δ_a.

The friction force direction depends on the sign of dP on that part. The various forces acting on a small element of length dx of zone "Z" is given in Figure 9. The differential equation for this is

$$F dx + EA\frac{d^2u}{d^2x} = 0. \quad (13)$$

The boundary conditions are found by equilibrium of forces at the bar end, and with a condition that u_a, and u_d must be of the same value, where zone Y and Z are connected. These boundary conditions are, as given in [8]:

$$x = 0, \quad P = -EA\frac{du_d}{dx}, \qquad x = \delta_d,\ u_d(\delta_a, \delta_d) = u(\delta_a, \delta_d). \quad (14)$$

Solving (13) by substituting (14) yields u_d. The displacement of the bar end will be a function of δ_a and δ_d.

The displacement function is given as

$$u_d(\delta_a, \delta_d) = \frac{\mu q_0\left(\delta_a^2 - \delta_d^2\right)}{2EA}. \quad (15)$$

The slip length δ_d is found by equilibrium of forces for the compressed region Z. By equilibrium of forces as shown in Figure 10 and solving one gets the force in terms of slip length. This is given by

$$P_d(\delta_a, \delta_d) = \mu q_0(\delta_a - 2\delta_d), \quad (16)$$

where δ_d is the slip length, when load is decreasing from P_{amp} to $-P_{\text{amp}}$.

(4) *Reloading Relations.* This is exactly opposite situation to the one in previous section. Zone Z, where tension will be seen in place of compression, will increase and extend as the force is increasing from $-P_{\text{amp}}$ to P_{amp}. The strain in the bar when force P is increasing is shown as the unbroken line in Figure 11, as given in [8].

The length of the stretched zone is denoted by the slip length δ_i. Zone Z will increase, and zone Y will decrease as P is increases. This will continue till $P = P_{\text{amp}}$. Zone Y is then eliminated, and δ_i equals δ_a. The differential equation in this case is the same as when the bar is initially loaded. The only difference is the boundary conditions. Here, the boundary conditions are, as given by [8]:

$$F dx - EA\frac{d^2u}{d^2x} = 0, \tag{17}$$

$$x = 0, \quad P = -EA\frac{du_i}{dx}, \qquad x = \delta_i, \; u_i(\delta_a, \delta_i) = u(\delta_a, \delta_i). \tag{18}$$

Solving (17) with (18) yields u_i at the bar end as a function of δ_a and δ_i:

$$\begin{aligned} u_i(\delta_i, \delta_a) &= \frac{\mu q_0\left(\delta_i^2 - \delta_a^2\right)}{2EA}, \\ P_d(\delta_i, \delta_a) &= \mu q_0(2\delta_i - \delta_a), \end{aligned} \tag{19}$$

where δ_i is slip length, when the load is increasing from $-P_{\text{amp}}$ to P_{amp}. The slip length δ_i is calculated by equilibrium of forces for the stretched region Y.

Once the relations of both force and displacement at the end of the bar for both unloading and reloading is established in terms of slip length based on initial loading relationships, the hysteretic curve can be generated. Figure 12 is the time domain representation of input force and Figure 13 is simulated displacement at the end of the bar. Using above parametric functions, the hysteretic curve is established and shown in Figure 14. The curve is built up with a starting part (portion OA) for initial loading, the lower curve (portion AB), for unloading $P_d(\delta_d)$ and $u_d(\delta_d)$, and an upper curve (portion BA) for reloading $P_i(\delta_i)$ and $u_i(\delta_i)$. Once the hysteretic curve is established, the damping energy is calculated. The work done per cycle is

$$W = \oint u dP. \tag{20}$$

(5) *Linearization.* The objective here is to transform the nonlinear properties of the friction damper into equivalent linear damping and stiffness. The hysteresis loop made by the force versus displacement function is thereby replaced with an equivalent elliptic loop. The linearized properties of the damper will then be used for forced response analysis. Linearization of the friction model is done by using Lazan's method [9].

The work done by equivalent viscous damping is given by

$$D = \pi\varpi C_{\text{eq}} u_{\text{amp}}^2, \tag{21}$$

by equating (20) and (21), we get

$$C_{\text{eq}} = \frac{W}{\pi\omega u_{\text{amp}}^2}. \tag{22}$$

Similarly the equivalent stiffness given by Lazan's method is

$$K_{\text{eq}} = \sqrt{\left(\frac{P_{\text{amp}}}{u_{\text{amp}}}\right)^2 - \left(\omega C_{\text{eq}}\right)^2}. \tag{23}$$

2.3. Reduced Model. The dynamically equivalent reduced blade model is represented using a two-degree freedom system simulating 1st flexure mode of blade under investigation, similar to that of given by [8], but additionally the material damping and the aerial damping are represented as a viscous damper. The reason for choosing a two-DOF system for simulations is that it is simple but still illustrates the effect of the damper. At first the response for blade model without damper insert is determined, and correlation is ensured for the first mode between design model and actual blade model. The calculations are performed for various excitation frequency and amplitude of excitation.

The 2-DOF system, shown in Figure 15, is built considering a mass less beam with two concentrated masses simulating aerofoil and the root portions of the blade and damper. The aerial and material damping is considered as a viscous damping and attached to the aerofoil portion of the blade. The friction damper is described by an equivalent viscous damper and spring, $C_{\text{eq}}(\delta_a)$ and $K_{\text{eq}}(\delta_a)$, respectively. Assuming harmonic force of excitation applied at the aerofoil portion of the blade. The displacements where the friction damper is attached and force applied are x_1 and x_2, respectively.

As the model is approximate, for obtaining the quantitative estimations, the model is identified using correction factors based on the statistical results obtained from testing the turbine blade and the modal analysis of a 3D model of the blade. Some of the parameters which are used for correcting the model are as follows.

The value of 1st natural frequency of the blade without a damper insert ($f1$) has provided the values of mass and stiffness of equivalent model.

The damping in the turbine blade without damper insert is in the range of $\zeta \approx 1\%$ as per the statistics provided by experiments and used as a viscous drag in the model.

The value of the excitation force is chosen such that the amplitude of varying stresses in the blade is in the range of 50–100 MPa, this value is estimated from the statistical data available for a typical class of turbine blades.

The friction coefficient is taken in the range of 0.2–0.4 considering the operating environment of the damper inserts.

The equation of motion for the system is

$$\begin{aligned} &M_1\ddot{x}_1(t) + C_{\text{eq}}(\delta_a)\dot{x}_1(t) + \left(K_1 + K_2 + K_{\text{eq}}(\delta_a)\right)x_1(t) \\ &\quad -K_2x_2(t) = 0, \\ &M_2\ddot{x}_2(t) + C\dot{x}_2(t) + K_2x_2(t) - K_2x_1(t) = P_a e^{j\omega t}. \end{aligned} \tag{24}$$

Assuming harmonic motion:

$$x_1(t) = X_1 e^{j\omega t}, \qquad x_2(t) X_2 e^{j\omega t}. \tag{25}$$

And defining the complex stiffness

$$K(\delta_a) = K_{\text{eq}}(\delta_a) + j\omega C_{\text{eq}}(\delta_a). \tag{26}$$

By substituting (26) and (25) in (24) yields algebraic equations:

$$-M_1\omega^2 X_1 + (K(\delta_a) + K_1 + K_2)X_1 - K_2X_2 = 0, \tag{27}$$

$$-M_2\omega^2 X_2 + C\omega X_2 + K_2X_2 - K_2X_1 = P_a. \tag{28}$$

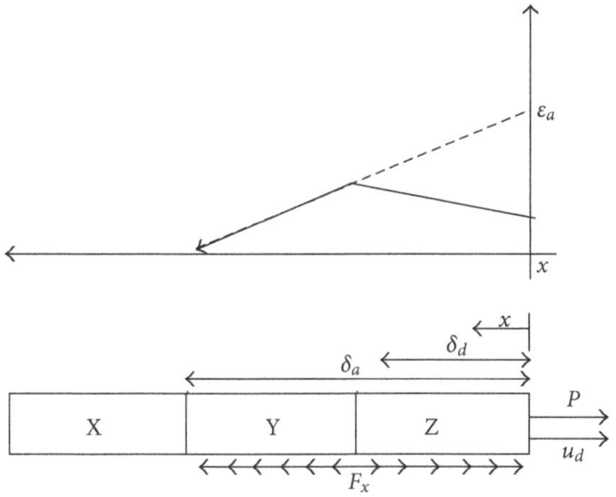

FIGURE 8: Force, displacement, and plot of strain function, when force is decreasing; zone X is totally struck and has zero strain, zone Y is struck and stretched; zone Z is slipping and compressed.

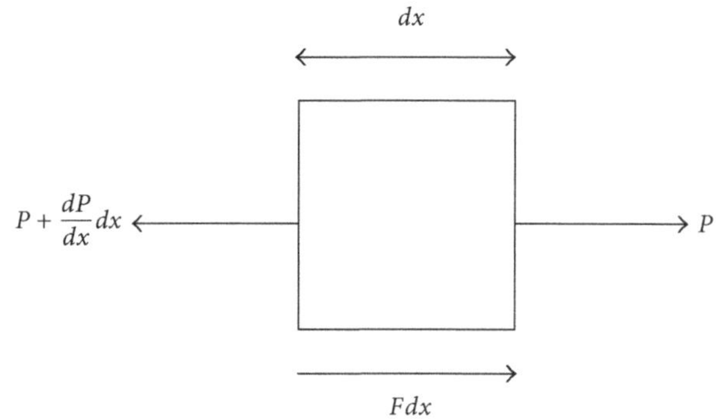

FIGURE 9: Forces acting on a small element in zone Z, when force is decreasing.

From (28), $X_2 = (P_a + K_2X_1)/(K_2 + jC\omega - M_2\omega^2)$. Substituting X_2 in (27) we get

$$X_1 = \frac{P_aK_2}{(K_2 + jC\omega - M_2\omega^2)((K_1 + K_2 + K(\delta_a) - M_1\omega^2) - K_2^2)},$$
$$X_1 = \frac{P_a}{(K_2 + jC\omega - M_2\omega^2)} + \frac{P_aK_2^2}{(K_2 + jC\omega - M_2\omega^2)((K_1 + K_2 + K(\delta_a) - M_1\omega^2) - K_2^2)}. \quad (29)$$

The response of the system is studied keeping the excitation force amplitude constant, and varying the normal load on the damper.

The solution procedure adopted in the frequency-domain is based on finding the response amplitudes iteratively. The excitation level is selected in such a way that the maximum stress in the blade is in the range of 100 MPa. The starting point being the response levels of the underlying linear system. The behaviour of the friction dampers is analyzed at a given relative response amplitude between the damper connection points and the individual dampers. The individual dampers are represented as equivalent complex stiffness, representing both restoring and energy dissipation characteristics as described above. The equivalent complex stiffness is then added to the otherwise linear system, and the response level of the modified system is calculated again. The procedure is repeated till convergence is achieved. The error between successive iterations should be below certain value. In the present case, it is defined as $1\text{E}10^{-5}$.

The response levels obtained at current frequency are used as initial guesses for the next frequency increment. Figure 16 shows the normalized responses verses the frequency ratio, under a given excitation for various normalized friction forces. When the friction force is increased the response levels will decrease and attain minimum at particular condition. Further increase in friction force will increase the response amplitude. Figure 17 is the normalized stress versus normalized friction force, and Figure 18 represents plot for damping coefficient verses normalized friction force. The damper design curve/performance curve given in Figure 19, is drawn between normalized response verses normalized excitation, the entire curve can be divided in to three zones, Zone 1 is a completely struck condition. This happens at very low excitation levels or at very high normal loads. In this zone, the system behaves as a linear system. Zone 2 is a slip condition at either low normal loads or at higher excitation level. Here again the system behaves as a linear system. Zone 3 is a slip-stick zone, where the system experiences both conditions in a given cycle, and the system is highly nonlinear in nature.

Figure 20 is the damper performance curve generated for various viscous damping levels. From this figure, it is clear that the damper performance curves depends on the damper properties rather than the normal load and excitation level, which means, for a given damper, one damper performance curve can be generated and that can be used for optimal selection of the damper. These observations are in line with observations made by [2, 3]. The advantage with this representation is that it provides a design point, which is independent of both excitation and viscous damping levels in the engine. The design point is selected corresponding to maximum allowable blade response and consequently, the maximum excitation that can be sustained by the present blade. If actual excitation exceeds this maximum value, then the entire blade must be redesigned since friction damping cannot keep the blade response below allowable limits. Using this approach provides a friction damper that is optimal. The stresses in the blade will be acceptable for as large an excitation as possible. Thus, the optimal design is independent of the excitation and is insensitive to variations in viscous damping.

3. Experimental Studies

The laboratory experiments were carried out in two phases [16]. The first phase establishes the dynamic characteristics of the blade. This includes estimation of natural frequencies, mode shapes, the logarithmic decrement/damping factors, and the relative stress distribution along the aerofoil for various conditions. The tests are performed for the free-free condition and the clamped condition. In the second phase, the friction damper is characterized. In the present paper, only the relevant and important results are discussed

3.1. Free-Free Condition. In this case, the blade is suspended using a thread and excited using an instrumented impact hammer. The damping estimation by this technique gives only the material damping of the blade (aerial damping due to oscillating blade can be neglected as it is proportionately very small). Knowing this material damping is a must in order to understand the amount of damping blades will experience due to blade fixity in clamped condition.

3.2. Clamped Condition. For this a customized fixture is made in order to accommodate the blade under investigation alongwith the dummy blades on either side with under platform damper inserts. An electrodynamic shaker is used to excite the blade assembly that is driven by the signal generator through the amplifier in the frequency range of interest. The amplitude of excitation is kept constant by using a feedback control loop. The response is measured by blade-mounted strain gages. While placing the strain gauges on the blades, the following aspects are considered:

(1) sensor operability is ensured throughout the operating conditions of the blade;

(2) instrumentation convenience;

(3) informative capability of the strain gauges for the mode of interest.

The first two conditions are usually satisfied by a strain gauge located on blade suction side near aerofoil root. The informative capability of strain gauge is ensured through the 3D FEM analysis and experimental strain survey. One will look for maximum variable stresses for a given oscillation mode. Figure 21 demonstrates the strain gauges pasted on the blade under test. Apart from these strain gages, various other transducers like eddy current probes and accelerometers are used for the rig operation.

The signals measured from strain gages mounted on blade are used for characterizing the blade. When studying frequency characteristics and the damping of fixed blades, it is necessary to identify frequencies using spectral analysis and then finds its attributes, namely, the one caused by blade's resonating frequencies and other caused by excitation of fixture subcomponents. When fixing blades in spring clamps,

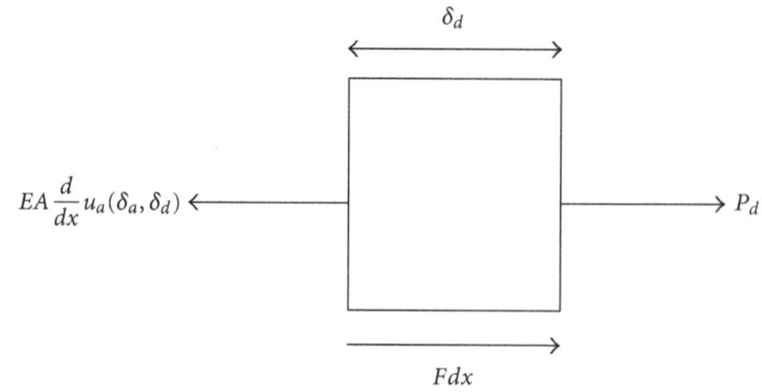

FIGURE 10: Forces acting on a small element in zone Z, when force is decreasing.

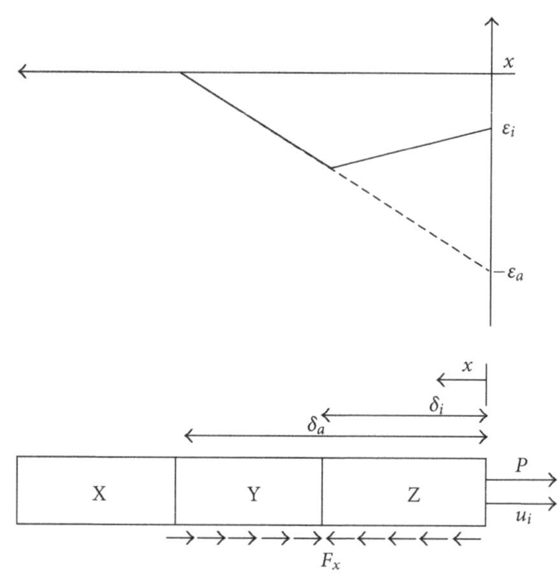

FIGURE 11: Force, displacement, and plot of strain function, when force is decreasing; zone X is totally struck and has zero strain, zone Y is struck and compressed; zone Z is slipping and stretched.

blade spectrums are investigated for various attachment designs, for different rigidity, mass, and squeezing efforts. Figure 22 shows the single blade setup under clamped condition, while Figure 23 shows the setup for dynamic characterization of blade damper assembly under clamped condition.

Table 3 gives measured natural frequency and corresponding damping value for both the fixity conditions. In both of these tests, it is clear that the first flexure mode is in the range of 1.01f1 and1.03f1. The logarithmic decrement is in the range of 0.5 to 1%, which is within the limits of scatter identified from the free-free test on the individual blade. From the test, it is apparent that the energy dissipation due to friction induced due to the clamping arrangement is very small. Certain divergence in frequency values can be attributed to difference in ways of fixation, scatter of blade geometric features within technological tolerance, and difference of real anisotropy of blade material

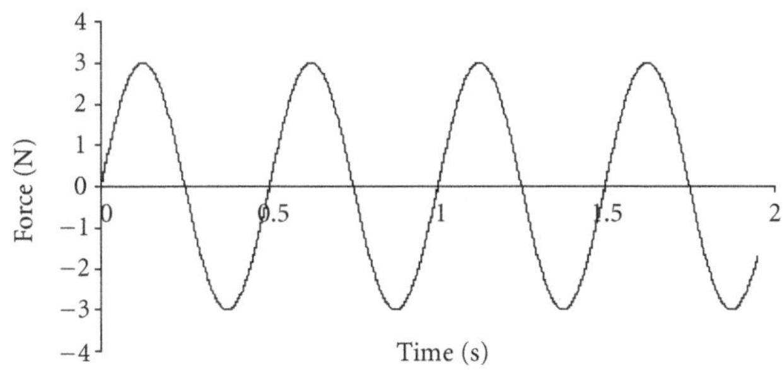

FIGURE 12: The applied force at the end of the bar.

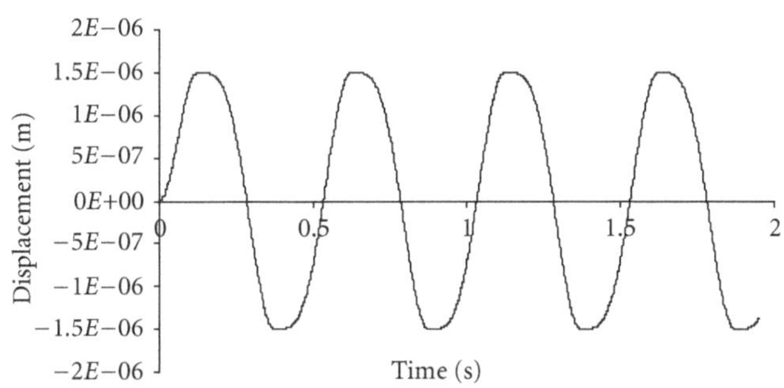

FIGURE 13: The simulated displacement at the end of the bar.

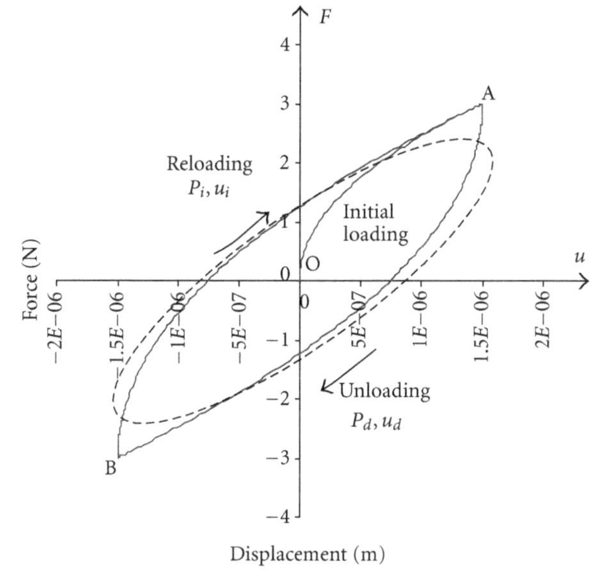

FIGURE 14: Hysteretic curve drawn between applied force and the simulated displacement at the end of the bar.

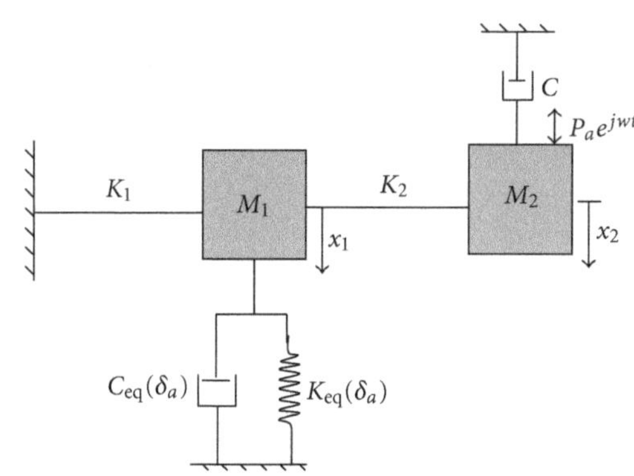

FIGURE 15: The equivalent blade model, representing two-dof system with friction damper.

TABLE 3: Measured modal parameters for 1st flexural mode at room temperature.

Blade fixity	Frequency (Hz)	Log decrement %
Spring clamping of root	1.01f1	0.47%
Three blades fixed in a fixture	1.03f1	0.58%

from orthotropic parameters set forth in calculations. Tables 4(a) and 4(b) indicate measured values of relative strain corresponding to first flexure mode of the blade under zero normal load on the damper. From the table, it is very clear that the most loaded area on the blade is near root, and the maximum displacement is measured at leading edge, pressure side of the blade. The relative deformation is

$$\overline{\varepsilon_i} = \frac{\varepsilon_i}{\varepsilon_{\max}}, \tag{30}$$

where i is the strain gauge number, $\overline{\varepsilon_i}$ is the relative deformation in blade aerofoil, ε_i is the deformation measured in aerofoil by the ith strain gauge, and $\varepsilon_{\max}$ is the maximum deformation measured by the any strain gauge on blade aerofoil.

3.3. Characterizing Friction Damper. The test setup for this case remains the same as above except for an additional customized feature to load the damper. The test is executed for various conditions of normal loads at controlled excitation level.

The force level is selected such that the maximum stress experienced by the blade is around 100 MPa. The frequency of excitation is selected around the natural frequency of the blade, and the responses were measured till the steady state achieved. The test is repeated for several damper load conditions. The values of resonance frequencies and its logarithmic decrement at different loads of the damper are given in Table 5. It is observed that the highest logarithmic decrement value is 5.08% at 500 N damper load. The natural frequency of the blade increases with increase in damper load, and to a greater extent in case of first bending mode.

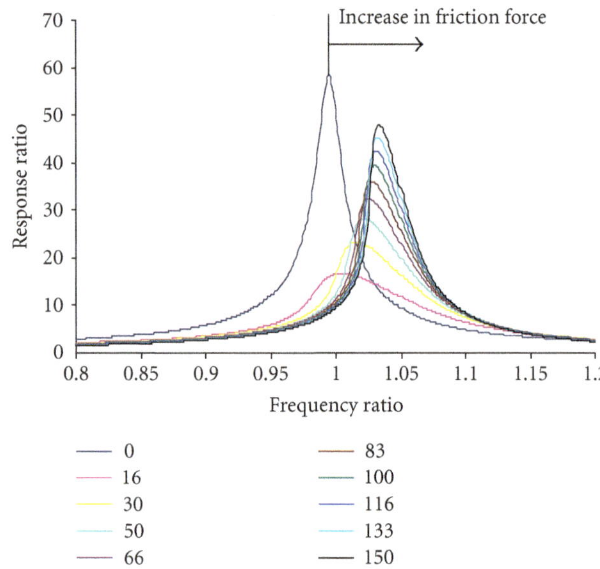

FIGURE 16: Response ratio versus frequency ratio for various values of normalized friction force.

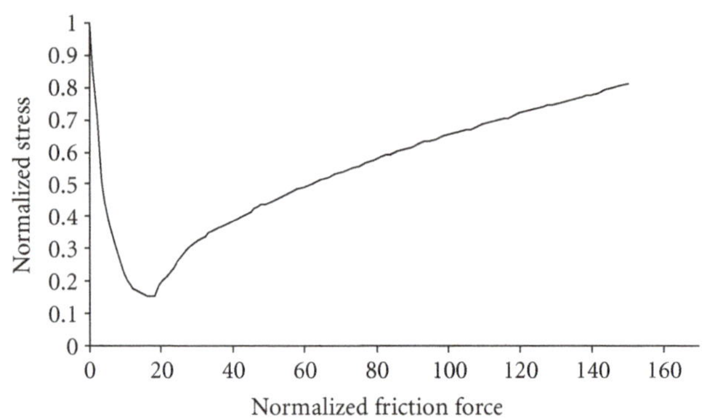

FIGURE 17: Normalised stress versus normalised friction force.

TABLE 4: Relative strain at various locations in 1st flexure mode for zero damper load.

(a)

Sl number	1	2	3	4	5	6	7	8
Gauge number	1	2	3	4	1′	2′	3′	4′
Relative strain	1	0.64	0.36	—	0.5	0.42	0.34	0.38

(b)

Sl number	9	10	11	12	13	14	15	16
Gauge number	5′	1″	2″	3″	4″	I	II	III
Relative strain	—	0.65	0.61	0.36	—	0.15	0.26	0.2

4. Estimation of Damper Efficiency

With the available experimental and theoretical data, the efficiency of the damper insert is evaluated [17].

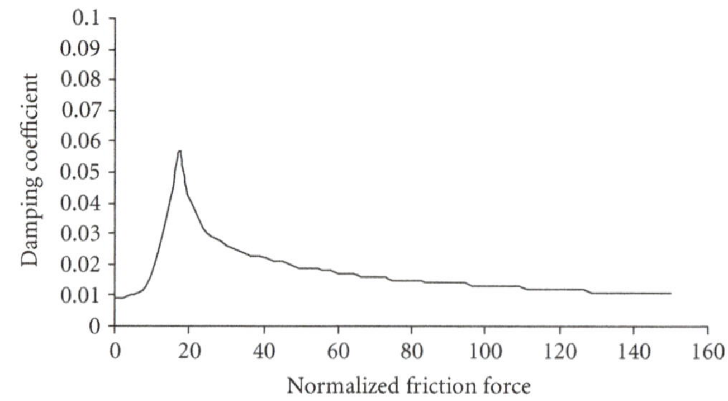

FIGURE 18: Damping coefficient versus normalized friction force.

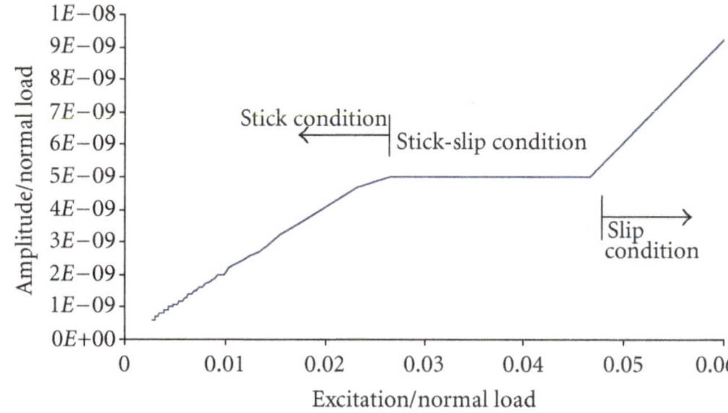

FIGURE 19: Damper design curve plotted between normalized response and the normalised force.

TABLE 5: Values of natural frequency and damping of 1st flexure mode of the blade under various damper loads.

Sl number	Damper load (N)	Frequency in Hz	Percentage δ
1	0	1.03f1	0.96
2	300	1.095f1	3.25
3	500	1.154f1	5.08
4	1000	1.170f1	2.54
5	2000	1.191f1	2.5

The normal load generated by the damper at a given operating speed is

$$N = m\omega^2 r,$$
$$m \times r = 1415\,\text{g} - \text{mm}, \tag{31}$$

where m is the mass of the damper in "g", and r is the radius at which damper located in "mm." Angular velocity ω in rads^{-1}.

Figure 24 is a Campbell plot for a blade. It shows two potential resonance conditions corresponding to first flexure mode, due to 6th- and 9th-order cross-overs, at 12000 rpm and 16000 rpm, respectively.

The normal load corresponding to 9th-order cross-over is $N = 1834\,\text{N}$.

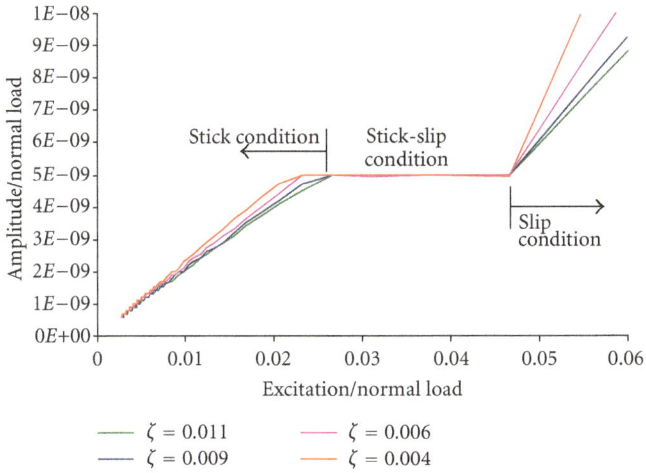

FIGURE 20: Damper design curve generated under various viscous damping levels.

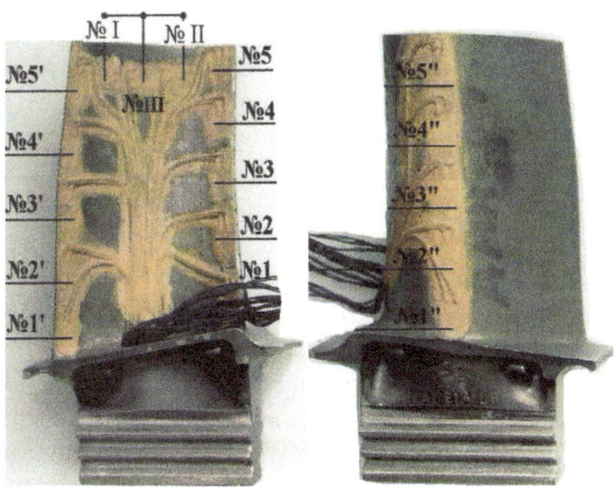

FIGURE 21: Indicating the strain gauges mounted on both pressure and section side of the blade.

A typical value of friction coefficient for metal-metal contact under slow slide is $\mu = 0.2$. This value under vibrating contact surfaces is known to reduce the contact and the friction ratios. Contrast to this, under high-frequency relative motion of the bodies, frictional seizure phenomena, due to heating of local zones can occur, leading to increase of friction coefficient. Therefore for calculations, a broad range of possible values for μ ranging from 0.2 to 0.4 is considered. The corresponding friction force F for this will be in the range of 180–734 N.

Using various factors generated from the finite element analysis the range of normalized friction force is identified. The ratio of tip displacement to resonant stresses is $k_{u\sigma} = 2.12e^{-3}\,\text{mm}^3/\text{N}$. Considering the resonance stresses in the range of 50–100 Mpa, the displacement of the blade tip is in the range of 0.11 mm–0.2 mm. Once the blade tip displacement is identified using $k_{uP} = 0.026$, the range of excitation P can be identified. Therefore, excitation P is in the range of 4–8 N.

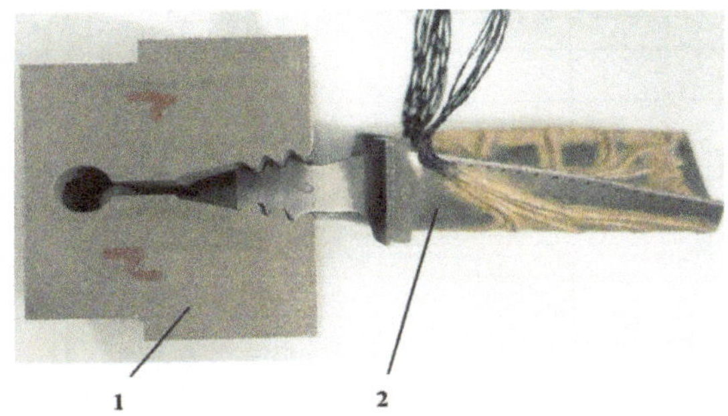

FIGURE 22: Blade fixation at the root in a spring clamp. (1) spring clamp and (2) blade under investigation.

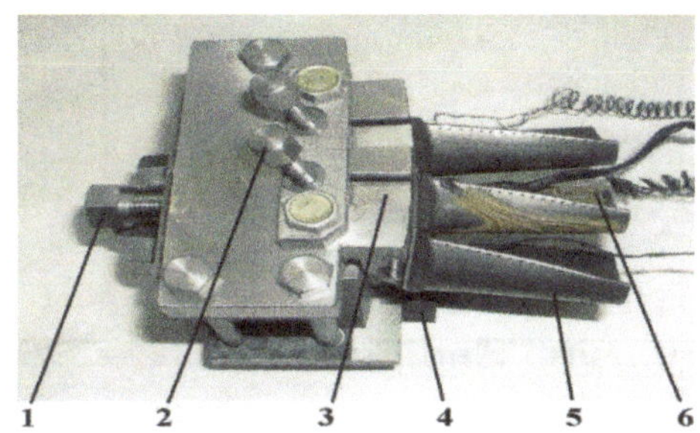

FIGURE 23: Tooling used for estimation of damper efficiency screw. (2) load screw, (3) lever, (4) support, (5) blade, and (6) strain gauges.

And the range of normalized friction force is calculated as

$$\left(\bar{F}\right)_{\text{cal}} = \frac{F}{P} \text{ is 22 to 174.} \tag{32}$$

Keeping all other parameters constant the $(\bar{F})$ is normalized friction force, proportional to the normal load applied on the damper.

Therefore, $(\bar{F})/N = \text{cont}'$.

The optimum range of normalized friction force corresponding to maximum rotational frequency can be found from

$$\left(\bar{F}\right)_{\text{opt}} = \frac{N_{\text{opt}}}{N}\left(\bar{F}\right)_{\text{cal}}, \tag{33}$$

where N_{opt} is the optimum damper load at which the response of the blade is minimum, that is, 500 N, and N is the normal load on the damper at a given rotational speed.

Therefore, the $(\bar{F})_{\text{opt}}$ is in the range of 6 to 47. In this range of friction forces significant reduction in stresses is expected. Further increase in normal load may increase the aerofoil stresses.

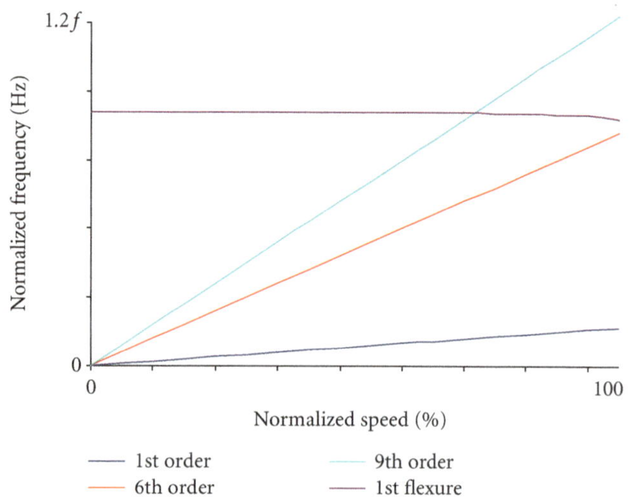

FIGURE 24: Campbell diagram, indicating 1st flexure mode with various excitation lines.

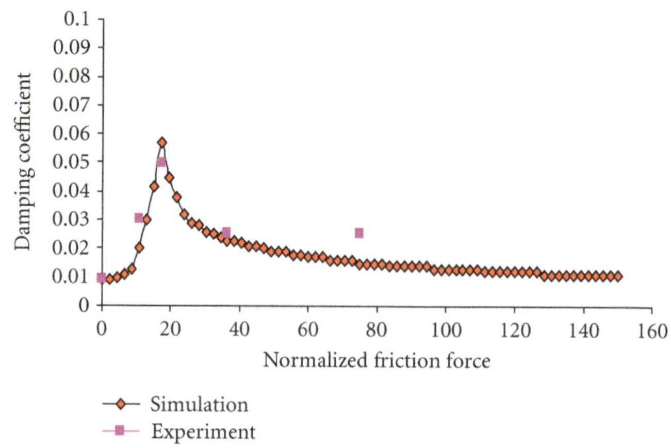

FIGURE 25: Comparison of damping coefficient value with normalized friction force.

5. Comparison of Analytical and Experimental Findings

The correlation studies is carried out in two phases, in phase one, the dynamic characteristics of the blade under investigation is carried out, which includes the natural frequency, mode shape, and the damping. Further it is ensured that the blade fixity will have negligible effect on the damping. The relative strain estimated along the aerofoil for the first flexure mode is in good correlation with experiments conducted without damper inserts. The damper effectiveness is measured on bench test is in good agreement with the analytical findings. The maximum logarithmic decrement value measured with damper insert is around 5.08%, in case of analytical estimation it is around 5.8%. From the experiments it is found that the optimal normalized friction force is in the range of 6–47, for the blade damper system under investigation, corresponding to the rotational speed of 16000 rpm, where the possibility of 1st flexure mode resonance exists. Similarly in the analytical solution, it is found to be in the range of 7–50, refer to Figure 16. It is clear that the range of normalized friction force for which the response amplitude is below 50% of resonance amplitude without a damper insert. Figure 25 indicates the comparison of both theoretically estimated and the experimentally measured damping coefficient.

6. Conclusions

Minimization of resonant stresses in turbine blades is a major concern in turbine engine. A detailed procedure is outlined for optimization of damper for turbine blade so that the maximum stress experienced by any blade should be below some designated maximum value. The procedure details the integration of analytical and experimental studies to ensure the correctness of the model developed and the friction damper design. A damper performance curve is established, which provides a design point that is independent of both damping and the excitation levels, these two quantities are very difficult to be determined for the new design of blades. At the same time, the design point optimizes the friction damper for as large an excitation as possible. A unique test fixture is developed for conducting the experiments, having a feature to setup various damper loads.

The predicted natural frequencies, stress distribution, and the displacements determined for the 1st flexure mode of the blade is in good correlation with the experimental results. The damper performance curves are generated for various levels of force and the damping values, which are used to arrive at optimal damper design. This can now be extended to the engine condition, namely, gas loads, and temperatures.

Nomenclature

F_x:	Friction force
μ:	Friction coefficient
δ_a:	Maximum slip corresponding to P_{amp} in "m"
u:	Displacement at the end of the bar in "m"
P:	Excitation at the end of the bar in "N"
N:	Total normal load acting on the interface in "N"
$q(x)$:	Normal load function
L:	Length of the bar = 0.02 "m"
EA:	Modulus of elasticity multiplied with area = 600000 "N"
ω:	Angular velocity rads^{-1}
P_{amp}:	Maximum value of the force "N"
u_{amp}:	Maximum displacement "m"
$P_a(\delta_a)$:	Force at the end of the bar corresponding to the initial loading

$u_a(\delta_a)$:	Displacement at the end of the bar corresponding to the initial loading
$P_d(\delta_a, \delta_d)$ and $u_d(\delta_a, \delta_d)$:	Force and displacement at the end of the bar for unloading
$P_i(\delta_a, \delta_i)$ and $u_i(\delta_a, \delta_i)$:	Force and displacement at the end of the bar for reloading
D:	Equivalent damping energy
C_{equ}:	Equivalent viscous damping in "Nsm^{-1}"
W:	Work done per cycle
K_{equ}:	Equivalent stiffness "Nsm^{-1}"
C:	Viscous damping coefficient Nsm^{-1}
dx:	Elemental length of the bar
du/dx:	Strain in the bar
δ:	Slip in the bar in "m".

Acknowledgment

The authors would like to thank the Director of GTRE, for his support throughout this work. And authors also would like to thank the Head of Mechanical Engineering Department, SV University, for his encouragement and guidance and Mr. Sreelal Sreedhar, Additional Director, GTRE, for his valuable inputs. Finally authors thank one and all whomever directly and indirectly involved and helped during this work.

References

[1] J. H. Griffin, "A review of friction damping of turbine blade vibration," *International Journal of Turbo and Jet Engines*, vol. 7, pp. 297–307, 1990.

[2] T. M. Cameron, J. H. Griffin, R. E. Kilb, and T. M. Hoosac, "An integrated approach for friction damper design," *The Role of Damping in Vibration and Noise Control*, vol. 5, pp. 205–211, ASME Booklet DE, 1987.

[3] R. Kilb, J. Griffin, and C. H. Menq, "Evaluation of a turbine blade damper using an integral approach," in *Proceedings of the 29th Structures, Structural Dynamics and Materials Conference AIAA/ASME/ASCE/AHS*, no. 88-2400, AIAA, Williamsburg, Va, USA, 1998.

[4] K. Y. Sanliturk, D. J. Ewins, R. Elliott, and J. S. Green, "Friction damper optimization: simulation of rainbow tests," *Journal of Engineering for Gas Turbines and Power*, vol. 123, no. 4, pp. 930–939, 2001.

[5] C. H. Menq, J. H. Griffin, and J. Bielak, "Influence of a variable normal load on the forced vibration of a frictionally damped structure," *Journal of Engineering for Gas Turbines and Power*, vol. 108, no. 2, pp. 300–305, 1986.

[6] C. H. Menq, J. Bielak, and J. H. Griffin, "The influence of microslip on vibratory response, part I: a new microslip model," *Journal of Sound and Vibration*, vol. 107, no. 2, pp. 279–293, 1986.

[7] C. H. Menq, J. Bielak, and J. H. Griffin, "The influence of microslip on vibratory response, Part II: a comparison with experimental results," *Journal of Sound and Vibration*, vol. 107, no. 2, pp. 295–307, 1986.

[8] G. Csaba, *Modelling microslip friction damping and its influence on turbine blade vibrations*, Dissertations no. 519, Linköping, Sweden, 1998.

[9] G. Csaba, "Forced response analysis in time and frequency domains of a tuned bladed disk with friction dampers," *Journal of Sound and Vibration*, vol. 214, no. 3, pp. 395–412, 1998.

[10] E. Cigeroglu and H. N. Ozguven, "Nonlinear vibration analysis of bladed disks with dry friction dampers," *Journal of Sound and Vibration*, vol. 295, no. 3–5, pp. 1028–1043, 2006.

[11] C. W. Schwingshackl et al., "Measured and estimated friction interface parameters in a nonlinear dynamic analysis," *Mechanical Systems and Signal Processing*, vol. 28, pp. 574–584, 2012.

[12] E. P. Petrov and D. J. Ewins, "Advanced modeling of underplatform friction dampers for analysis of bladed disk vibration," *Journal of Turbomachinery*, vol. 129, no. 1, pp. 143–150, 2007.

[13] J. Szwedowicz, C. Gibert, T. P. Sommer, and R. Kellerer, "Numerical and experimental damping assessment of a thin-walled friction damper in the rotating setup with high pressure turbine blades," *Journal of Engineering for Gas Turbines and Power*, vol. 130, no. 1, Article ID 012502, 2008.

[14] M. Allara, "A model for the characterization of friction contacts in turbine blades," *Journal of Sound and Vibration*, vol. 320, no. 3, pp. 527–544, 2009.

[15] C. M. Firrone, S. Zucca, and M. M. Gola, "The effect of under platform dampers on the forced response of bladed disks by a coupled static/dynamic harmonic balance method," *International Journal of Non-Linear Mechanics*, vol. 46, no. 2, 2011.

[16] Vibration Engineering Group, "Experimental investigation on turbine blade damper," Internal Report, GTRE, Bangalore, India.

[17] Vibration Engineering Group, "Turbine blade damper optimization studies," Internal Document, GTRE, Bangalore, India.

12

The Effects of Inlet Box Aerodynamics on the Mechanical Performance of a Variable Pitch in Motion Fan

A. G. Sheard

Fläkt Woods Limited, Axial Way, Colchester, Essex CO4 5ZD, UK

Correspondence should be addressed to A. G. Sheard, geoff.sheard@flaktwoods.com

Academic Editor: Mohammad Tawfik

This paper describes research involving an in-service failure of a "variable pitch in motion" fan's blade bearing. Variable pitch in motion fans rotate at a constant speed, with the changing blade angle varying the load. A pitch-change mechanism facilitates the change in blade angle. A blade bearing supports each blade enabling it to rotate. The author observed that as the fan aerodynamic stage loading progressively increased, so did the rate of blade-bearing wear. The reported research addressed two separate, but linked, needs. First, the ongoing need to increase fan pressure development capability required an increase in fan loading. This increase was within the context of an erosive operating regime which systematically reduced fan pressure development capability. The second need was to identify the root cause of blade-bearing failures. The author addressed the linked needs using a computational analysis, improving the rotor inflow aerodynamic characteristics through an analysis of the inlet box and design of inlet guide vanes to control flow nonuniformities at the fan inlet. The results of the improvement facilitated both an increase in fan-pressure-developing capability and identification of the root cause of the blade-bearing failures.

1. Introduction

In induced draft applications, the fan extracts combustion products from a boiler. Current legislation requires boilers to be fitted with emission control systems. These systems reduce the sulphur-dioxide, nitrox-oxide and, in some cases, mercury emissions from the power plant. They also have an associated pressure loss, which the induced draft fan must overcome.

Increasingly stringent legislation has resulted in emission control systems becoming progressively more complex over the last decade. This results in a market requirement to upgrade in-service fans to cope with the associated increased system resistance. Although not reported in the open literature, the author was aware that the blade bearings situated under each blade of the variable pitch in motion induced draft fans were wearing more rapidly with each increase in fan pressure developing capability. Following the most recent upgrade, it had been necessary to replace worn out blade bearings before the scheduled five year major overhaul. The author speculated that a further increase in fan pressure developing capability was likely to exacerbate an already unacceptable situation.

The cost to upgrade an existing power plants emission control system to meet current and foreseen future regulatory requirements can be as high as $500 million. Existing coal fired power plants have a finite life, and in order for the upgrade to make commercial sense, the value of generated electricity must be greater than the cost of upgrading and operating the power plant. The largest single cost when operating a coal fired power plant is the cost of coal. Perhaps counterintuitively, upgrading a power plants emission control system facilitates a move to lower quality coal. Without an emission control system or with a low specification emission control system, a power plant must burn higher quality low-sulphur coal to remain within existing emission limits. Once a power plant fits or upgrades an emission control system, it can burn lower quality, high-sulphur coal as the emission control system will clean the exhaust gas.

The business case for fitting or upgrading an emission control systems is linked to a move from higher cost high quality coal to lower cost low quality coal. Low quality coal includes not only more sulphur, but also more silica. Silica is hard, resulting in increased wear in the ball-mills that grind the coal before it enters the boiler. Wear in the boiler ball-mills results in larger carbon particles that are more likely to pass through the boiler and into the induced draft fan. Consequently, the shift to low quality coal results in a shift towards an increase in the silica and unburnt carbon passing though induced draft fans. As such, an unwanted consequence of switching from high to low quality coal is to increase the rate at which fan blades erode, decreasing the fan's pressure developing capability.

The addition or upgrading of emission control equipment increases the system resistance against which an induced draft fan must operate. An increase in the fan blades' erosion rate results in the need for a further increase in fan pressure developing capability to ensure that as blades erode over the fans service life stall margin does not reduce to a point at which the fan stalls in-service. A way to increase an axial flow fan's pressure developing capability is to increase fan loading by fitting inlet guide vanes. Increased fan loading is, in practice, at the expense of fan stall margin and is therefore an inherently risky approach to increasing fan pressure developing capability in an erosive application.

Concern for maintaining a safe operating stall margin is more pressing when one places the fan downstream of an intricate duct system, as is typically the case with induced draft fans. The inlet flow distortion associated with an intricate duct system results in an "installation effect" that reduces fan stall margin. When examining the consequence of installation effect, the bulk of published works has focused on noise generation [1, 2]. Installation effects invariably degrade cooling fans' acoustic performance; however, acoustic considerations are not of primary concern for fans in induced draft application. We can lag induced draft fans to reduce acoustic emissions. The primary consideration for an induced draft fan is the aerodynamic impact of a non-asymmetric inflow, with a particular concern for reduced fan efficiency and stall margin. Some researchers, notably Roslyng [3] and Staiger and Stetter [4], have studied the aerodynamic impact of fan installation effects. Despite the undoubted quality of the reported research, neither was able to generalize their results in a way that facilitated application beyond their work.

The impact of erosion on induced draft fan performance has also received relatively little attention in the literature. However, Corsini et al. [5] have studied the impact of erosion on induced-draft fan leading edge geometry. They concluded that three factors dominate the erosion process: (i) the inflow's incidence angle, (ii) the flow deviation correlated to the blade profile, and (iii) the blade tip speed. Corsini et al. [6] continue their work, developing a computational method to predict erosion rate and the time it takes to erode the induced-draft fan blades to the point where they were no longer able to generate the necessary pressure rise to avoid the fan stalling at its design point. Corsini et al. [7] reported the eroded blade profile at different blade angles, and were able to computationally establish the characteristics of both the datum fan, with blades as per the design intent and eroded blades. The erosion occurred over approximately 12,000 hours in-service operation, corresponding to less than two years continuous operation. Corsini's research team removed the eroded blades when the fan efficiency fell so far that the motor driving the fan was no longer able to supply sufficient power to maintain the required fan duty point.

In the present study, the author models the inlet duct, inlet guide vanes, fan rotor, and stator. He initiated this in response to an in-service issue. A facility operator had incorrectly specified a 1767 mm diameter single stage induced draft fan with too low a pressure developing capability. The blade bearings also suffered premature failure. The fan's installed inlet configuration was complex, leading to the inlet flow's significant distortion which reduced fan performance. In combination with the fan's under-specification, the performance reduction that occurred as a consequence of the poorly conditioned inlet flow resulted in the fan limiting the power station power output. The author initiated the computational program of work in an effort to both increase the fan pressure developing capability and identify the root cause of the premature blade bearing failure.

We can configure induced draft fans with a rectangular inlet box that accepts flow coming vertically down. The box turns the flow ninety degrees and into a round and horizontal fan inlet. The study used computational analysis as a "numerical laboratory". This provided insight into the fan's actual inlet flow field. The objective was to facilitate inlet guide vane design improvement. The desired outcomes were first, an increase in fan pressure development capability and, second, an understanding of the aerodynamic mechanism that was driving the fan blade bearings' premature failure.

2. Computational Analysis

The author analysed the geometry of a 1767 mm single stage induced-draft coal fired boiler fan as installed in a power plant, but with a standardized air inlet system. Greenzweig et al. [8] experimentally studied the effect of actual in-service air inlet system installations on induced draft coal fired boiler fan performance and concluded that the effect of inlet flow distortion on fan performance was significant, with the effect varying with the rating point's position along the fan curve. They found distortion effects greater at higher flow conditions and lower at lower flow conditions. This observation led Greenzweig et al. [8] to conclude that the actual impact of inlet distortion is dependent on the relationship between the fan design parameters, and further that engineers currently do not understand the relationship between these parameters.

The relationship between fan design parameters and the impact of inlet flow distortion on fan performance is a subject in its own right. In the current program of work the author's focus was not an experimental research program aimed at the elucidation of physical mechanisms underpinning the effect of flow distortion on fan performance. The author's focus was developmental, aimed at applying computational methods to improve the performance of an existing in-service fan. In addition, the author was able to use the computational results to provide insight into the circumferential variations in the flow field that then assisted him in identifying the root cause of the observed

Table 1: In-service duty point and specification of the PFS fan studied in the reported research.

Fan in-service duty point		
Pressure (Pa)		2245
Flow (m^3/s)		54.3
Blade angle (degrees)		56.0
Fan rotor data		
Blade number		18
Hub-to-casing diameter ratio		0.7
Tip diameter (mm)		1767
Rotor tip clearance (% span)		6
Rated rotational frequency (rpm)		890
Fan blade geometry	Hub	Tip
Chord (mm)	487	446
Solidity	1.48	0.96
Stagger angle (degrees)	38	60
Camber angle (degrees)	45	36

in-service premature blade bearing failure. As such, the author chose to study the in-service fan with a standardized inlet. This standardized inlet was both used in model testing to establish the fan's characteristics and modeled in the author's computational code when predicting the fan's characteristics.

The fan that is the subject of this research is designated Pan Fläkt Single (PFS), where "pan fläkt" in Swedish means "boiler fan" and "single" designates a single stage fan, Table 1. The author derived the studied fans characteristics, Figure 1, from data that he obtained using a low speed 1250 mm aerodynamic fan model. The aerodynamic model had the correct tip solidity and hub ratio, with model test data scaled to the studied fan speed and diameter using the fan laws. The author selected the studied fan for induced draft application in a power plant, with a duty point that is typical for the application, Figure 1 and Table 1. The duty point falls below the fan's operating characteristic peak efficiency region as induced draft applications are classically an erosive application. Fan blade erosion reduces the fan's pressure developing capability, and therefore one must select the fan conservatively to accommodate a reduction in stall margin over the fan's service life.

The author modeled the studied fan with a standardized inlet box, rotor, stator, and diffuser, plus a section of an expansion chamber, Figure 2. The inlet box comprised a square section vertical intake angled through ninety degrees and converted into a circular inlet. A shaft fairing traversed the inlet duct's lower part and expanded in a conical form onto the fan annulus. Following Teixeira et al.'s method [9], the author placed a fan with outlet guide vanes downstream of the inlet box discharging into a conical diffuser.

In the original configuration, the author fitted five long chord sheet metal stays connecting the fan's hub surface to the outer casing, Figure 3. This original configuration was symmetrical with respect to a vertical plane through the fan center line, with the stays' primary purpose as the provision

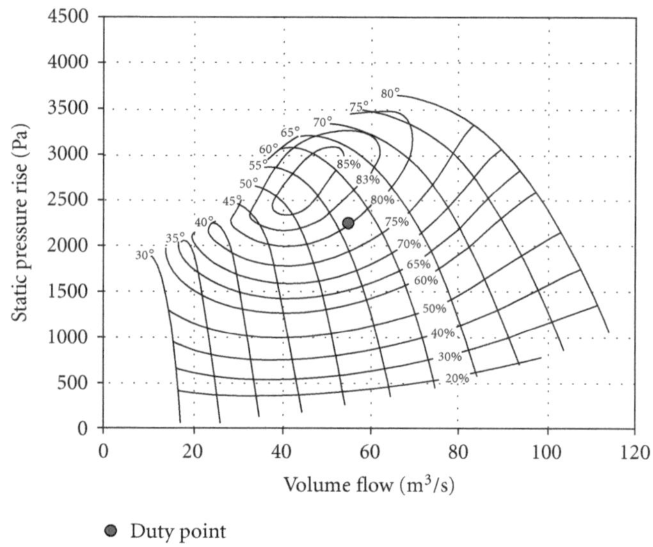

Figure 1: The performance characteristics of the studied 1767 mm diameter PFS fan. This fan chart has been scaled from experimental data which the author obtained using a 1250 mm low speed aerodynamic model, scaled for speed and diameter using the fan laws.

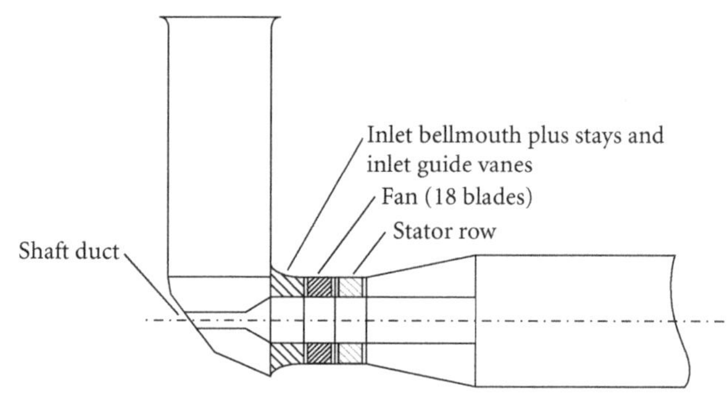

Figure 2: Side elevation of the fan installation arrangement.

of structural strength. The stays themselves were intended to present as near a neutral aerodynamic effect as possible.

The author studied multiple variations of stay and inlet guide vane arrangements. He obtained the greatest increase in fan pressure developing capability and reduction in circumferential variation in pressure when a further three stays and eight short chord inlet guide vanes supplemented the original five stays, Figure 3. This improved eight stay plus eight inlet guide vane configuration was circumferentially symmetrical, with the additional stays and guide vanes requiring no original stay modification. In this way the author could retrofit the improved configuration to the original configuration without compromising the fan's structural integrity. The computational results, Figure 3, are for the original five stay and improved eight stay plus eight inlet guide vane configurations.

2.1. Aerodynamic Details. The inlet box introduced the initial aerodynamic complexity, where vertical inlet flow turned through 90 degrees from a square-section duct into an

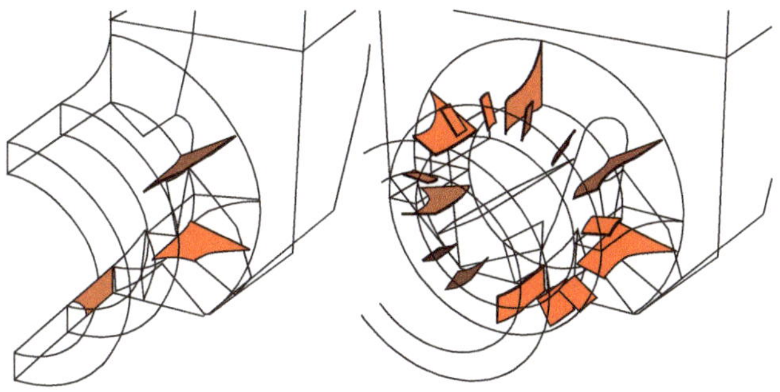

Figure 3: The original five stay configuration, left, and improved configuration with eight stays plus eight guide vanes, right.

annular intake, while accommodating a sizeable shaft fairing. The annular intake incorporates the original five stays. The improved configuration supplements the original five stays with an additional three cambered stays, plus cambered chord-wise extensions to the original five stays. The eight additional inlet guide vanes associated with the improved configuration were also cambered. The resultant effect was that the improved configuration preswirled the inlet flow into the fan and increased fan aerodynamic loading. Additionally, by varying the camber on each stay and inlet guide vane, the author aimed to minimize circumferential variations in the fan rotor inlet flow which the inlet box configuration introduced.

For the fan, an effect of additional loading is a shift of the characteristic upwards and to the right. Hence, for a given flow rate, the fan produces a higher pressure by operating further along its characteristic. This last point implies an unavoidable reduction in stall margin. The additional loading will also result in some decrease in fan efficiency. However, for a well-designed blade that was, in practice, operating at a duty point below peak efficiency, the increased loading may bring the actual fan operating point closer to the peak efficiency operating point. Operating closer to the fans peak efficiency will go some way towards offsetting the efficiency decrease associated with increased fan loading.

2.2. Objective of the Simulations. The study's primary objective was to safely maximize the fan loading through the addition of stays and inlet guide vanes. The author sought evidence of separation throughout the blade span, as well as information about rotor inlet flow conditions' possible pitch-wise variations. Additionally, the author used the simulations to gain a deeper understanding of the flow field in the inlet box and ultimately to provide information to enable elimination or, at least, minimization of sources of inefficiencies which he identified in the computational results. The author obtained solutions for four flow conditions: duty point, 10% and 20% above and 5% below the duty point condition.

The author computed two forms of the inlet box, the original five stay configuration and the improved eight stay, plus eight inlet guide vane configurations. The original configuration was symmetrical around a vertical plane through the fan center line, and therefore the author was able to model only half the inlet box. This simplifying assumption is not applicable to the improved configuration. The author varied the camber that was associated with each stay and inlet guide vane in an attempt to minimize the circumferential variation in outlet flow from the stays and inlet guide vanes. Consequently, each stay and inlet guide vane had a different camber, and therefore required the author to model the full annulus.

3. Computational Tool

The computational fluid dynamics code that the author employed in the present study was CFX-TASCflow, Version 2.11.1. The author used a strongly conservative finite element-based finite volume discretisation method to partition the physical domain into a set of hexahedral control volumes. Shape function usage introduces the finite element connection which the code employs to define subelemental geometry coordinates which are connected with the assembly of the control volumes. The author carried out all computations with the standard k-ε turbulence model with standard wall functions.

Out of the possible discretisation schemes available, the current simulations employed the Modified Linear Profile Scheme (MLPS). The MLPS gives second order error reduction in most instances [10]. For steady-state computations, the code employs a transient formulation as a convenient means of introducing relaxation into the iterative nonlinear solution [11]. If only the steady-state solution is of interest, one can use the time step Δt as a free parameter through which it optimizes the convergence rate. This corresponds to the use of pseudotime which one employs solely to advance the solution in an iterative manner.

The relaxation scheme which one employs in the code belongs to the Lower-Upper (LU) factorization family. The code makes use of a multigrid solution acceleration technique designed to reduce the global linear residuals in between the solution of the nonlinear terms. The rationale underpinning multigrid acceleration is rooted in the observation that iterative solvers tend to reduce errors which correlate with short grid spacing much faster than they do to those which correlate with longer lengths.

When running the inlet box simulations, it was appropriate to displace the outlet boundary conditions location away from the bell-mouth discharge and for this purpose the author added an annular ducting grid section to the model. The grid that the author employed in the second set of simulations includes rotor and stator blocks, a conical diffuser and a long expansion chamber, Figure 4. Although the mesh's extent is large in this last component, it is useful to have the outflow boundary located far downstream of a backward step, thereby preventing recirculation flow occurrence in the boundary's proximity since this can act as a source of overall convergence problems. The grid sizes are about 600,500 nodes for the original configuration of five stays and 1,200,000 nodes for the improved eight stay plus eight inlet guide vane configuration. The number of nodes that the author needed for the improved configuration was approximately double that of the original as a consequence of the requirement to extend the original configuration's

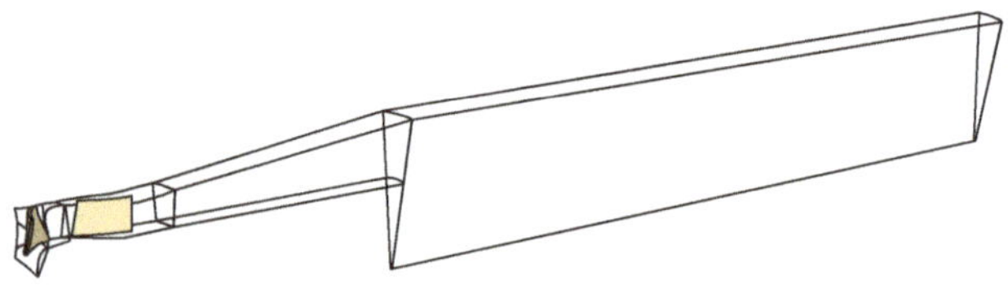

FIGURE 4: Rotor, stator, diffuser, and expansion chamber computational domains.

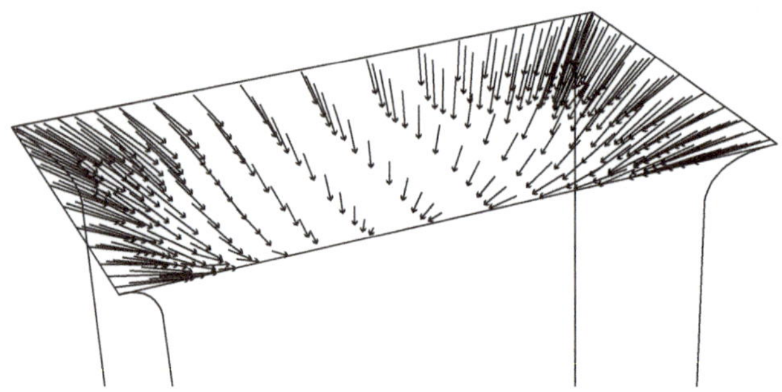

FIGURE 5: Prescribed flow direction vectors at the fan duct inlet.

half-annulus model to a full-annulus model. The author augmented both simulations with an inlet box grid that comprised 325,000 nodes.

4. Simulation Parameters

The author specified the boundary conditions applicable to the two simulations as total pressure and flow direction at inlet and mass flow at outlet. The author ran both the original and improved inlet box configuration simulations with prescribed inlet profiles. The prescribed inlet profiles' function was strictly that of aligning the flow with the local geometric features given that earlier runs had shown that a prescribed uniform direction, taking no account of these features, induces the appearance of lengthy low momentum flow structures along the inlet duct walls and corners. With a prescribed inlet profile the flow direction's local distribution in the form of a vector plot is indicative of a well-conditioned inlet, Figure 5.

The author performed the computations in this study on six 1.5 GHz twin-processor PCs networked as a Linux cluster. In this platform, and employing two CPUs in parallel, each computation of the original five stays configuration took approximately 45 CPU hours to perform and 1700 iterations to reach a maximum root mean square residual error of 3E-06.

5. Inlet Box Simulation Results

An analysis of simulation results of the original five stay configuration illustrated that the flow was generally well behaved without any significant areas of separation and that downstream of the stay outflow plane the flow is predominantly axial, notwithstanding the presence in places of a small amount of tangential secondary flow. This translates into a two to four degree departure from a zero rotor inflow angle

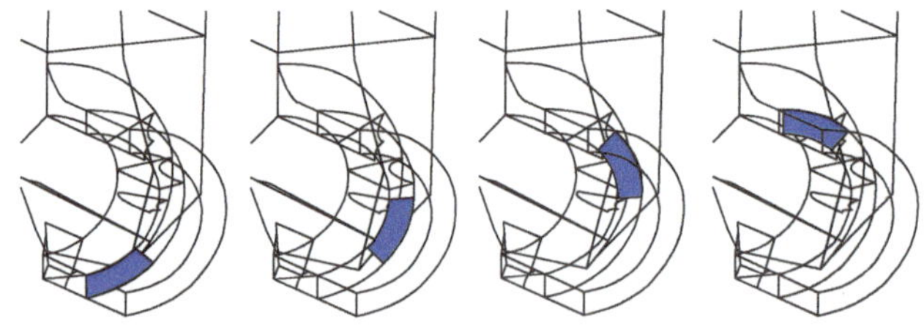

FIGURE 6: Flow averaging angular sectors for the original five stay configuration.

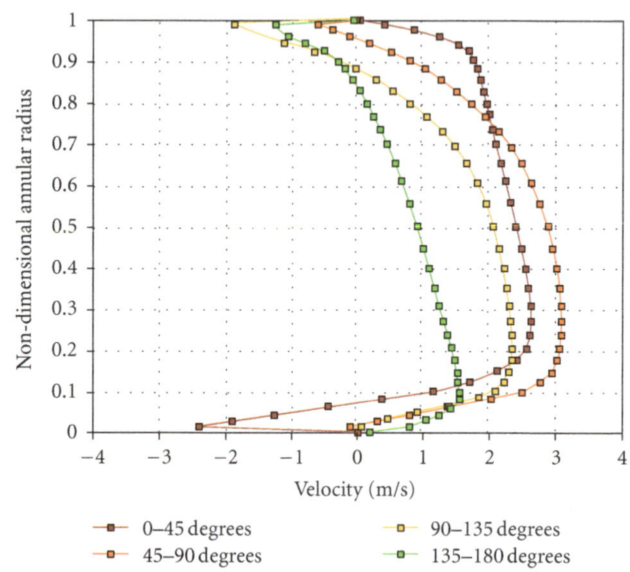

FIGURE 7: Tangential velocity radial distribution at the inlet box discharge for the original five stay configuration at the duty point condition.

at given pitch-wise locations. For the purposes of supplying the corresponding fan assembly model with an inlet profile, the author divided the inlet box outflow into four 45-degree sectors, Figure 6. Figure 7 presents a comparative plot of the sector-averaged tangential velocity. The flow's positive direction is in a clockwise sense when looking upstream of the discharge plane.

The author applied a similar postprocessing method to the improved eight stays plus eight inlet guide vane configuration and therefore the equivalent tangential velocity plot shows results for eight pitch wise sectors of 45 degrees, Figure 8. The origin of angular displacement measurement is as for the five stays case.

The effect of fitting the cambered stays and inlet guide vanes to the improved configuration is apparent, but the comparisons with the original configuration are limited to equivalent sectors alone. Unlike the complete inlet box arrangement, the tangential flow distribution in the five stays geometry is valid only for the side of the box which the author computed. Under the flow symmetry assumption, the opposite side will present a similar distribution, but flowing in the opposite direction. One must then superimpose the inlet box's "handed" effect in its original configuration onto the overall spinning motion which the added turning effects

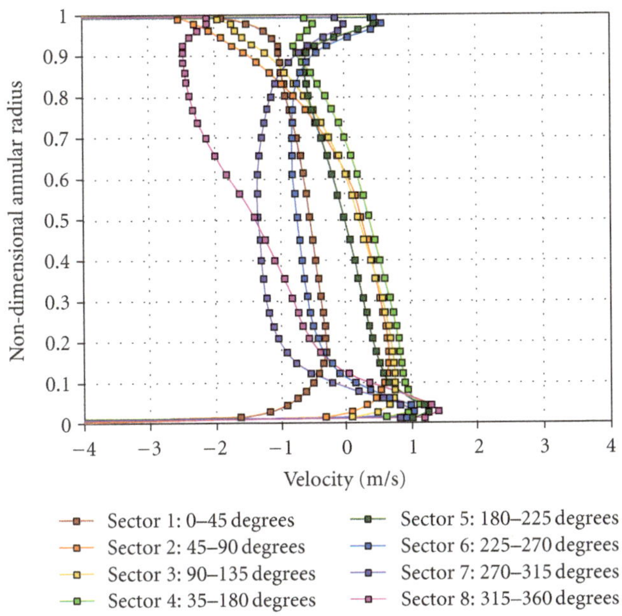

Figure 8: Tangential velocity radial distribution at the inlet box discharge for the improved eight stay plus eight inlet guide vane configuration at the duty point condition.

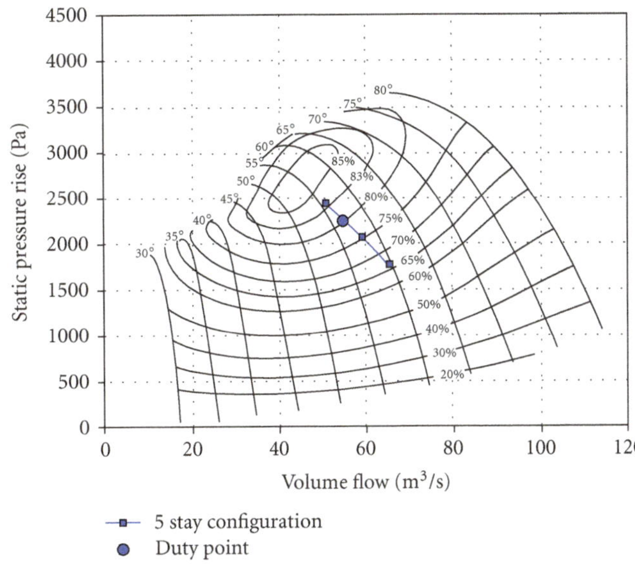

Figure 9: Experimentally derived characteristics for the studied 1767 mm diameter PFS fan, with simulation results for the original five stay configuration at four flow conditions, duty point, 10% and 20% above and 5% below the duty point condition.

induce, but it is not immediately obvious what this should be locally.

A closer analysis of the results, Figure 8, can help clarify the issue associated with the handed effect since, in general, the velocity distribution in related sectors for the two plots is broadly similar and corresponds mostly to the highest values. Conversely, the lowest values are those which would match the geometric opposite's sectors of the computed five stays domain confirming, therefore, the notion that the flow's right hand side in the inlet box contributes towards the swirling motion, while that of the left hand side tends to oppose it.

6. Fan Simulation Results

The second simulation set addressed the combined geometries of the rotor, stator, diffuser, and downstream channel with the inflow boundary conditions supplied to the simulations which the author obtained from the inlet box computations at the appropriate flow rates. The original five stays configuration computation involved the study of four flow conditions corresponding to the pitch-wise sectors described above, while the author carried out that of the eight stay plus eight inlet guide vanes for the corresponding points at the eight circumferential positions.

The author presented the fan simulation results for the original five stay configuration on the fan characteristic derived from model test data, Figure 9. The duty point pressure is slightly over predicted, and the slope of the simulation results is lower than the model test data. The over prediction of the duty point pressure is small enough for us to consider within the combined uncertainty of the model test data and simulation results. The lower slope of the simulation results implies that the simulation is under predicting the fan's peak pressure developing capability and is therefore conservative and consequently acceptable.

The author neither model tested nor performance tested in-service the improved eight stay plus eight guide vane configuration. Consequently, the author presented the improved configuration fan simulation results in a comparative manner between the simulation results for the original and improved configurations. The author compared the rotor relative inflow angles for the two configurations for equivalent angular sectors at the duty point, Figure 10. The author predicted that the mid-height rotor relative inflow angle would increase by approximately five degrees with the change from original to improved configuration. The author estimates that in-service, the fan's motor had a margin on peak power sufficient to allow inflow angle to increase by up to seven degrees, with five representing a conservative maximum.

The effect of increasing inflow angle on fan pressure rise is to increase predicted duty point pressure from 2320 Pa to 3300 Pa, Figure 11. It is apparent that besides the expected increase in pressure with the change from original to improved configuration, there is a marked change in the characteristic lines gradient which is consistent with a movement towards stall. Despite moving towards stall, the simulation results do not predict that the improved configuration will stall at the duty point. As the original configuration did predict the fan's stall margin conservatively, the improved configuration's stall margin was acceptable.

An analysis of the simulations adiabatic efficiency results for both the original and improved configurations indicates that the improved configuration has a duty point efficiency approximately 1% less than the original configuration, Figure 12. This is consistent with the principle that for a well-designed blade increasing the rotor relative inflow angle

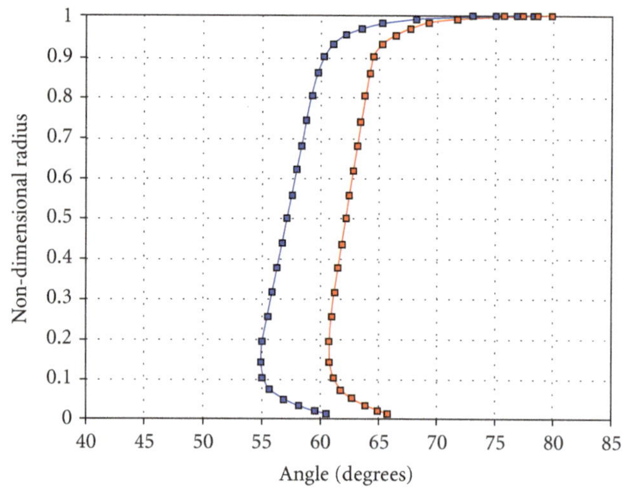

FIGURE 10: Radial distribution of the blade relative inflow angle for the original five stay and improved eight stay plus eight inlet guide vane configurations at the duty point condition.

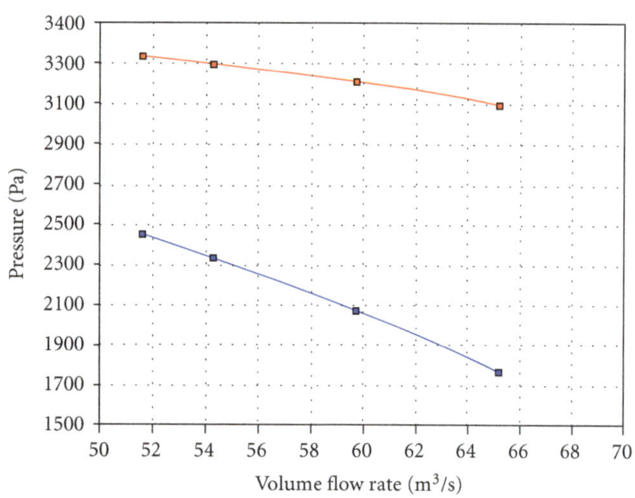

FIGURE 11: Total pressure rise for the original five stay and improved eight stay plus eight inlet guide vane configurations for four flow conditions: duty point, 10% and 20% above and 5% below.

should equate to some local increase in losses. Although the simulation results for the duty point confirm increase in losses, the simulation results indicate that at higher flow, the improved configuration is able to maintain efficiency more effectively than the original configuration. The improved configuration's ability to maintain efficiency at higher flow suggests that when considering the rotor in isolation, there is a small reward for operation with a higher relative inflow angle.

The improved eight stay plus eight inlet guide vane configuration increases the fan aerodynamic loading. Increasing fan aerodynamic loading also increased fan acoustic emissions [12]. When we analyze the improved eight stay

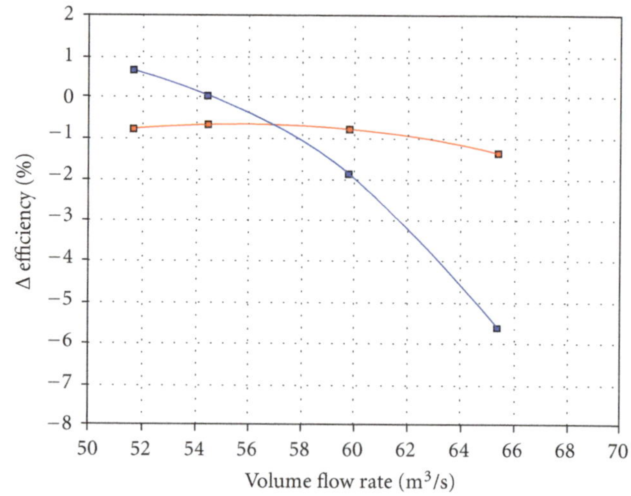

FIGURE 12: Normalized efficiency variation relative to the duty point for the original five stay and improved eight stay plus eight inlet guide vane configurations for four flow conditions: duty point, 10% and 20% above and 5% below.

plus eight inlet guide vane configuration using Pascu et al.'s method [12], we can predict an overall increase in fan acoustic emissions of between 3 and 5 dB. In practice, the author installed the original configurations and then later retrofitted the improved configuration into the same fan. The fan was located next to the boiler of a coal fired power plant, and in close proximity to other heavy duty rotating equipment. This proved impossible in practice to accurately measure on-site fan acoustic emissions with either the original or improved configuration. As such, the author was not able to ascertain that the change in stay and guide vane geometry resulted in an increase in fan acoustic emissions; however, it is likely that if the background noise in the fan's vicinity had been lower, an increase would have been apparent.

7. Mechanical Impact of Fan Simulation Results

The author undertook the computational analysis in response to an in-service issue with an induced draft fan. The induced draft fan was not generating the required pressure. The inability to produce the required pressure was a consequence of the fan's under-specification for the application in the first place. It was subject to a distorted inlet flow and also suffered significant erosion as a consequence of a poorly installed and maintained air filtration system upstream of the fan. The improved stay and inlet guide vane configuration increased the fan pressure developing capability, and additionally reduced the circumferential variation in velocity entering the fan blades. The author considered this reduction in circumferential velocity as important as a consequence of a second in-service issue with the induced draft fans generally: premature blade bearing failure.

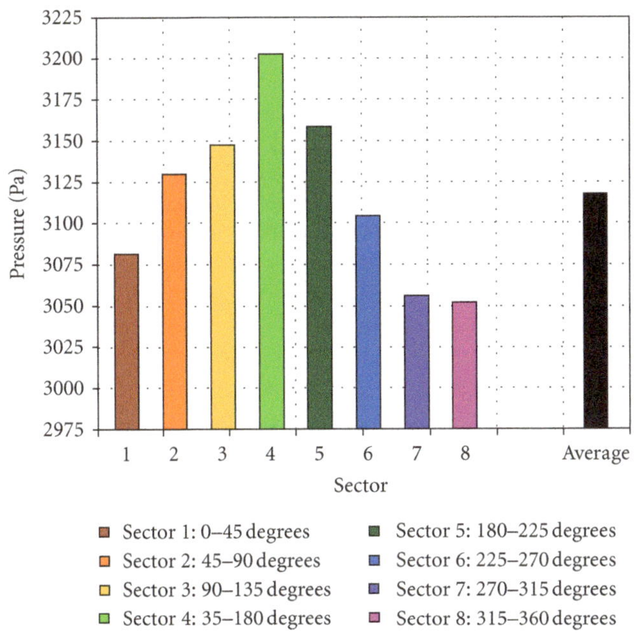

Figure 13: Total pressure variation with pitch-wise sector for the improved eight stay plus eight inlet guide vane configuration at the duty point condition.

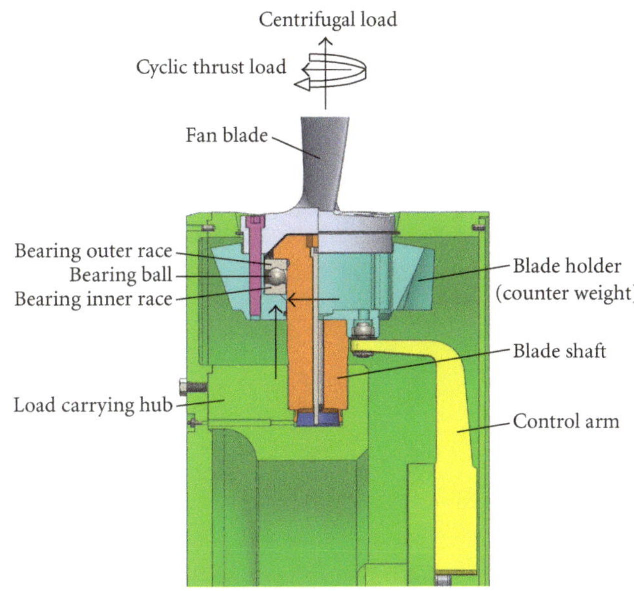

Figure 14: Mechanical loads induced in the blade bearing by the pitch-wise total pressure variation.

Collected data downstream of the rotor for the improved eight stay plus eight inlet guide vane configuration, Figure 13, shows the predicted cyclic variation in total pressure due to the uneven pressure distribution emanating from the inlet box. The graph's scale exaggerates the true magnitude of the dissimilarity between different sector values. The maximum deviation from the average value is 2%. This deviation, due to the inlet box conditions, is small enough for us to regard as aerodynamically negligible.

Despite that observation that the uneven pressure deviation's aerodynamic effect due to the inlet box conditions may be neglected, we must consider it within the context of the studied fan: a constant speed variable pitch in motion fan. A blade bearing that facilitates change in blade angle supports each fan blade to enable fan-load adjustment. The deviation in aerodynamic load results in a once-per-revolution cyclic load and moment on each blade which then changes the load distribution on the blade bearing, Figure 14. This once-per-revolution load results in the bearing balls' cyclic contact angles on the bearing raceway. The cyclic twisting moment causes an infinitesimal rotation in pitch change direction, which then results in bearing ball orbital motion which is different for each ball.

A consequence of the ball orbital motion is that bearing balls tend to migrate towards one location on the bearing raceway [13]. The balls push until they remove all clearance, with the resultant ball to ball separator forces becoming large. This ball to ball separator force results in increased bearing friction due to the friction between bearing balls and ball separator. This results in accelerated ball separator wear, Figure 15, and consequently, bearing ball wear. The result is the bearing's rapid mechanical deterioration, and consequently, it rapidly increases forces to change the blade angle.

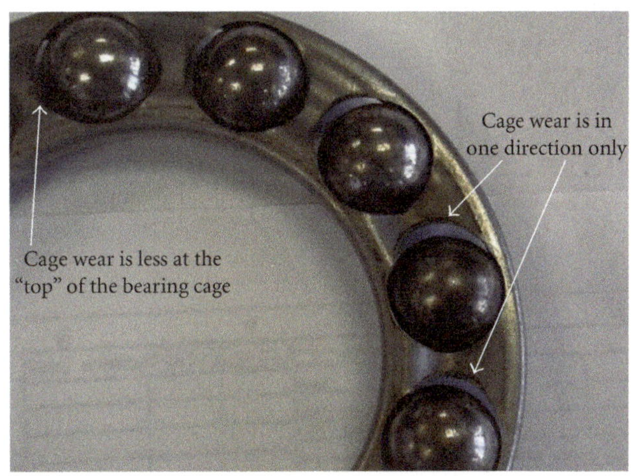

Figure 15: Wear in a traditional "cage" bearing separator that occurs as a consequence of mechanical loads on blade bearing balls. Pitch-wise total pressure variations induce the mechanical loads.

Reducing the cyclic variation in total pressure due to the uneven pressure distribution emanating from the inlet box below 2% would be difficult to achieve in practice in an industrial fan application. Rather than attempting to eliminate the uneven pressure distribution, Sheard and Rhoden [14] proposed a novel bearing separator comprising individual bearing separator elements, Figure 16, manufactured from a low friction carbon impregnated Teflon. The resulting bearing configuration, Figure 17, has the same imposed mechanical loads that previously resulted in bearing separator wear (Figure 15). However, the use of separate bearing separator elements results in the imposed loads inducing rotation in each separator that then transports lubricant into the bearing ball to bearing separator contact zone.

FIGURE 16: Sheard and Rhoden's bearing separator concept [14].

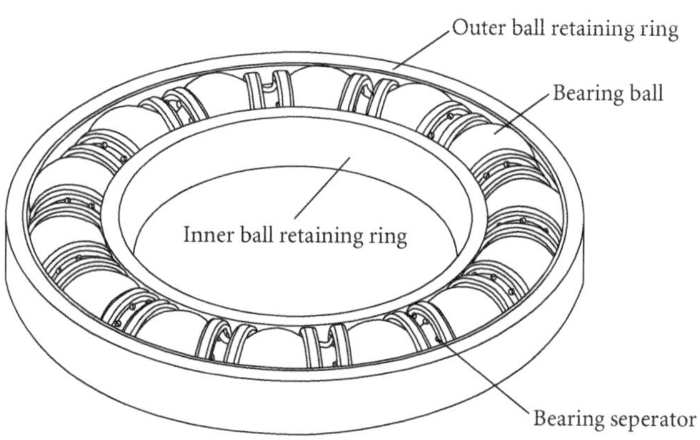

FIGURE 17: A blade bearing, fitted with Sheard and Rhoden's bearing separator [14].

Sheard and Rhoden's bearing separator design [14] has demonstrated good in-service reliability, withstanding the imposed loads that arise with the cyclic variation in total pressure due to the uneven pressure distribution emanating from the inlet. An investigation after 22,000 in-service hours indicated that approximately 50% of bearing life had been consumed. The bearing itself was still well packed with lubrication that showed good persistence in all contact areas within the bearing. The bearing balls, separator, and raceway showed no visible signs of damage.

8. Conclusions

The author studied an inlet box in its five stays original form and an improved configuration with eight stays plus eight inlet guide vanes, designed to pre-swirl the flow before discharge onto a fan stage. The two inlet box simulations enabled the identification of handed flow effects in the original configuration which are still present in the improved configuration in the form of small, but detectable tangential speed variations. The corollary is that for complex inlet assembly geometries, particularly for higher speed applications, there is a need to account for the cumulative flow effects in order to avoid discharged flow pitch-wise nonuniformities. We can best meet this objective through some form of computational method which is able to resolve the three dimensional flow-field. For the case under study, the fan's stall margin in its original configuration is adequate to accommodate the additional loading associated with the improved configurations additional stays and inlet guide vanes. Nevertheless, the fan remains susceptible to erosion effects.

The analysis of the improved eight stays plus eight inlet guide vanes rotor-stator simulations over the eight pitch-wise sectors enabled the identification of a 2% cyclic variation in pressure due to the uneven pressure distribution emanating from the inlet box. Although the aerodynamic impact is small, there is a mechanical consequence for the studied variable pitch in motion fan. The resultant mechanical loads on the blade-bearing result in the bearing ball separator's rapid wear and consequently, premature blade bearing failure. The author developed a novel bearing separator design that has proven capable of withstanding the imposed mechanical loads and thus eliminating the need to reduce further the cyclic variations in total pressure. After 22,000 hours in-service operation, the author estimated that the blade bearings fitted with the novel bearing separator had approximately 50% bearing life remaining. As such, the novel bearing separator design facilitates an estimated five year bearing life, self-consistent with the major service intervals of variable pitch in motion fans in induced draft application.

References

[1] A. Guédel, "Prediction of the noise installation effect induced by a bend at the inlet of an axial flow fan," in *Proceedings of the 2nd International Symposium Fan Noise*, Senlis, France, September 2003.

[2] S. Bianchi, A. Corsini, and A. G. Sheard, "Installed acoustic performance of cooling axial fans fitted with end-plates," *Noise Control Engineering Journal*, vol. 60, no. 5, pp. 519–527.

[3] O. Roslyng, "Installation effect on axial flow fan caused by swirl and non-uniform velocity distribution," in *Proceedings of the Installation Effects in Ducted Fan Systems Conference*, Institution of Mechanical Engineers, Power Industries Division, London, UK, May 1984, AIAA paper 96-0297.

[4] M. Staiger and H. Stetter, "Aerodynamic response of axial fan bladings to non-uniform inlet flow fields," in *Institution of Mechanical Engineers Seminar on Installation Effects in Fan Systems*, London, UK, 1993.

[5] A. Corsini, A. egiani, S. Minotti, and A. G. Sheard, "Numerical investigation on the aerodynamic influence of eroded leading-edge geometry on boiler fan performance," in *Proceedings of the 9th European Conference on Turbomachinery, Fluid Dynamics & Thermodynamics*, pp. 879–892, Istanbul, Turkey, March 2011.

[6] A. Corsini, A. Marchegiani, F. Rispoli, A. G. Sheard, and P. Venturini, "Predicting blade leading edge erosion in an axial induced draft fan," *Journal of Engineering for Gas Turbines and Power*, vol. 134, article 042601, 9 pages.

[7] A. Corsini, F. Rispoli, A. G. Sheard, and P. Venturini, "The impact of blade erosion on the induced-draft boiler fan aerodynamic performance and stall margin ventilation," in *Proceedings of the 57th American Society of Mechanical*

Engineers Turbine and Aeroengine Congress, Copenhagen, Denmark, June 2012, paper no. GT2012-69048.

[8] J. T. Greenzweig, R. Henry, and T. Holm, "The effect on inlet flow distortion on the performance of centrifugal fans for utility applications," in *Proceedings of the 56th American Society of Mechanical Engineers Turbine and Aeroengine Congress*, Vancouver, Canada, June 2011, paper no. GT2011-45422.

[9] J. A. Teixeira, A. Tourlidakis, P. C. Ivey, A. G. Sheard, J. A. Molin, and I. R. Kinghorn, "Computational analysis of an industrial axial-flow fan including installation effects," in *Proceedings of the IMechE Advances of CFD in Fluid Machinery Design Seminar*, London, UK, November 2003.

[10] "TASCflow documentation," ANSYS Canada, Ontario, Canada, 2003.

[11] P. F. Galpin and G. D. Raithby, "Numerical solution of problems in incompressible fluid flow: treatment of the temperature-velocity coupling," *Journal of Numerical Heat Transfer*, vol. 10, no. 2, pp. 105–129, 1986.

[12] M. Pascu, M. Miclea, P. Epple, A. Delgado, and F. Durst, "Analytical and numerical investigation of the optimum pressure distribution along a low-pressure axial fan blade," *Journal of Mechanical Engineering Science*, vol. 223, no. 3, pp. 643–658, 2009.

[13] R. H. Perkinson, M. C. Soule, and H. J. Holl, "Separator for bearing assemblies with cyclic loads," patent number US8167501, May 2012.

[14] A. G. Sheard and A. Rhoden, "Variable pitch fans," patent no. GB2485634, September 2012.

Nonsmooth Modeling of Vibro-Impacting Euler-Bernoulli Beam

C. P. Vyasarayani,[1] Sukhpreet Singh Sandhu,[2] and John McPhee[3]

[1] *Department of Mechanical Engineering, Indian Institute of Technology Hyderabad, Ordnance Factory Estate, Yeddumailaram 502205, India*
[2] *MSC Software, 101 North Brand Boulevard, Suite 900 Glendale, Santa Ana, CA 91203, USA*
[3] *Systems Design Engineering, University of Waterloo, Waterloo, ON, Canada N2L 3G1*

Correspondence should be addressed to C. P. Vyasarayani, vcprakash@iith.ac.in

Academic Editor: Abul Azad

A new technique to simulate nonsmooth motions occurring in vibro-impacting continuous systems is proposed. Sticking motions that are encountered during vibro-impact simulation are imposed exactly using a Lagrange multiplier, which represents the normal reaction force between the continuous system and the obstacle. The expression for the Lagrange multiplier is developed in closed form. The developed theory is demonstrated by numerically simulating the forced response of a pinned-pinned beam impacting a point-like rigid obstacle.

1. Introduction

The dynamics of continuous systems like beams, strings, and rods subjected to impacts caused by motion-limiting constraints have been studied extensively. A representative set of studies can be found in [1–17]. The impact of these systems is modelled using two different methods.

(1) Force integration method (penalty method) [2, 4–6, 10–14]: in this method, a contact force proportional to the penetration of the beam at the contact point is applied at the contact location. This method can be used to model linear or nonlinear stiffness in the obstacles.

(2) Nonsmooth method (coefficient of restitution (CoR) approach): this approach treats the obstacle as rigid and uses the CoR assumption to relate the states before and after impacting neglecting the change in configuration during the contact. A modal form of CoR rule for continuous systems was proposed by Wagg and Bishop [1, 7–9] and further developed by Vyasarayani et al. [18] to make the formulation energy-consistent.

The penalty method is best suited for simulating impacts that are compliant but, when used to simulate rigid impacts, they suffer from long simulation times due to the abrupt discontinuity in stiffness arising at the event of impact [19], requiring special numerical treatment [20, 21]. The COR method is an attractive method for simulating the impact with rigid point-like obstacles due to its several advantages. The modeling process is simple because when the impact occurs one simply uses the CoR rule to obtain the post-impact modal states from the pre-impact modal states. An energy-consistent modal form of CoR rule has been proposed by Vyasarayani et al. [18] and was shown to perform better in terms of simulation time than the penalty approach. In the work of Vyasarayani et al. [18], the authors have not discussed the modeling of sticking motions.

The aim of this work is to highlight and address the issues associated with modeling sticking motions that can occur when a continuous system impacts a rigid obstacle. During the impact of a continuous system with a rigid obstacle, before a sticking motion occurs, it is preceded by a chatter sequence. In a chatter sequence [22] the beam impacts the obstacle repeatedly, and the time between the two successive impacts ($\Delta\tau_i$) continuously decreases. Simulating the chatter sequence without assuming sticking when $\Delta\tau_i \ll 1$ can be computationally prohibitive. For lumped parameter systems, Cusumano and Bai [23] proposed to monitor $\Delta\tau_i$ in a chatter sequence and, if this time falls below a certain tolerance,

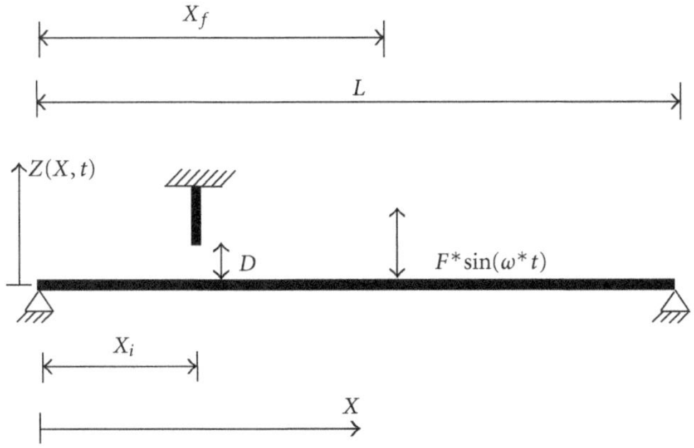

FIGURE 1: Schematic of the physical system.

the structure is assumed to stick with the obstacle. The reaction at the interface of the stuck mass and obstacle is monitored and when it crosses zero the contact phase is assumed to be over and the mass is released. The sticking motion can also be interpreted as motions occurring with zero coefficient of restitution in the impact model. However, for continuous systems the method proposed by Cusumano and Bai [23] cannot be directly applied.

Sticking motions in continuous systems are usually simulated by using mode-switching method [24, 25], wherein when the beam is assumed to be stuck to the obstacle the basis functions of the nonimpacting motion are changed to a different basis, usually the mode shapes of the beam with a pin at the location of the obstacle. The reaction at the pin is monitored to see when it crosses through zero and then the basis is again changed from contact phase to noncontact phase. In this paper, we propose a new method to simulate the beam motion during the sticking phase which does not require the reevaluation of mode shapes with a pin at the location of impact, thus enabling the efficient handling of sticking motions within the CoR method. This is done by deriving closed form expressions for the reaction force at the pin such that the gap constraint is exactly enforced. A suitable numerical example will be shown to demonstrate this method on a vibroimpacting simply supported beam.

2. Mathematical Modeling

We consider a simply supported beam with a motion-limiting constraint. The beam-stop system is shown in Figure 1. The beam is subjected to a uniformly distributed harmonic force $F^* \sin(\omega^* t)$. The stop that limits the motion of the beam is located at X_i. The governing partial differential equation (PDE) for the beam using Euler-Bernoulli beam theory, without considering the impact event, is as follows:

$$EI\frac{\partial^4 Z}{\partial X^4} + C\frac{\partial Z}{\partial t} + \rho A\frac{\partial^2 Z}{\partial t^2} = F^* \sin(\omega^* t), \quad Z(X_i, t) \le D, \tag{1}$$

where E is the Young's modulus, I is the area moment of inertia, ρ is the mass density, A is the cross-sectional area, C is the viscous damping coefficient per unit length, and t is time. The pinned-pinned boundary conditions are enforced as follows:

$$Z(0,t) = Z(L,t) = 0, \qquad \frac{\partial^2 Z(0,t)}{\partial X^2} = \frac{\partial^2 Z(L,t)}{\partial X^2} = 0. \tag{2}$$

We scale the equation of motion by introducing the following parameters:

$$x = \frac{X}{L}, \quad d = \frac{D}{L}, \quad w = \frac{Z}{L}, \quad \tau = t\sqrt{\frac{EI}{\rho A L^4}},$$
$$\omega = \omega^*\sqrt{\frac{\rho A L^4}{EI}}, \quad F = \frac{F^* L^3}{EI}, \quad c = CL\sqrt{\frac{EI}{\rho A L^4}}. \tag{3}$$

The equation of motion in scaled form is as follows:

$$\frac{\partial^4 w}{\partial x^4} + c\frac{\partial w}{\partial t} + \frac{\partial^2 w}{\partial \tau^2} = F\sin(\omega\tau), \quad w(x_i,\tau) \le d. \tag{4}$$

The boundary conditions are

$$w(0,\tau) = w(1,\tau) = 0, \qquad \frac{\partial^2 w(0,\tau)}{\partial x^2} = \frac{\partial^2 w(1,\tau)}{\partial x^2} = 0. \tag{5}$$

The initial conditions are as follows:

$$w(x,0) = w_0(x), \qquad \dot{w}(x,0) = \dot{w}_0(x). \tag{6}$$

Using the expansion theorem, a solution to (4) is assumed to be of the following form:

$$w(x,\tau) = \sum_{j=1}^{\infty} W_j(x) q_j(\tau), \tag{7}$$

where $W_j(x) = C\sin(j\pi x)$ and $q_j(\tau)$ correspond to the jth undamped mode shape and the corresponding modal coordinate for the beam, respectively. Substituting the above form of solution into (4), we get

$$\sum_{j=1}^{\infty} \frac{d^4 W_j(x)}{dx^4} q_j(\tau) + \sum_{j=1}^{\infty} W_j(x)\dot{q}_j(\tau) + \sum_{j=1}^{\infty} W_j(x)\ddot{q}_j(\tau) = F\sin(\omega\tau). \tag{8}$$

The above equation is multiplied by $W_k(x)$ and then integrated over the spatial domain, which yields the following expression:

$$\sum_{j=1}^{\infty} q_j(\tau)\int_0^1 \frac{d^4 W_j(x)}{dx^4} W_k(x)dx + \sum_{j=1}^{\infty} \dot{q}_j(\tau)\int_0^1 W_j(x)W_k(x)dx,$$
$$\sum_{j=1}^{\infty} \ddot{q}_j(\tau)\int_0^1 W_j(x)W_k(x)dx = F\sin(\omega\tau)\int_0^1 W_k(x)dx, \quad k = 1,2,\ldots,\infty. \tag{9}$$

We now evaluate the following integral:

$$\int_0^1 W_j(x) W_k(x) dx = C^2 \int_0^1 \sin(j\pi x)\sin(k\pi x) dx = \frac{C^2}{2}\delta_{jk}, \quad (10)$$

where δ_{jk} is Kronecker's delta, defined as:

$$\delta_{jk} = \begin{cases} 1, & \text{if } j = k, \\ 0, & \text{if } j \neq k. \end{cases} \quad (11)$$

Since the amplitudes of the mode shapes are arbitrary, we choose the amplitude of mode shapes (C) to be $\sqrt{2}$. This procedure of choosing the amplitudes of mode shapes such that the right-hand side of (10) that becomes δ_{jk} is called mass-normalization. By using the mass-normalized mode shapes, it can be shown that:

$$\int_0^1 \frac{d^4 W_j(x)}{dx^4} W_k(x) dx = \omega_j^2 \delta_{jk} = (j\pi)^4 \delta_{jk}. \quad (12)$$

The coupled ordinary differential equations (ODEs) given by (9) can be decoupled using the orthogonality relations ((10) and (12)). The uncoupled ODEs, after truncating the system to N modes, are as follows:

$$\ddot{q}_j(\tau) + 2\zeta_j \omega_j \dot{q}_j + \omega_j^2 q_j(\tau) = \alpha_j F \sin(\omega\tau), \quad j = 1, 2, \ldots, N, \quad (13)$$

where $\alpha_j = \int_0^1 W_k(x) dx$ and $\zeta_j = c/(2\omega_j)$. To obtain the modal initial conditions, we substitute the series solution given by (7) into (6), which gives the following relations for displacement and velocity initial conditions:

$$w(x,0) = \sum_{j=1}^{\infty} W_j(x) q_j(0),$$
$$\dot{w}(x,0) = \sum_{j=1}^{\infty} W_j(x) \dot{q}_j(0). \quad (14)$$

Multiplying both sides of (14) by $W_k(x)$, and then integrating over the domain, results in the following modal initial conditions:

$$q_j(0) = \int_0^1 w(x,0) W_j(x) dx,$$
$$\dot{q}_j(0) = \int_0^1 \dot{w}(x,0) W_j(x) dx. \quad (15)$$

2.1. Impact Modeling. Equation (13) when integrated with appropriate initial conditions can simulate the beam motion until the impact. Let τ_- be the time just before the impact (pre-impact time). Pre-impact states, $w(x,\tau_-)$ and $\dot{w}(x,\tau_-)$, are known. We have to find the states immediately after impact (post-impact), $w(x,\tau_+)$ and $\dot{w}(x,\tau_+)$. Using the classical Newton CoR assumption for rigid impacts, the configuration of the beam is assumed not to change during impact [1] (as rigid impacts are assumed to occur in an infinitesimal time), we have

$$w(x,\tau_+) = w(x,\tau_-). \quad (16)$$

The post-impact modal displacement initial conditions are

$$q_j(\tau_+) = q_j(\tau_-), \quad j = 1, 2, \ldots, N. \quad (17)$$

We also know that the beam velocity at the location of impact is altered according to the classical CoR rule:

$$\dot{w}(x_i,\tau_+) = -R\dot{w}(x_i,\tau_-), \quad (18)$$

where R is the coefficient of restitution. We must now obtain the post-impact velocity distribution, $\dot{w}(x,\tau_+)$, satisfying (18). The procedure for obtaining the post-impact velocity distribution is explained in detail in Vyasarayani et al. [18]; we present a brief explanation here. Let us assume that at the instant of impact (τ_-) an impulse of magnitude P acts on the beam at the impact location. The equation of motion of the beam can be written as:

$$\frac{\partial^4 w}{\partial x^4} + c\frac{\partial w}{\partial t} + \frac{\partial^2 w}{\partial \tau^2} = P\delta_d(x - x_i)\delta_d(\tau - \tau_-). \quad (19)$$

We assume the same pinned-pinned boundary conditions for the above PDE. Substituting the series solution given by (7) into the above PDE, we have

$$\sum_{j=1}^{N} \frac{d^4 W_j(x)}{dx^4} q_j(\tau) + \sum_{j=1}^{N} W_j(x)\dot{q}_j(\tau) + \sum_{j=1}^{N} W_j(x)\ddot{q}_j(\tau) = P\delta_d(x - x_i)\delta_d(\tau - \tau_-). \quad (20)$$

Integrating the above equation between τ_- and τ_+, we get

$$\sum_{j=1}^{N} \frac{d^4 W_j(x)}{dx^4} \int_{\tau_-}^{\tau_+} q_j(\tau) d\tau + \sum_{j=1}^{N} W_j(x) \int_{\tau_-}^{\tau_+} \dot{q}_j(\tau) d\tau + \sum_{j=1}^{N} W_j(x) \int_{\tau_-}^{\tau_+} \ddot{q}_j(\tau) d\tau = P\delta_d(x - x_i) \int_{\tau_-}^{\tau_+} \delta_d(\tau - \tau_-) d\tau. \quad (21)$$

The post-impact displacement distribution will remain the same as the pre-impact distribution, as the impact is assumed to occur in an infinitesimal amount of time, that is, $w(x,\tau_+) = w(x,\tau_-)$. Hence we have $q_j(\tau_+) = q_j(\tau_-)$ and thus the first integral in (21) becomes zero. Integrating the second term with respect to time, and from the definition of Dirac's delta function, $\int_{\tau_-}^{\tau_+} \delta_d(\tau - \tau_-) d\tau = 1$, (21) becomes

$$\sum_{j=1}^{N} W_j(x) \left(\dot{q}_j(\tau_+) - \dot{q}_j(\tau_-) \right) dt = P\delta_d(x - x_i). \quad (22)$$

Now substituting for $\dot{w}(x,\tau) = \sum_{j=1}^{N} W_j(x)\dot{q}_j(\tau)$, we get:

$$\dot{w}(x,\tau_+) = \dot{w}(x,\tau_-) + P\delta_d(x - x_i). \quad (23)$$

From the above equation we can say that the effect of applying an impulse on the Euler-Bernoulli beam results in a sudden change in velocity of the beam, immediately after the application of impact. The resulting velocity distribution after impact is nothing but a Dirac's function in space added to the pre-impact velocity distribution. The scaling parameter P (magnitude of impulse) in (23) is to be found such that the post-impact velocity at the impact location is as described by (18). Multiplying both sides of (23) by $W_k(x)$ and integrating over the domain results in:

$$\sum_{j=1}^{N} \dot{q}_j(\tau_+) \int_0^1 W_j(x) W_k(x) dx = \sum_{j=1}^{N} \dot{q}_j(\tau_-) \int_0^1 W_j(x) W_k(x) dx + P \int_0^1 W_k(x) \delta_d(x - x_i) dx. \quad (24)$$

Making use of (10), the above equation reduces to:

$$\dot{q}_j(\tau_+) = \dot{q}_j(\tau_-) + P W_j(x_i), \quad (25)$$

we now substitute (25) into (18):

$$\sum_{j=1}^{N} W_j(x_i) \left[\dot{q}_j(\tau_-) + P W_j(x_i) \right] = -R \sum_{j=1}^{N} W_j(x_i) \dot{q}_j(\tau_-), \quad (26)$$

which can be simplified to obtain the following expression for P:

$$P = -\frac{(1+R) \sum_{j=1}^{N} W_j(x_i) \dot{q}_j(\tau_-)}{\sum_{j=1}^{N} W_j(x_i)^2}. \quad (27)$$

The post-impact modal velocities can be obtained by substituting (27) into (25), which results in the following expression:

$$\dot{q}_j(\tau_+) = \dot{q}_j(\tau_-) - \frac{(1+R) \dot{w}(x_i, \tau_-)}{\sum_{j=1}^{N} W_j(x_i)^2} W_j(x_i), \quad j = 1, 2, \ldots, N. \quad (28)$$

2.2. Simulating Sticking Motions. During a chatter sequence the beam repeatedly impacts the obstacle and the time between two successive impacts continuously drops. The pre-impact velocity also continually drops and approaches zero.To switch to the sticking motion phase the time between two successive impacts is monitored and when this time falls below a certain threshold, the beam is assumed to stick with the obstacle. When the reaction between the obstacle and the beam vanishes the sticking motion is assumed to end.

Figure 2 illustrates the method of simulating a chatter sequence using a CoR method with Lagrange multiplier. In the figure, $\Delta\tau_i$ represents the time between successive impacts. When this time duration falls below a threshold value, ϵ, the beam is assumed to stick to the obstacle. To exactly satisfy the condition of sticking, as shown in Figure 2, when a chatter sequence is recognized the beam is brought to rest at the point of impact by applying an impulse using CoR = 0. The post-impact modal position and velocities can now be obtained from (17) and (28) using CoR = 0. The sticking

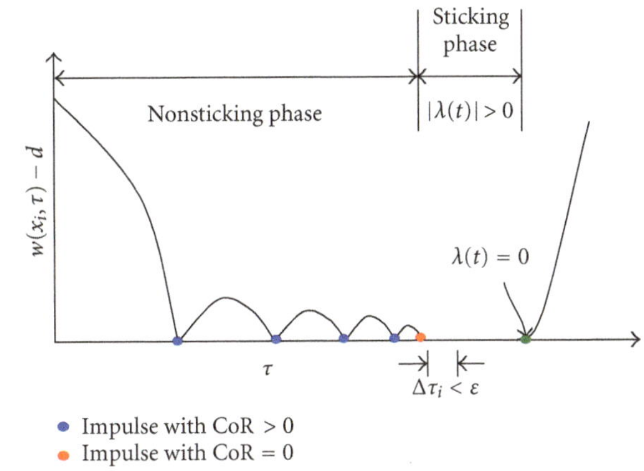

FIGURE 2: Simulating a typical chatter sequence using CoR method with Lagrange multiplier.

phase is now simulated by applying a reaction force at the point of impact such that the sticking constraint is satisfied. The sticking phase ends when the reaction force changes its sign. We now derive the expression for this reaction force.

Let $\lambda(\tau)$ be the reaction force at the location of the impact during sticking. The modal equation, (13), is modified as follows by the addition of the reaction force:

$$\ddot{q}_j(\tau) + 2\zeta_j \omega_j \dot{q}_j + \omega_j^2 q_j = \alpha_j(x_f) F \sin(\omega\tau) + W_j(x_i) \lambda(\tau), \quad j = 1, 2, \ldots, N. \quad (29)$$

The constraint equation resulting from the sticking condition is given by:

$$w(x_i, t) = \sum_{j=1}^{N} W_j(x_i) q_j(\tau) = d, \quad (30)$$

we differentiate (30) twice with respect to time:

$$\sum_{j=1}^{N} W_j(x_i) \ddot{q}_j(\tau) = 0, \quad (31)$$

substituting (29) into (31) we get:

$$\sum_{j=1}^{N} W_j(x_i) \left(-\omega_j^2 q_j(\tau) - 2\zeta_j \omega_j \dot{q}_j + \alpha_j F \sin(\omega\tau) + W_j(x_i) \lambda(t) \right) = 0. \quad (32)$$

Solving the above equation for $\lambda(t)$ we get

$$\lambda(t) = -\frac{\sum_{j=1}^{N} W_j(x_i) \alpha_j F \sin(\omega\tau)}{\sum_{j=1}^{N} W_j(x_i)^2} - \frac{\sum_{j=1}^{N} W_j(x_i) \left(\omega_j^2 q_j(\tau) + 2\zeta_j \omega_j \dot{q}_j \right)}{\sum_{j=1}^{N} W_j(x_i)^2}, \quad (33)$$

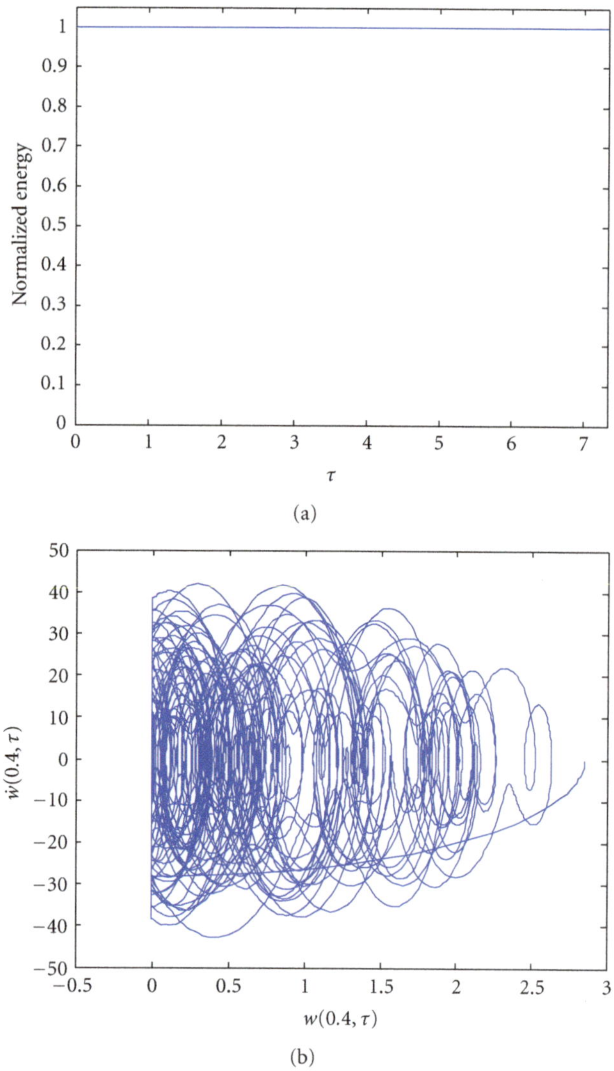

FIGURE 3: Free vibration response of the beam with $w(x,0) = 3\sin(\pi x)$, $R = 1$, and $x_i = 0.4$: (a) total energy of the beam, and (b) phase plot at the location of impact.

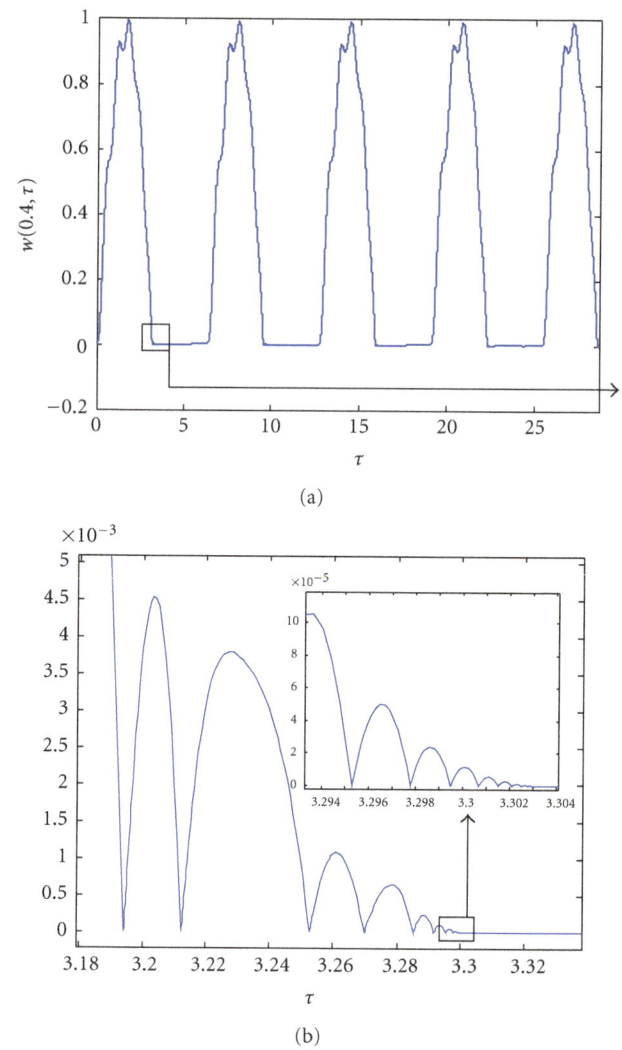

FIGURE 4: Predicted motion of the beam for $F = 76.8$, $\omega = 0.1\omega_1$, $R = 0.7$, $d = 0$, $\zeta_j = 0.05$, $x_i = 0.4$, and $\epsilon = 1 \times 10^{-6}$: (a) beam displacement at the location of impact showing chatter and (b) close-up view near the end of chatter and the beginning of sticking phase.

where $\lambda(t)$ is the dynamic reaction at the location of the obstacle during sticking, which enforces the position constraint (30). Once the reaction approaches zero, the sticking motion ends and the modal equations of motion will be integrated with $\lambda(t) = 0$. The process of simulating the vibro-impacting motion of a beam with sticking motions using Lagrange multiplier can now be summarized as below.

(1) Use the modal equations, (13), to integrate the states, $\{q(\tau), \dot{q}(\tau)\}$, forward in time.

(2) Check for impact: if impact is detected use (28) to model the impact.

(3) Check for sticking: if time between two successive impacts is less than threshold, ϵ, the sticking phase begins, go to step 4. Otherwise continue with step 1.

(4) Integrate the modal equations, (32) which include the Lagrange multiplier (which represent the reaction at the pin) in the modal equations. The Lagrange multiplier, $\lambda(\tau)$, can be calculated using (33).

(5) Check for contact separation by monitoring $\lambda(\tau)$: if $\lambda(\tau)$ crosses zero sticking phase ends go to step 1; otherwise, continue with step 4.

Before presenting numerical simulations we summarize the developments in this paper. In Section 2 a mathematical model based on modal formulation has been developed for simulating nonimpacting motions of the beam. A modal form of CoR rule for simulating impacts was discussed in Section 2.1. The simulation of sticking motions using Lagrange multiplier is proposed in Section 2.2 within the framework of CoR rule.

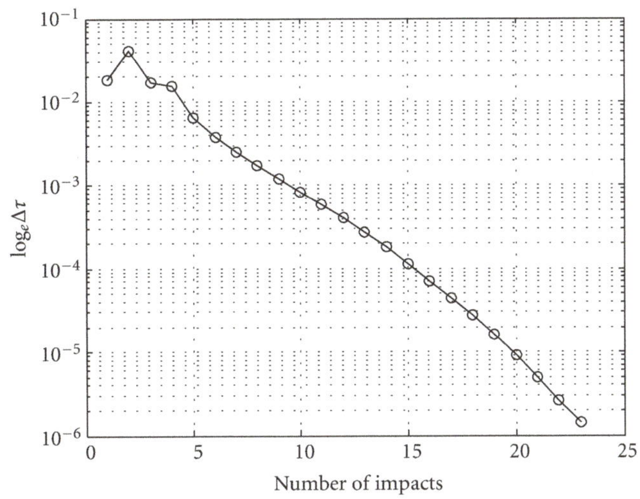

FIGURE 5: Impact duration between successive impacts for typical chatter sequence. The following parameters are used during the simulation: $F = 76.8$, $\omega = 0.1\omega_1$, $R = 0.7$, $d = 0$, $\zeta_j = 0.05$, $x_i = 0.4$, and $\epsilon = 1 \times 10^{-6}$.

3. Numerical Simulation and Discussion

A numerical code was developed in Matlab for simulating a pinned-pinned vibro-impacting beam (four modes were considered in the simulation). Matlab's built-in integrator (ode15s) for solving ordinary differential equations along with the built-in event detection algorithm for detecting impacts was employed. We used an absolute and relative tolerance of $1e^{-6}$ in the numerical simulations. For simulating the sticking motions the proposed Lagrange multiplier method within the framework of CoR rule was used. A threshold time for detecting chatter was chosen to be $\epsilon = 1e^{-6}$ (later we show how this value can be selected). To validate our numerical code we first studied the free vibration response of the beam with CoR = 1. The initial conditions for the beam were chosen to be $w(x, 0) = 3\sin(\pi x)$, which corresponds to the first mode shape and $\dot{w}(x, 0) = 0$. The beam is allowed to vibrate against an obstacle located at $x_i = 0.4$. The gap of $d = 0$ was selected. The total energy of the beam is shown in Figure 3(a) and is conserved during the entire simulation (the beam impacts the obstacle 40 times during the simulation time), as required for perfectly elastic impacts. Figure 3(b) shows the phase plot of the beam and shows that the constraint $(w(0.4, \tau_\pm) - d = 0)$ is exactly satisfied at every impact.

3.1. Choosing the Value ϵ. Figure 4(a) shows a typical chatter sequence of the beam under external distributed harmonic forcing. Multiple impacts with reducing amplitudes are seen in Figure 4(b); the motion is similar to that of a bouncing ball falling under gravity on a perfectly rigid ground with $R < 1$, where an infinite number of impacts occur before the ball comes to rest. The logarithm of time difference between the successive impacts ($\log_e \Delta\tau$) corresponding to Figure 4(b) is shown in Figure 5. It can be seen that before the onset of sticking motions the time between successive impacts drops exponentially or linearly (almost) in the logarithmic scale. Now what if we stick the obstacle to the beam immediately when the chatter is detected (with $\epsilon = 1e - 6$, the beam impacts the obstacle 19 times before it is made to stick); does it cause any significant change in the subsequent dynamics of the beam? To answer this question we have simulated the beam motion by choosing $\epsilon = 1e - 3$ and $\epsilon = 1e - 4$. These values are within the region of chatter sequence (see Figure 5). The phase plot of the beam motion at the location of impact for different values of ϵ is shown in Figure 6 and is evident from

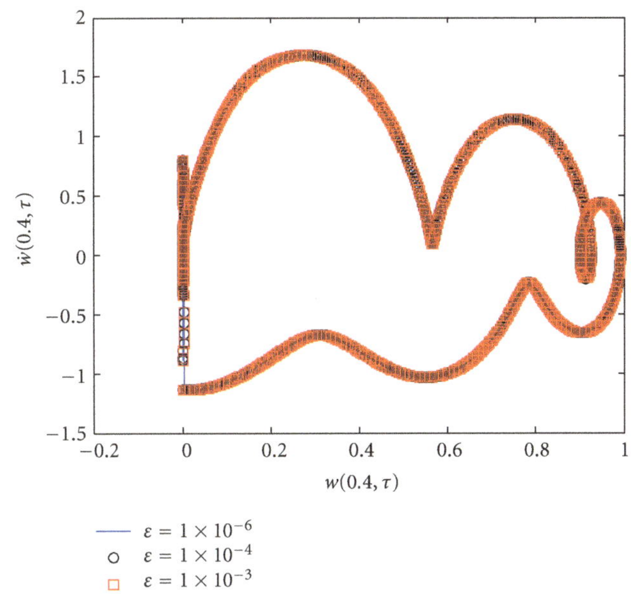

(a)

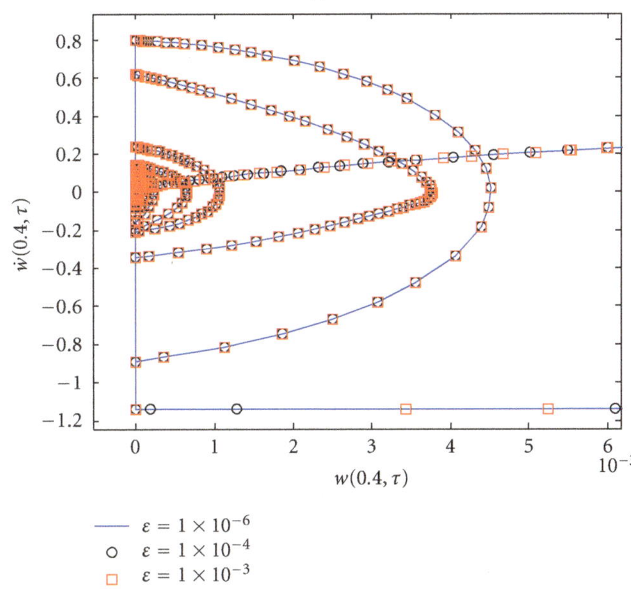

(b)

FIGURE 6: Response of the beam for three different values of ϵ: (a) phase-plot at the impact location and (b) phase plot showing a close-up view near the impact.

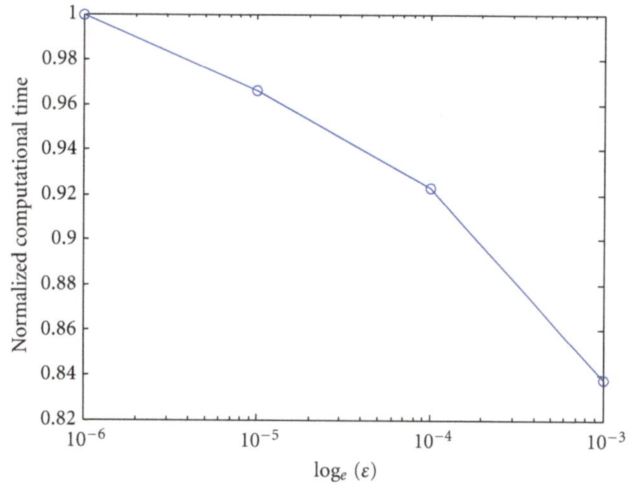

FIGURE 7: Comparison of computational time (for 10 units of simulation time) for different values of ϵ.

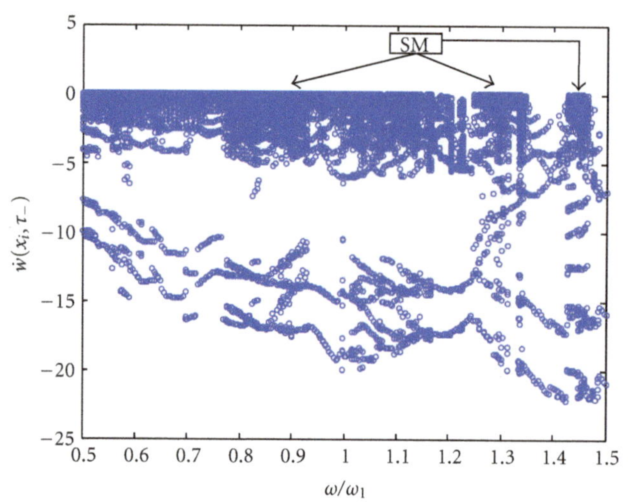

FIGURE 8: Poincaré section at the location of impact for $F = 72$, $R = 0.7$, $d = 0$, $\zeta_j = 0.05$, $x_i = 0.4$, and $\epsilon = 1 \times 10^{-3}$ in the frequency range of $0.5 \leq \omega/\omega_1 \leq 1.5$.

the figure that the overall motion of the beam (poststicking phase) remains the same irrespective of what epsilon value we choose, except the number of impacts the beam makes before it sticks to the obstacle are different for each ϵ. Once the chatter is detected (exponential decay of impact times between successive impacts) the beam can be made to stick to the obstacle without compromising on poststicking response behavior. The amount of computational time required to simulate the beam motion for different values of ϵ is shown in Figure 7. The comparison was made for 10 units of simulation time (the computational time was calculated by taking average computational time for 20 simulations). By using a lower value of $\epsilon = 1 \times 10^{-3}$ instead of $\epsilon = 1 \times 10^{-6}$, we were able to gain 16% of computational time. In order to characterize the nonlinear vibro-impacting continuous systems, several long-time integrations need to be performed while varying the detuning parameter, so any computational

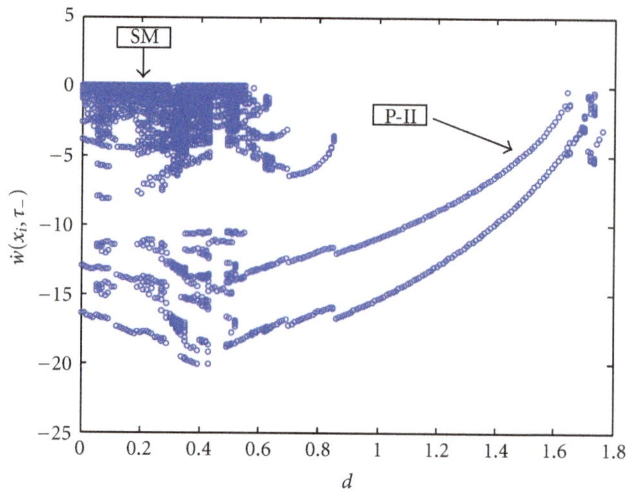

FIGURE 9: Poincaré section at the location of impact for $F = 72$, $R = 0.7$, $\zeta_j = 0.05$, $x_i = 0.4$, and $\epsilon = 1 \times 10^{-3}$, $\omega = 0.7\omega_1$ and $0 \leq d \leq 1.8$.

time gained will be beneficial while performing parametric studies.

3.2. Regions of Sticking Motions. The sticking motions primarily occur due to wave phenomena in continuous systems. To emphasize the point consider this example; when the beam is subjected to impact at a location, the beam is abruptly brought to rest, while the rest of the beam continaes to move in the direction of the forcing. The impulse generated at the impact initiates a bending wave which will travel to the boundary and get reflected, thus causing further multiple impacts. Several parameters, like CoR, forcing amplitude, forcing frequency, and gap, can initiate sticking motions. As the intention of the paper is to demonstrate the capability of the developed theory for simulating vibro-impacting continuous systems, we will not discuss the specifics of motion, like studying the bifurcations and periodicity of motion, but rather we focus solely on sticking motions. Figure 8 shows the Poincaré section at the location of impact with zero gap. In the frequency range of interest $0.5 \leq \omega/\omega_1 \leq 1.5$, we found a large region where sticking motions (SMs) were present for the selected parameter values. Sticking motions are identified as zero rebound velocity impacts with finite time duration. It is evident from Figure 8 that the developed theory can be used to successfully simulate the vibro-impacting motion of the beam for a wide range of physical parameters.

3.3. Control of Sticking Motions. The sticking motions can be controlled by changing different physical parameters, but we choose to vary the gap d for illustration. In this study, the same parameters that were used to generate Figure 8 are used, except we fix the frequency at $\omega = 0.7\omega_1$. Figure 9 shows the Poincaré section at the location of impact for different values of gap. As we increase the gap the sticking motions disappear around $d \simeq 0.6$. Stable periodic double impact solutions (P-II), showing two impacts for every cycle of motion, were

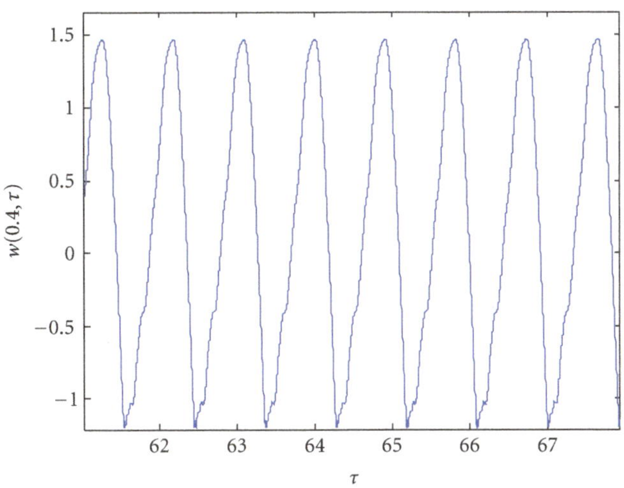

(a)

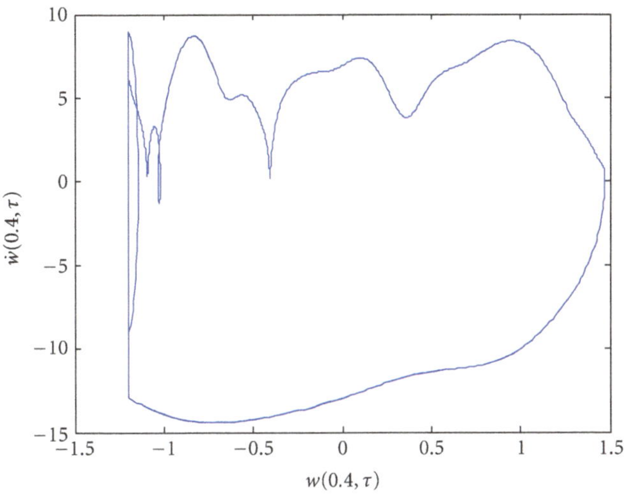

(b)

FIGURE 10: Response of the beam for $F = 72$, $R = 0.7$, $\omega = 0.7\omega_1$, $d = 1.2$, $\zeta_j = 0.05$, $x_i = 0.4$, and $\epsilon = 1 \times 10^{-3}$: (a) displacement at impact location and (b) phase plot at impact location.

found over a wide range of gaps. Figure 10(a) shows the motion of the beam at the location of impact for a double impact motion for $d = 1.2$. The corresponding phase space is shown in Figure 10(b). Other kinds of motions are also visible in Figure 9, like period-I, period-III, and also some bifurcations can be seen during sticking motions. A detailed parametric study of these motions in itself is an involved problem and is beyond the scope of the present work. The emphasis of this study was to show the capability of the developed theory to simulate a wide variety of impacting motions.

4. Conclusions

A new procedure for modeling sticking motions in vibro-impacting continuous systems has been proposed in the framework of the CoR method. The method does not require prior information of the mode shapes of the beam in sticking phase (when the beam is in contact with the obstacle) unlike mode-switching method. The sticking constraints are imposed exactly by calculating the closed-form expression for the constraint reactions using the Lagrange multiplier. A systematic method for selecting the sticking tolerance ϵ based on time between successive impacts during a chatter sequence was proposed. A pinned-pinned vibro-impacting beam was numerically simulated using the developed theory to demonstrate the usefulness of the method.

Disclosure

Some parts of this paper has appeared in "Effective Modelling of Sticking Motions in Vibro-Impacting Continuous Systems," by C. P. Vyasarayani, S. S. Sandhu, and J. McPhee, Paper Number IMECE2010-39485, Volume 8: Dynamic Systems and Control, Parts A and B. Permission was obtained from ASME for reproducing the published material in this manuscript.

Acknowledgments

Funding of this research by the Natural Sciences and Engineering Research Council of Canada is gratefully acknowledged. Thanks to Dr. Kingshook Bhattacharyya of IIT Kharagpur for his technical inputs during the course of this study.

References

[1] D. J. Wagg and S. R. Bishop, "Application of non-smooth modelling techniques to the dynamics of a flexible impacting beam," *Journal of Sound and Vibration*, vol. 256, no. 5, pp. 803–820, 2002.

[2] Z. Bao, S. Goyal, L. J. Leu, and S. Mukherjee, "The role of beam flexibility and ground contact model in the clattering of deformable beams," *Journal of Dynamic Systems, Measurement and Control, Transactions of the ASME*, vol. 126, no. 2, pp. 421–425, 2004.

[3] W. Schiehlen and R. Seifried, "Three approaches for elastodynamic contact in multibody systems," *Multibody System Dynamics*, vol. 12, no. 1, pp. 1–16, 2004.

[4] H. C. Tsai and M. K. Wu, "Methods to compute dynamic response of a cantilever with a stop to limit motion," *Computers and Structures*, vol. 58, no. 5, pp. 859–867, 1996.

[5] C. Wang and J. Kim, "New analysis method for a thin beam impacting against a stop based on the full continuous model," *Journal of Sound and Vibration*, vol. 191, no. 5, pp. 809–823, 1996.

[6] C. Wang and J. Kim, "The dynamic analysis of a thin beam impacting against a stop of general three-dimensional geometry," *Journal of Sound and Vibration*, vol. 203, no. 2, pp. 237–249, 1997.

[7] D. J. Wagg, "A note on coefficient of restitution models including the effects of impact induced vibration," *Journal of Sound and Vibration*, vol. 300, no. 3–5, pp. 1071–1078, 2007.

[8] D. J. Wagg, "A note on using the collocation method for modelling the dynamics of a flexible continuous beam subject to impacts," *Journal of Sound and Vibration*, vol. 276, no. 3-5, pp. 1128–1134, 2004.

[9] D. J. Wagg, "A note on modelling multi-degree-of-freedom vibroimpact systems using coefficient of restitution models," *Journal of Sound and Vibration*, vol. 236, no. 1, pp. 176–184, 2000.

[10] E. L. B. Van De Vorst, M. F. Heertjes, D. H. Van Campen, A. De Kraker, and R. H. B. Fey, "Experimental and numerical analysis of the steady state behaviour of a beam system with impact," *Journal of Sound and Vibration*, vol. 212, no. 2, pp. 321–336, 1998.

[11] P. Metallidis and S. Natsiavas, "Vibration of a continuous system with clearance and motion constraints," *International Journal of Non-Linear Mechanics*, vol. 35, no. 4, pp. 675–690, 2000.

[12] J. Knudsen and A. R. Massih, "Vibro-impact dynamics of a periodically forced beam," *Journal of Pressure Vessel Technology, Transactions of the ASME*, vol. 122, no. 2, pp. 210–221, 2000.

[13] J. A. Turner, "Non-linear vibrations of a beam with cantilever-Hertzian contact boundary conditions," *Journal of Sound and Vibration*, vol. 275, no. 1-2, pp. 177–191, 2004.

[14] W. Lin, N. Qiao, and H. Yuying, "Bifurcations and chaos in a forced cantilever system with impacts," *Journal of Sound and Vibration*, vol. 296, no. 4-5, pp. 1068–1078, 2006.

[15] E. K. Ervin and J. A. Wickert, "Experiments on a beam-rigid body structure repetitively impacting a rod," *Nonlinear Dynamics*, vol. 50, no. 3, pp. 701–716, 2007.

[16] E. K. Ervin and J. A. Wickert, "Repetitive impact response of a beam structure subjected to harmonic base excitation," *Journal of Sound and Vibration*, vol. 307, no. 1-2, pp. 2–19, 2007.

[17] K. J. L. Fegelman and K. Grosh, "Dynamics of a flexible beam contacting a linear spring at low frequency excitation: experiment and analysis," *Journal of Vibration and Acoustics, Transactions of the ASME*, vol. 124, no. 2, pp. 237–249, 2002.

[18] C. P. Vyasarayani, J. McPhee, and S. Birkett, "Modeling impacts between a continuous system and a rigid obstacle using coefficient of restitution," *Journal of Applied Mechanics, Transactions ASME*, vol. 77, no. 2, Article ID 021008, 2010.

[19] K. D. Murphy and T. M. Morrison, "Grazing instabilities and post-bifurcation behavior in an impacting string," *Journal of the Acoustical Society of America*, vol. 111, no. 2, pp. 884–892, 2002.

[20] Q. H. Qin and X. Q. He, "Variational principles, FE and MPT for analysis of non-linear impact-contact problems," *Computer Methods in Applied Mechanics and Engineering*, vol. 122, no. 3-4, pp. 205–222, 1995.

[21] I. Huněk, "On a penalty formulation for contact-impact problems," *Computers and Structures*, vol. 48, no. 2, pp. 193–203, 1993.

[22] C. Budd and F. Dux, "Chattering and related behaviour in impact oscillators," *Philosophical Transactions: Physical Sciences and Engineering*, vol. 347, no. 1683, pp. 365–389, 1994.

[23] J. P. Cusumano and B. Y. Bai, "Period-infinity periodic motions, chaos, and spatial coherence in a 10 degree of freedom impact oscillator," *Chaos, Solitons and Fractals*, vol. 3, no. 5, pp. 515–535, 1993.

[24] F. C. Moon and S. W. Shaw, "Chaotic vibrations of a beam with non-linear boundary conditions," *International Journal of Non-Linear Mechanics*, vol. 18, no. 6, pp. 465–477, 1983.

[25] S. W. Shaw, "Forced vibrations of a beam with one-sided amplitude constraint: theory and experiment," *Journal of Sound and Vibration*, vol. 99, no. 2, pp. 199–212, 1985.

14

Frequency Domain Compressive Sampling for Ultrasound Imaging

Céline Quinsac, Adrian Basarab, and Denis Kouamé

University of Toulouse, IRIT UMR 5505 CNRS, 31062 Toulouse Cedex 9, France

Correspondence should be addressed to Adrian Basarab, basarab@irit.fr

Academic Editor: Erdal Oruklu

Compressed sensing or compressive sampling is a recent theory that originated in the applied mathematics field. It suggests a robust way to sample signals or images below the classic Shannon-Nyquist theorem limit. This technique has led to many applications, and has especially been successfully used in diverse medical imaging modalities such as magnetic resonance imaging, computed tomography, or photoacoustics. This paper first revisits the compressive sampling theory and then proposes several strategies to perform compressive sampling in the context of ultrasound imaging. Finally, we show encouraging results in 2D and 3D, on high- and low-frequency ultrasound images.

1. Introduction

Ultrasound (US) imaging acquisition, like all other imaging modalities, relies on Shannon's theorem. This theorem states that, in order to reconstruct perfectly a signal, its sampling frequency must be at least twice the highest-frequency component present in the signal. Often, US devices use a sampling rate that is at least four times the central frequency (wideband), to guarantee Shannon's theorem. However, the volume of data obtained is large, especially in 3D imaging, and can impair imaging in real-time or data transfer [1–4].

Compressed sensing (CS) or compressive sampling is a novel theory aiming to reduce the volume of data acquired, below the one dictated by Shannon's theorem. First introduced by Candès et al. and Donoho in 2006 [5, 6], CS guarantees the reconstruction of a signal from far fewer samples than usually necessary. CS has led to many applications, including medical imaging (particularly in magnetic resonance imaging (MRI) and tomography), sampling the spatial or frequency domains [7–9]. In ultrasound imaging, a few groups proposed very recently preliminary works for adapting the compressive sampling framework to ultrasound imaging [10–15], to ultrasound Doppler [16, 17], or to photoacoustic tomography [18].

The success of this reconstruction lies on two principles inherent to CS: sparsity and incoherence.

Sparsity reflects the ability of a signal to be compressed. A signal that has a sparse representation in a given basis will have most of its coefficients (or entries) null or very close to zero in this representation. Hence, by suppressing those negligible coefficients, the signal can be compressed, that is, reconstructed from relatively few samples [19]. Of course, during data acquisition, there is no knowledge about which coefficients are significant and which are not. CS overcomes this issue using a sampling basis incoherent with the sparsity basis.

Incoherence in CS expresses the idea that signals that are sparse in a given basis cannot be sampled in this basis but in another where the signal is dense. This property guarantees that the samples acquired contain the same amount of information. If the sparse basis was sampled, there would be a risk of acquiring negligible coefficients, not participating to the signal reconstruction.

The challenge of CS is to design a sampling protocol that will capture the information contained on the relatively few coefficients of the sparse basis. This sampling protocol will also suit any type of images within a specific application, here US imaging, without prior knowledge on the signal or image to sample.

By means of optimization methods, the original signal can then be recovered from those few measurements and the reconstruction quality will be similar to the one obtained respecting Shannon's theorem.

In summary, CS is a simple acquisition method where only a few samples of a signal are blindly measured. The full signal is later retrieved using reconstruction methods.

The purpose of this paper is twofold: first, make CS familiar to US imaging, and second, show the mechanisms involved in a successful CS reconstruction. The structure of the paper includes a reminder of the theory of sampling. Then, an overview of the CS theory and its components (essentially sparsity, incoherence, and optimization) will be given in detail and illustrated in a US context. Then, an application to US imaging will be proposed and results of US image reconstruction from usually 25%, 33%, and 50% of samples acquired will be shown. Finally, perspectives for CS in US will be drawn and future work will be described.

2. Sampling a Signal

The general framework of sampling can be summarized by measuring linear combinations of an analog signal $f(t)$, possibly considered as projections on a given basis:

$$y_k = \langle \varphi_k, f \rangle, \qquad \text{for } k = 1, \ldots, m, \tag{1}$$

where $\langle \cdot, \cdot \rangle$ denotes an inner product, y_k are the measurements, φ_k are sampling vectors, and m is the number of measurements. The most common sampling protocol consists of vectors φ_k of Diracs at equal time laps (ideal sampling). The measurements obtained represent then a simple discretization of $f(t)$.

However, if the sampling vectors φ_k are complex exponentials, then the measurements y_k are Fourier coefficients. This sampling protocol is used in MRI, for example.

In CS, the number of measurements m is far below the criteria established by Shannon's theorem for a given signal duration. If f is a digital signal of size n (respecting Shannon's theorem), then $m \ll n$. This situation can rise with slow imaging devices, for example, or when the number of sensors is limited. When it is possible to sample signals respecting Shannon's theorem, like in US imaging, it might be more advantageous to reduce the volume of data or the acquisition time.

However, when the number of measurements is smaller than the signal size, then we are facing an ill-posed inverse problem. If Φ is a matrix of size $m \times n$, concatenating the sampling vectors φ_k, then $y = \Phi f$. When we want to recover the signal $\hat{f}$ corresponding to the measurements y, then there is an infinity of solutions possible.

CS shows that it is possible to recover $\hat{f}$, provided that it has a *sparse* representation in a given basis and that the measurements are *incoherent* with that basis [20]. The following two sections explain those two concepts.

3. The Concept of Sparsity

Sparsity is the idea that signals may have a concise representation in a given basis. For example, a signal composed of three sinusoids will be sparse in the Fourier domain as its representation in this domain is very concise: namely 12 coefficients (6 symmetrical magnitudes and phases). Hence a dense signal in the time domain can be coded with only a few samples.

Other examples include photographic images. On the image, almost all the pixels have a nonzero value. However, in the wavelet domain, these images are sparse; that is, they contain a majority of null or very small coefficients. By discarding those negligible samples, an approximation of the original image can be obtained, with minimal loss of information. Usually, this loss of information is not noticeable: that is, the concept of JPEG2000 compression [19].

Mathematically, it translates as follows:

$$f(t) = \sum_{i=1}^{n} x_i \psi_i(t), \tag{2}$$

where $f(t)$ is the original signal, x_i are the coefficients of the signal in the sparse basis, and $\psi_i(t)$ is an orthonormal basis (Fourier or wavelets e.g.). The S largest coefficients x_i are noted x_S, and the corresponding signal $f_S(t)$. If $f(t)$ is sparse in the basis Ψ composed of the vectors ψ_i, then $f = \Psi x$ and the error $\|f - f_S\|_2$ is small.

Figure 1 illustrates the sparsity of radio-frequency (RF) US signals. Because the sparse representation of the RF US signal (here, its Fourier transform) concentrates the information on a few coefficients, it is possible to reconstruct almost perfectly the signal from only 30% of its largest Fourier coefficients (keeping only 30% of the largest x_i in (2)).

This concept of concentrated information is also visible when plotting the Fourier coefficients in order of magnitude (Figure 2). If they decay rapidly, then the compressed signal $f_S(t)$ including the S largest coefficients will be close to the original signal $f(t)$.

Sparsity therefore leads to the compressive nature of a signal: if a signal has a sparse representation, then the information coding that signal can be compressed on a few coefficients. A reconstruction from those few coefficients can be obtained with minimal loss compared to the original signal. Note however that CS and compression are different in that when sampling a signal, it is impossible to directly acquire the significant coefficients as their positions are not known *a priori*. CS overcomes this issue via an incoherent sampling.

4. Incoherent Sampling

The term incoherent sampling conveys the idea that the sampling protocol φ_k in (1) has to be as little correlated as possible with the sparse representation ψ_i in (2). This requirement prevents from the risk of sampling insignificant bits of information (the close-to-zero coefficients described in Section 3). Instead, the idea of an incoherent sampling is to introduce noise-like interferences to the signal to recover.

The mathematical definition of incoherence is [21, 22]:

$$\mu(\Phi, \Psi) = \sqrt{n} \max_{1 \le k, j \le n} \left| \left\langle \phi_k, \psi_j \right\rangle \right|, \tag{3}$$

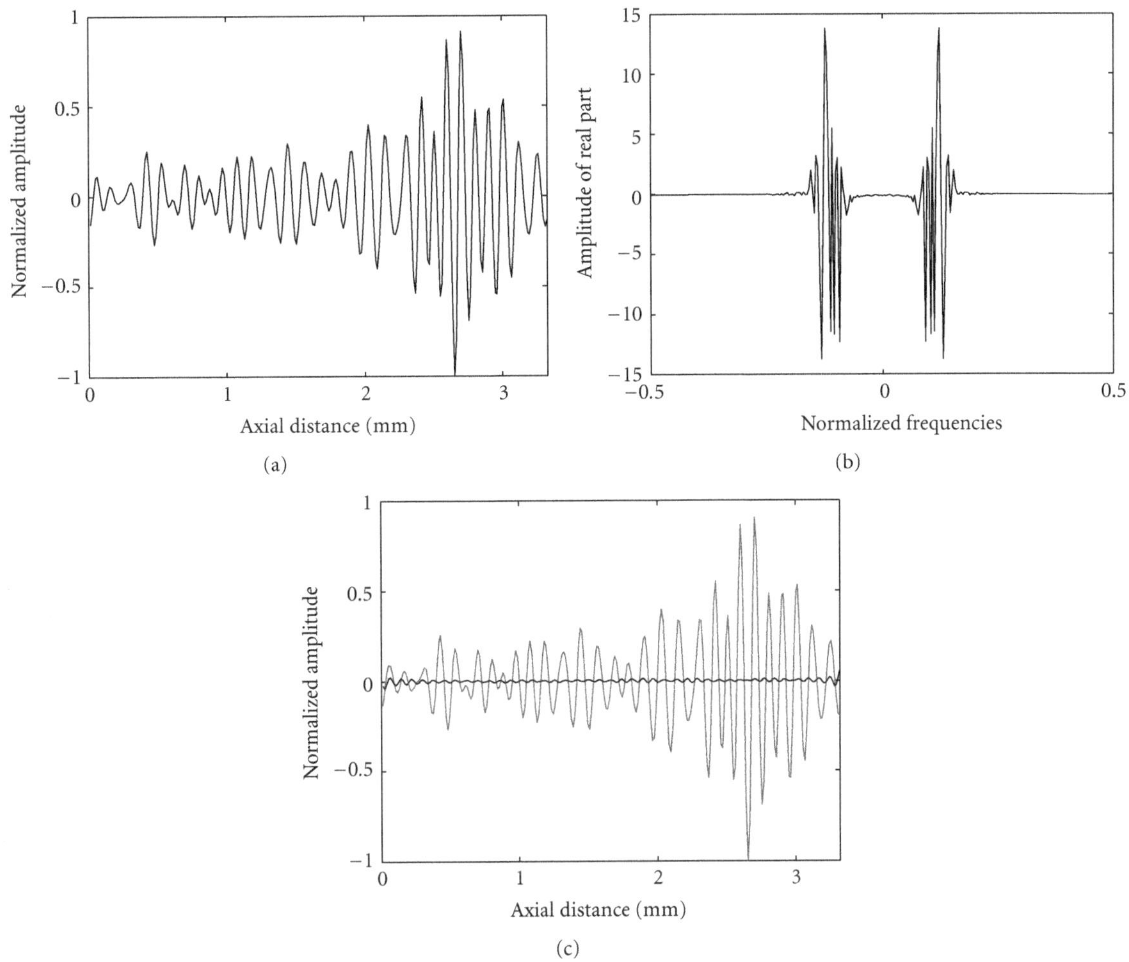

FIGURE 1: (a) A full US RF signal and (b) its sparse representation via Fourier transform. Most of the coefficients are equal or close to zero. (c) Compressed US RF signal (gray), corresponding to 30% of the largest Fourier coefficients, the rest of them being set to zero. The difference between the full and compressed US RF signal (black) is minimal.

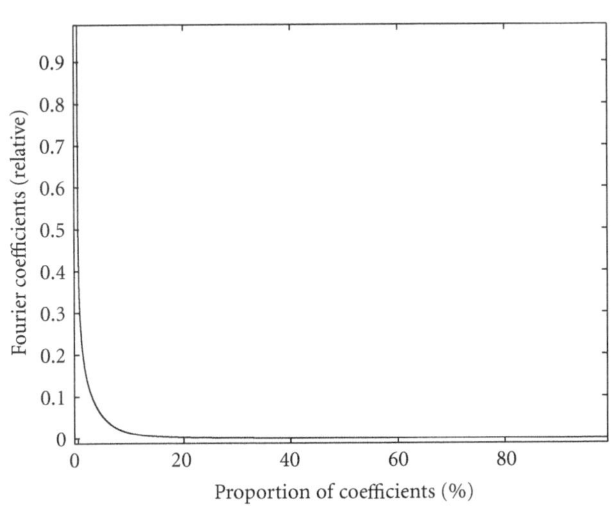

FIGURE 2: Ordered relative values of a thyroid *in vivo* US image Fourier magnitude coefficients (see Figure 13). They decay very rapidly, indicating a sparse representation of the image by Fourier transform.

where Φ is the sampling basis and Ψ is the sparsifying basis. According to (3), if the two bases are strongly correlated, then μ will be close to $\sqrt{n}$, and if they are not correlated at all, then it will be close to 1. CS requires a low coherence between the bases. In other words, incoherence is guaranteed provided that the two bases are not correlated.

Pairs of bases with minimum coherence include, for example, a basis of Diracs associated with a Fourier basis (a spatial Dirac contains information about all the frequencies). In addition, if the sampling basis is completely random, then it will be maximally incoherent with any sparsifying fixed basis (wavelets, curvelets, etc.) [21].

Figure 3 shows an example of coherent and incoherent samplings of a US RF signal. When the sampling is incoherent with the sparsifying basis (Figures 3(c) and 3(d)), then the measurements (namely, in practical situations) in that basis are dense (by opposition to sparse). The original sparse signal is polluted by noise-like interferences and can be reconstructed by optimization. However, when the sampling basis and the sparsifying basis are coherent (Figures 3(e) and 3(f)), the measurements in the sparsifying basis are

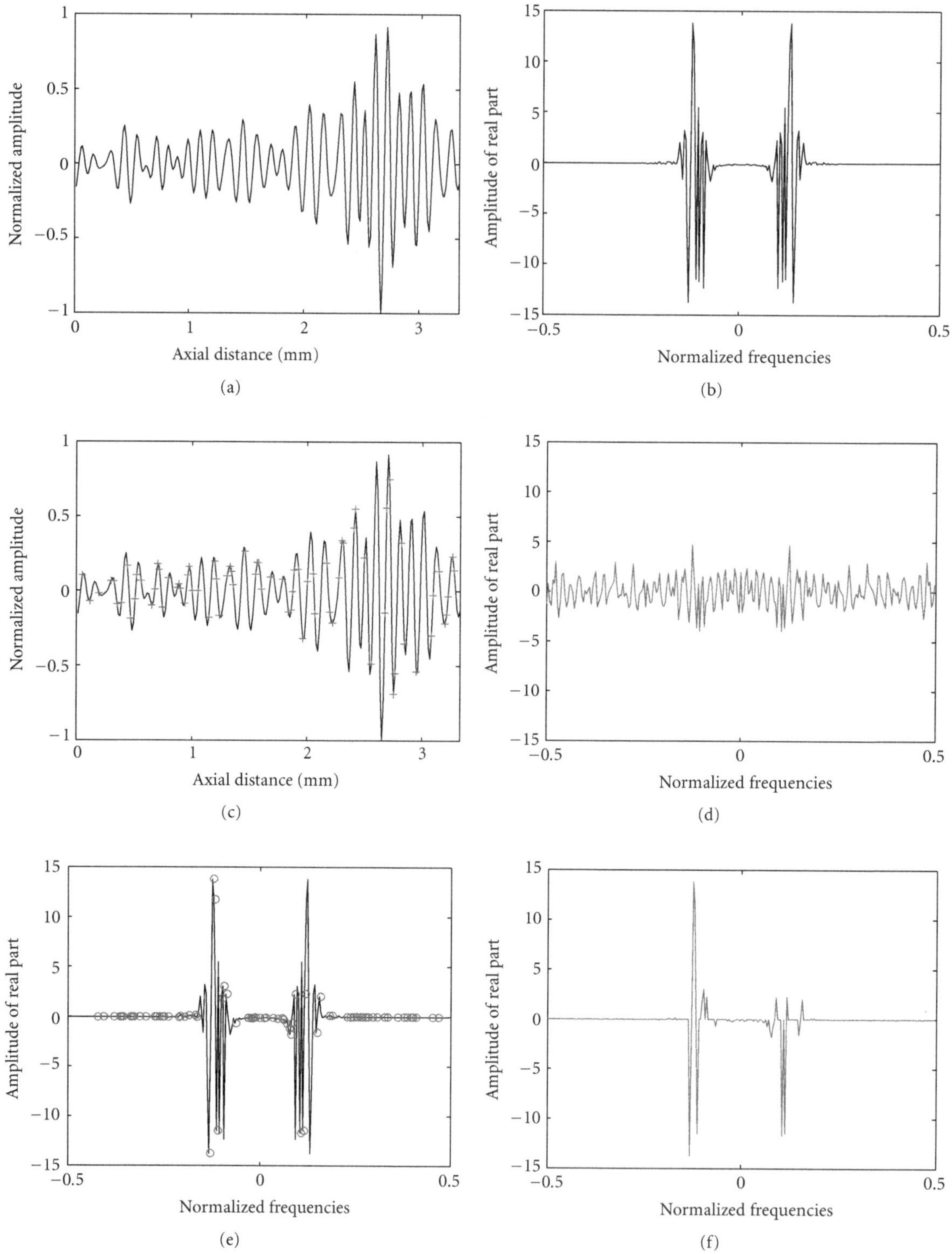

FIGURE 3: (a) A full US RF signal and (b) its sparse representation via Fourier transform. (c) Sampling 30% of the US RF signal using a basis of random Diracs (crosses), maximally incoherent with the Fourier basis, leads to (d) a dense signal in the sparsifying basis (Fourier transform). However, (e) sampling 30% of the US RF signal using a basis of random sinusoids (circles), coherent with the sparsifying basis, is equivalent to sample random Fourier coefficients of the signal. (e) The resulting measurements in the sparsifying basis (Fourier transform) are sparse as well.

themselves sparse. There is significant information (large Fourier coefficients) missing and CS will not be able to recover the original signal.

From those incoherent measurements and knowing the sparsifying basis, CS theory states that it is possible to recover the original signal using an optimization routine.

5. Signal Reconstruction through Optimization

Knowing that the signal to recover $\hat{f}$ has a sparse representation x in a given basis Ψ, it is possible to reconstruct it from the incomplete incoherent measurements y obtained using the sampling basis Φ. This reconstruction is performed via a convex optimization program:

$$\min \|\hat{x}\|_1 \quad \text{subject to } y = \Phi f = \Phi\Psi\hat{x}, \tag{4}$$

where $\hat{x}$ is the reconstructed sparse signal and $\|\cdots\|_1$ denotes the ℓ_1 norm.

This optimization searches amongst all the signals that verify the measurements y, the one with the smallest ℓ_1 norm, that is, the sparsest. The choice of the ℓ_1 norm (sum of magnitudes) over the ℓ_0 norm (size of support) is mainly practical: while solving of the ℓ_0 norm minimization is computationally infeasible, the ℓ_1 norm minimization can easily be recast as a linear program. The ℓ_2 norm (sum of magnitudes squared) is unsuited to CS because the minimization would not recover the sparsest signal [23].

In other words, the optimization routine (4) removes the interferences caused by the incoherent undersampling from the sparse representation of the measurements (as in Figure 3(d)). Figure 4 illustrates the process of ℓ_1 norm minimization applied to the Fourier transform of an RF signal: the significant coefficients tend to be amplified while the others, corresponding to the interferences, are attenuated.

In this example, at the last iteration, the recovered signal is exactly equal to the original signal. Generally speaking, recovering the signal is true with overwhelming probability if the number of measurements m follows:

$$m \geq C \cdot \mu^2(\Phi, \Psi) \cdot S \cdot \log n, \tag{5}$$

where C is a positive constant, μ is the coherence as defined in (3), S is the degree of sparsity, and n is the signal size. From (5) it follows that the number of measurements depends on the sparsity S of the signal (the sparsest, the best) in a given basis and the coherence μ of the sampling protocol with that basis.

In practice, many researchers observed that accurate reconstructions can be achieved if the number of measurements m is roughly 2 to 5 times the sparsity S of the signal [24–26]. The work herein matches this statement. For an offline reconstruction, plots in Figure 2 could be used to determine the degree of sparsity and consequently the minimum number of measurements required. For online reconstruction, priors on the US imaging device bandwidth could be exploited. Classic US scanners bandwidth ranges from 50% to 100% or more (depending on the scanner) and is practically estimated at 3 or 6 dB attenuations. An example of an experimental PSF together with its Fourier transform showing the bandwidth at 6 dB is given in Figure 5. Thus, we can observe that, taking into account the device bandwidth, the k-space may be considered even sparser than the impression given by Figure 2. The sparsity could then be set as the number of significant coefficients in the practical bandwidth, or two times this number.

The optimization routine utilizes no prior knowledge about the positions or amplitudes of the sparse coefficients or about the signal to recover.

In practical situations, the measurements y are often corrupted by noise η, originating from the instrumentation. Therefore, the term guaranteeing data consistency in (4) has to be relaxed, so that $y = Ax+\eta$. In addition, the signal might not have an exact sparse representation but an approximate sparse representation where very small but not exactly nil coefficients will be neglected. Again, this approximation will introduce some noise.

In those cases, the CS method will still allow a reconstruction of the signal, provided that the CS matrix A ($A = \Phi\Psi$) respects the Restricted Isometry Property (RIP) [21, 27]:

$$(1 - \delta_S)\|x\|_2^2 \leq \|Ax\|_2^2 \leq (1 + \delta_S)\|x\|_2^2, \tag{6}$$

where S are integers and δ_S is the isometry constant. A obeys (6) when the smallest δ_S that verifies (6) for all S-sparse signals x is not too close to 1. In other words, if A obeys the RIP, then the Euclidean lengths, or norms, will be preserved in A: this is the isometry. This property basically ensures that a sparse signal x will not fall in a null space in A, where it would be impossible to recover.

If (6) is true, then the minimization will allow an accurate reconstruction of the signal. More precisely, the reconstruction of an approximately sparse signal will approach the corresponding compressed signal.

Interestingly, random matrices obey the RIP with overwhelming probability if $m \geq C \cdot S \cdot \log(n/S)$ with C being a constant.

The minimization in the case of noisy data is as follows:

$$\min \|\hat{x}\|_1 \quad \text{subject to } \left\|A\hat{x}_y\right\|_2 \leq \varepsilon, \tag{7}$$

where the fidelity of the measurement constraint is relaxed to take into account the level of noise ε. This is again a convex minimization, computationally feasible.

6. Sampling Protocols in Ultrasound Imaging

The sampling protocols in US imaging are designed to fulfill both the requirements of CS and of the US instrumentation. The CS theory has been described in the previous section and, from this perspective, the sampling basis mainly has to be incoherent with the sparsifying basis. The US imaging devices have physical constraints that limit the sampling strategies one can adopt for CS.

The data acquisition in US imaging is performed in the image space (spatial domain), unlike MRI, for example

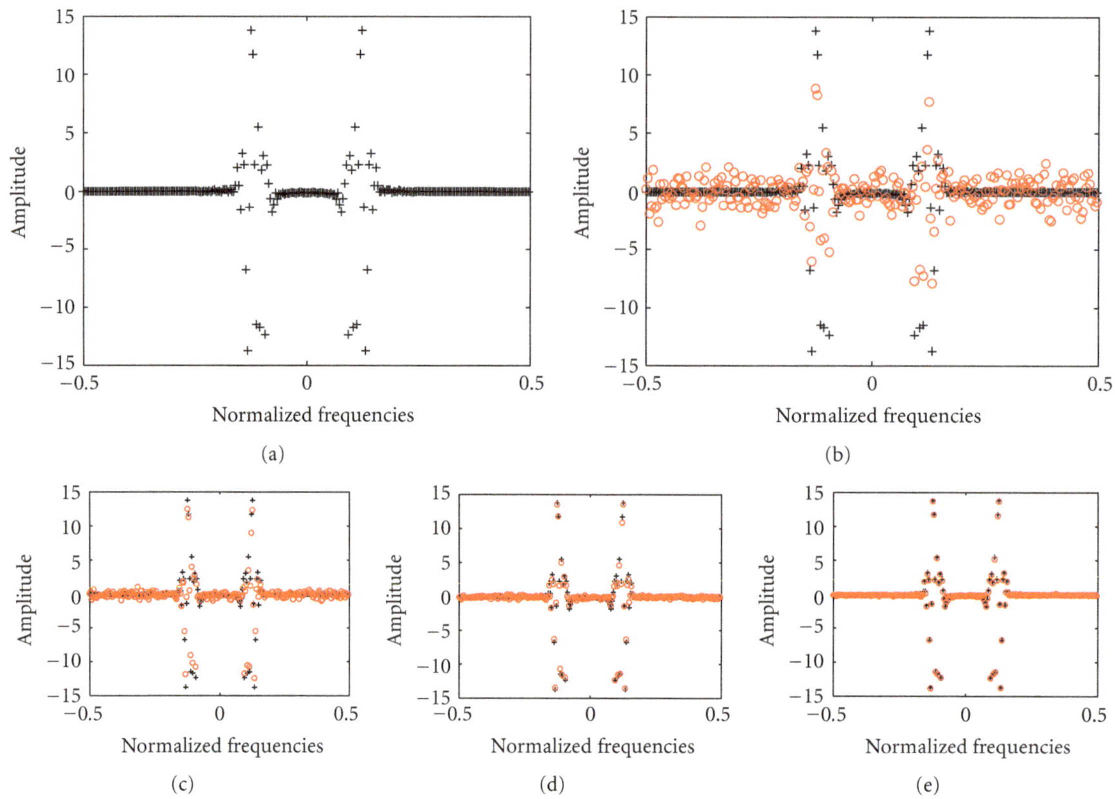

FIGURE 4: (a) Fourier transform (real part) of an RF signal (256 points) and (b) the initial guess (circles) of the ℓ_1 norm minimization for 128 random measurements. (c) Evolution of the reconstructed sparse signal using a primal-dual algorithm for linear programming at iterations 3, (d) 5 and (e) 11. After 11 iterations the reconstructed signal corresponds exactly to the original.

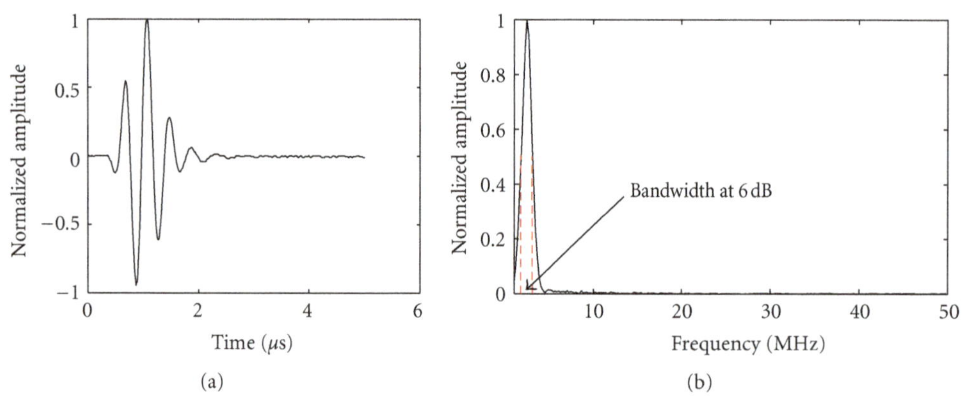

FIGURE 5: (a) Experimental PSF. (b) Fourier transform of the PSF (only positive frequencies are plotted), showing the bandwidth at 6 dB and reflecting the sparsity of the ultrasound k-space.

[7, 24]. There are several possible sampling protocols adapted to US imaging and incoherent with the sparsifying basis. They all consist in taking samples of that image at more or less random locations. This is equivalent to taking samples at specific times on the RF signals or taking RF lines at specific locations.

In this paper, eight different sampling protocols are proposed: three 2D masks and five 3D masks. In two dimensions, the CS uniformly random mask, denoted Φ_1 and shown in Figure 6(a), will be studied on different types of RF US images. This sampling protocol is maximally incoherent with the Fourier transform (considered as the sparse decomposition basis in this paper) and therefore should give the best results.

However, switching rapidly from one position to the next in this kind of sampling pattern might be difficult from the instrumentation point of view. Consequently, two other sampling masks denoted by Φ_2 and Φ_3 where studied

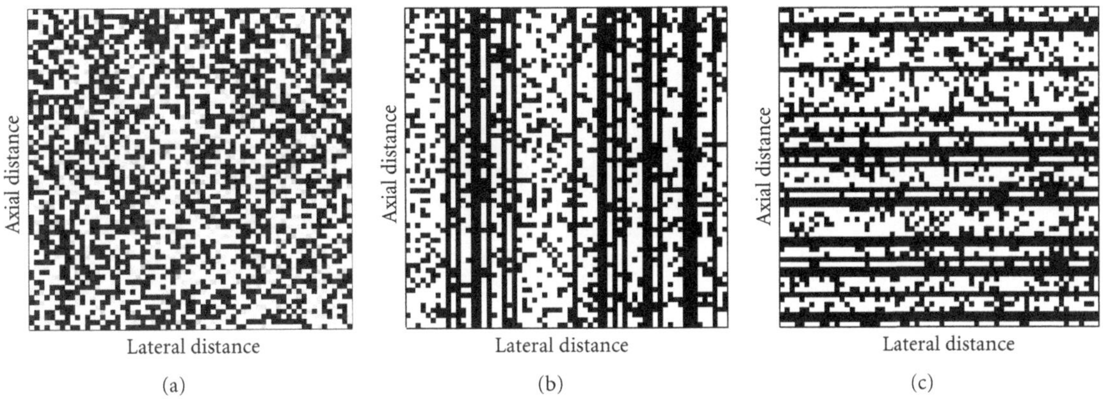

FIGURE 6: Sampling masks Φ_1 (a), Φ_2 (b), and Φ_3 (c) adapted to a spatial sampling of the US images. The white pixels correspond to the samples used for CS. The proportion of samples here is 50% of the original image.

were whole lines or columns of the images are not sampled at all (Figures 6(b) and 6(c), resp.). The sampled lines or columns are chosen in a uniformly random fashion. On the remaining lines or columns, random points are sampled. The total number of points sampled and chosen was the same as for the sampling mask Φ_1 to be able to compare the quality of the CS reconstructions. The sampling masks Φ_2 and Φ_3 are slightly less incoherent than Φ_1 (due to a certain coherence in the direction that is not sampled at all), so the results are expected to be worse than for Φ_1. However, these two strategies could translate as a gain of time from the instrumentation point of view as some lines (resp., columns) of the image will not be acquired at all.

For 3D US datasets, five sampling strategies are proposed. The first, similar to Φ_1, and maximizing incoherence is a uniformly random mask in three directions, denoted Θ_1 (Figure 7(a)).

The other four, Θ_2 to Θ_5, are inspired by Φ_2 and Φ_3, that is, sampling only certain RF lines. Whereas Θ_2 and Θ_3 consist in sampling different RF lines or columns on each slice of the azimuthal direction (see Figures 7(b) and 7(c)), with Θ_4 and Θ_5, the set of unsampled RF lines or lateral profiles is always the same in each slice (see Figures 7(d) and 7(e)). Consequently, with Θ_4 (resp., Θ_5), some whole axial-azimuthal (resp., lateral-azimuthal) plans of the volume are not sampled.

7. Reconstructions of Ultrasound Images and Volumes

In US imaging, the acquisition consists in taking samples of the image (or of the RF signals). This sampling protocol is similar to a basis of Diracs (sampling mask). To guarantee the success of the CS reconstruction, a basis incoherent with Diracs and where the US images are sparse is needed. The basis chosen in this paper is the Fourier basis as it is maximally incoherent with Diracs and because the US image k-space is sufficiently sparse. The function to minimize is

$$\arg\min_{M} \|AM - y\|_2 + \lambda |M|_1, \qquad (8)$$

where M is the k-space of the US RF image m ($M = \mathcal{F}m$), and A is the sampling scheme ($A = \Phi\mathcal{F}^{-1}$ here, where Φ corresponds to Φ_k ($k = 1,\ldots,3$), the RF random sample locations in 2D, or to Θ_k ($k = 1,\ldots,5$) in 3D). $\mathcal{F}^{-1}$ stands for the inverse Fourier transform, y are the RF US image measurements and λ is a coefficient weighting for sparsity.

Other bases of sparsity such as the wavelet transform of the US image k-space have been investigated in previous work and give similar results to those presented here [12].

The first term of (8) represents the fidelity of the measurements and the second term guarantees the signal sparsity in the Fourier basis. The balance between those two terms is given by λ. The choice of λ is crucial to a good reconstruction as it corresponds to a threshold for the recovery of significant coefficients in the sparse basis.

In this paper, the optimal λ chosen was the one giving the minimum errors of reconstruction. However, finding the optimal λ by trials and errors is obviously not possible during acquisition where no comparison with the real signal can be performed.

One solution to minimize the fluctuations in the CS reconstructions due to a poor choice of lambda is to add an elastic-net regularization, that is, ℓ_2 minimization term on the sparse coefficients [28]. The resulting reconstruction errors around the optimal λ were however quite large compared to the minimum errors found with only the ℓ_1 minimization.

Another method, based on an adapting λ and called reweighted ℓ_1 minimization, has been developed [29] to address that issue. It consists in performing (8) with an initial λ (equal to one e.g.). Then, (8) is reiterated using a new λ, which value is calculated from the results of the first iteration of (8). Thus, the optimization performed is given as follows:

$$\arg\min_{M} \|AM - y\|_2 + \sum_i \left(w_i^{(\ell)} |M_i|_1 \right). \qquad (9)$$

Namely, the weighted λ at each iteration is $w_i^{(\ell+1)} = 1/(|M_i^{(\ell)}| + \varepsilon)$ where ℓ is the iteration number, $M_i^{(\ell)}$ are the sparse coefficients estimated after ℓ iterations, and ε is a coefficient ensuring stability (that should be slightly

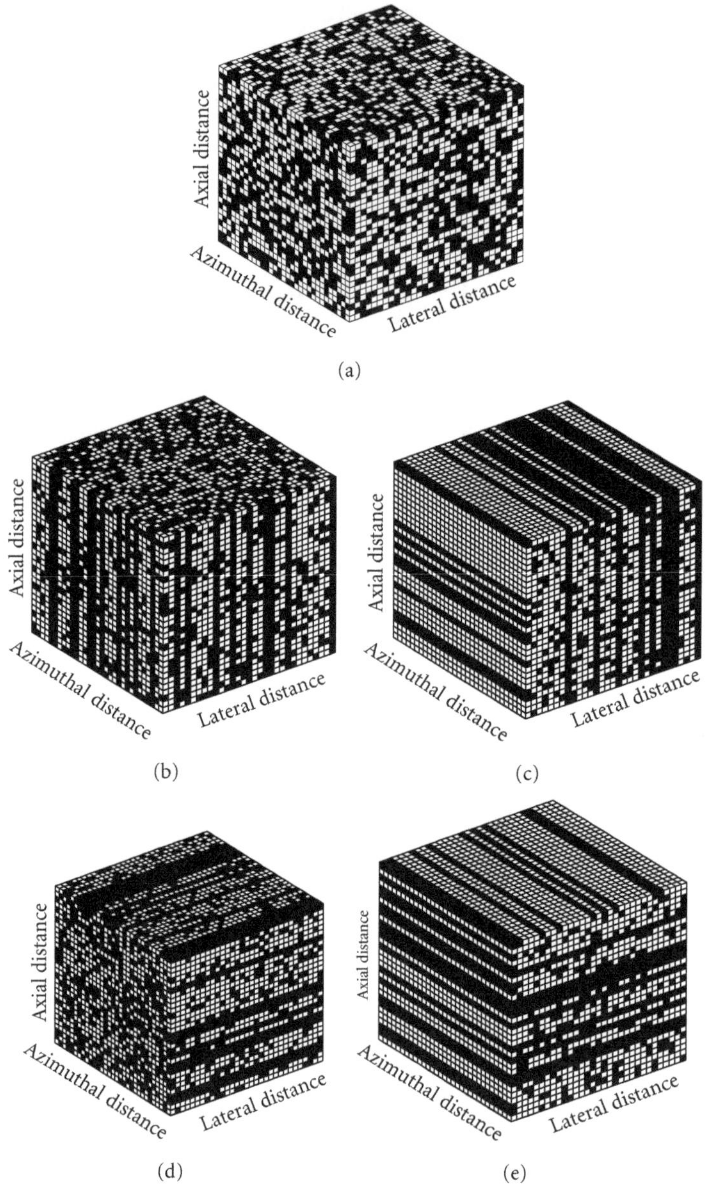

FIGURE 7: Sampling masks Θ_1 (a), Θ_2 (b), Θ_3 (c), Θ_4 (d), and Θ_5 (e) adapted to a spatial sampling of the 3D US volumes. The white pixels correspond to the samples used for CS. The proportion of samples here is 50% of the original volume.

smaller than the smallest nonnil sparse coefficient). At the first iterations, all w_i coefficients are set to 1 ($w_i^{(0)} = 1$, for all i). In this setting, the choice of lambda is replaced by the choice of ε. However, we observed that the results of the CS reconstructions were a lot less dependant on ε than λ. A major drawback of the reweighted ℓ_1 minimization method is its iterative process. Thus, at least two classic optimizations are performed (corresponding to a minimum of two iterations) in order to get an accurate result.

In Section 8, the two techniques of CS reconstruction: using a fixed optimal λ and the reweighted ℓ_1 minimization were compared for the sampling patterns described in Section 6. In both cases, a nonlinear conjugate gradient algorithm was used for numerical optimization. This algorithm is particularly suited to large-scale data and is used for similar convex optimization problems (see, e.g., [30]).

In addition, different undersampling ratios were tested.

8. Results on a 2D Simulation Image

The CS strategy described in (8) was used to reconstruct the k-space of an RF image simulated using the Field II US simulation program [31]. The parameters and the example scatterer map of a kidney used were as follows Transducer centre frequency = 5 MHz, sampling frequency = 20 MHz, number of scatterers = 50,000, number of RF lines generated = 256, size of the object = 88 × 147 mm. The image was then cropped to a 1792 by 128 size matrix.

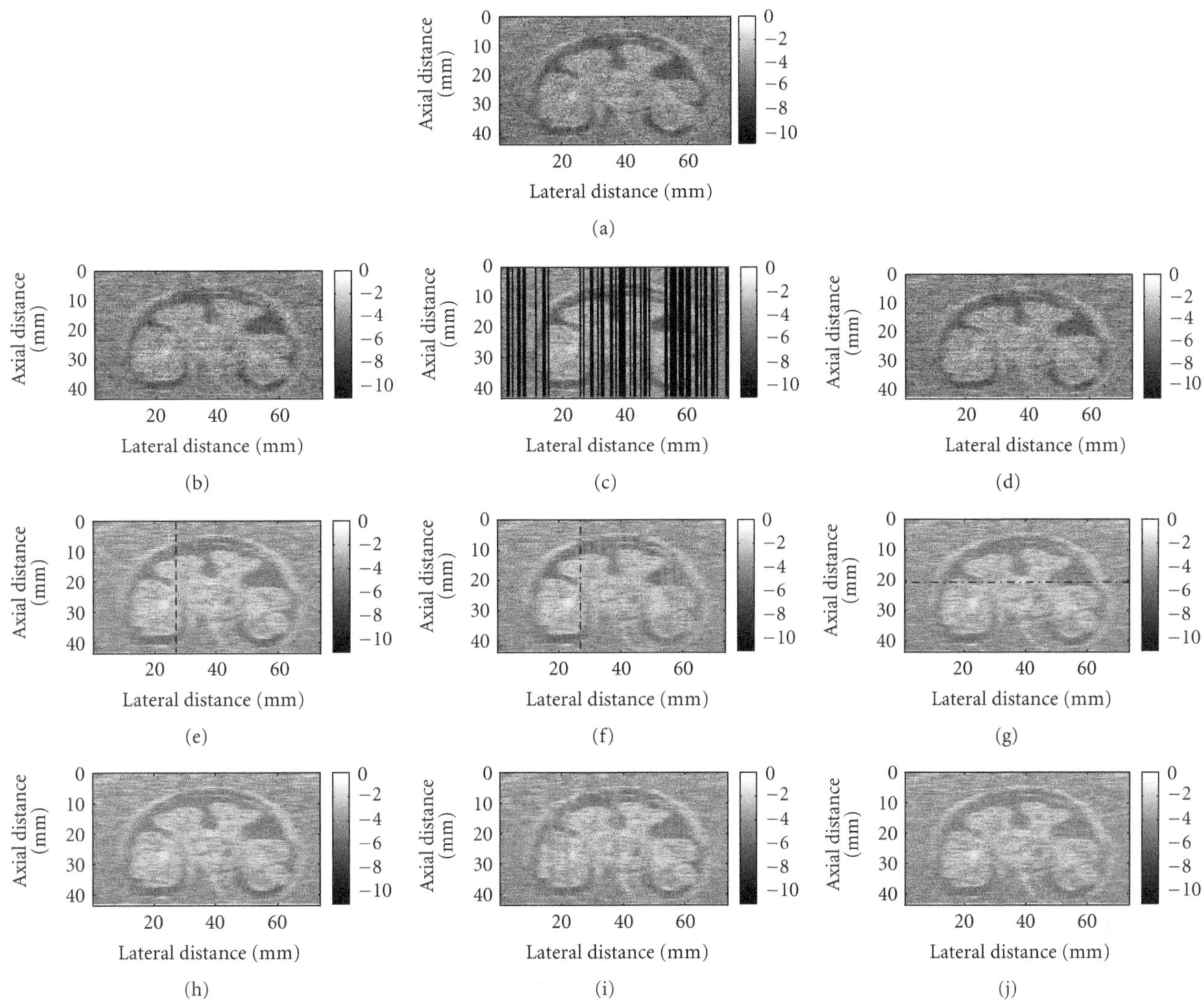

FIGURE 8: CS on a simulated US image (a) using a two-dimensional random sampling pattern Φ_1 (b) and two alternative sampling mask Φ_2 (c) and Φ_3 (d) using 33% of the samples. The B-mode images of the RF random measurements are shown on the second row. Those from the CS reconstructed k-spaces with the optimal $\lambda = 0.005$ are on the third row: (e) to (g), and those from the CS-reconstructed k-spaces with the reweighted ℓ_1 minimization on the fourth row: (h) to (j). The dotted lines correspond to the position of the RF signals plotted in Figures 9, 10, and 11.

The three different schemes of sampling, Φ_1, Φ_2, and Φ_3 (Figure 6), were studied to compare the CS reconstructions, using a fixed optimal λ set to 0.005 and the reweighted ℓ_1 minimization. The results are shown in Figure 8 for the three sampling patterns (using 33% of the samples).

For the classic 2D random sampling pattern Φ_1, one RF line of the reconstructed signal was plotted against the corresponding RF lines of the original signal and of the random measurements (Figure 9).

For the sampling pattern Φ_2, one RF line of the reconstructed signal was displayed in Figure 10, corresponding to a line that was not sampled at all (shown with the dash-dotted line on Figure 8(f)). RF lines that were partially sampled are very similar to the ones shown in Figure 9.

Figure 11 shows one lateral profile that was not sampled at all with the sampling mask Φ_3 (denoted by a dash-dotted line on Figure 8(g)). Again, the reconstruction quality of lateral profiles that were partially sampled was similar to Figure 9.

Table 1 shows the errors of CS reconstructions for the three sampling masks Φ_1, Φ_2, and Φ_3 at three undersampling ratios: 25%, 33%, and 50%.

For the reconstruction from Φ_1 (Figure 8(e)), the parts of the image that contained less signal, shown in black on the B-mode images, were less successfully reconstructed, but the diagnostic information was maintained. On the RF signals (Figures 9, 10, and 11), the amplitudes of the reconstructed signals were sometimes reduced. However, the differences in amplitude were constant and the timing information of the signal was maintained. Consequently, the CS-reconstructed images would give very close visualizations in B-mode and could also be used for tissue motion estimation or tissue

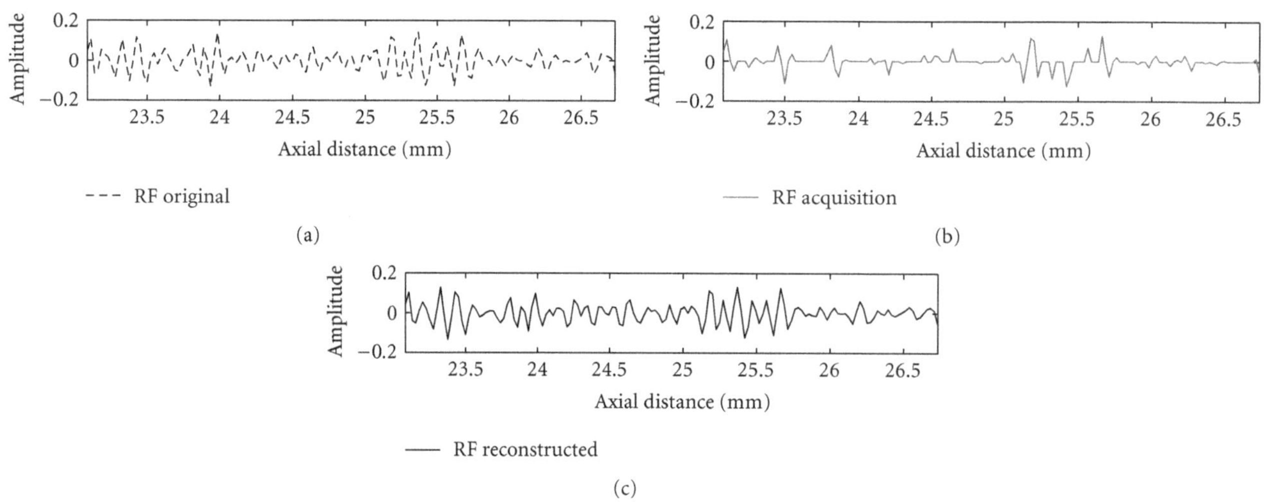

FIGURE 9: An example of a local region of an RF line after CS reconstruction using Φ_1 sampling pattern, corresponding to the dotted line in Figure 8(e).

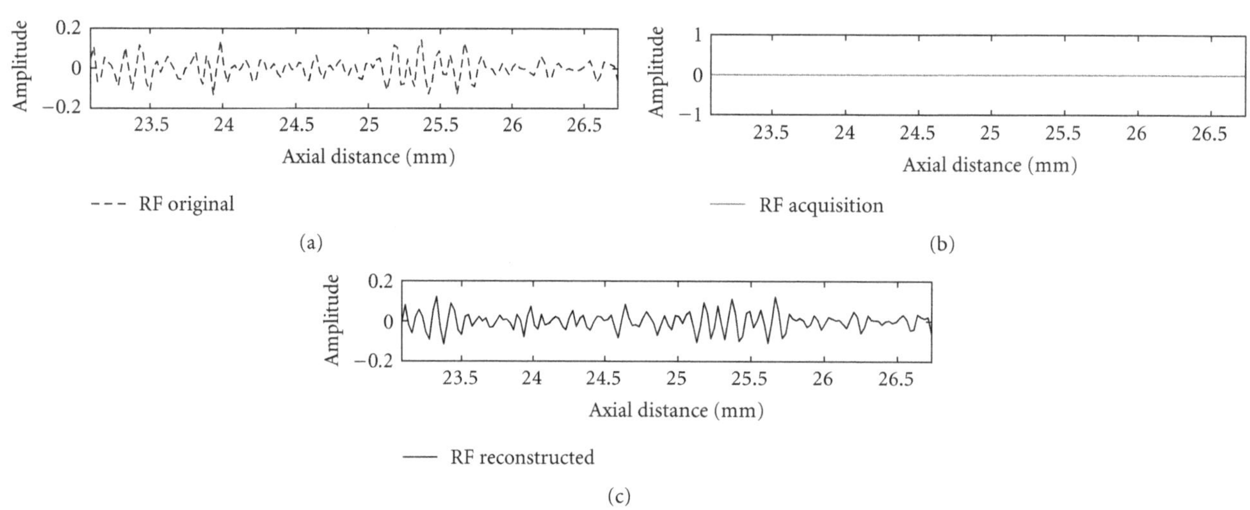

FIGURE 10: An example of a local region of an RF line after CS reconstruction using Φ_2 sampling mask, corresponding to the dotted line in Figure 8(f). This line was not sampled at all.

characterization. The errors of reconstruction increased with a smaller undersampling ratio as one would expect. In addition, the errors were always reduced when Φ_1 was used, due to a greater incoherence.

With Φ_2, the reconstruction of partially sampled RF lines was again very close to the original signal. For the RF lines that were not sampled at all (Figure 10), a good reconstruction was performed as well, showing the potential of CS for US imaging with reduced pulse emissions. The visualizations in B-mode were again very satisfactory in terms of diagnostic power.

When the sampling mask Φ_3 was used, partially sampled and unsampled lateral lines were well reconstructed (Figure 11). The overall CS reconstruction displayed in B-mode did not exhibit any line artifact.

Results obtained from the reweighted minimization (i.e., with an adaptive λ) were similar to those with the optimal λ, found experimentally.

9. Results on *In Vivo* 2D Images

In this section, US CS is performed using high- and low-frequency ultrasound images. These images were sampled *a posteriori* using CS.

First, results of a CS reconstruction using method (8) on *in vivo* images of the skin are shown. The central frequency was 20 MHz and the sampling frequency 100 MHz (ATYS Medical). Results from three different sampling patterns Φ_1, Φ_2, and Φ_3 (Figure 6) are shown in Figure 12.

The same CS method (8) was used to reconstruct a US image of the right lobe of a normal human thyroid.

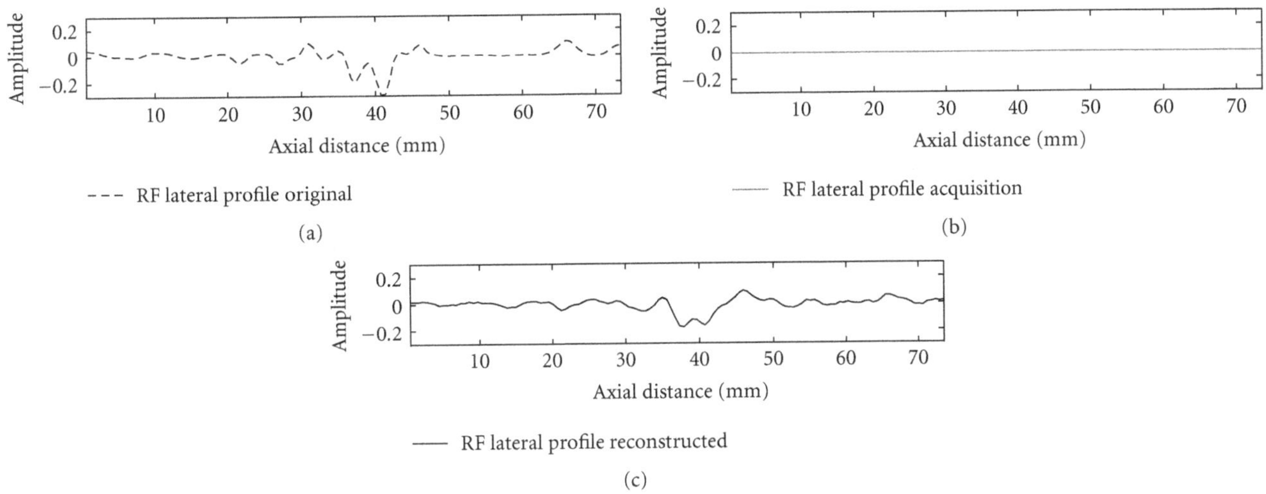

FIGURE 11: An example of a lateral profile of aN RF image after CS reconstruction using Φ_3 sampling mask, corresponding to the dotted line in Figure 8(g). This lateral profile was not sampled at all.

TABLE 1: NRMSE between the CS-reconstructed RF US image and the original simulated image of a kidney for different sampling ratios and patterns.

		Φ_1	Φ_2	Φ_3
Classic ℓ_1 minimization ($\lambda = 0.005$)	25%	0.48	0.58	0.6
	33%	0.33	0.45	0.49
	50%	0.15	0.28	0.29
Reweighted ℓ_1 minimization	25%	0.51	0.56	0.64
	33%	0.36	0.47	0.49
	50%	0.17	0.24	0.3

The imaging was performed using a clinical scanner that was modified for research with a 7.5 MHz linear probe (Sonoline Elegra, Siemens Medical Systems, Issaquah, WA, USA). The sampling frequency was adjusted to 40 MHz. For the image presented here, $\lambda = 0.01$ and 33% of the samples were measured *a posteriori*. The three different sampling patterns Φ_1, Φ_2 and Φ_3 (Figure 6) were used. The reconstructed US images of the thyroid are shown in Figure 13.

On *in vivo* US images, similarly to simulation images, the CS reconstruction using both sampling patterns was very good. The tissue structures were restored and the diagnostic information was maintained. Note that the results do not depend on the US frequency used. Tables 2 and 3 show the reconstruction errors for both *in vivo* images and for the different sampling schemes proposed.

10. Results on an *In Vivo* 3D Volume

In vivo US volumes of mouse embryos, acquired on anaesthetised mice, were reconstructed using the sampling masks described in Section 6 (Figure 7).

A single element-high-resolution scanner SHERPA, developed and commercialized by Atys Medical (Lyon, France) where RF data was available, was used (central frequency 22 MHz, frame rate 10 images per second, scanning width 16 mm, sampling frequency 80 M samples/second, emission frequency 20 MHz, exploration depth 7.8 mm). The volume was then cropped to a 128^3 size volume for illustration purposes.

The CS reconstruction of the volume was performed using (8) and $\lambda = 0.4$.

Figure 14 shows the CS reconstruction obtained from the sampling masks Θ_1, Θ_2, Θ_3, Θ_4, and Θ_5 for a 50% undersampling factor. Figure 14(a) represents the original volume and Figures 14(g), 14(h), 14(i), 14(j), and 14(k) the CS reconstructions obtained from each mask, whereas Figures 14(b), 14(c), 14(d), 14(e), and 14(f) are the measurements obtained from Θ_1 to Θ_5 masks, respectively.

The first observation to make is that for all the sampling masks, the CS method (8) provided good reconstructions of the whole volume from only 50% of the samples. The plan that was best reconstructed in each case was always the axial-lateral plan, where the 2D masks were applied (and then repeated along the azimuthal direction). However this setting could easily be changed for other applications where another plan is more crucial.

When the coherence increased, that is, from Θ_3 to Θ_2 to Θ_1 and from Θ_5 to Θ_4 to Θ_1, the reconstructions were degraded, as expected. This is particularly visible on the axial-azimuthal plans of Figures 14(g), 14(h), and 14(i). However, considering that absolutely no samples were kept for the axial-azimuthal plan visible on Figure 14(i), the result

TABLE 2: NRMSE between the CS-reconstructed RF US image and the original *in vivo* image of the skin for different sampling ratios and patterns.

		Φ_1	Φ_2	Φ_3
Classic ℓ_1 minimization ($\lambda = 0.01$)	25%	0.64	0.71	0.64
	33%	0.48	0.57	0.35
	50%	0.28	0.35	0.35
Reweighted ℓ_1 minimization	25%	0.65	0.67	0.73
	33%	0.51	0.65	0.61
	50%	0.38	0.44	0.43

TABLE 3: NRMSE between the CS-reconstructed RF US image and the original *in vivo* image of a human thyroid right lobe for different sampling ratios and patterns.

		Φ_1	Φ_2	Φ_3
Classic ℓ_1 minimization ($\lambda = 0.01$)	25%	0.34	0.63	0.38
	33%	0.22	0.53	0.26
	50%	0.12	0.38	0.13
Reweighted ℓ_1 minimization	25%	0.36	0.64	0.39
	33%	0.24	0.56	0.26
	50%	0.15	0.40	0.16

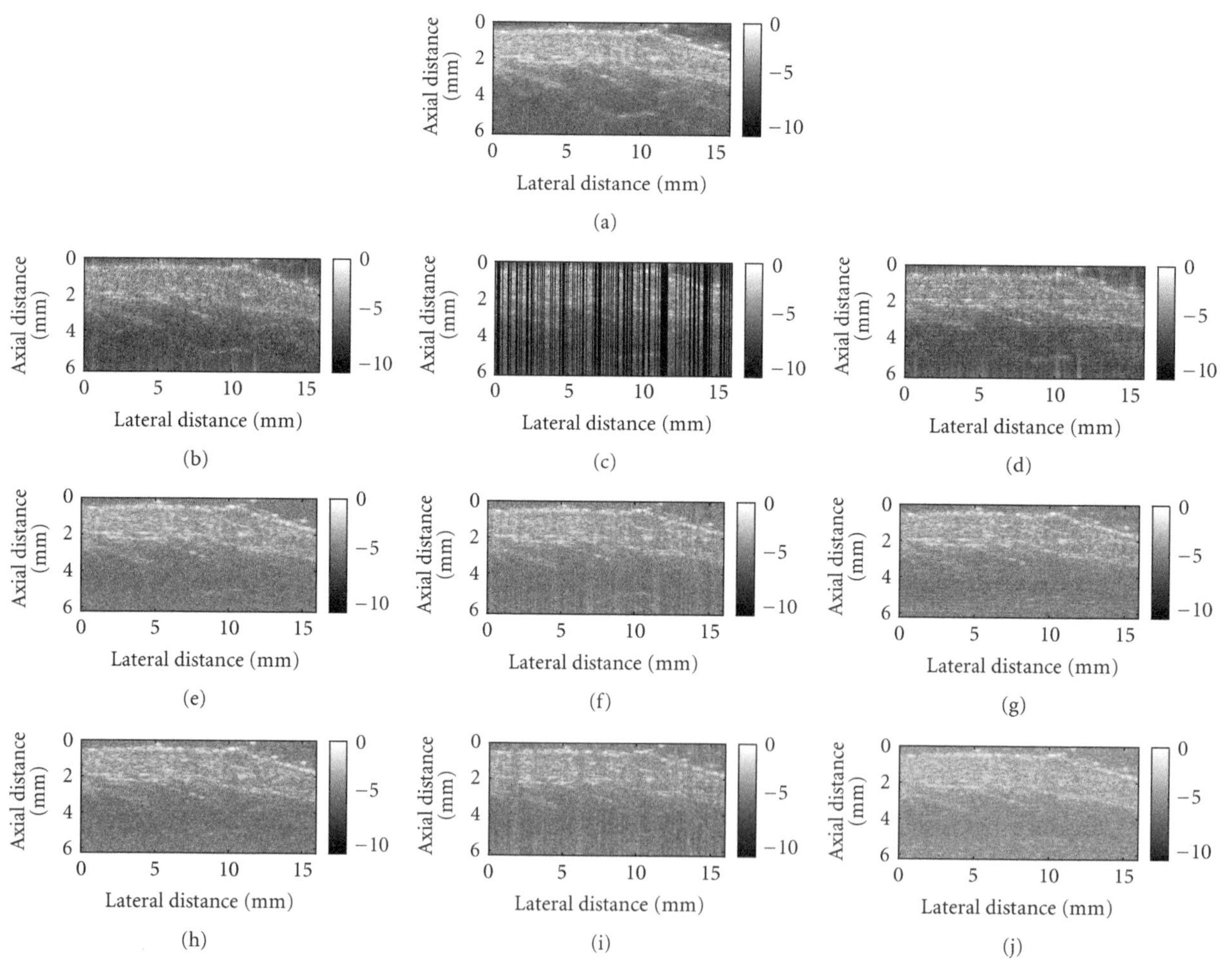

FIGURE 12: CS reconstruction on a US image of the skin (a) using a two-dimensional random sampling pattern Φ_1 (b) and two alternative sampling masks Φ_2 (c) and Φ_3 (d) and 33% of samples. The B-mode images of the samples are shown on the second row from (b) to (d). The B-mode images of the CS-reconstructed k-spaces with the optimal λ are shown on the third rows (e) to (g). Those from the CS-reconstructed k-spaces using the reweighted ℓ_1 minimization are on the fourth row: (h) to (j).

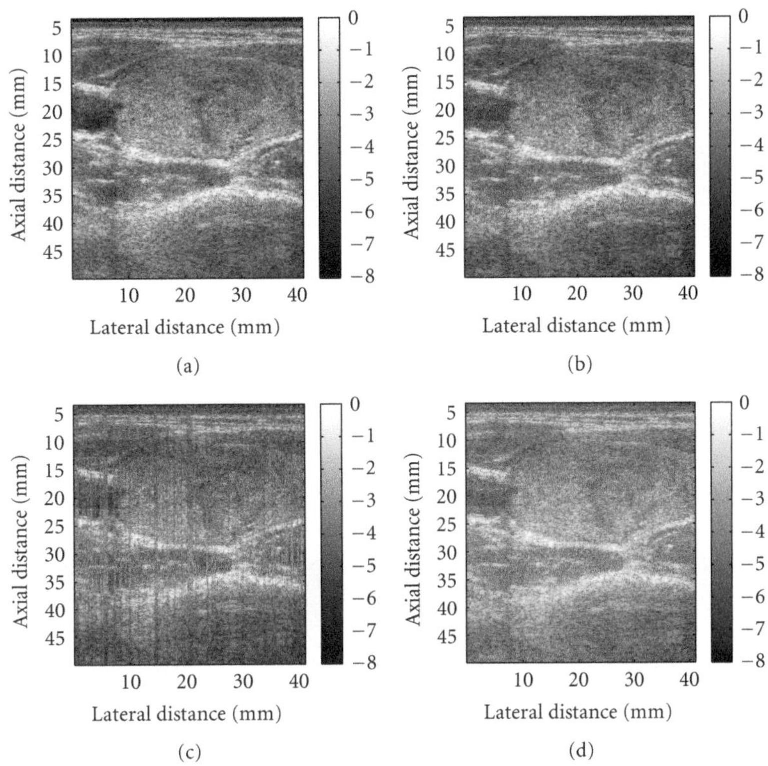

FIGURE 13: CS reconstruction on a US image of a human thyroid right lobe (a) using a two-dimensional random sampling pattern Φ_1 (b) and two alternative sampling masks Φ_2 (c) and Φ_3 (d) and 33% of samples. The B-mode images of the CS-reconstructed k-spaces with the optimal $\lambda = 0.01$ are shown on the second row from (b) to (d).

TABLE 4: NRMSE between the reconstructed RF US volume and the original volume for different sampling ratios and patterns.

	Sampling	25%	33%	50%
	Θ_1	0.61	0.49	0.27
	Θ_2	0.69	0.57	0.40
Cubic spline interpolation	Lateral regular	1.09	1.02	0.83

is still quite impressive. This setting could be used in a situation where the speed of imaging would prevail on the quality of the reconstruction. In addition, despite being less sharp, the image still exhibits the tissue structure and might be sufficient in many applications.

For puropse of illustration, Table 4 shows the normalised root mean squared errors of reconstruction between the RF original and reconstructed US volumes for the two sampling patterns Θ_1 and Θ_2 and different rates of decimation (25%, 33%, and 50%). In addition, the NRMSE of the CS reconstructions was compared with results from an interpolation reconstruction method, based on a 2D cubic spline interpolation. The decimation used for the interpolation was a regular lateral undersampling corresponding to the sampling ratios 33% (no axial decimation). As expected, CS outperforms interpolated regular subsampling.

11. Some Hints for Practical Implementation

As shown previously, one of the key points for compressive sampling success is the incoherence of the acquired samples. For this, random sampling schemes are necessary, in both axial and lateral directions.

Regarding the axial direction for the sampling masks Φ_1 and Φ_2, one way to incoherently sample one RF line is to fix a constant sampling frequency f_e, to consider a vector of random integers n_r and to only acquire samples situated at n_r/f_e. This can be achieved by programming specific acquisition devices such as FPGA or CPLD. If the mask Φ_3 is considered, the same n_r random vector is repeated on each RF line.

Alternatively, for all the masks proposed in the paper, the whole RF line can be acquired respecting Shannon's theorem and random samples subsequently discarded, in order to speed up the transfer to the scanner memory.

In the lateral direction, for RF lines selection, the acquisition scheme depends on the type of scanner. For multiple element transducers, each element can be randomly set to be active or not. For single-element transducers, each RF line is acquired separately. Thus, RF lines at random lateral positions can be omitted.

Table 5: Qualitative assessment of the speed of acquisition, the quality of reconstruction, and the ease of a practical implementation of different sampling masks and undersampling ratios (+++ denotes high-speed), high-quality reconstruction and easy implementation.

	Speed	Reconstruction	Implementation
Φ_1	+	+++	+
Φ_2	+++	++	+++
Φ_3	++	++	++

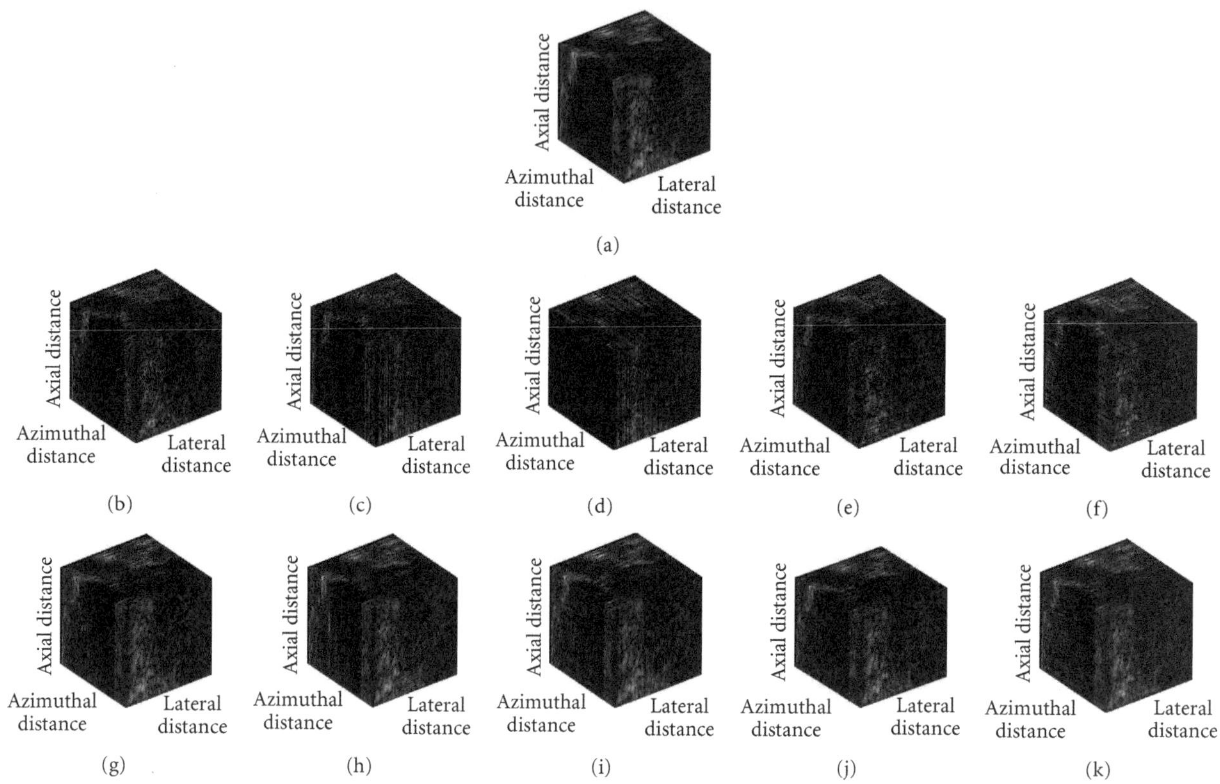

Figure 14: 3D CS reconstructions following (8) on an *in vivo* US volume of a mouse kidney (a) using 50% of the samples and the sampling masks Θ_1, Θ_2, Θ_3, Θ_4, and Θ_5. The B-mode volumes of the RF random measurements are shown in (b), (c), (d), (e), and (f) and of CS reconstructed US volume in (g), (h), (i), (j), and (k).

This methodology could be suitable to all the sampling strategies presented in this paper.

Of course, one could also implement direct random acquisitions through random triggering acquisition board. But this possibility needs more sophisticated electronic design.

12. Summary

Table 5 summarizes the different results presented so far and qualitatively compares the speed of the acquisition, the quality of the reconstruction in term of errors, and the ease of a practical implementation of the different masks and undersampling ratios tested in this paper.

13. Discussion and Conclusion

This paper investigates the feasibility of CS in ultrasound imaging. From this presentation many questions are opened.

13.1. Sparsity of Ultrasound Signals. Sparsity is key point in the success of CS. We dealt here with the fact that the Fourier transform of US is sparse. This may of course be questionable.

US signals exhibit bandpass characteristics and thus are sparse in frequency domain. Consequently, a highly oversampled version of the ultrasound signal could be reconstructed from fewer regular samples. However, it is well known that sampling using high sampling rate is neither easy nor cost-effective particularly in high-frequency US applications. The interest of CS lies in the ability of allowing under sampling from Nyquist limits. Indeed when sampled at Nyquist (which is cost-effective), by taking the demodulated I/Q signal, which remains sparse, CS allows correct reconstruction, whereas no reconstruction is possible after under sampling from Nyquist rate, according to the regular sampling theory.

In addition CS allows skipping RF line in lateral and azimuthal directions.

13.2. Real-Time Nature. In this paper, we showed the powerful potential of CS to reduce data volume and speed up acquisitions at the price of a reconstruction using the l_1 norm. However, using dedicated circuits (GPU type) for the CS reconstruction could allow a great improvement in processing times and overall increase the imaging rate, keeping the real-time nature of US imaging.

In addition, various sampling protocols suited to US imaging were proposed here where the RF signals can be sampled at random times to provide measurements of the final image k-space. Through the l_1 minimization, the original k-space can be reconstructed and the RF US image subsequently recovered with minimal loss of information.

The method presented here differs from inpainting methods as the reconstruction is performed in another domain than the image itself.

Future work will include the identification of optimal conditions as well as an investigation of several optimization routines and better sparsity basis. Additional knowledge about the US images will be inserted in the reconstruction process (statistics of the signal, attenuation). The aim is to reach the fastest and most reliable reconstruction from as little samples as possible. Various applications will also be considered (multidimensional Doppler and tissue characterization).

Acknowledgments

This work was supported by the French National Agency of Research (ANR) under the SURFOETUS Grant. The authors would like to thank Jean-Marc Gregoire and Hugues Herault for ultrasound imaging software development.

References

[1] A. Elen, H. F. Choi, D. Loeckx et al., "Three-dimensional cardiac strain estimation using spatio-temporal elastic registration of ultrasound images: a feasibility study," *IEEE Transactions on Medical Imaging*, vol. 27, no. 11, pp. 1580–1591, 2008.

[2] W. Lees, "Ultrasound imaging in three and four dimensions," *Seminars in Ultrasound CT and MRI*, vol. 22, no. 1, pp. 85–105, 2001.

[3] T. R. Nelson and D. H. Pretorius, "Three-dimensional ultrasound imaging," *Ultrasound in Medicine and Biology*, vol. 24, no. 9, pp. 1243–1270, 1998.

[4] A. Fenster and D. B. Downey, "3-D ultrasound imaging: a review," *IEEE Engineering in Medicine and Biology Magazine*, vol. 15, no. 6, pp. 41–51, 1996.

[5] E. J. Candès, J. Romberg, and T. Tao, "Robust uncertainty principles: exact signal reconstruction from highly incomplete frequency information," *IEEE Transactions on Information Theory*, vol. 52, no. 2, pp. 489–509, 2006.

[6] D. L. Donoho, "Compressed sensing," *IEEE Transactions on Information Theory*, vol. 52, no. 4, pp. 1289–1306, 2006.

[7] M. Lustig, D. L. Donoho, J. M. Santos, and J. M. Pauly, "Compressed sensing MRI: a look at how CS can improve on current imaging techniques," *IEEE Signal Processing Magazine*, vol. 25, no. 2, pp. 72–82, 2008.

[8] H. Yu, G. Cao, L. Burk et al., "Compressive sampling based interior reconstruction for dynamic carbon nanotube micro-CT," *Journal of X-Ray Science and Technology*, vol. 17, no. 4, pp. 295–303, 2009.

[9] J. Provost and F. Lesage, "The application of compressed sensing for photo-acoustic tomography," *IEEE Transactions on Medical Imaging*, vol. 28, no. 4, pp. 585–594, 2009.

[10] D. Friboulet, H. Liebgott, and R. Prost, "Compressive sensing for raw RF signals reconstruction in ultrasound," in *Proceedings of the IEEE International Ultrasonics Symposium (IUS '10)*, pp. 367–370, October 2010.

[11] A. Achim, B. Buxton, G. Tzagkarakis, and P. Tsakalides, "Compressive sensing for ultrasound RF echoes using α-stable distributions," in *Proceedings of the 32nd Annual International Conference of the IEEE Engineering in Medicine and Biology Society (EMBC '10)*, pp. 4304–4307, September 2010.

[12] C. Quinsac, A. Basarab, J. M. Girault, and D. Kouame, "Compressed sensing of ultrasound images: sampling of spatial and frequency domains," in *Proceedings of the IEEE Workshop on Signal Processing Systems (SiPS '10)*, pp. 231–236, October 2010.

[13] C. Quinsac, A. Basarab, J. M. Gregoire, and D. Kouame, "3D compressed sensing ultrasound imaging," in *Proceedings of the IEEE International Ultrasonics Symposium (IUS '10)*, San Diego, Calif, USA, October 2010.

[14] C. Quinsac, F. de Vieilleville, A. Basarab, and D. Kouame, "Compressed sensing of ultrasound single-orthant analytical signals," in *Proceedings of the IEEE International Ultrasonics Symposium (IUS '10)*, San Diego, Calif, USA, October 2010.

[15] R. Tur, Y. C. Eldar, and Z. Friedman, "Innovation rate sampling of pulse streams with application to ultrasound imaging," *IEEE Transactions on Signal Processing*, vol. 59, no. 4, pp. 1827–1842, 2011.

[16] J. Richy, H. Liebgott, R. Prost, and D. Friboulet, "Spectral doppler using compressive sensing," in *Proceedings of the IEEE International Ultrasonics Symposium (IUS '10)*, October 2010.

[17] S. M. S. Zobly and Y. M. Kakah, "Compressed sensing: doppler ultrasound signal recovery by using non-uniform sampling and random sampling," in *Proceedings of the 28th National Radio Science Conference (NRSC '11)*, pp. 1–9, April 2011.

[18] Z. Guo, L. Song, and L. Wang, "Compressed sensing in photoacoustic tomography in vivo," *Journal of Biomedical Optics*, vol. 15, no. 2, Article ID 021311, 2010.

[19] A. Skodras, C. Christopoulos, and T. Ebrahimi, "The JPEG 2000 still image compression standard," *IEEE Signal Processing Magazine*, vol. 18, no. 5, pp. 36–58, 2001.

[20] E. Candès and J. Romberg, "Sparsity and incoherence in compressive sampling," *Inverse Problems*, vol. 23, no. 3, pp. 969–985, 2007.

[21] E. J. Candès and M. B. Wakin, "An introduction to compressive sampling: a sensing/sampling paradigm that goes against the common knowledge in data acquisition," *IEEE Signal Processing Magazine*, vol. 25, no. 2, pp. 21–30, 2008.

[22] D. L. Donoho and X. Huo, "Uncertainty principles and ideal atomic decomposition," *IEEE Transactions on Information Theory*, vol. 47, no. 7, pp. 2845–2862, 2001.

[23] J. Romberg, "Imaging via compressive sampling: Introduction to compressive sampling and recovery via convex programming," *IEEE Signal Processing Magazine*, vol. 25, no. 2, pp. 14–20, 2008.

[24] M. Lustig, D. Donoho, and J. M. Pauly, "Sparse MRI: the application of compressed sensing for rapid MR imaging," *Magnetic Resonance in Medicine*, vol. 58, no. 6, pp. 1182–1195, 2007.

[25] Y. Tsaig and D. L. Donoho, "Extensions of compressed sensing," *Signal Processing*, vol. 86, no. 3, pp. 549–571, 2006.

[26] E. Candès and J. Romberg, "Signal recovery from random projections," in *IS and T Electronic Imaging—Computational Imaging III*, vol. 5674 of *Proceedings of SPIE*, pp. 76–86, January 2005.

[27] R. Baraniuk, M. Davenport, R. DeVore, and M. Wakin, "A simple proof of the restricted isometry property for random matrices," *Constructive Approximation*, vol. 28, no. 3, pp. 253–263, 2008.

[28] C. de Mol, E. de Vito, and L. Rosasco, "Elastic-net regularization in learning theory," *Journal of Complexity*, vol. 25, no. 2, pp. 201–230, 2009.

[29] E. J. Candès, M. B. Wakin, and S. P. Boyd, "Enhancing sparsity by reweighted $\ell 1$ minimization," *Journal of Fourier Analysis and Applications*, vol. 14, no. 5-6, pp. 877–905, 2008.

[30] J. Nocedal and S. J. Wright, *Numerical Optimization*, Springer, New York, NY, USA, 2nd edition, 2006.

[31] J. A. Jensen and N. B. Svendsen, "Calculation of pressure fields from arbitrarily shaped, apodized, and excited ultrasound transducers," *IEEE Transactions on Ultrasonics, Ferroelectrics, and Frequency Control*, vol. 39, no. 2, pp. 262–267, 1992.

15

Dynamical Analysis of Long Fiber-Reinforced Laminated Plates with Elastically Restrained Edges

Liz G. Nallim,[1] Facundo J. Bellomo,[1] Ricardo D. Quinteros,[1] and Sergio Oller[2]

[1] *INIQUI-CONICET, Facultad de Ingeniería, Universidad Nacional de Salta, Avenue Bolivia 5150, 4400 Salta, Argentina*
[2] *International Center for Numerical Method in Engineering, (CIMNE) Technical University of Catalonia-Barcelona Tech (UPC), Edif. C1, Campus Nord, Jordi Girona 1-3, 08034 Barcelona, Spain*

Correspondence should be addressed to Liz G. Nallim, lnallim@unsa.edu.ar

Academic Editor: Kok Keong Choong

This paper presents a variational formulation for the free vibration analysis of unsymmetrically laminated composite plates with elastically restrained edges. The study includes a micromechanics approach that allows starting the study considering each layer as constituted by long unidirectional fibers in a continuous matrix. The Mori-Tanaka method is used to predict the mechanical properties of each lamina as a function of the elastic properties of the components and of the fiber volume fraction. The resulting mechanical properties for each lamina are included in a general Ritz formulation developed to analyze the free vibration response of thick laminated anisotropic plates resting on elastic supports. Comprehensive numerical examples are computed to validate the present method, and the effects of the different mechanical and geometrical parameters on the dynamical behavior of different laminated plates are shown. New results for general unsymmetrical laminates with elastically restrained edges are also presented. The analytical approximate solution obtained in this paper can also be useful as a basis to deal with optimization problems under, for instance, frequency constraints.

1. Introduction

Fiber-reinforced composite laminated plates are extensively used in many engineering applications. The free vibration analysis of these plates plays a very important role in the design of civil, aerospace, mechanical, and marine structures. In addition to the favorable high specific strength and high specific stiffness, fiber-reinforced composite laminates offer the possibility of optimal design through the variation of stacking pattern, angle of fiber orientation, fiber content, and so forth, known as composite tailoring. All these mechanical and geometrical characteristics, as well as the various coupling effects that take place, must be considered in the prediction of the laminates dynamical response to assure that this is reliable, accurate, and adequate to the design requirements.

It is well known that laminated composite plates have relatively low transverse shear stiffness, playing the shear deformation an important role in the global and local behavior of these structures. Among the numerous theories used for laminated plates that include the transverse shear strain, the first-order shear deformation theory (FSDT) [1, 2] is adequate for the computation of global responses (such as natural frequencies) and simultaneously has some advantages due to its simplicity and low computational cost. Many investigations have been reported for free vibration analysis of moderately thick composite laminates using the FSDT kinematics (see for instance [3–13]). However, the results are, in most cases, limited to certain lamination schemes and boundary conditions. As far as the study of thick plates with elastically restrained edges is concerned, most of the previous works are limited to isotropic ones ([14–19] among others). But, limited information is found for the case of thick anisotropic laminated plates resting on elastic supports. For instance, Setoodeh and Karami [20] implemented a layer-wise laminated plate theory linked with three-dimensional elasticity approach for vibration and buckling of symmetric and antisymmetric fiber-reinforced composite plates having elastically restraint edges support and results for cross-ply laminates are presented, whereas Karami et al. [21] applied the differential quadrature method for the free vibration analysis of moderately thick symmetric laminated plates with elastically restrained edges. For the same boundary conditions, semianalytical solutions for the free vibration of angle-ply symmetrically laminated plates were presented by Ashour [22]. Nallim and Grossi [23] also studied the vibration of symmetric laminated plates resting on elastic support employing the Ritz method and beam orthogonal polynomials as approximated functions. These kind of approximate functions (in one or two variables) have been used by many authors to the free vibration analysis of, both homogeneous and nonhomogeneous, plates (Chakraverty et al. [24–26] and Chow et al. [27], among others).

In this paper, a general Ritz formulation for the free vibration analysis of anisotropic laminated plates is developed. All kind of boundary conditions including elastically restrained edges are considered enhancing the study. This feature allows a more realistic analysis of some structural problems. The analysis includes a micromechanical approach (according to the classification of Altenbach et al. [28]), where the average mechanical properties of each anisotropic lamina are estimated from the known characteristics of the fibers and the matrix materials taking into account the fiber volume ratio and the fiber-packing arrangement. At structural level, the dynamic response of the unsymmetrical laminated plate, with elastically restrained edges, is analyzed using the first-order shear deformation theory and the Ritz method with beam orthogonal polynomials as coordinate functions. The approximate analytical solution developed here is very useful to understand, both qualitatively and quantitatively, the behavior of complex laminated plates.

2. Formulation

2.1. Effective Elastic Moduli of Long Fiber-Reinforced Laminae. The micromechanics-based Mori-Tanaka method [29] is used in this section to predict the elastic mechanical properties of the orthotropic unidirectional laminae. This method may be viewed as the simplest mean field approach for inhomogeneous materials that encompass the full physical range of phase volume fraction.

Eshelby's results [33] show that if an elastic homogeneous ellipsoidal inclusion in an infinite linear elastic matrix is subjected to an eigenstrain $\boldsymbol{\varepsilon}^T$, uniform strain states $\boldsymbol{\varepsilon}^C$ is induced, and it is related to the eigenstrain by the expression

$$\boldsymbol{\varepsilon}^C = \mathbf{S}^E : \boldsymbol{\varepsilon}^T, \quad (1)$$

where $\mathbf{S}^E$ is the Eshelby tensor, which depends on the reinforcement dimensions and the Poisson ratio of the matrix ν_m. The components of this tensor for a circular, cylindrical inclusion with an infinite length-to-diameter ratio parallel to the 1-axis (parallel to the fiber direction, Figure 1) are

$$S_{1111} = S_{1133} = S_{1122} = 0, \qquad S_{3333} = S_{2222} = \frac{5-4\nu_m}{8(1-\nu_m)},$$
$$S_{3322} = S_{2233} = \frac{4\nu_m - 1}{8(1-\nu_m)}, \qquad S_{3311} = S_{2211} = \frac{\nu_m}{2(1-\nu_m)},$$
$$S_{3232} = \frac{3-4\nu_m}{8(1-\nu_m)}, \qquad S_{1313} = S_{1212} = \frac{1}{4}. \quad (2)$$

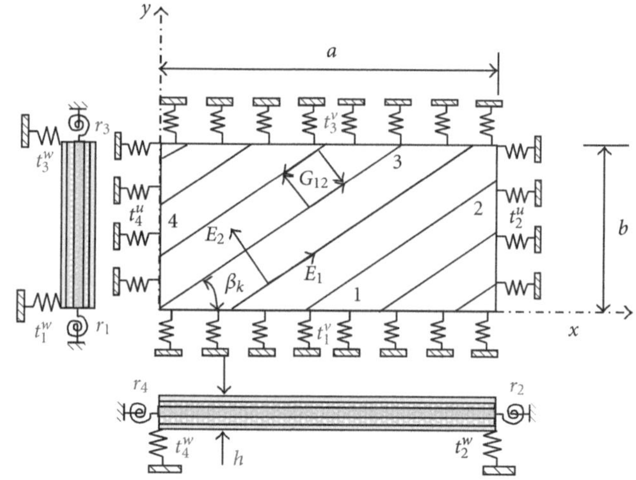

FIGURE 1: General laminated plate resting on elastic supports.

The transformations strains are obtained considering the equivalent homogeneous inclusion for inhomogeneous inclusions developed by Eshelby [33] together with the interaction effects of Mori-Tanaka [29]. These transformations strains are used to equate the total stresses in the inhomogeneities and their equivalent inclusions, as described in the following equation:

$$\mathbf{C}_f : \left(\boldsymbol{\varepsilon}^a + \boldsymbol{\varepsilon}^{\text{int}} + \boldsymbol{\varepsilon}^C\right) = \mathbf{C}_m : \left(\boldsymbol{\varepsilon}^a + \boldsymbol{\varepsilon}^{\text{int}} + \boldsymbol{\varepsilon}^C - \boldsymbol{\varepsilon}^T\right), \quad (3)$$

where $\mathbf{C}_f$ and $\mathbf{C}_m$ are the stiffness tensors of fiber and matrix, respectively, $\boldsymbol{\varepsilon}^a$ is the uniform far field strain applied to the domain at infinity, and $\boldsymbol{\varepsilon}^{\text{int}}$ is the average elastic strain defined by Mori-Tanaka which is given by

$$\boldsymbol{\varepsilon}^{\text{int}} = -k_f\left(\boldsymbol{\varepsilon}^C - \boldsymbol{\varepsilon}^T\right), \quad (4)$$

where k_f is the fiber volume fraction.

Finally, the stiffness tensor $\mathbf{C}$ for different unidirectional laminae can be obtained from energy considerations [34] and (1) to (4) as

$$\mathbf{C} = \mathbf{C}_m \times \left\{\mathbf{I} - k_f\left[\left(\mathbf{C}_f - \mathbf{C}_m\right)\left(\mathbf{S}^E - k_f\left(\mathbf{S}^E - \mathbf{I}\right) + \mathbf{C}_m\right)\right]^{-1} \times \left(\mathbf{C}_f - \mathbf{C}_m\right)\right\}^{-1}, \quad (5)$$

where $\mathbf{I}$ is the fourth order identity tensor.

Using this method the mechanical properties of unidirectional carbon/epoxy laminae are found considering various fiber volume fractions, and they are depicted in Table 1. These properties, for each unidirectional lamina, are then used in the next section to obtain the reduced constitutive matrix.

2.2. General Laminated Plate Resting on Elastic Supports. Let us consider a rectangular fiber-reinforced composite laminated plate, of dimension $a \times b$ and total thickness h

TABLE 1: Mechanical properties of unidirectional laminae (AS4-3501-6), obtained using Mori-Tanaka method. Fiber and matrix properties $E_{f1} = 225\,\text{GPa}$, $E_{f2} = 15\,\text{GPa}$, $G_{f12} = 15\,\text{GPa}$, $G_{f23} = 7\,\text{GPa}$, $\nu_{f12} = 0.20$; $E_m = 4.2\,\text{GPa}$, $\nu_m = 0.34$ ([30]).

k_f	0.1	0.2	0.3	0.4	0.5	0.6	0.7	0.8
E_1	26.29	48.38	70.46	92.54	114.62	136.70	158.78	180.86
E_2	5.11	5.69	6.31	7.01	7.81	8.76	9.87	11.23
$G_{12} = G_{13}$	1.84	2.17	2.57	3.07	3.70	4.54	5.68	7.35
G_{23}	1.75	1.97	2.22	2.52	2.88	3.32	3.88	4.61
n_{12}	0.32	0.31	0.29	0.28	0.27	0.25	0.24	0.23

($h = \sum h_k$, h_k represents the thickness of a layer). The laminated plate is composed of an arbitrary number of N_c orthotropic layers and fibre orientation β_k $(k = 1, 2, \dots, N_c)$ as shown in Figure 1. A rectangular Cartesian coordinate system (x, y, z) is used to represent the plate geometry and the strain–displacement relations. The x-y plane coincides with the middle plane of the plate. The displacement field of the first-order shear deformation theory is assumed to be of the form [1, 2]

$$\begin{aligned} u(x,y,z,t) &= u_0(x,y,t) + z\phi_x(x,y,t), \\ v(x,y,z,t) &= v_0(x,y,t) + z\phi_y(x,y,t), \\ w(x,y,z,t) &= w_0(x,y,t), \end{aligned} \quad (6)$$

where t is the time dimension u_0, v_0 and w_0 denote the mid-surface translational displacements along the x, y, and z axes, and ϕ_x, ϕ_y are the rotations about y- and x-axes, respectively. The displacement model (6) yields the following kinematic relations:

$$\{\varepsilon\} = \begin{bmatrix} \{\varepsilon_0\} \\ \{\varepsilon_0^*\} \end{bmatrix} + z \begin{bmatrix} \{\varepsilon_1\} \\ \{0\} \end{bmatrix}, \quad (7)$$

where

$$\begin{aligned} \{\varepsilon\}^T &= \{\varepsilon_x, \varepsilon_y, \gamma_{xy}, \gamma_{yz}, \gamma_{xz}\}, \\ \{\varepsilon_0\}^T &= \left\{ \frac{\partial u_0}{\partial x}, \frac{\partial v_0}{\partial y}, \frac{\partial u_0}{\partial y} + \frac{\partial v_0}{\partial x} \right\}, \\ \{\varepsilon_0^*\}^T &= \left\{ \frac{\partial w_0}{\partial y} + \phi_y, \frac{\partial w_0}{\partial x} + \phi_x \right\}, \\ \{\varepsilon_1\}^T &= \left\{ \frac{\partial \phi_x}{\partial x}, \frac{\partial \phi_y}{\partial y}, \frac{\partial \phi_x}{\partial y} + \frac{\partial \phi_y}{\partial x} \right\}. \end{aligned} \quad (8)$$

The stress-strain relation of each layer is given by the generalized Hooke's law as follows:

$$\begin{pmatrix} \sigma_x \\ \sigma_y \\ \tau_{xy} \\ \tau_{yz} \\ \tau_{xz} \end{pmatrix}^{(k)} = \begin{bmatrix} \overline{Q}_{11} & \overline{Q}_{12} & \overline{Q}_{16} & 0 & 0 \\ \overline{Q}_{12} & \overline{Q}_{22} & \overline{Q}_{26} & 0 & 0 \\ \overline{Q}_{16} & \overline{Q}_{26} & \overline{Q}_{66} & 0 & 0 \\ 0 & 0 & 0 & \overline{Q}_{44} & \overline{Q}_{45} \\ 0 & 0 & 0 & \overline{Q}_{45} & \overline{Q}_{55} \end{bmatrix}^{(k)} \begin{bmatrix} \varepsilon_x \\ \varepsilon_y \\ \gamma_{xy} \\ \gamma_{yz} \\ \gamma_{xz} \end{bmatrix}, \quad (9)$$

where $\overline{Q}_{ij}$ are the components of the plane-stress reduced constitutive matrix [35] which are function of the elastic constant determined in Section 2.1 and the ply angle β_k.

2.3. Energy Functional Components. Taking into account (7) and (9), the strain energy due to the laminated plate deflection can be written as

$$\begin{aligned} U_p = \frac{1}{2}\iint_R \Big[&\{\varepsilon_0\}[A]\{\varepsilon_0\}^T + \{\varepsilon_0\}[B]\{\varepsilon_1\}^T + \{\varepsilon_1\}[B]\{\varepsilon_0\}^T \\ &+ \{\varepsilon_1\}[D]\{\varepsilon_1\}^T + \{\varepsilon_0^*\}[A^*]\{\varepsilon_0^*\}^T \Big] dx\,dy, \end{aligned} \quad (10)$$

where R is the mid-surface area (Figure 1) and the stiffness coefficients [35, 36] are given by $(A_{ij}, B_{ij}, D_{ij}) = \int_{-h/2}^{h/2} \overline{Q}_{ij}(1, z, z^2)dz$ $(i, j = 1, 2, 6)$, $A_{ij}^* = \int_{-h/2}^{h/2} k_{ij}\overline{Q}_{ij}dz$ $(i, j = 4, 5)$ k_{ij} being the shear correction factors.

The strain energy corresponding to the elastic edge restraints is given by

$$\begin{aligned} U_t = &\frac{1}{2}\int_0^a \Big[t_1^w\left(w_0|_{y=0}\right)^2 + t_3^w\left(w_0|_{y=b}\right)^2 + t_1^v\left(v_0|_{y=0}\right)^2 \\ &+ t_3^v\left(v_0|_{y=b}\right)^2 + r_1\left(\phi_y|_{y=0}\right)^2 + r_3\left(\phi_y|_{y=b}\right)^2 \Big] dx \\ &+ \frac{1}{2}\int_0^b \Big[t_2^w(w_0|_{x=a})^2 + t_4^w(w_0|_{x=0})^2 + t_2^u(v_0|_{x=a})^2 \\ &+ t_4^u(v_0|_{x=0})^2 + r_2(\phi_x|_{x=a})^2 + r_4\left(\phi_x|_{y=0}\right)^2 \Big] dy, \end{aligned} \quad (11)$$

where $t_i^\bullet$ $(i = 1, \dots, 4$ and $\bullet = u, v, w)$ are the elastic translational coefficients and r_i $(i = 1, \dots, 4)$ are the elastic rotational coefficients.

The kinetic energy is expressed as

$$\begin{aligned} T = \frac{1}{2}\iint_R \Big[&I_0(\dot{u}_0^2 + \dot{v}_0^2 + \dot{w}_0^2) + 2I_1\left(\dot{u}_0\dot{\phi}_x + \dot{v}_0\dot{\phi}_y\right) \\ &+ I_2\left(\dot{\phi}_x^2 + \dot{\phi}_y^2\right) \Big] dx\,dy, \end{aligned} \quad (12)$$

being I_i $(i = 0, 1, 2)$ the mass inertias of the plate defined as [35]

$$I_i = \int_{-h/2}^{h/2} \rho^{(k)} z^i dz, \quad (i = 0, 1, 2), \quad (13)$$

where $\rho^{(k)}$ is the material density of the k-th layer.

3. Application of the Ritz Method for the Free Vibration Analysis

The Ritz method is applied to determine analytical approximate solutions for dynamical behavior of arbitrarily laminated plates resting on elastic supports. During free vibration, the displacements components are assumed split in

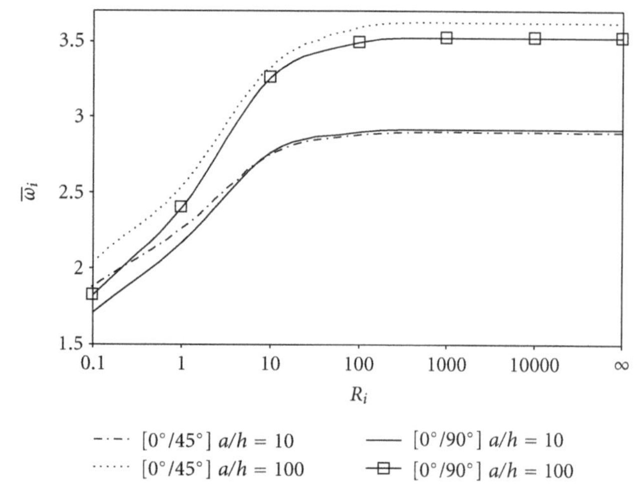

FIGURE 2: Variation of the fundamental frequency coefficient $\overline{\omega}$ with the rotational restraint parameter R_i, $T_i^w = T_i^u = T_i^v = \infty$ ($i = 1, \ldots, 4$) for square [0°/45°] and [0°/90°] carbon-epoxy (Table 1), with $k_f = 0.6$.

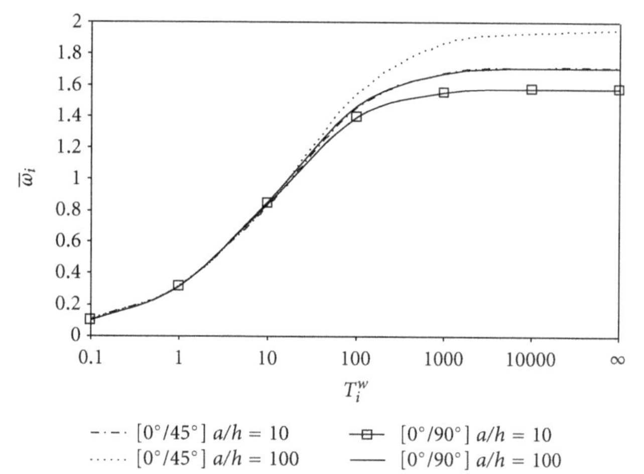

FIGURE 3: Variation of the fundamental frequency coefficient $\overline{\omega}$ with the translational restraint parameter T_i^w, $T_i^u = T_i^v = R_i = 0$ ($i = 1, \ldots, 4$) for square [0°/45°] and [0°/90°] carbon-epoxy (Table 1), with $k_f = 0.6$.

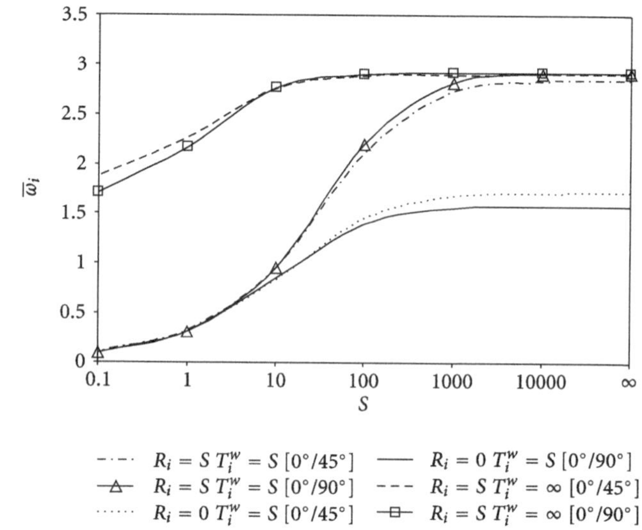

FIGURE 4: Variation of the fundamental frequency coefficient $\overline{\omega}$ with the translational and the rotational restraint parameters $T_i^w = R_i$, $T_i^u = T_i^v = 0$ ($i = 1, \ldots, 4$) for square [0°/45°] and [0°/90°] carbon-epoxy (Table 1), with $k_f = 0.6$ and $a/h = 10$.

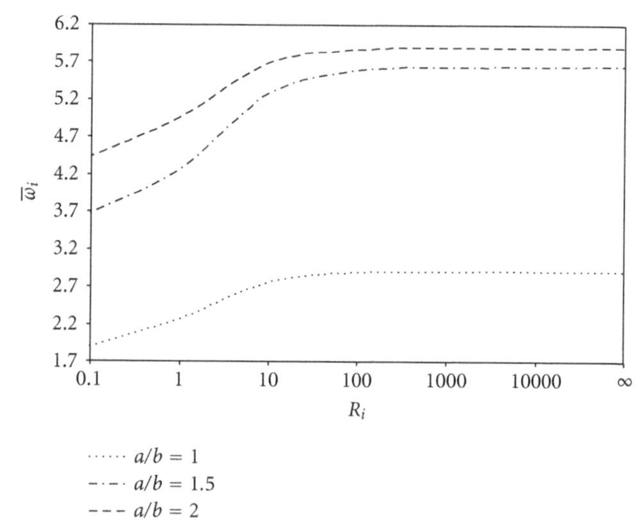

FIGURE 5: Variation of the fundamental frequency coefficient $\overline{\omega}$ with the rotational restraint parameter R_i, $T_i^w = T_i^u = T_i^v = \infty$ ($i = 1, \ldots, 4$) for different aspect ratios [0°/45°] carbon-epoxy (Table 1), with $k_f = 0.6$.

the spatial and temporal parts, being the last one periodic in time; that is,

$$\begin{aligned} u_0(x,y,t) &= U(x,y)\sin\omega t, \\ v_0(x,y,t) &= V(x,y)\sin\omega t, \\ w_0(x,y,t) &= W(x,y)\sin\omega t, \\ \phi_x(x,y,t) &= \Phi_x(x,y)\sin\omega t, \\ \phi_y(x,y,t) &= \Phi_y(x,y)\sin\omega t, \end{aligned} \quad (14)$$

where ω is the natural frequency in radian.

Putting these displacements into the energy functional components ((10) to (12)) the maximum values of the kinetic energy ($T_{\max}$) and the strain energies ($U_{p,\max}, U_{t,\max}$) are derived. Then, the energy functional for free vibration of the laminated plate is given by

$$\Pi = U_{p,\max} + U_{t,\max} - T_{\max}, \quad (15)$$

which is to be minimized according to the Ritz principle.

3.1. Boundary Conditions and Approximating Functions. There are some options when choosing the unknown functions of displacement components to apply the Ritz method. Particularly, the use of orthogonal polynomials as coordinate

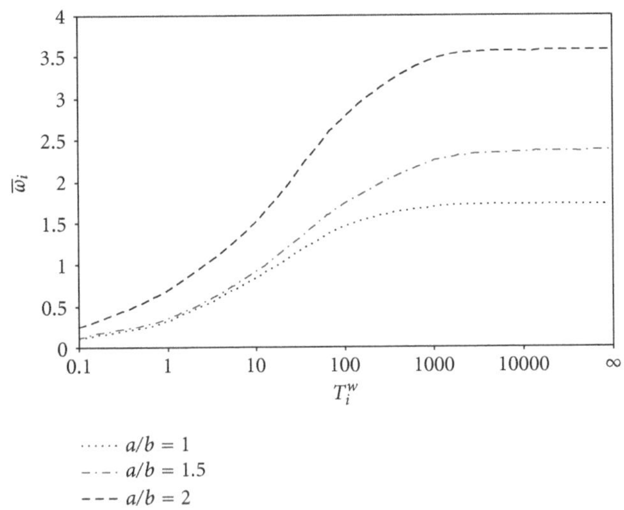

FIGURE 6: Variation of the fundamental frequency coefficient $\overline{\omega}$ with the translational restraint parameter T_i^w, $T_i^u = T_i^v = R_i = 0$ ($i = 1,\ldots,4$) for different aspect ratios [0°/45°] carbon-epoxy (Table 1), with $k_f = 0.6$.

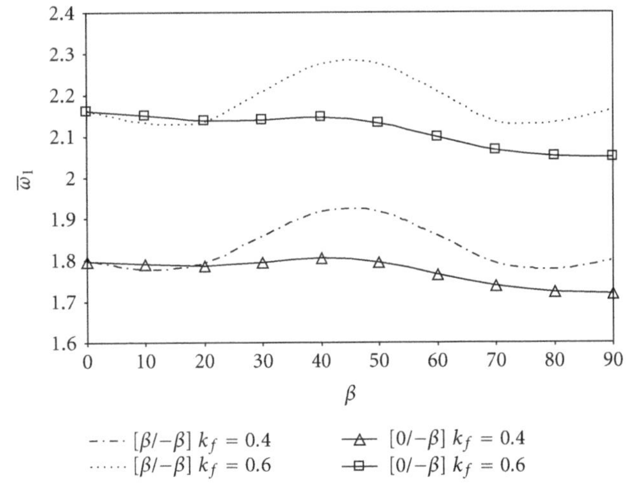

FIGURE 7: Effect of the fiber orientation on the first vibration frequency coefficient $\overline{\omega}_1$, for two different fiber volume fraction k_f with $C_1C_1C_1C_1$ boundary condition and $a/h = 10$.

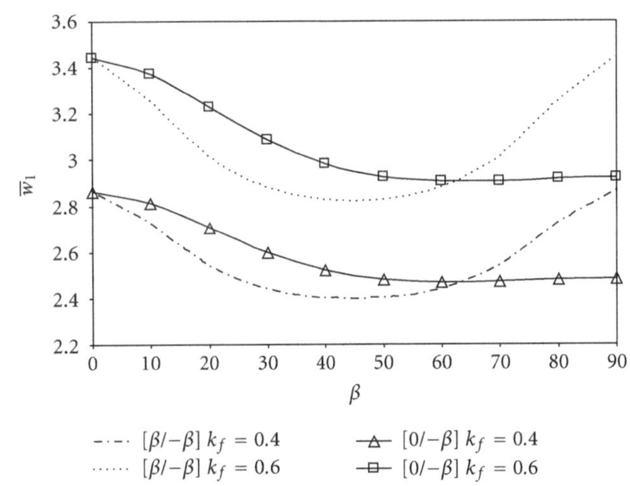

FIGURE 8: Effect of the fiber orientation on the first vibration frequency coefficient $\overline{\omega}_1$, for two different fiber volume fraction k_f with $S_1S_1S_1S_1$ boundary condition and $a/h = 10$.

functions has important advantages related to numerical stability and fast convergence as has been demonstrated in previous works [23, 37, 38], even for plates with complicated boundary conditions and high degree of anisotropy. For these reasons, in this work, the displacement components are expressed by sets of beam characteristic orthogonal polynomials $\{p_i^{(\cdot)}(x)\}$, $\{q_j^{(\bullet)}(y)\}$, $(\bullet) = u, v, w, \phi_x, \phi_y$, resulting in

$$U(x,y) \approx U_{MN}(x,y) = \sum_{i=1}^{M}\sum_{j=1}^{N} c_{ij}^{(u)} p_i^{(u)}(x)\, q_j^{(u)}(y),$$

$$V(x,y) \approx V_{MN}(x,y) = \sum_{i=1}^{M}\sum_{j=1}^{N} c_{ij}^{(v)} p_i^{(v)}(x)\, q_j^{(v)}(y),$$

$$W(x,y) \approx W_{MN}(x,y) = \sum_{i=1}^{M}\sum_{j=1}^{N} c_{ij}^{(w)} p_i^{(w)}(x)\, q_j^{(w)}(y),$$

$$\Phi_x \approx \Phi_{xMN}(x,y) = \sum_{i=1}^{N}\sum_{j=1}^{M} c_{ij}^{(\phi_x)} p_i^{(\phi_x)}(x)\, q_j^{(\phi_x)}(y),$$

$$\Phi_y \approx \Phi_{yMN}(x,y) = \sum_{i=1}^{N}\sum_{j=1}^{M} c_{ij}^{(\phi_y)} p_i^{(\phi_y)}(x)\, q_j^{(\phi_y)}(y),\tag{16}$$

where $c_{ij}^{(u)}$, $c_{ij}^{(v)}$, $c_{ij}^{(w)}$, $c_{ij}^{(\phi_x)}$, $c_{ij}^{(\phi_y)}$ are the unknown coefficients, and M, N are the numbers of polynomials in each coordinate.

The procedure for the construction of the orthogonal polynomials has been developed by Bhat [39]. The first members of the sets, $p_1^{(\bullet)}(x)$ and $q_1^{(\bullet)}(y)(\bullet) = u, v, w, \phi_x, \phi_y$ are obtained as the simplest polynomials that satisfy all the geometrical boundary conditions of the plate in their respective x and y directions. The higher members of each set are constructed by employing the Gram-Schmidt orthogonalization procedure. The coefficients of the polynomials are chosen in such a way as to make the polynomials orthonormal. However, the functions $p_k^{(\bullet)}(x)$ and $q_k^{(\bullet)}(y)$ for $(\bullet) = \phi_x, \phi_y$ are obtained from relative rotation conditions starting from polynomials of an order lower than the chosen for the transversal displacements and then applying the sequence of Gram-Schmidt orthogonalization procedure. This particular choice is made to avoid the overestimation of the rate of elastic energy due to the shear respect to the rate due to the bending. This concept has been applied by Auciello and Ercolano [40], to Timoshenko beams, to avoid

TABLE 2: Notations for various combinations of boundary conditions, in which n and s indicate the directions normal and tangential to the respective plate edges.

Transverse constraints	In-plane constraints			
	$u_n = 0, u_s = 0$	$N_n = 0, u_s = 0$	$u_n = 0, N_{ns} = 0$	$N_n = 0, N_{ns} = 0$
Clamped: $w = 0; \phi_n = 0$	C_1	C_2	C_3	C_4
Simply supported: $w = 0; M_n = 0; \phi_s = 0$	S_1	S_2	S_3	S_4
Free: $M_n = 0; M_{ns} = 0; Q_n = 0$	F_1	F_2	F_3	F_4

TABLE 3: Convergence study of frequencies $\omega_i^* = \omega_i a^2\sqrt{\rho/(E_2h^2)}$ for a two-layered [0°/45°] square plate. $E_1/E_2 = 25$, $G_{12} = G_{13} = 0.5E_2$, $G_{23} = 0.2E_2$, $\nu_{12} = 0.25$.

a/h	M, N ($M = N$)	Mode sequence number 1	2	3	4	5	6
10	4	15.487	23.582	30.182	35.243	42.217	49.098
	5	15.480	23.401	30.005	33.514	38.099	44.303
	6	15.478	23.359	29.960	33.217	37.825	43.331
	7	15.478	23.356	29.957	33.112	37.691	42.997
	8	15.478	23.356	29.956	33.104	37.679	42.901
	9	15.478	23.356	29.956	33.103	37.676	42.889
	10	15.478	23.356	29.956	33.103	37.675	42.887
	Shi et al. [5]	15.504	23.399	29.991	33.170	37.740	42.973
20	4	19.248	30.947	42.338	50.788	71.618	81.985
	5	19.227	30.415	41.788	46.288	54.225	64.750
	6	19.221	30.279	41.586	45.434	53.531	62.080
	7	19.219	30.271	41.572	45.047	53.090	61.032
	8	19.219	30.268	41.569	45.016	53.047	60.655
	9	19.218	30.267	41.569	45.008	53.032	60.600
	10	19.218	30.267	41.568	45.007	53.030	60.587
	Shi et al. [5]	19.350	30.490	41.769	45.400	53.385	61.035
100	4	21.348	36.366	52.130	65.773	298.228	304.052
	5	21.314	35.002	50.893	56.783	68.251	86.403
	6	21.291	34.676	50.155	54.927	66.919	79.733
	7	21.289	34.660	50.124	53.854	65.602	76.972
	8	21.288	34.655	50.107	53.792	65.504	75.800
	9	21.288	34.654	50.106	53.757	65.449	75.631
	10	21.288	34.654	50.106	53.754	65.444	75.553
	Shi et al. [5]	21.802	35.692	51.304	55.298	67.257	77.843

TABLE 4: Comparison of fundamental frequency coefficient $\omega_i^* = \omega_i a^2\sqrt{\rho/(E_2h^2)}$ for a four layered [45°/−45°/45°/−45°] plate with different aspect ratios a/b. $E_1/E_2 = 40$, $G_{12} = 0.6E_2$, $G_{13} = G_{23} = 0.5E_2$, $\nu_{12} = 0.25$.

a/h		a/b 0.2	0.6	0.8	1	1.2	1.6	2
10	Present	9.013	13.02	15.74	18.62	21.59	27.66	34.57
	Alibeigloo et al. [31]	8.559	12.565	15.187	17.983	20.895	27.031	33.634
	Redy [32]	8.724	12.965	15.712	18.609	21.567	27.736	34.247
30	Present	9.965	15.409	19.293	23.638	28.377	39.062	51.480
	Alibeigloo et al. [31]	9.420	14.790	18.487	22.637	27.200	37.534	49.499
	Redy [32]	9.667	15.385	19.304	23.676	28.381	38.940	51.132
50	Present	10.056	15.66	19.70	24.25	29.25	40.70	54.25
	Alibeigloo et al. [31]	9.5016	15.0261	18.8586	23.195	28.003	39.05	52.686
	Redy [32]	9.816	15.689	19.759	24.343	29.321	40.653	53.989

TABLE 5: (a) Frequency parameters $\overline{\omega}_i$ for [0°/45°]and [0°/90°] carbon-epoxy AS4-3501-6 (Table 1), with different translational restraint parameter T_i^w, $T_i^u = T_i^v = R_i = 0$ $(i = 1,\ldots,4)$, and $a/h = 10$. (b) Frequency parameters $\overline{\omega}_i$ for [0°/45°] and [0°/90°] carbon-epoxy AS4-3501-6 (Table 1), with different translational restraint parameter T_i^w, $T_i^u = T_i^v = R_i = 0$ $(i = 1,\ldots,4)$, and $a/h = 100$.

(a)

k_f	Mode	Translational restraint parameter T_i^w						
		0.1	1	10	100	1000	10000	1.00E + 10
				$a/h = 10$				
	[0°/45°]							
0.2	1	0.060	0.187	0.520	0.969	1.161	1.191	1.195
	2	0.085	0.266	0.803	1.827	2.380	2.470	2.481
	3	0.085	0.267	0.822	2.079	2.930	3.063	3.078
	4	0.754	0.810	1.224	2.609	3.776	4.004	4.032
0.4	1	0.083	0.256	0.689	1.219	1.419	1.448	1.452
	2	0.117	0.367	1.083	2.284	2.829	2.912	2.922
	3	0.117	0.368	1.125	2.723	3.638	3.767	3.782
	4	0.890	0.978	1.590	3.325	4.530	4.735	4.759
0.6	1	0.101	0.310	0.824	1.443	1.671	1.705	1.709
	2	0.142	0.445	1.303	2.685	3.290	3.382	3.393
	3	0.142	0.447	1.364	3.274	4.342	4.492	4.509
	4	1.040	1.151	1.903	3.936	5.292	5.517	5.543
	[0°/90°]							
0.2	1	0.060	0.188	0.528	0.942	1.084	1.104	1.106
	2	0.085	0.267	0.814	1.970	2.657	2.756	2.768
	3	0.085	0.267	0.814	1.970	2.657	2.756	2.768
	4	0.628	0.697	1.164	2.642	3.708	3.888	3.909
0.4	1	0.083	0.258	0.705	1.181	1.321	1.340	1.342
	2	0.117	0.368	1.110	2.546	3.251	3.344	3.355
	3	0.117	0.368	1.110	2.546	3.251	3.344	3.355
	4	0.750	0.858	1.540	3.424	4.519	4.675	4.675
0.6	1	0.101	0.313	0.847	1.395	1.553	1.574	1.577
	2	0.142	0.446	1.343	3.035	3.829	3.932	3.944
	3	0.142	0.446	1.343	3.035	3.829	3.932	3.944
	4	0.910	1.040	1.864	4.089	5.329	5.514	5.536

(b)

k_f	Mode	Translational restraint parameter T_i^w						
		0.1	1	10	100	1000	10000	1.00E + 10
				$a/h = 100$				
	[0°/45°]							
0.2	1	0.060	0.187	0.525	1.014	1.276	1.329	1.340
	2	0.085	0.268	0.813	1.928	2.708	2.881	2.908
	3	0.085	0.269	0.832	2.214	3.489	3.771	3.815
	4	0.804	0.858	1.266	2.774	4.519	5.027	5.098
0.4	1	0.083	0.256	0.697	1.292	1.577	1.633	1.646
	2	0.117	0.369	1.099	2.445	3.269	3.440	3.468
	3	0.117	0.370	1.142	2.952	4.444	4.746	4.794
	4	0.947	1.032	1.642	3.588	5.531	6.011	6.073
0.6	1	0.101	0.310	0.835	1.531	1.859	1.925	1.940
	2	0.142	0.447	1.324	2.880	3.804	3.995	4.028
	3	0.143	0.450	1.385	3.555	5.304	5.654	5.711
	4	1.104	1.210	1.962	4.255	6.457	6.980	7.046

(b) Continued.

k_f	Mode	Translational restraint parameter T_i^w 0.1	1	10	100	1000	10000	1.00E + 10
	[0°/90°]							
	1	0.060	0.188	0.533	0.981	1.161	1.190	1.194
0.2	2	0.085	0.268	0.823	2.086	3.043	3.212	3.233
	3	0.085	0.268	0.823	2.086	3.043	3.212	3.233
	4	0.666	0.732	1.197	2.805	4.365	4.732	4.782
	1	0.083	0.258	0.714	1.239	1.422	1.449	1.453
0.4	2	0.117	0.370	1.126	2.738	3.783	3.948	3.969
	3	0.117	0.370	1.126	2.738	3.783	3.948	3.969
	4	0.797	0.901	1.583	3.699	5.418	5.774	5.821
	1	0.101	0.313	0.857	1.464	1.674	1.706	1.710
0.6	2	0.143	0.449	1.363	3.271	4.453	4.638	4.661
	3	0.143	0.449	1.363	3.271	4.453	4.638	4.661
	4	0.967	1.093	1.917	4.427	6.392	6.797	6.851

the shear locking effect and is extended here for laminated plates.

The classical boundary conditions considered in this study are depicted in Table 2. By keeping in mind that in the Ritz method only the geometric boundary conditions need to be satisfied, it is possible to work with any sets of required edge boundary condition and also is very simple the consideration of elastically restrained edges where there are not essential boundary conditions to satisfy.

Upon inserting the displacement forms (16) into the energy functional of the system (15), the minimization with respect to the coefficients of the displacement functions is given by

$$\frac{\partial \Pi}{\partial c_{ij}^{(u)}} = 0, \qquad \frac{\partial \Pi}{\partial c_{ij}^{(v)}} = 0, \qquad \frac{\partial \Pi}{\partial c_{ij}^{(w)}} = 0,$$
$$\frac{\partial \Pi}{\partial c_{ij}^{(\phi_x)}} = 0, \qquad \frac{\partial \Pi}{\partial c_{ij}^{(\phi_y)}} = 0. \tag{17}$$

From (17) a set of algebraic simultaneous equations is obtained. The number of these equations becomes $5 \times M \times N$. The algebraic equations obtained are given as follows, in the form of the generalized eigenvalue problem:

$$(\mathbf{K} - \omega^2 \mathbf{M})\{\mathbf{C}\} = \{\mathbf{0}\}, \tag{18}$$

where **K** and **M** are stiffness and inertia matrices, respectively (their expressions are given in the Appendix, $\{\mathbf{C}\}$ contains the unknown coefficients of (16).

For a nontrivial solution, the eigenvalues which make the determinant equal to zero, correspond to the free vibration frequencies.

4. Verification of the Formulation and Numerical Applications

4.1. General Description. The variational algorithm developed in this paper was programmed in Fortran language and used for the free vibration analysis of generally laminated thin and moderately thick laminated plates having different geometric parameters, stacking sequences, material properties, fiber volume fractions, and boundary conditions. The examples considered in this study are confined to laminates with layers of equal thickness, even though the procedure was formulated for plies with arbitrary thickness. In all cases the shear correction factor was taken a 5/6.

Let us introduce the terminology to be used throughout the remainder of the paper for describing the boundary conditions of the considered plates. The designation $C_iS_iF_iS_i$, for example, identifies a plate with edges (1) clamped, (2) simply supported, (3) free, and (4) simply supported (see Figure 1) the subscript i ($i = 1, \ldots, 4$) identifies the in-plane constraints according to Table 2. When the edges are elastically restrained against rotation or translation, the following nondimensional restraint parameters are used

$$T_i^{\bullet} = \frac{a^3 t_i^{\bullet}}{D_0}, \quad (i = 1, \ldots, 4 \text{ and } \bullet = u, v, w),$$
$$R_i = \frac{r_i a}{D_0}, \quad (i = 1, \ldots, 4), \tag{19}$$

where $D_0 = E_1 h^3 / 12(1 - \nu_{12}\nu_{21})$.

The main purposes of the numerical applications presented in this section are twofold. One is to demonstrate the accuracy, the flexibility, and the efficiency of the proposed method and the other is to produce some results which may be regarded as benchmark solutions for other academic research workers and design engineers.

4.2. Validation and Convergence Studies. The accuracy and reliability of the results obtained with the present approach are next demonstrated by comparing them with some selected values published by Shi et al. [5] for moderately thick ($a/h = 10$, 20) and thin ($a/h = 100$) arbitrarily clamped laminated plates. The comparison presented in Table 3 authenticates the validity of the present method for arbitrarily laminated plates. Very close agreement for the first

TABLE 6: (a) Frequency parameters $\overline{\omega}_i$ for [0°/45°] and [0°/90°] carbon-epoxy AS4-3501-6 (Table 1), with different rotational restraint parameter R_i, $T_i^w = T_i^u = T_i^v = \infty$ ($i = 1,..,4$), and $a/h = 10$. (b) Frequency parameters $\overline{\omega}_i$ for [0°/45°] and [0°/90°] carbon-epoxy AS4-3501-6 (Table 1), with different rotational restraint parameter R_i, $T_i^w = T_i^u = T_i^v = \infty$ ($i = 1,..,4$), and $a/h = 100$.

(a)

k_f	Mode	Rotational restraint parameter R_i						
		0.1	1	10	100	1000	10000	1.00E + 10
				$a/h = 10$				
	[0°/45°]							
0.2	1	1.461	1.623	1.943	2.056	2.071	2.072	2.072
	2	2.709	2.903	3.268	3.390	3.406	3.407	3.407
	3	3.350	3.497	3.861	4.010	4.030	4.032	4.032
	4	4.322	4.477	4.818	4.951	4.969	4.971	4.971
0.4	1	1.837	2.029	2.371	2.483	2.497	2.498	2.498
	2	3.249	3.493	3.875	3.988	4.002	4.003	4.003
	3	4.112	4.302	4.709	4.854	4.872	4.874	4.874
	4	5.129	5.333	5.700	5.823	5.838	5.840	5.840
0.6	1	2.182	2.408	2.803	2.931	2.947	2.949	2.949
	2	3.799	4.090	4.527	4.654	4.669	4.671	4.671
	3	4.884	5.119	5.611	5.783	5.804	5.806	5.807
	4	5.987	6.239	6.672	6.812	6.830	6.832	6.832
	[0°/90°]							
0.2	1	1.386	1.562	1.916	2.034	2.048	2.050	2.050
	2	2.901	3.110	3.561	3.719	3.738	3.740	3.741
	3	2.901	3.110	3.561	3.719	3.738	3.740	3.741
	4	4.052	4.287	4.764	4.970	4.994	4.996	4.996
0.4	1	1.755	1.969	2.352	2.468	2.482	2.483	2.483
	2	3.532	3.814	4.321	4.473	4.491	4.493	4.493
	3	3.532	3.814	4.321	4.473	4.491	4.493	4.493
	4	4.932	5.218	5.745	5.949	5.971	5.973	5.973
0.6	1	2.090	2.340	2.776	2.905	2.920	2.922	2.922
	2	4.167	4.511	5.102	5.272	5.292	5.294	5.295
	3	4.167	4.511	5.102	5.272	5.292	5.294	5.295
	4	5.833	6.179	6.790	7.021	7.045	7.047	7.048

(b)

k_f	Mode	Rotational restraint parameter R_i						
		0.1	1	10	100	1000	10000	1.00E + 10
				$a/h = 100$				
	[0°/45°]							
0.2	1	1.598	1.814	2.318	2.534	2.564	2.567	2.567
	2	3.108	3.410	4.115	4.415	4.457	4.461	4.462
	3	4.161	4.421	5.275	5.759	5.831	5.839	5.840
	4	5.366	5.689	6.611	7.100	7.173	7.181	7.181
0.4	1	2.044	2.316	2.901	3.135	3.166	3.169	3.170
	2	3.783	4.185	4.997	5.311	5.353	5.358	5.358
	3	5.265	5.627	6.699	7.240	7.317	7.325	7.326
	4	6.450	6.911	8.018	8.523	8.594	8.601	8.602
0.6	1	2.428	2.748	3.423	3.689	3.725	3.729	3.729
	2	4.418	4.896	5.821	6.171	6.217	6.222	6.223
	3	6.250	6.692	7.967	8.593	8.682	8.691	8.692
	4	7.498	8.058	9.337	9.897	9.974	9.983	9.983

(b) Continued.

k_f	Mode	Rotational restraint parameter R_i						
		0.1	1	10	100	1000	10000	1.00E + 10
	[0°/90°]							
0.2	1	1.493	1.719	2.223	2.413	2.437	2.440	2.440
	2	3.396	3.720	4.583	4.960	5.010	5.016	5.016
	3	3.396	3.720	4.583	4.960	5.010	5.016	5.016
	4	5.036	5.390	6.396	6.966	7.039	7.047	7.048
0.4	1	1.917	2.203	2.780	2.976	3.001	3.003	3.003
	2	4.202	4.665	5.725	6.122	6.172	6.178	6.178
	3	4.202	4.665	5.725	6.122	6.172	6.178	6.178
	4	6.173	6.688	7.937	8.560	8.634	8.641	8.642
0.6	1	2.286	2.621	3.277	3.494	3.521	3.523	3.524
	2	4.952	5.515	6.745	7.185	7.241	7.247	7.247
	3	4.952	5.515	6.745	7.185	7.241	7.247	7.247
	4	7.297	7.918	9.364	10.060	10.141	10.149	10.150

sixth nondimensional frequencies $\omega_i^* = \omega_i a^2\sqrt{\rho/(E_2h^2)}$ is obtained for all cases and display monotonic convergence tendency to constant values. For thick plates, as shown in Table 3, as number of N and M is increased from 7 to 10, the frequency parameter decreases merely 0.002% for the first mode and 0.26% for the sixth. For thin plates the relative decreases of the frequency parameters are 0.004% for the first mode and 1.88% for the sixth as the numbers of polynomials M, N are increased from 7 to 10, exhibiting slower convergence rate than that of moderately thick plates. Consequently the number of beam characteristic polynomials used in the following computations for thin and thick plates is chosen as $M = N = 7$.

The validation of the proposed methodology for different aspect ratios (a/b) is presented in Table 4, showing a good agreement with Alibeigloo et al. [31] and Reddy [32].

4.3. Numerical Results and Discussion. Several examples including new results for arbitrarily laminated plates with elastically restrained edges are presented in this section. The elastic properties of the composite materials used here are those shown in Table 1. The influence of different values of fiber volume ratios (k_f) is analyzed in several figures and tables.

Values of the first four frequency parameters $\overline{\omega}_i = \omega_i(a^2/h)\sqrt{\rho/E_{f1}}$ for square thick $(a/h = 10)$ and thin $(a/h = 100)$ unsymmetric laminated plates are shown for increasing values of the translational restraint parameter T_i^w, in Tables 5(a) and 5(b). Moreover, the influence of rotational restraint parameter R_i in the free vibration frequency coefficients is shown in Tables 6(a) and 6(b).

In Figures 2–4 the fundamental frequency coefficients $\overline{\omega}$ corresponding to two laminated square plates are plotted against the restraint parameters R_i and T_i^w. Figure 2 shows the variation of $\overline{\omega}$ for various values of the rotational restraint R_i, while Figure 3 shows the variation of $\overline{\omega}$ for various values of the translational restraint T_i^w. A major increase of frequency occurs when the elastic restraint values are in the interval 0.1–50. Figure 4 shows the variation of $\overline{\omega}$ for various values of the rotational and translational restraint parameters: (a) $R_i = 0$, $T_i^w = S$; (b) $R_i = S$, $T_i^w = \infty$, and (c) $R_i = T_i^w = S$. The obtained curves illustrate the restraint parameters intervals for which the frequency coefficient $\overline{\omega}$ is sensitive to R_i and T_i^w.

To asses the influence of the aspects ratio a/b in the laminated plate response, values of the first four frequency parameters $\overline{\omega}_i = \omega_i(a^2/h)\sqrt{\rho/E_{f1}}$ for rectangular thick $(a/h = 10)$ unsymmetric laminated plates are shown, for increasing values of the translational restraint parameter T_i^w (Table 7(a)) and the rotational restraint parameter R_i (Table 7(b)) considering $a/b = 1.5$ and $a/b = 2$.

Figure 5 shows the variation of $\overline{\omega}$ for various values of the rotational restraint R_i, while Figure 6 shows the variation of $\overline{\omega}$ for various values of the translational restraint T_i^w for rectangular laminated plates.

To evaluate the effect of different fiber orientation angles (β) and fiber volume fraction on the dynamic properties of the laminates, the variation of the first free vibration coefficient $\overline{\omega}_1$ is plotted in Figures 7 and 8, considering two lamination stacking sequences, $[\beta/-\beta]$ and $[0/\beta]$. Two boundary conditions have been included, $C_1C_1C_1C_1$ in Figure 7 and $S_1S_1S_1S_1$ in Figure 6. It is observed that the $[\beta/-\beta]$ laminate is more sensitive to the fiber orientation angle than $[0/\beta]$ lamination scheme. The adimensional frequency parameter is noticeable higher as the fiber volume fraction k_f increases and as the boundary conditions become clamped.

Finally, the first four free vibration coefficients are presented in Table 8 to illustrate the influence of various fiber volume fractions and boundary conditions on the dynamical behavior of an unsymmetric [0°/45°] laminated plate.

5. Concluding Remarks

A Ritz approach for free vibration analysis of general laminated plates with edges elastically restrained against translation and rotation is presented in this work. The study

TABLE 7: (a) Frequency parameters $\overline{\omega}_i$ for [0°/45°] carbon-epoxy AS4-3501-6 (Table 1), with different translational restraint parameter T_i^w, $T_i^u = T_i^v = R_i = 0$ $(i = 1,..,4)$, and $a/h = 10$. (b) Frequency parameters $\overline{\omega}_i$ for [0°/45°] carbon-epoxy AS4-3501-6 (Table 1), with different rotational restraint parameter R_i, $T_i^w = T_i^u = T_i^v = \infty$ $(i = 1,\ldots,4)$, and $a/h = 10$.

(a)

k_f	Mode	Translational restraint parameter T_i^w						
		0.1	1	10	100	1000	10000	1.00E + 10
				$a/b = 2$				
0.2	1	0.158	0.464	1.043	2.045	2.778	2.903	2.917
	2	0.184	0.574	1.625	3.592	4.520	4.704	4.726
	3	0.255	0.793	2.167	3.948	6.915	7.139	7.149
	4	1.652	1.823	2.838	5.127	7.177	7.419	7.449
0.4	1	0.199	0.578	1.268	2.392	3.046	3.144	3.156
	2	0.233	0.723	1.999	4.175	5.024	5.169	5.186
	3	0.322	0.999	2.649	4.774	7.759	8.125	8.148
	4	1.832	2.071	3.374	6.021	7.973	8.173	8.196
0.6	1	0.235	0.681	1.487	2.757	3.455	3.557	3.569
	2	0.275	0.853	2.347	4.837	5.763	5.915	5.933
	3	0.380	1.179	3.115	5.589	8.864	9.309	9.342
	4	2.104	2.394	3.945	6.992	9.186	9.408	9.434
				$a/b = 1.5$				
0.2	1	0.111	0.341	0.904	1.724	2.239	2.345	2.364
	2	0.141	0.443	1.335	3.326	4.739	5.023	5.067
	3	0.171	0.539	1.627	3.705	5.946	6.580	6.673
	4	1.423	1.542	2.285	4.768	7.832	8.845	8.980
0.4	1	0.140	0.429	1.108	2.012	2.501	2.598	2.617
	2	0.178	0.559	1.664	3.964	5.396	5.649	5.686
	3	0.216	0.681	2.037	4.480	6.688	7.229	7.313
	4	1.587	1.731	2.729	5.680	8.857	9.749	9.869
0.6	1	0.165	0.506	1.296	2.316	2.847	2.954	2.976
	2	0.210	0.660	1.957	4.587	6.194	6.469	6.509
	3	0.256	0.804	2.400	5.243	7.625	8.198	8.292
	4	1.815	1.989	3.182	6.582	10.085	11.034	11.163

(b)

k_f	Mode	Rotational restraint parameter $R_i^{\phi_x,\phi_y}$						
		0.1	1	10	100	1000	10000	1.00E + 10
				$a/b = 2$				
0.2	1	3.444	3.858	4.588	4.806	4.833	4.835	4.836
	2	5.467	5.742	6.278	6.450	6.472	6.474	6.474
	3	8.161	8.349	8.750	8.890	8.908	8.909	8.910
	4	8.357	8.741	9.485	9.722	9.751	9.754	9.754
0.4	1	3.854	4.314	4.989	5.183	5.183	5.185	5.186
	2	6.061	6.362	6.855	7.009	7.009	7.011	7.011
	3	8.971	9.187	9.552	9.678	9.678	9.680	9.680
	4	8.987	9.437	10.142	10.37	10.347	10.349	10.349
0.6	1	4.421	4.946	5.682	5.866	5.887	5.889	5.889
	2	6.961	7.309	7.858	8.006	8.024	8.026	8.026
	3	10.160	10.584	10.999	11.124	11.139	11.140	11.141
	4	10.352	10.720	11.512	11.709	11.732	11.734	11.735

(b) Continued.

k_f	Mode	Rotational restraint parameter $R_i^{\phi_x,\phi_y}$ 0.1	1	10	100	1000	10000	1.00E + 10
		$a/b = 1.5$						
0.2	1	2.764	3.203	4.114	4.438	4.479	4.483	4.484
	2	5.686	6.122	7.270	7.771	7.839	7.846	7.847
	3	7.047	7.669	9.272	9.946	10.035	10.045	10.046
	4	9.588	10.107	11.673	12.471	12.586	12.598	12.599
0.4	1	3.188	3.703	4.631	4.925	4.961	4.965	4.966
	2	6.470	7.016	8.260	8.739	8.801	8.807	8.808
	3	7.847	8.632	10.324	10.927	11.003	11.011	11.012
	4	10.679	11.329	13.007	13.742	13.843	13.853	13.854
0.6	1	3.663	4.252	5.272	5.585	5.624	5.628	5.628
	2	7.394	8.040	9.448	9.973	10.040	10.047	10.048
	3	8.920	9.842	11.721	12.360	12.440	12.448	12.449
	4	12.113	12.873	14.748	15.537	15.643	15.654	15.655

Table 8: Frequency parameters $\overline{\omega}_i$ for [0°/45°] carbon-epoxy AS4-3501-6 (Table 1), with different fiber volume fraction k_f and boundary conditions. $a/h = 10$.

Boundary conditions	k_f	Mode sequence number 1	2	3	4
$S_1S_1S_1S_1$	0.1	1.198	2.379	0.193	2.872
	0.2	1.435	2.683	0.166	3.337
	0.3	1.628	2.951	0.154	3.723
	0.4	1.804	3.212	0.148	4.092
	0.5	1.973	3.475	0.145	4.463
	0.6	2.143	3.754	0.145	4.859
	0.7	2.319	4.054	0.146	5.290
	0.8	2.489	4.344	0.148	5.724
$C_1C_1C_1C_1$	0.1	1.826	3.091	3.560	4.502
	0.2	2.073	3.407	4.032	4.961
	0.3	2.287	3.702	4.452	5.387
	0.4	2.498	4.003	4.874	5.824
	0.5	2.715	4.321	5.317	6.290
	0.6	2.949	4.671	5.807	6.809
	0.7	3.207	5.062	6.359	7.396
	0.8	3.466	5.449	6.934	7.982
$S_2S_2C_1C_1$	0.1	1.446	2.825	3.037	3.975
	0.2	1.651	3.137	3.469	4.621
	0.3	1.830	3.417	3.847	5.045
	0.4	2.002	3.697	4.213	5.470
	0.5	2.173	3.987	4.587	5.912
	0.6	2.351	4.303	4.989	6.395
	0.7	2.541	4.651	5.430	6.933
	0.8	2.724	4.989	5.876	7.464

includes the effective elastic moduli of each lamina obtained using the Mori-Tanaka mean field theory, which allows taking into account the influence of the fiber volume ratios and the elastic properties of the components (fiber and matrix) into the vibration behavior. The formulation is based on the first-order shear deformation theory, and the generalized displacements are approximate using sets of characteristic orthogonal polynomials generated by the Gram-Schmidt procedure. The consideration of all possible rotational and translational restraints allows generating any classical boundary condition, only approaching the corresponding spring parameter to zero or infinity. The algorithm is computationally efficient, and the solutions are stables and convergent. Close agreement with existing results in the literature is shown and new results are presented in tables and figures which could be useful for design and optimization problems of general long fiber-reinforced laminated plates.

Appendix

The matrices K and M in (18) are given by

$$[\mathbf{K}] = \begin{bmatrix} [K_{ijkh}^{uu}] & [K_{ijkh}^{uv}] & [K_{ijkh}^{uw}] & [K_{ijkh}^{u\phi_x}] & [K_{ijkh}^{u\phi_y}] \\ & [K_{ijkh}^{vv}] & [K_{ijkh}^{vw}] & [K_{ijkh}^{v\phi_x}] & [K_{ijkh}^{v\phi_y}] \\ & & [K_{ijkh}^{ww}] & [K_{ijkh}^{w\phi_x}] & [K_{ijkh}^{w\phi_y}] \\ & & & [K_{ijkh}^{\phi_x\phi_x}] & [K_{ijkh}^{\phi_x\phi_y}] \\ \text{sym} & & & & [K_{ijkh}^{\phi_y\phi_y}] \end{bmatrix}, \tag{A.1}$$

where

$$\begin{aligned} 3K_{ijkh}^{uu} ={}& A_{11}\iint_R p_i'^{(u)} p_k'^{(u)} q_j^{(u)} q_h^{(u)} dx\,dy \\ &+ A_{16}\iint_R \left(p_i'^{(u)} p_k^{(u)} q_j^{(u)} q_h'^{(u)} + p_i^{(u)} p_k'^{(u)} q_j'^{(u)} q_h^{(u)}\right) dx\,dy \\ &+ A_{66}\iint_R p_i^{(u)} p_k^{(u)} q_j'^{(u)} q_h'^{(u)} dx\,dy \\ &+ t_2^u \int_0^b \left[p_i^{(u)} p_k^{(u)}\right]_{x=1} q_j^{(u)} q_h^{(u)} dy \\ &+ t_4^u \int_0^b \left[p_i^{(u)} p_k^{(u)}\right]_{x=0} q_j^{(u)} q_h^{(u)} dy, \end{aligned}$$

$$\begin{aligned} K_{ijkh}^{uv} ={}& A_{12}\iint_R p_i'^{(u)} p_k^{(v)} q_j^{(u)} q_h'^{(v)} dx\,dy \\ &+ A_{16}\iint_R p_i'^{(v)} p_k'^{(u)} q_j^{(v)} q_h^{(u)} dx\,dy \\ &+ A_{26}\iint_R p_i^{(v)} p_k^{(u)} q_j'^{(v)} q_h'^{(u)} dx\,dy \\ &+ A_{66}\iint_R p_i'^{(v)} p_k^{(u)} q_j^{(v)} p_h'^{(u)} dx\,dy, \end{aligned}$$

$$K_{ijkh}^{uw} = 0, \qquad K_{ijkh}^{vw} = 0,$$

$$\begin{aligned} K_{ijkh}^{u\phi_x} ={}& B_{11}\iint_R p_i'^{(u)} p_k'^{(\phi_x)} q_j^{(u)} q_h^{(\phi_x)} dx\,dy \\ &+ B_{16}\iint_R p_i'^{(u)} p_k^{(\phi_x)} q_j^{(u)} q_h'^{(\phi_x)} dx\,dy \\ &+ B_{16}\iint_R p_i^{(u)} p_k'^{(\phi_x)} q_j'^{(u)} q_h^{(\phi_x)} dx\,dy \\ &+ B_{66}\iint_R p_i^{(u)} p_k^{(\phi_x)} q_j'^{(u)} p_h'^{(\phi_x)} dx\,dy, \end{aligned}$$

$$\begin{aligned} K_{ijkh}^{u\phi_y} ={}& B_{12}\iint_R p_i'^{(u)} p_k^{(\phi_y)} q_j^{(u)} q_h'^{(\phi_y)} dx\,dy \\ &+ B_{16}\iint_R p_i'^{(u)} p_k'^{(\phi_y)} q_j^{(u)} q_h^{(\phi_y)} dx\,dy \\ &+ B_{26}\iint_R p_i^{(u)} p_k^{(\phi_y)} q'^{(u)}_j q'^{(\phi_y)}_h dx\,dy \\ &+ B_{66}\iint_R p_i^{(u)} p_k'^{(\phi_y)} q'^{(u)}_j q_h^{(\phi_y)} dx\,dy, \end{aligned}$$

$$\begin{aligned} K_{ijkh}^{vv} ={}& A_{22}\iint_R p_i^{(v)} p_k^{(v)} q_j'^{(v)} q_h'^{(v)} dx\,dy \\ &+ A_{26}\iint_R \left(p_i^{(v)} p_k'^{(v)} q_j'^{(v)} q_h^{(v)} + p_i'^{(v)} p_k^{(v)} q_j^{(v)} q_h'^{(v)}\right) dx\,dy \\ &+ A_{66}\iint_R p_i'^{(v)} p_k'^{(v)} q_j^{(v)} q_h^{(v)} dx\,dy \\ &+ t_1^v \int_0^a p_i^{(v)} p_k^{(v)} \left[q_j^{(v)} q_h^{(v)}\right]_{y=0} dx \\ &+ t_3^v \int_0^b p_i^{(v)} p_k^{(v)} \left[q_j^{(v)} q_h^{(v)}\right]_{y=1} dx, \end{aligned}$$

$$\begin{aligned} K_{ijkh}^{v\phi_x} ={}& B_{12}\iint_R p_i^{(v)} p_k'^{(\phi_x)} q_j'^{(v)} q_h^{(\phi_x)} dx\,dy \\ &+ B_{16}\iint_R p_i'^{(v)} p_k'^{(\phi_x)} q_j^{(v)} q_h^{(\phi_x)} dx\,dy \\ &+ B_{26}\iint_R p_i^{(v)} p_k^{(\phi_x)} q_j'^{(v)} q_h'^{(\phi_x)} dx\,dy \\ &+ B_{66}\iint_R p_i'^{(v)} p_k^{(\phi_x)} q_j^{(v)} q_h'^{(\phi_x)} dx\,dy, \end{aligned}$$

$$\begin{aligned} K_{ijkh}^{v\phi_y} ={}& B_{22}\iint_R p_i^{(v)} p_k^{(\phi_y)} q_j'^{(v)} q_h'^{(\phi_y)} dx\,dy \\ &+ B_{26}\iint_R p_i^{(v)} p_k'^{(\phi_y)} q_j'^{(v)} q_h^{(\phi_y)} dx\,dy \\ &+ B_{26}\iint_R p_i'^{(v)} p_k^{(\phi_y)} q_j^{(v)} q_h'^{(\phi_y)} dx\,dy \\ &+ B_{66}\iint_R p_i'^{(v)} p_k'^{(\phi_y)} q_j^{(v)} q_h^{(\phi_y)} dx\,dy, \end{aligned}$$

$$
\begin{aligned}
K_{ijkh}^{ww} &= A_{44} \iint_R p_i^{(w)} p_k^{(w)} q_j'^{(w)} q_h'^{(w)} dx\,dy \\
&\quad + A_{55} \iint_R p_i'^{(w)} p_k'^{(w)} q_j^{(w)} q_h^{(w)} dx\,dy \\
&\quad + A_{45} \iint_R \left(p_i'^{(w)} p_k^{(w)} q_j^{(w)} q_h'^{(w)} \right. \\
&\qquad \left. + p_i^{(w)} p_k'^{(w)} q_j'^{(w)} q_h^{(w)} \right) dx\,dy \\
&\quad + t_1^w \int_0^a p_i^{(w)} p_k^{(w)} \left[q_j^{(w)} q_h^{(w)} \right]_{y=0} dx \\
&\quad + t_2^w \int_0^b \left[p_i^{(w)} p_k^{(w)} \right]_{x=1} q_j^{(w)} q_h^{(w)} dy \\
&\quad + t_3^w \int_0^a p_i^{(w)} p_k^{(w)} \left[q_j^{(w)} q_h^{(w)} \right]_{y=1} dx \\
&\quad + t_4^w \int_0^b \left[p_i^{(w)} p_k^{(w)} \right]_{x=0} q_j^{(w)} q_h^{(w)} dy, \\
K_{ijkh}^{w\phi_x} &= K \left(A_{45} \iint_R p_i^{(w)} p_k^{(\phi_x)} q_j'^{(w)} q_h^{(\phi_x)} dx\,dy \right. \\
&\qquad \left. + A_{55} \iint_R p_i'^{(w)} p_k^{(\phi_x)} q_j^{(w)} q_h^{(\phi_x)} dx\,dy \right), \\
K_{ijkh}^{w\phi_y} &= K \left(A_{44} \iint_R p_i^{(w)} p_k^{(\phi_y)} q_j'^{(w)} q_h^{(\phi_y)} dx dy \right. \\
&\qquad \left. + A_{45} \iint_R p_i'^{(w)} p_k^{(\phi_y)} q_j^{(w)} q_h^{(\phi_y)} dx\,dy \right), \\
K_{ijkh}^{\phi_x\phi_x} &= K A_{55} \iint_R p_i^{(\phi_x)} p_k^{(\phi_x)} q_j^{(\phi_x)} q_h^{(\phi_x)} dx\,dy \\
&\quad + D_{11} \iint_R p_i'^{(\phi_x)} p_k'^{(\phi_x)} q_j^{(\phi_x)} q_h^{(\phi_x)} dx\,dy \\
&\quad + D_{16} \iint_R \left(p_i'^{(\phi_x)} p_k^{(\phi_x)} q_j^{(\phi_x)} q_h'^{(\phi_x)} \right. \\
&\qquad \left. + p_i^{(\phi_x)} p_k'^{(\phi_x)} q_j'^{(\phi_x)} q_h^{(\phi_x)} \right) dx\,dy \\
&\quad + D_{66} \iint_R p_i^{(\phi_x)} p_k^{(\phi_x)} q_j'^{(\phi_x)} q_h'^{(\phi_x)} dx\,dy \\
&\quad + r_2 \int_0^b \left[p_i^{(\phi_x)} p_k^{(\phi_x)} \right]_{x=a} q_j^{(\phi_x)} q_h^{(\phi_x)} dy \\
&\quad + r_4 \int_0^b \left[p_i^{(\phi_x)} p_k^{(\phi_x)} \right]_{x=0} q_j^{(\phi_x)} q_h^{(\phi_x)} dy, \\
K_{ijkh}^{\phi_x\phi_y} &= K A_{45} \iint_R p_i^{(\phi_x)} p_k^{(\phi_y)} q_j^{(\phi_x)} q_h^{(\phi_y)} dx\,dy \\
&\quad + D_{12} \iint_R p_i'^{(\phi_x)} p_k^{(\phi_y)} q_j^{(\phi_x)} q_h'^{(\phi_y)} dx\,dy \\
&\quad + D_{16} \iint_R p_i'^{(\phi_x)} p_k'^{(\phi_y)} q_j^{(\phi_x)} q_h^{(\phi_y)} dx\,dy \\
&\quad + D_{26} \iint_R p_i^{(\phi_x)} p_k^{(\phi_y)} q_j'^{(\phi_x)} q_h'^{(\phi_y)} dx\,dy \\
&\quad + D_{66} \iint_R p_i^{(\phi_x)} p_k'^{(\phi_y)} q_j'^{(\phi_x)} q_h^{(\phi_y)} dx\,dy, \\
K_{ijkh}^{\phi_y\phi_y} &= K A_{44} \iint_R p_i^{(\phi_y)} p_k^{(\phi_y)} q_j^{(\phi_y)} q_h^{(\phi_y)} dx\,dy \\
&\quad + D_{22} \iint_R p_i^{(\phi_y)} p_k^{(\phi_y)} q_j'^{(\phi_y)} q_h'^{(\phi_y)} dx\,dy \\
&\quad + D_{26} \iint_R \left(p_i^{(\phi_y)} p_k'^{(\phi_y)} q_j'^{(\phi_y)} q_h^{(\phi_y)} \right. \\
&\qquad \left. + p_i'^{(\phi_y)} p_k^{(\phi_y)} q_j^{(\phi_y)} q_h'^{(\phi_y)} \right) dx\,dy \\
&\quad + D_{66} \iint_R p_i'^{(\phi_y)} p_k'^{(\phi_y)} q_j^{(\phi_y)} q_h^{(\phi_y)} dx\,dy \\
&\quad + r_1 \int_0^a p_i^{(\phi_y)} p_k^{(\phi_y)} \left[q_j^{(\phi_y)} q_h^{(\phi_y)} \right]_{y=0} dx \\
&\quad + r_3 \int_0^a p_i^{(\phi_y)} p_k^{(\phi_y)} \left[q_j^{(\phi_y)} q_h^{(\phi_y)} \right]_{y=b} dx.
\end{aligned}
\tag{A.2}
$$

with $K = 5/6$

$$
[\mathbf{M}] = \begin{bmatrix}
\left[M_{ijkh}^{uu}\right] & 0 & 0 & 0 & 0 \\
 & \left[M_{ijkh}^{vv}\right] & 0 & 0 & 0 \\
 & & \left[M_{ijkh}^{ww}\right] & 0 & 0 \\
 & & & \left[M_{ijkh}^{\phi_x\phi_x}\right] & 0 \\
\text{sym} & & & & \left[M_{ijkh}^{\phi_y\phi_y}\right]
\end{bmatrix},
$$

$$
\begin{aligned}
M_{ijkh}^{uu} &= I_0 \iint_R p_i^{(u)} p_k^{(u)} q_j^{(u)} q_h^{(u)} dx\,dy, \\
M_{ijkh}^{vv} &= I_0 \iint_R p_i^{(v)} p_k^{(v)} q_j^{(v)} q_h^{(v)} dx\,dy, \\
M_{ijkh}^{ww} &= I_0 \iint_R p_i^{(w)} p_k^{(w)} q_j^{(w)} q_h^{(w)} dx\,dy, \\
M_{ijkh}^{\phi_x\phi_x} &= I_2 \iint_R p_i^{(\phi_x)} p_k^{(\phi_x)} q_j^{(\phi_x)} q_h^{(\phi_x)} dx\,dy, \\
M_{ijkh}^{\phi_y\phi_y} &= I_2 \iint_R p_i^{(\phi_y)} p_k^{(\phi_y)} q_j^{(\phi_y)} q_h^{(\phi_y)} dx\,dy.
\end{aligned}
\tag{A.3}
$$

Acknowledgment

The authors wish to thank the economic support of CONICET (PIP no. 0105/2010) and CIUNSa.

References

[1] E. Reissner, "On transverse bending of plates, including the effect of transverse shear deformation," *International Journal of Solids and Structures*, vol. 11, no. 5, pp. 569–576, 1975.

[2] R. D. Mindlin, "Influences of rotatory inertia and shear inflexural motion of isotropic, elastic plates," *Journal of Applied Mechanics*, vol. 18, pp. 1031–1036, 1951.

[3] H. R. H. Kabir, "On the frequency response of moderately thick simply supported rectangular plates with arbitrary lamination," *International Journal of Solids and Structures*, vol. 36, no. 15, pp. 2285–2301, 1999.

[4] K. M. Liew, "Solving the vibation of thick symmetric laminates by Reissner/Mindlin plate theory and the p-Ritz method," *Journal of Sound and Vibration*, vol. 198, no. 3, pp. 343–360, 1996.

[5] J. W. Shi, A. Nakatani, and H. Kitagawa, "Vibration analysis of fully clamped arbitrarily laminated plate," *Composite Structures*, vol. 63, no. 1, pp. 115–122, 2004.

[6] M. Aydogdu and T. Timarci, "Vibration analysis of cross-ply laminated square plates with general boundary conditions," *Composites Science and Technology*, vol. 63, no. 7, pp. 1061–1070, 2003.

[7] L. Liu, L. P. Chua, and D. N. Ghista, "Mesh-free radial basis function method for static, free vibration and buckling analysis of shear deformable composite laminates," *Composite Structures*, vol. 78, no. 1, pp. 58–69, 2007.

[8] G. R. Liu, X. Zhao, K. Y. Dai, Z. H. Zhong, G. Y. Li, and X. Han, "Static and free vibration analysis of laminated composite plates using the conforming radial point interpolation method," *Composites Science and Technology*, vol. 68, no. 2, pp. 354–366, 2008.

[9] J. Wang, K. M. Liew, M. J. Tan, and S. Rajendran, "Analysis of rectangular laminated composite plates via FSDT meshless method," *International Journal of Mechanical Sciences*, vol. 44, no. 7, pp. 1275–1293, 2002.

[10] K. S. Numayr, R. H. Haddad, and M. A. Haddad, "Free vibration of composite plates using the finite difference method," *Thin-Walled Structures*, vol. 42, no. 3, pp. 399–414, 2004.

[11] W. Lanhe, L. Hua, and W. Daobin, "Vibration analysis of generally laminated composite plates by the moving least squares differential quadrature method," *Composite Structures*, vol. 68, no. 3, pp. 319–330, 2005.

[12] P. Ribeiro, "First-order shear deformation, p-version, finite element for laminated plate nonlinear vibrations," *AIAA Journal*, vol. 43, no. 6, pp. 1371–1379, 2005.

[13] A. J. M. Ferreira and G. E. Fasshauer, "Analysis of natural frequencies of composite plates by an RBF-pseudospectral method," *Composite Structures*, vol. 79, no. 2, pp. 202–210, 2007.

[14] P. Malekzadeh and S. A. Shahpari, "Free vibration analysis of variable thickness thin and moderately thick plates with elastically restrained edges by DQM," *Thin-Walled Structures*, vol. 43, no. 7, pp. 1037–1050, 2005.

[15] F. Ohya, M. Ueda, T. Uchiyama, and M. Kikuchi, "Free vibration analysis by the superposition method of rectangular Mindlin plates with internal columns resting on uniform elastic edge supports," *Journal of Sound and Vibration*, vol. 289, no. 1-2, pp. 1–24, 2006.

[16] D. Zhou, "Vibrations of Mindlin rectangular plates with elastically restrained edges using static Timoshenko beam functions with the Rayleigh-Ritz method," *International Journal of Solids and Structures*, vol. 38, no. 32-33, pp. 5565–5580, 2001.

[17] K. Khorshidi, "Vibro-acoustic analysis of Mindlin rectangular plates resting on an elastic foundation," *Scientia Iranica*, vol. 18, no. 1, pp. 60–69, 2001.

[18] L. H. Wu and Y. Lu, "Free vibration analysis of rectangular plates with internal columns and uniform elastic edge supports by pb-2 Ritz method," *International Journal of Mechanical Sciences*, vol. 53, no. 7, pp. 494–504, 2011.

[19] G. Y. Jin, H. Chen, J. T. Du, T. J. Yang, and W. Y. Li, "The Influence of edge restraining stiffness on the transverse vibrations of rectangular plate structures," *Journal of Marine Science and Application*, vol. 9, no. 4, pp. 393–402, 2010.

[20] A. R. Setoodeh and G. Karami, "A solution for the vibration and buckling of composite laminates with elastically restrained edges," *Composite Structures*, vol. 60, no. 3, pp. 245–253, 2003.

[21] G. Karami, P. Malekzadeh, and S. R. Mohebpour, "DQM free vibration analysis of moderately thick symmetric laminated plates with elastically restrained edges," *Composite Structures*, vol. 74, no. 1, pp. 115–125, 2006.

[22] A. S. Ashour, "Vibration of angle-ply symmetric laminated composite plates with edges elastically restrained," *Composite Structures*, vol. 74, no. 3, pp. 294–302, 2006.

[23] L. G. Nallim and R. O. Grossi, "Vibration of angle-ply symmetric laminated composite plates with edges elastically restrained," *Composite Structures*, vol. 81, no. 1, pp. 80–83, 2007.

[24] S. Chakraverty and M. Petyt, "Natural frequencies for free vibration of nonhomogeneous elliptic and circular plates using two-dimensional orthogonal polynomials," *Applied Mathematical Modelling*, vol. 21, no. 7, pp. 399–417, 1997.

[25] B. Singh and S. Chakraverty, "Flexural vibration of skew plates using boundary characteristic orthogonal polynomials in two variables," *Journal of Sound and Vibration*, vol. 173, no. 2, pp. 157–178, 1994.

[26] S. Chakraverty, R. B. Bhat, and I. Stiharu, "Free vibration of annular elliptic plates using boundary characteristic orthogonal polynomials as shape functions in the Rayleigh-Ritz method," *Journal of Sound and Vibration*, vol. 241, no. 3, pp. 524–539, 2001.

[27] S. T. Chow, K. M. Liew, and K. Y. Lam, "Transverse vibration of symmetrically laminated rectangular composite plates," *Composite Structures*, vol. 20, no. 4, pp. 213–226, 1992.

[28] H. Altenbach, J. Altenbach, and W. Kisssing, *Mechanics of Composite Structural Elements*, Springer, Berlin, Germany, 2004.

[29] T. Mori and K. Tanaka, "Average stress in matrix and average elastic energy of materials with misfitting inclusions," *Acta Metallurgica*, vol. 21, no. 5, pp. 571–574, 1973.

[30] P. D. Soden, M. J. Hinton, and A. S. Kaddour, "Lamina properties, lay-up configurations and loading conditions for a range of fibre-reinforced composite laminates," *Composites Science and Technology*, vol. 58, no. 7, pp. 1011–1022, 1998.

[31] A. Alibeigloo, M. Shakeri, and M. R. Kari, "Free vibration analysis of antisymmetric laminated rectangular plates with distributed patch mass using third-order shear deformation theory," *Ocean Engineering*, vol. 35, no. 2, pp. 183–190, 2008.

[32] J. N. Reddy, "A simple higher-order theory for laminated composite plates," *ASME Journal of Applied Mechanics*, vol. 51, no. 4, pp. 745–752, 1984.

[33] J. D. Eshelby, "The determination of the elastic field of an ellipsoidal inclusion and related problems," *Proceedings of the Royal Society*, vol. 241, pp. 376–396, 1957.

[34] G. P. Tandon and G. J. Weng, "The effect of aspect ratio of inclusions on the elastic properties of unidirectionally aligned composites," *Polymer Composites*, vol. 5, no. 4, pp. 327–333, 1984.

[35] J. N. Reddy, *Mechanics of Laminated Composite Plates and Shells: Theory and Analysis*, CRC Press, Boca Raton, Fla, USA, 2nd edition, 2003.

[36] J. M. Whitney, *Structural Analysis of Laminated Anisotropic Plates*, Technomic Publishing Co., Lancaster, Pa, USA, 1987.

[37] L. G. Nallim and R. O. Grossi, "On the use of orthogonal polynomials in the study of anisotropic plates," *Journal of Sound and Vibration*, vol. 264, no. 5, pp. 1201–1207, 2003.

[38] L. G. Nallim and S. Oller, "An analytical-numerical approach to simulate the dynamic behaviour of arbitrarily laminated

composite plates," *Composite Structures*, vol. 85, no. 4, pp. 311–325, 2008.

[39] R. B. Bhat, "Plate deflection using orthogonal polynomials," *Journal of Engineering Mechanics*, vol. 101, no. 11, pp. 1301–1309, 1985.

[40] N. M. Auciello and A. Ercolano, "A general solution for dynamic response of axially loaded non-uniform Timoshenko beams," *International Journal of Solids and Structures*, vol. 41, no. 18-19, pp. 4861–4874, 2004.

16

Three-Dimensional Acoustic Source Mapping with Different Beamforming Steering Vector Formulations

Ennes Sarradj

Department of Technical Acoustics, Brandenburg University of Technology, Siemens-Halske-Ring 14, 03046 Cottbus, Germany

Correspondence should be addressed to Ennes Sarradj, ennes.sarradj@tu-cottbus.de

Academic Editor: Mario Kupnik

Acoustic source mapping techniques using acoustic sensor arrays and delay-and-sum beamforming techniques suffer from bad spatial resolution at low-aperture-based Helmholtz numbers. This is especially a problem for three-dimensional map grids, when the sensor array is not arranged around the region spanned by the grid but on only one side of it. Then, the spatial resolution of the result map in the direction pointing away from the array is much worse than in the other lateral directions. Consequently, deconvolution techniques need to be applied. Some of the most efficient deconvolution techniques rely on the properties of the spatial beamformer filters used. As these properties are governed by the steering vectors, four different steering vector formulations from the literature are examined, and their theoretical background is discussed. It is found that none of the formulations provide both the correct location and source strength. As a practical example the CLEAN-SC deconvolution methodology is applied to simulated data for a three-source scenario. It is shown that the different steering vector formulations are not equally well suited for three-dimensional application. The two preferred formulations enable the correct estimation of the source location at the cost of a negligible error in the estimated source strength.

1. Introduction

In the context of acoustic measurements, methods based on acoustic sensor arrays can be used to locate acoustic sources and to estimate their strength [1]. In most cases, these methods are adopted to produce acoustic source maps. In general, such maps can be thought of as an image of the spatial distribution of an indicator quantity of source strength.

Acoustic source mapping techniques using beamforming methods have been widely applied for the study of acoustic sources (e.g., for trains [2], aeroacoustic testing [3, 4], airframe noise [5, 6], noise source characterization at a helicopter [7], and jet noise [8]). These methods use the signals from an array of acoustic sensors (mostly microphones) to filter out the signal from a source at an assumed location. Such a spatial filter behaves like a directional sound receiver with a directional characteristic that favours sound emanating from the assumed source location [9]. If several of such spatial filters are applied in parallel for a number of different assumed source locations, an acoustic source mapping may be generated from the filter outputs. Usually the assumed source locations are arranged in some sort of grid and the amplitude of the output from each individual filter is mapped on the respective grid location. Thus, the filters constitute a mapping device.

The map produced by this device is an image of the spatial source distribution. Ideally, it shall have the following properties: if a certain assumed source location in the grid coincides with an actual source location, the map shows a higher value at this location. If there is no coincidence of assumed and actual source location, the map shows a lower value. Moreover, stronger sources shall result in higher values in the map. Thus, the map provides information about the location of the source, and it allows to estimate the strength of the source.

The feasibility of this approach depends on the properties of the beamformer mapping device as given by the point spread function. The point spread function is the spatial impulse response of the beamformer. It can be thought of being the map that is produced if only one single point source is present at a certain location. It shows the image of the source as a spot at the source location (main lobe) that is accompanied by a number of spots (side lobes) at other locations and lower in level. This mapping is imperfect for two reasons. First, the width of the main lobe limits the spatial resolution because sources that are too close to each other will produce a mapping very similar to that of a single source. Second, the images of weaker sources may be masked by the side lobes of a stronger source. The point spread function depends on a number of factors: the number and the geometrical layout of the array microphones, the array aperture, the frequency, and the type and properties of the filter. It also depends on the source location.

A number of deconvolution techniques have been developed to recover the true spatial source distribution from the beamformer result by removing the influence of the point spread function. Some rely on precalculated point spread functions (e.g., DAMAS [10]), other approaches assess the point spread function from the acoustic data recorded during the measurement (e.g., CLEAN-SC [11]). A critical point, especially for the latter techniques, is that they require the maximum in the map to coincide with an actual source position.

The beamforming source mapping approach is typically applied using a planar two-dimensional grid. In this case, all sources are mapped into one plane regardless of their actual position. In situations where the acoustic sources under test are not in a common plane (e.g., for complex machinery parts, engines, and some aeroacoustic sources, such as, landing gear and pantographs), this leads to an erroneous source mapping. Therefore, a three-dimensional mapping is desirable that allows source localisation in three dimensions. In principle, the three-dimensional application of beamforming techniques is straightforward and can be easily realised by using a three-dimensional grid [12–20]. However, there appear to be some practical problems in the application.

First, the resolution in the third dimension (depth-wise) is much worse than in the other dimensions unless the microphone array arrangement encloses the source region to be mapped. The second problem is the larger number of points in a three-dimensional grid. While for a two-dimensional mapping some thousand grid points may be sufficient, a three-dimensional grid can easily have some hundred thousand points. Because one filter per grid point is required to calculate the result, the computational effort increases considerably. That is why most applications are using only several ten thousand grid points.

As with two-dimensional map grids, the resolution can be improved using deconvolution techniques also for the third dimension. The computational effort connected with these techniques is generally high and depends on the number of grid points. Those deconvolution techniques that require precalculated point spread functions (e.g., DAMAS [10]) also require to solve a system of equations with as many unknowns as there are grid points. The effort for the estimation of the point spread function increases with the fourth power of the number of grid points in the map grid. The effort to solve the system of equations itself grows with the second to third power. Thus, the large number (hundreds of thousands or more) of grid points required for a three-dimensional mapping with fine spatial resolution renders the application of these techniques inefficient. A possible solution that has been proposed [21] is to assume that the point spread function is assumed to be shift invariant. Thus, the effort would reduce substantially and it would increase then only with the second power of the number of grid points for the estimation of the point spread function and with somewhat less than the second power for the solution of the system of equations based on the fast fourier transform. However, this approach is limited to cases where the source region is small compared with its distance to the array [21]. Thus, it is not applicable to the general case of a larger source region.

Computationally less demanding deconvolution techniques, such as, CLEAN-SC that do not require solving a huge system of equations in turn need to find sources by maximums in the three-dimensional map. Therefore, a final problem in the three-dimensional application of beamforming mapping techniques is to find spatial filters that have the desired properties also in three dimensions to provide these maximums. While this problem arises specifically for deconvolution methods that require maximums in the map to coincide with acoustic sources, it is relevant because the computational cost increases roughly linearly with the number of grid points. This makes these techniques most appropriate for three-dimensional application and allows for practical application with several hundred thousand grid points [17–19].

The problem of desired beamformer filter properties shall be considered here by comparing different spatial filter characteristics given by different steering vectors and analysing their properties with regard to three-dimensional beamforming with deconvolution. In the remainder of this contribution the theoretical basis for the beamforming methods is briefly presented, and four different spatial filter characteristics are discussed. Their properties are demonstrated using a single-source scenario as an example. Finally, a slightly more realistic case is considered. Simulated data for three-dimensional source mapping is analysed using the different beamformers and the CLEAN-SC deconvolution approach. The results are compared with results from two-dimensional source mapping.

2. Theory

First, the analysis of a single sound source located at $\mathbf{x}_s$ using an array of N microphones is assumed. The complex-valued sound pressure p at the i-th microphone at $\mathbf{x}_i$ is

$$p(\mathbf{x}_i) = a(\mathbf{x}_i, \mathbf{x}_0, \mathbf{x}_s) q(\mathbf{x}_s). \quad (1)$$

The strength of the source is characterised by the sound pressure q at the reference location $\mathbf{x}_0$ due to that source. Though in principle $\mathbf{x}_0$ can be freely chosen, for the purpose of the following analysis it is set to the array centre at $(1/N) \sum_{i=1}^{N} \mathbf{x}_i$. The transfer function a depends on the type of the source, its location $\mathbf{x}_s$, and the environmental conditions.

If a monopole source under free-field conditions is assumed and no flow is present, the transfer function is given by:

$$a(\mathbf{x}_i, \mathbf{x}_0, \mathbf{x}_s) = \frac{r_{s,0}}{r_{s,i}} e^{-jk(r_{s,i}-r_{s,0})}, \tag{2}$$

with $r_{s,i} = |\mathbf{x}_s - \mathbf{x}_i|$ and $r_{s,0} = |\mathbf{x}_s - \mathbf{x}_0|$ indicating the distance between the source and the microphone location and the distance between the source and the array centre location, respectively, and k is the wave number. The vector of sound pressures at the microphones due to a source at $\mathbf{x}_s$ is given by $\mathbf{p} = \mathbf{a}(\mathbf{x}_0, \mathbf{x}_s)q(\mathbf{x}_s)$. The transfer vector $\mathbf{a}(\mathbf{x}_0, \mathbf{x}_s)$ contains all respective transfer functions and accounts for the individual time delays and attenuations of the sound that travels from the source to the microphones.

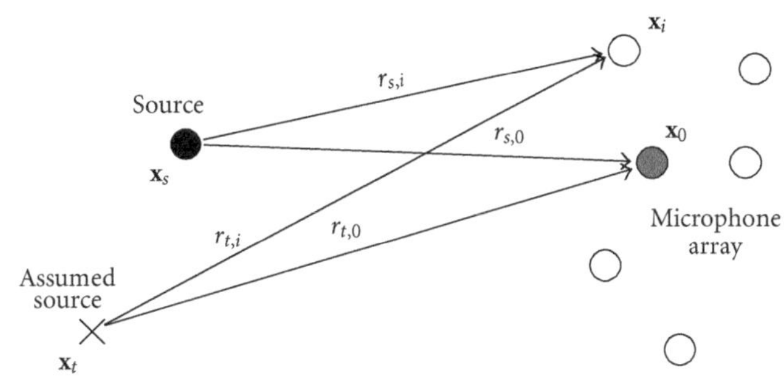

FIGURE 1: Microphone array and distances to actual and assumed source location.

The beamformer filter is realised by calculating the weighted sum of the microphone sound pressures using complex-valued weight factors. The vector $\mathbf{h}(\mathbf{x}_t)$ of these factors is called the steering vector and depends on an assumed source location $\mathbf{x}_t$. The filter output is then:

$$p_F(\mathbf{x}_t) = \mathbf{h}(\mathbf{x}_t)^H \mathbf{p}, \tag{3}$$

where the superscript H denotes the hermitian transpose. Instead of p_F, the real-valued autopower spectrum B of the filter output can be used as a quantity to construct a source map. Using the cross spectral matrix of microphone signals $\mathbf{G}$, it can be written as

$$\begin{aligned} B(\mathbf{x}_t) &= E\{p_F(\mathbf{x}_t)p_F^*(\mathbf{x}_t)\} = \mathbf{h}^H(\mathbf{x}_t)E\{\mathbf{p}\mathbf{p}^H\}\mathbf{h}(\mathbf{x}_t) \\ &= \mathbf{h}^H(\mathbf{x}_t)\mathbf{G}\mathbf{h}(\mathbf{x}_t), \end{aligned} \tag{4}$$

with $E\{\}$ denoting the expectation operator and the superscript $*$ denoting the complex conjugate.

Two properties of the beamformer filter are desirable for a successful application in the case of acoustic source mapping. First, the filter should provide maximum output power when the assumed and the actual source position are the same:

$$B(\mathbf{x}_t = \mathbf{x}_s) > B(\mathbf{x}_t \neq \mathbf{x}_s). \tag{5}$$

This property is essential to have a map showing the peak value at the source position. Second, if the assumed and the actual source coincide, the filter output should be a measure of the source strength. This is true, if

$$B(\mathbf{x}_t = \mathbf{x}_s) = CE\{qq^*\} \tag{6}$$

holds, where C is an arbitrary constant.

The properties of the beamformer filter are governed by the steering vector. This vector depends on the assumed source location x_t also referred to as steering location. While not further considered here, it can be noted that it may also depend on the measured data itself and adapt the filter properties to the data. Sometimes the distance between array and source is very large and a plane wave propagation can be assumed. In this case, x_t is replaced by the direction of arrival and the distance between array and source is set to be practically infinite. However, for three-dimensional application this approach is not feasible.

An in-depth examination of the literature on acoustic beamforming reveals that there are at least four choices available regarding the formulation of the steering vector elements h_i under these circumstances. In the following, these formulations are presented and analysed regarding the desirable properties stated in (5) and (6).

2.1. Formulation I. The most basic idea is to simply compensate for the phase delay [22] between assumed source and the individual microphone. The steering vector elements are then derived from the phase part of the transfer vector elements for the assumed source location:

$$h_i^{\mathrm{I}} = \frac{1}{N}\frac{a_i(\mathbf{x}_0, \mathbf{x}_t)}{|a_i(\mathbf{x}_0, \mathbf{x}_t)|} = \frac{1}{N}e^{-jk(r_{t,i}-r_{t,0})}. \tag{7}$$

The latter part of this equation holds if the transfer function from (2) is assumed and $r_{t,i} = |\mathbf{x}_t - \mathbf{x}_i|$ and $r_{t,0} = |\mathbf{x}_t - \mathbf{x}_0|$ (see Figure 1). Given this steering vector formulation that shall be referred to here as "formulation I", the output power of the beamformer filter can be derived for the case $\mathbf{x}_t = \mathbf{x}_s$:

$$B(\mathbf{x}_t = \mathbf{x}_s) = \left(\frac{1}{N}\sum_{i=1}^{N}\frac{r_{s,0}}{r_{s,i}}\right)^2 E\{qq^*\}. \tag{8}$$

Because $\sum_{i=1}^{N}(r_{s,0}/r_{s,i}) \approx N$, the output is an estimate of the source strength. However, it becomes obvious that the condition (6) is met only approximately.

Condition (5) requires a local maximum of $B(\mathbf{x}_t)$ at $\mathbf{x}_t = \mathbf{x}_s$. A necessary condition for this is that all partial derivatives of B with respect to the elements of $\mathbf{x}_t$ are zero. For this, it is sufficient to have

$$\left.\frac{\partial}{\partial r_{t,i}}B(\mathbf{x}_t)\right|_{\mathbf{x}_t=\mathbf{x}_s} = 0, \qquad \left.\frac{\partial}{\partial r_{t,0}}B(\mathbf{x}_t)\right|_{\mathbf{x}_t=\mathbf{x}_s} = 0. \tag{9}$$

This holds for the partial derivatives that can be computed when the steering vector given by (7) is applied in (4). Thus, the necessary condition to have a maximum response at the source location is met in this case.

2.2. Formulation II. Another formulation of the steering vector that is frequently used in the literature (e.g., [3, 5]) aims at compensating also for the amplitude:

$$h_i^{\mathrm{II}} = \frac{1}{N}\frac{a_i(\mathbf{x}_0,\mathbf{x}_t)}{a_i(\mathbf{x}_0,\mathbf{x}_t)a_i^*(\mathbf{x}_0,\mathbf{x}_t)} = \frac{1}{N}\frac{r_{t,i}}{r_{t,0}}e^{-jk(r_{t,i}-r_{t,0})}. \quad (10)$$

This formulation assures that condition (6) is met because $B(\mathbf{x}_t = \mathbf{x}_s) = E\{qq^*\}$. However, this comes at the cost that the derivatives in (9) do not vanish. There is no maximum at $\mathbf{x}_t = \mathbf{x}_s$, and consequently condition (5) is not met.

2.3. Formulation III. The third formulation of the steering vector [23] that should be discussed is based on the idea that signals from the assumed source position should pass undistorted through the filter, so that $B(\mathbf{x}_t = \mathbf{x}_s) = E\{qq^*\}$. At the same time signals from all other positions should be attenuated as much as possible. This is equivalent to minimising the filter response to spatially white noise [24]. By solving this optimisation problem, the formulation

$$h_i^{\mathrm{III}} = \frac{a_i(\mathbf{x}_0,\mathbf{x}_t)}{\mathbf{a}^H(\mathbf{x}_0,\mathbf{x}_t)\mathbf{a}(\mathbf{x}_0,\mathbf{x}_t)} = \frac{1}{r_{t,0}r_{t,i}\sum_{j=1}^{N}\left(1/r_{t,j}^2\right)}e^{-jk(r_{t,i}-r_{t,0})} \quad (11)$$

arises. The steering vector here is parallel to $\mathbf{a}(\mathbf{x}_0,\mathbf{x}_s)$ when $\mathbf{x}_t = \mathbf{x}_s$. Again, this formulation does not meet condition (5).

2.4. Formulation IV. Formulation IV introduces a steering vector that is also parallel to $\mathbf{a}(\mathbf{x}_0,\mathbf{x}_s)$ when $\mathbf{x}_t = \mathbf{x}_s$ but uses a normalisation to ensure that $h_i^* h_i$ remains constant. This formulation is usually developed via a least square minimisation of the error between modelled and measured sound pressures at the microphones (e.g., [4, 8]). Using the normalisation $h_i^* h_i = 1/N$, the formulation reads

$$h_i^{\mathrm{IV}} = \frac{1}{\sqrt{N}}\frac{a_i(\mathbf{x}_0,\mathbf{x}_t)}{\sqrt{\mathbf{a}^H(\mathbf{x}_0,\mathbf{x}_t)\mathbf{a}(\mathbf{x}_0,\mathbf{x}_t)}} = \frac{1}{r_{t,i}\sqrt{N\sum_{j=1}^{N}\left(1/r_{t,j}^2\right)}}e^{-jk(r_{t,i}-r_{t,0})}. \quad (12)$$

In this case condition (5) is met, but the response for $\mathbf{x}_t = \mathbf{x}_s$ is only an approximate measure of the source strength:

$$B(\mathbf{x}_t = \mathbf{x}_s) = \frac{1}{N}\sum_{i=1}^{N}\frac{r_{s,0}^2}{r_{s,i}^2}E\{qq^*\}. \quad (13)$$

2.5. Comparison of Formulations. It can be concluded that in theory neither of this four formulations of the steering vector has both properties desirable for the application of the beamformer to acoustic source mapping. The first formulation (7) and the fourth formulation (12) provide the correct location but produce an error in the source strength and both the second formulation (10) and the third formulation (6) provide the correct source strength, but the maximum does not coincide with the correct location. However, it should be borne in mind that for practical application it suffices to get approximate estimates of both location and source strength with acceptable accuracy.

In contrast to the theoretical model used up to here, in a practical scenario there is often more than one source. The beamformer output is then a superposition of contributions from the individual sources. Thus, the presence of multiple sources has an impact on the performance of the beamformer filter (see [23]). If the signals from these sources are mutually uncorrelated, $B(\mathbf{x}_t)$ is the sum of their nonnegative contributions. In such a case the results will deviate somewhat from those of the mathematical analysis presented for the one source scenario. Any conclusions that were derived for the properties of the different formulations regarding conditions (5) and (6) are not rigorously valid, but only approximately. This becomes especially important when sources are closely spaced with distances less than the wavelength. As a consequence, the low-frequency application of source-mapping beamforming techniques leads to results of limited value. Appropriate deconvolution techniques may improve the results in this case and are frequently applied for this reason.

While for all four formulations of the steering vector meaningful practical results have been reported for acoustic source mapping applications, there are no results available yet that compare the mappings produced by the application of these different steering vectors. Moreover, the vast majority of these applications use two-dimensional mapping grids and assume a priori that all relevant sources are located in the mapping plane. A three-dimensional mapping grid does not need this assumption but will usually require a deconvolution technique to deal with the otherwise inadequate depth-wise resolution.

The large number of points in a three-dimensional mapping grid calls for an deconvolution technique that is computationally efficient. Computationally less demanding deconvolution techniques, like CLEAN-SC, rely on the correct estimation of the source location from the maximum in the map. Because of this requirement, it is important to assess the quality of the results that follow the application of the different formulations of the steering vector for three-dimensional source mapping. In what follows, the effect of using different steering vector formulations shall be analysed for both two- and three-dimensional acoustic source mapping on the basis of simulated measurements.

3. Results

Most practical applications are concerned with sources that are not compact but spatially extended. A usual assumption is to assume that the sources can be seen as spatial distributions of uncorrelated point sources. Thus, in order to analyze the different steering vectors, no extended source is considered here but a simple point source scenario. Results

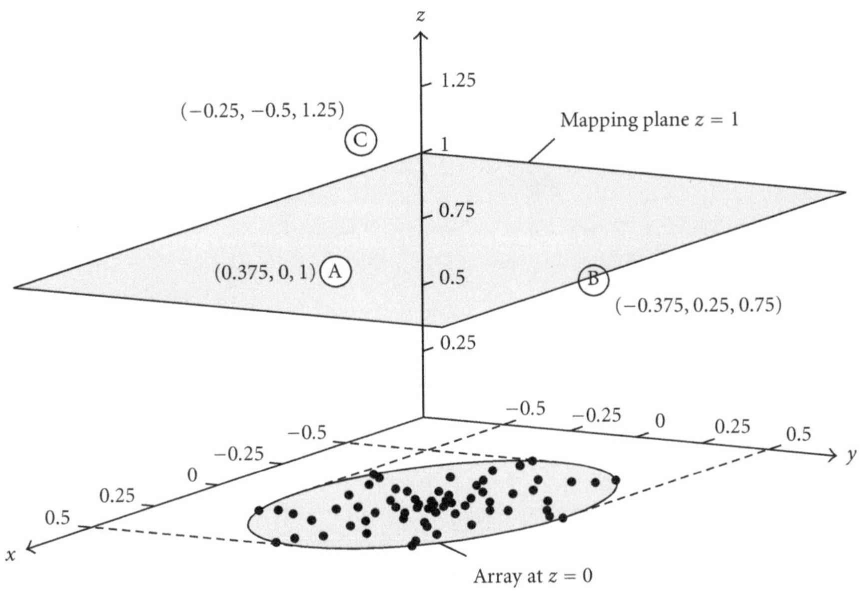

FIGURE 2: Set-up used for the simulations: 64-channel microphone array and three uncorrelated sources A, B, C with identical source strength, coordinates, and lengths are nondimensionalised by the array aperture.

for practical cases with extended sources that use three-dimensional mapping can be found elsewhere [18, 19, 25].

The source mappings that should be discussed here are based on simulated data that was generated using the set-up shown in Figure 2. It uses a 64 microphone array to analyse three point sources. The microphone array is planar and has a layout that consists of 7 spiral arms that contain 9 logarithmically spaced microphones each and an additional microphone in the centre. The aperture d of the array, defined by the diameter of the smallest circle containing all microphones, is used as a scaling parameter in the analysis. All coordinates and lengths in the analysis are nondimensionalised by the aperture. Frequencies are given in nondimensional form as Helmholtz numbers $\mathrm{He} = d/\lambda = fd/c$ with λ being the wavelength and c being the speed of sound.

The positions of the sources are chosen randomly to span a region with lateral dimensions comparable to the array itself. The distance to the array varies between 0.75 and 1.25 apertures. This choice was made because the application of threedimensional source mapping is most interesting when distances of the sources to the array are somewhat different. However, very large ratios of the distances (i.e., one source very close and another source far away) are not likely to appear in a practical application. Altogether the scenario is somewhat representative of a situation where the object under test has dimensions comparable to the aperture. Typically this would allow for the beamforming analysis at frequencies with wavelengths much smaller than the object. An aperture much larger than the object is desirable for the analysis at lower frequencies. When the application of large aperture arrays is not practicable, the analysis requires deconvolution methods. As results from deconvolution are of special interest here, a scenario representative for this case was chosen as illustrative example.

The simulated microphone signals were calculated using a transfer function similar to (2). The point sources were driven by simulated white noise signals from different gaussian random processes to ensure that they are not coherent. All three sources had the same power. Nevertheless, because of the different distances to the array plane and therefore also to the array centre $\mathbf{x}_0$, the relative sound pressure levels at $\mathbf{x}_0$ due to the individual sources were different: 0 dB, 1.7 dB, and −2.2 dB for source A, B, and C, respectively.

The array was placed in the plane $z = 0$ and two different grids were used. The first grid used for the three-dimensional source mapping covered a block-shaped region with $-1 < x < 1$, $-1 < y < 1$, and $0.125 < z < 2$. It had a uniform grid spacing of 1/32 aperture and the overall number of grid points was 257, 725. The second grid for two-dimensional source mapping in a plane parallel to the array had the same spacing and had the same extent $-1 < x < 1$, $-1 < y < 1$, but for $z = 1$. The overall number of points in this case was 4225. The simulated microphone signals were sampled at a rate that corresponds to $\mathrm{He} = 64$. A fast Fourier transform with prior von Hann weighting was applied for every channel to 1000 consecutive, 50% overlapping blocks of 1024 samples each. All 64^2 cross spectra were calculated and averaged over the 1000 blocks to produce the cross spectral matrix.

The quality of an acoustic source mapping can be determined by evaluating the errors in the source levels and source locations that are estimated using the mapping result. According to the definition in (1), the source level,

$$L_s = 10\log_{10}\frac{E\{qq^*\}}{p_{\mathrm{ref}}^2}\mathrm{dB}, \tag{14}$$

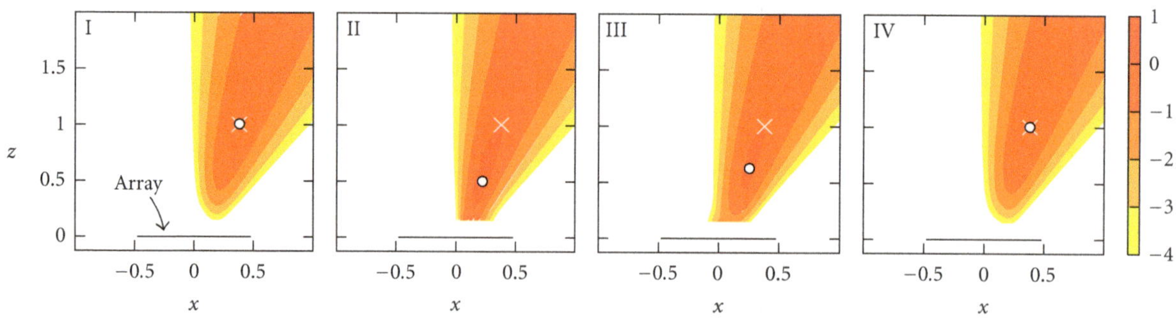

FIGURE 3: Maps of sound pressure level contributions at the array centre (in dB relative to the correct value) for steering vector formulations I–IV, Helmholtz number He = 2, $y = 0$, white dot: location of the maximum, cross: true source location.

of a certain source is defined as the sound pressure level calculated from the sound pressure q at the array centre caused by that source (p_{ref} is the reference sound pressure). The estimated source level is then

$$L_e = 10\log_{10}\frac{B}{p_{\mathrm{ref}}^2}\,\mathrm{dB}, \quad (15)$$

and the error in this level is given by $\Delta L = L_e - L_s$, where L_s is taken to be the true level.

The source location is a vector quantity and involves three components. While the error here could be given as the distance between the true and the estimated location, this quantity is always positive and contains no information on the spatial arrangement. Instead, the error in the estimated source location shall be defined here as an error of the distance r between source and the array centre. It is given by $\Delta r = r_e - r_s$, where r_e and r_s are the estimated and true distances, respectively. As it is reasonable to assume that this error will also increase with the distance, it is used here in the normalised form $\Delta r/r_s$. As the errors in source location and level also depend on the frequency, they are discussed here regarding their dependence on the Helmholtz number.

3.1. Single Source. In the first test case, only source A was operated. In this simple single source scenario it is feasible to use the classic beamforming approach (4) and to do without a deconvolution technique. In Figure 3, the results for all four formulations of the steering vector (I–IV) are compared for He = 2. The source mapping itself is three-dimensional. However, for clarity of presentation only a slice of the mapping along the plane $y = 0$ (perpendicular to the array plane) is shown which contains the true source location. The four formulations obviously lead to different mapping results. In agreement with the theoretical analysis, both formulations I and IV meet the condition (5) that the maximum in the map coincides with the actual source position. For both formulations II and III the maximums in the map are situated somewhere between the actual source position and the array centre. Thus, it would not be possible to estimate the exact source position from the maximum in the source mapping. The actual source position is located on the 0 dB contour for both formulations II and III. In agreement with theory, this shows that condition (6) is met.

While for formulations I and IV the error in source location as shown in Figure 4(a) is zero regardless of He, for formulations II and III; this error becomes less than 5% only above He $\approx$ 5. However, the estimated distance between array centre and source is never larger than the actual distance.

In the present case, the only option to estimate the source level is to use the maximum in the map. Figure 4(b) shows the error in comparison for all four formulations. Formulations I and IV show constant, small errors that follow the theoretical analysis in (8) and (13), respectively. For low Helmholtz numbers, the formulations II and III lead to larger errors because of the error in the estimated source location (the value at the location of the maximum in the map taken as source level). The error vanishes for larger Helmholtz numbers which is in agreement with the theoretical analysis.

3.2. Three Sources, Two-Dimensional Mapping. The second test case where all three sources are operated is a slightly more realistic scenario. In this case, the strength and the location of all three sources are of interest. If a classic beamforming approach with the mapping plane parallel to the array is used, the two-dimensional beamforming maps (Figure 5) show only minor differences for the four formulations. While only source A is actually located in the mapping plane at $z = 1$, contributions from all three sources appear in the maps. Thus, without any further analysis the maps suggest that all sources are located within this plane.

With the exception of source A, the maximums do not coincide with the projected source positions for both Helmholtz numbers shown. The reason is that sources B and C are not located in the mapping plane. The results show the tendency to map the sources nearer to the array more into the direction of the projected array center, while sources with a distance greater than that of the mapping plane are mapped to a an apparent position further away from the center. If there is no information available about the true distance between source and array, there is no way to estimate the exact source positions from the mapping

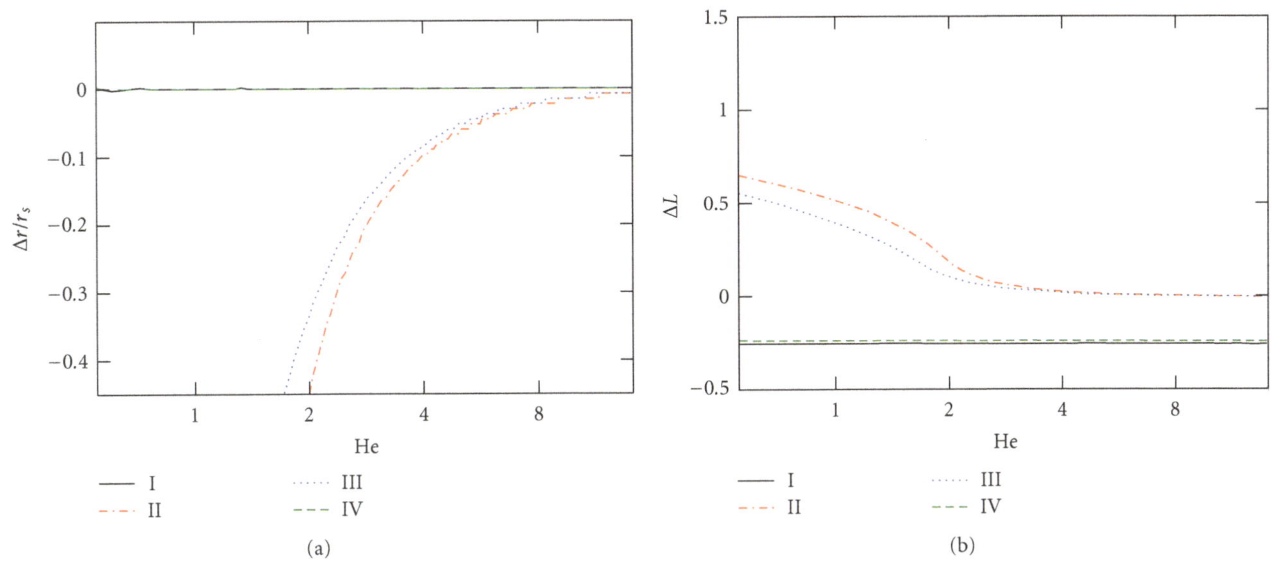

Figure 4: Errors in (a) source location and (b) source strength for formulations I–IV as a function of the Helmholtz number He.

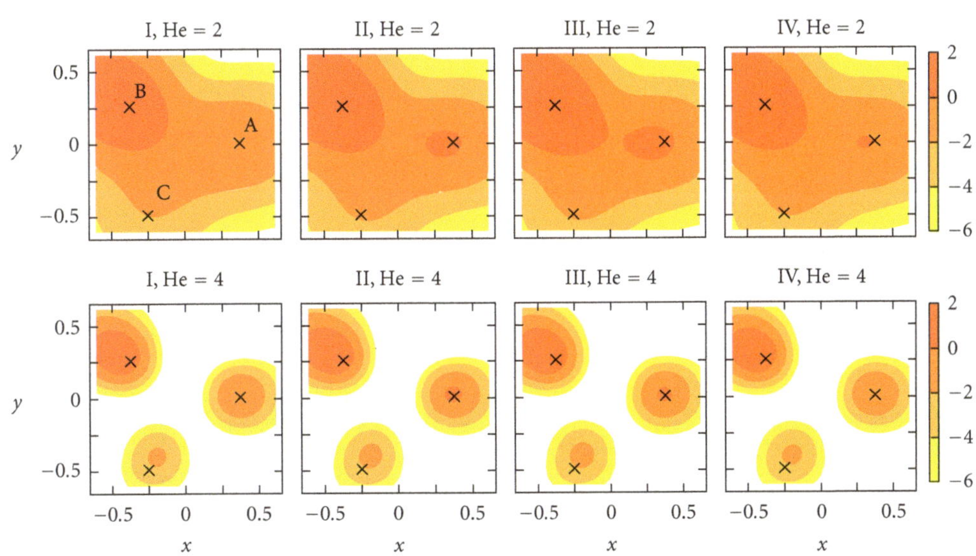

Figure 5: Maps of relative sound pressure level contributions at the array centre (correct values are 0 dB, 1.7 dB, and −2.2 dB for source A, B, and C, resp.) using classic beamforming without deconvolution for steering vector formulations I–IV, Helmholtz numbers He = 2 and 4, $z = 1$, crosses mark the projected position of the sources.

result. For the same reason, it is not feasible to estimate the source strength or to rank the sources using the result from the two-dimensional mapping. While all sources have the same source strength, the result shows the respective sound pressure level contribution at the array centre. If all sources are assumed to be within the mapping plane, B then appears to be the strongest and C appears to be the weakest source.

At the lower frequency shown (He = 2), the sources are less clearly distinguishable because of the large main lobe width at this frequency. If the source spacing would be smaller, the same would happen even at higher frequencies. Thus, the result can be improved if the real beamformer filter properties are taken into account by using a deconvolution technique. In the present case, the deconvolution method CLEAN-SC was applied (see [11] for more details) to the beamforming map. The result is a map that shows nonzero entries only at those grid points where a source is found. This corresponds to a negligible main lobe width regardless of frequency.

When applied to the two-dimensional beamforming results from Figure 5, CLEAN-SC delivers maps that allow for an easy separation of the sources even at the lower

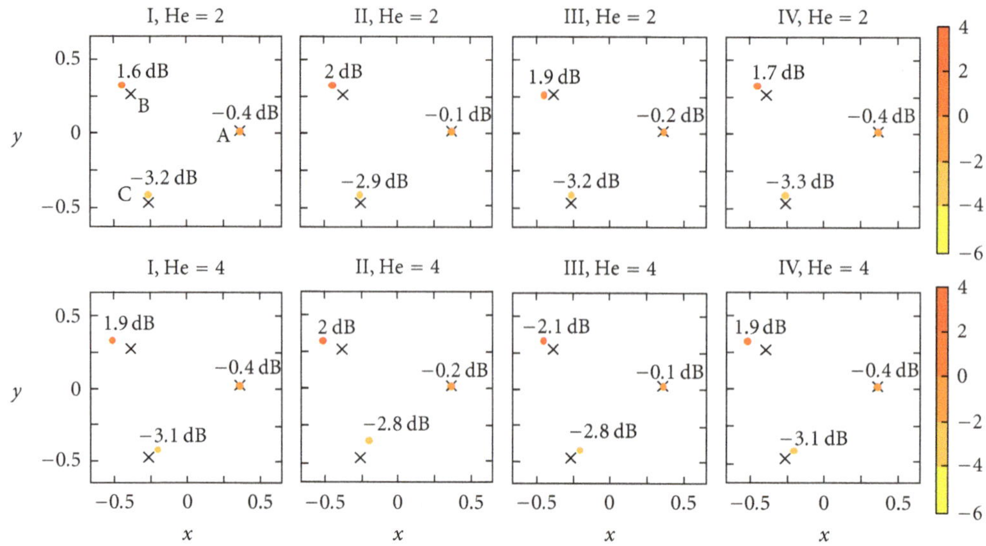

FIGURE 6: Maps of relative sound pressure level contributions at the array centre (correct values are 0 dB, 1.7 dB, and −2.2 dB for source A, B, and C, resp.) using CLEAN-SC on a two-dimensional grid at $z = 1$, for steering vector formulations I-IV, Helmholtz numbers He = 2 and 4, crosses mark the projected position of the sources.

frequency (Figure 6). Similar to the result from classic beamforming, these maps show all sources as they where situated in the mapping plane. The estimated locations of sources B and C again do not coincide with the projected source positions. The four different steering vector formulations lead to some differences in the estimated source level but show no divergent effects otherwise. To summarise, these results allow to conclude that there are at least three sources, but the information about the (projected) location is limited and a ranking of the sources is not possible.

The effect on the estimated source level as a function of frequency can also be estimated. If the source positions are known, the source level can be estimated from the map by simply taking the values at the grid points that are located at source positions. To allow for small errors in the estimated source positions, in the present case the source level was estimated by integrating over small square regions of the map. These regions were centred at the nominal source positions and had a side length of 0.1 array apertures.

The results in Figures 7(a), 7(c), and 7(e) show that the errors of the estimated source levels tend to be larger at very low Helmholtz numbers and become smaller for He > 2. However, for the sources B and C that are not situated within the mapping plane, the error again increases above He = 8, obviously as a result of the wrong mapping. Similar to the result shown in Figure 6, the four different steering vector formulations show only small differences, with formulations I and IV giving somewhat smaller levels compared to formulations II and III.

3.3. Three Sources, Three-Dimensional Mapping. More information can be gathered when the deconvolution is applied to a three-dimensional beamforming map. To study the results in comparison to two-dimensional beamforming, the analysis of a slice from the three-dimensional result is one option. Figure 8 shows such a slice at $z = 1$ that is equivalent to the mapping plane shown in Figure 2. Sources B and C that are not situated within this plane do not appear in any of the maps. Source A is within the plane but appears only for formulations I and IV and in case of formulation III for the higher frequency (He = 4). The reason for the absence of any source in the remaining maps is that the sources are mapped at the wrong position in the z direction. This becomes obvious in the two-view orthographic projections (Figure 9) of the three-dimensional map that is another option for the graphical representation of the result.

The projections along the y-axis (x-z-plane) reveal that the sources are indeed mapped much closer to the array as they really are for formulations II and III and low frequencies (He = 2). For formulations I and IV, the error in the location is much smaller at this frequency but not zero ($\Delta r/r_s = \pm 0.1 \cdots 0.2$). This effect is present only for the multisource scenario. It can be attributed to the fact that the location of the maximum in the beamforming map for a certain source is slightly shifted by the influence of other sources. For He = 4, no error is visible in the maps with exception of source A and formulation II. This small error also vanishes at even higher frequencies. While the estimated source levels are somewhat different for the different formulations, the errors are small, and no formulation seems to produce distinctly smaller errors than the others.

Finally, the error of the estimated source level shall be examined. Again, the source level was estimated by integrating over regions with a side length of 0.1 array apertures

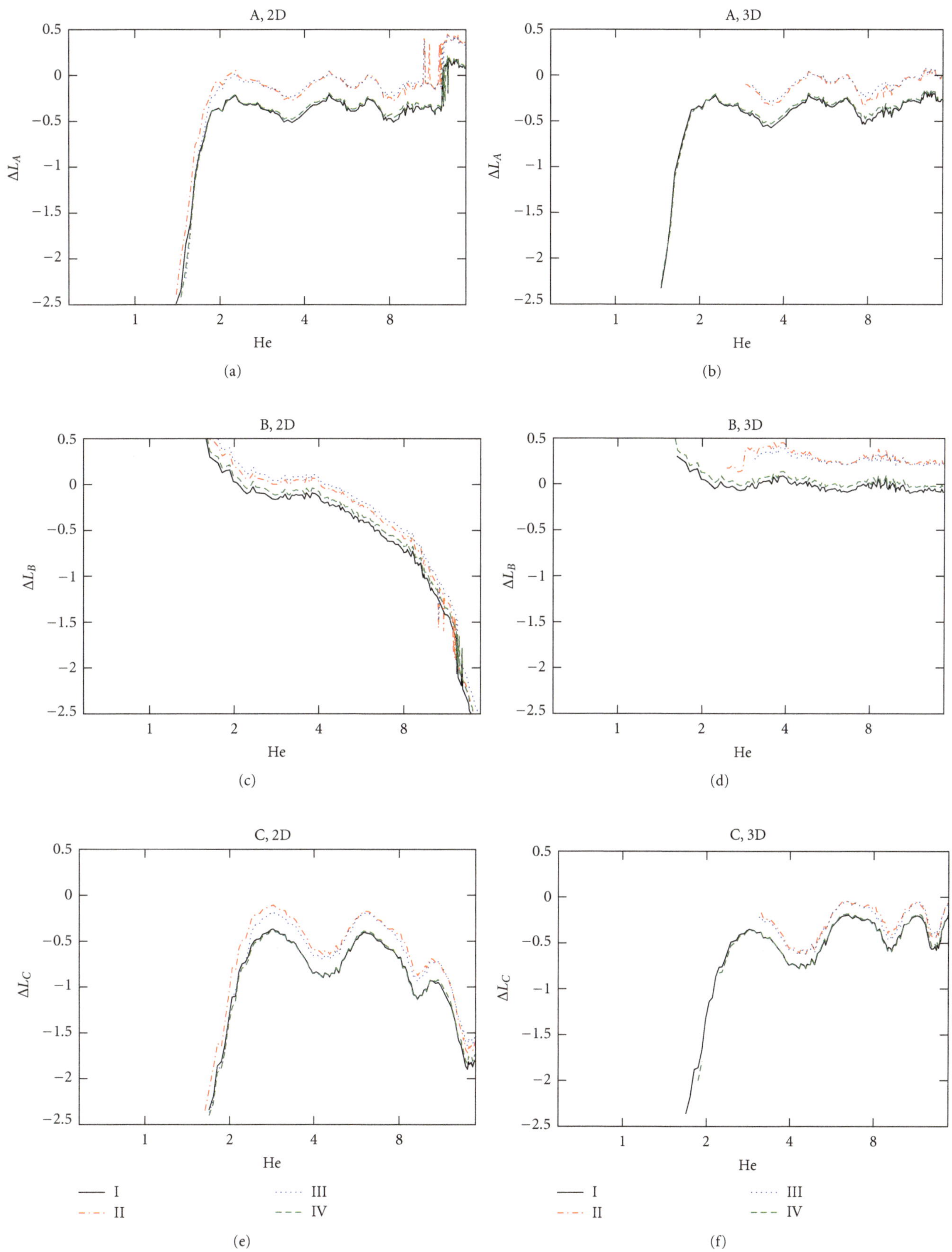

FIGURE 7: Errors (in dB) in the estimated sound pressure levels for formulations I–IV: (a, b) source A, (c, d) source B, and (e, f) source C.

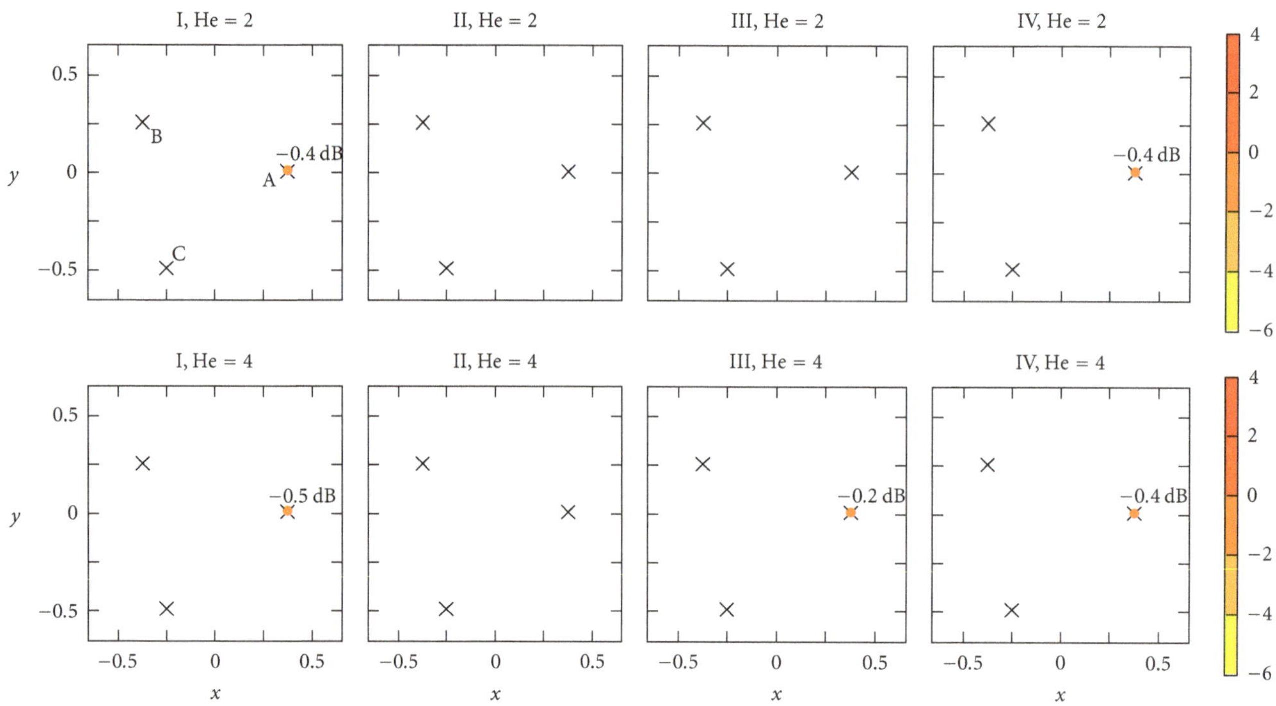

FIGURE 8: Slices at $z = 1$ of three-dimensional maps of relative sound pressure level contributions at the array centre (correct values are 0 dB, 1.7 dB, and −2.2 dB for source A, B, and C, resp.) using CLEAN-SC on a three-dimensional grid for steering vector formulations I–IV, Helmholtz numbers He = 2 and 4, crosses mark the projected position of the sources.

centred at the source position, but this time the regions were cubic shaped. Figures 7(b), 7(d), and 7(f) show the errors of the estimated source levels for all three sources. The results for formulations I and IV are again very similar and show a small negative error over a wide range of Helmholtz numbers. This error is consistent with the results from the theoretical analysis in (8) and (13) and is negligible for most practical applications. In contrast to the theoretical analysis, the error does not vanish completely for formulations II and III, though it is also negligibly small. For lower Helmholtz numbers, the beamforming and deconvolution method maps the source to a location outside the sector used for the integration. Thus, the estimated source level becomes infinitely small and consequently $\Delta L = -\infty$. While this is the case for He < 2 for all three sources and formulations I and IV, the estimated source level for formulations II and III vanishes already below He $\approx 3 \cdots 4$. The error does not increase for higher Helmholtz numbers as it is the case for the levels estimated from two-dimensional mapping.

It can be concluded that the principal theoretical findings regarding the different formulations I–IV in a single-source scenario remain true for the multiple source case: formulations I and IV deliver the correct source locations already for low Helmholtz numbers but show a small systematic error in the estimated source level. Formulations II and III deliver a slightly less erroneous level, but only for higher Helmholtz numbers, when the error in the estimated source location is small enough to place the source within the integration sector. Thus, for the given scenario the practical differences between the different formulations are generally small for higher Helmholtz numbers. However, in applications where small Helmholtz numbers arise, only formulations I and IV can be applied.

Because the errors in source level are very small for all formulations, formulations I and IV seem to be preferable for the practical application of three-dimensional acoustic source mapping using a beamforming approach. Moreover, once the correct position of a source is estimated, the systematic errors in the source levels for formulations I and IV can be corrected for by taking into account the factors in (8) and (13), respectively. Finally, formulation I has the extraadvantage that the calculation of the steering vectors requires less arithmetic operations.

4. Conclusion

A crucial element for the three-dimensional application of beamforming source mapping techniques using a microphone array is the formulation of the steering vectors. It was shown here that four different formulations found in the literature lead to different results. In theory, no formulation produces both correct source location and strength. Two formulations lead to the correct location at the cost of a

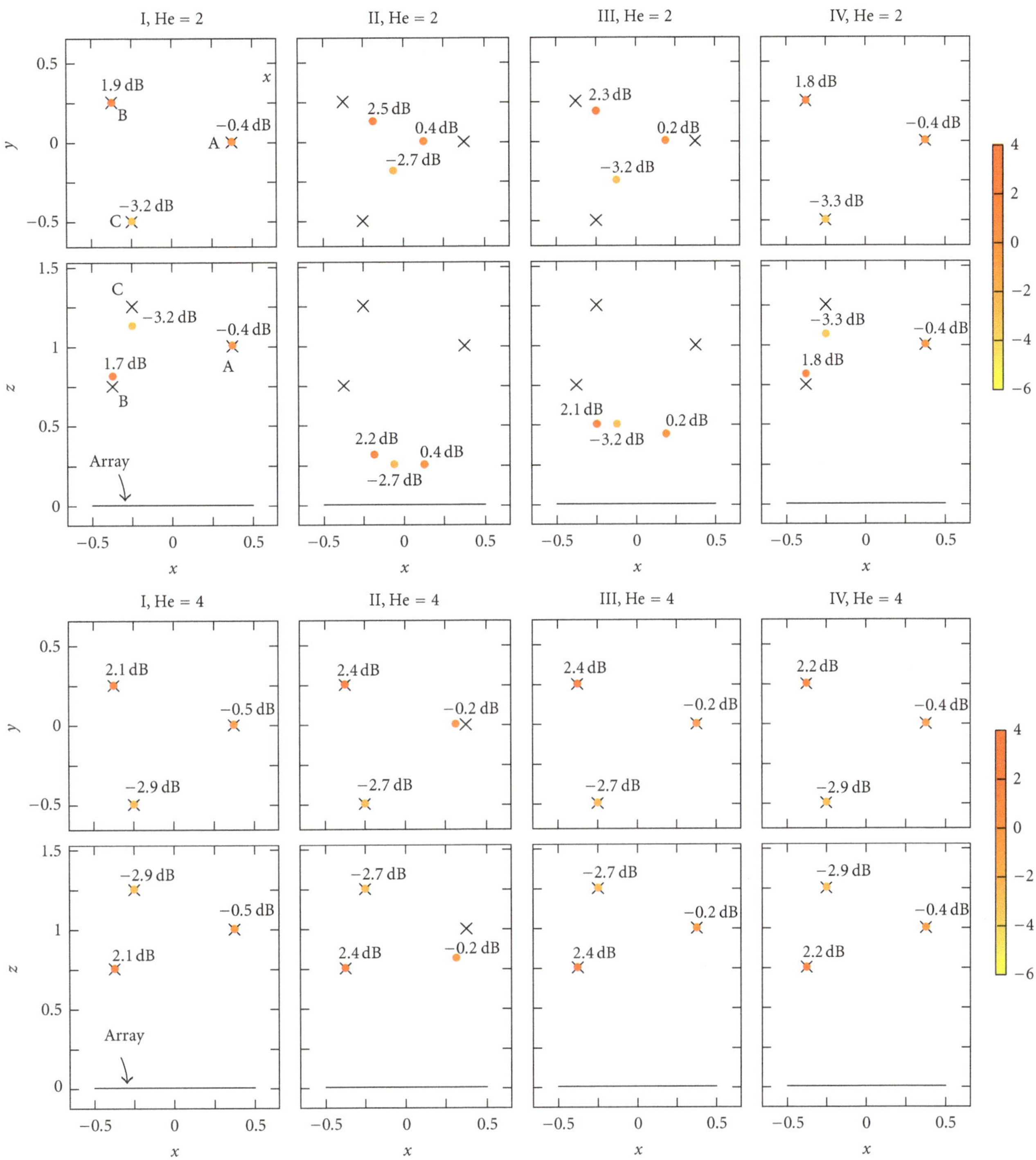

FIGURE 9: Projections of three-dimensional maps of relative sound pressure level contributions at the array centre (correct values are 0 dB, 1.7 dB, and −2.2 dB for source A, B, and C, resp.) using CLEAN-SC on a three-dimensional grid for steering vector formulations I–IV, Helmholtz numbers He = 2 and 4, crosses mark the projected position of the sources.

small error in the estimated source strength. The other two formulations estimate the correct strength but show an error in the estimated location of the source. Using simulated measurement data, it was shown that this error is relevant especially at low Helmholtz numbers based on the array aperture. In a simulated three-source scenario with CLEAN-SC deconvolution, all four formulations lead to small errors in the estimated strength of the sources. Unlike the systematic errors in source location, the systematic errors in the level can be corrected for in principle. Thus, the major conclusion is that for three-dimensional source mapping those steering vector formulations are preferable that enable the best estimation of the source location, for example, formulations I or IV.

References

[1] P. Nelson, "Source identification and location," in *Advanced Applications in Acoustics, Noise, and Vibration*, F. Fahy and J. Walker, Eds., pp. 100–153, Taylor & Francis, 2004.

[2] B. Barsikow, "Experiences with various configurations of microphone arrays used to locate sound sources on railway trains operated by the DB AG," *Journal of Sound and Vibration*, vol. 193, no. 1, pp. 283–293, 1996.

[3] M. Mosher, "Phased arrays for aerocoustic testing: theoretical development," in *Proceedings of the 2nd AIAA/CEAS Aeroacoustics Conference*, State College, Philadelphia, Pa, USA, May 1996, AIAA-1996-1713.

[4] R. Dougherty, "Beamforming in acoustic testing," in *Aeroacoustic Measurements*, T. Mueller, Ed., pp. 62–97, Springer, Berlin, Germany, 2002.

[5] T. Brooks and W. Humphreys, "Effect of directional array size on the measurement of airframe noise components," AIAA paper 99-1958, 1999.

[6] S. Oerlemans, L. Broersma, and P. Sijtsma, "Quantification of airframe noise using microphone arrays in open and closed wind tunnels," *International Journal of Aeroacoustics*, vol. 6, no. 4, pp. 309–333, 2007.

[7] D. Blacodon, G. Elias, D. Papillier, and J. Prieur, "Noise source localization on a dauphin helicopter in flight," in *Proceedings of the 25th European Rotorcraft Forum*, Roma, Italy, September 1999.

[8] T. Suzuki, "A review of diagnostic studies on jet-noise sources and generation mechanisms of subsonically convecting jets," *Fluid Dynamics Research*, vol. 42, no. 1, Article ID 014001, 2010.

[9] B. D. van Veen and K. M. Buckley, "Beamforming: a versatile approach to spatial filtering.," *IEEE ASSP magazine*, vol. 5, no. 2, pp. 4–24, 1988.

[10] T. F. Brooks and W. M. Humphreys, "A deconvolution approach for the mapping of acoustic sources (DAMAS) determined from phased microphonearrays," AIAA paper 2004-2954, 2004.

[11] P. Sijtsma, "CLEAN based on spatial source coherence," AIAA paper 2007-3436, 2007.

[12] J. Hileman, B. Thurow, and M. Samimy, "Development and evaluation of a 3-D microphone array to locate individual acoustic sources in a high-speed jet," *Journal of Sound and Vibration*, vol. 276, no. 3–5, pp. 649–669, 2004.

[13] T. Brooks and W. Humphreys, "Three-dimensional application of DAMAS methodology for aeroacoustic noise source definition," AIAA paper 2005-2960, 2005.

[14] S. N. Y. Gerges, W. D. Fonseca, and R. P. Dougherty, "State of the art beamforming software and hardware for applications," in *Proceedings of the 16th International Congress on Sound and Vibration*, Krakow, Poland, July 2009, keynote lecture 2.

[15] R. P. Dougherty, "Jet noise beamforming with several techniques," in *Proceedings on CD of the 3rd Berlin Beamforming Conference*, February 2010, BeBeC-2010-17.

[16] K. Oakley, "Noisecam—using 3d beamforming to better localise noise sources on hovercraft," in *Proceedings on CD of the 3rd Berlin Beamforming Conference*, February 2010, BeBeC-2010-22.

[17] H. Brick, T. Kohrs, E. Sarradj, and T. Geyer, "Noise from high-speed trains: experimental determination of the noise radiation of the pantograph," in *Proceedings of the Forum Acusticum*, Aalborg, Denmark, 2011.

[18] T. Geyer, E. Sarradj, and J. Giesler, "Application of a beamforming technique to the measurement of airfoil leading edge noise," *Advances in Acoustics and Vibration*, vol. 2012, Article ID 905461, 2012.

[19] E. Sarradj, "Three-dimensional acoustic source mapping," in *Proceedings on CD of the 4th Berlin Beamforming Conference*, February 2012, BeBeC 2012-11.

[20] M. Legg and S. Bradley, "Comparison of clean-sc for 2d and 3d scanning surfaces," in *Proceedings on CD of the 4th Berlin Beamforming Conference*, February 2012, BeBeC 2012-06.

[21] R. Dougherty, "Extensions of DAMAS and benefits and limitations of deconvolution in beamforming," AIAA paper 2005-2961, 2005.

[22] D. H. Johnson and D. E. Dudgeon, *Array Signal Processing: Concepts and Techniques*, Simon & Schuster, 1992.

[23] E. Sarradj, "A fast signal subspace approach for the determination of absolute levels from phased microphone array measurements," *Journal of Sound and Vibration*, vol. 329, no. 9, pp. 1553–1569, 2010.

[24] P. Stoica and R. Moses, *Introduction to Spectral Analysis*, Prentice Hall, New York, NY, USA, 1997.

[25] T. Geyer, *Trailing edge noise generation of porous airfoils [Ph.D. thesis]*, Brandenburg University of Technology at Cottbus, 2011.

Acoustic Predictions in Industrial Spaces Using a Diffusion Model

Alexis Billon,[1] Judicaël Picaut,[2] Vincent Valeau,[3] and Anas Sakout[1]

[1] *LEPTIAB, University of La Rochelle, 17042 La Rochelle, France*
[2] *Département IM UR EASE, Ifsttar Nantes, Route de Bouaye, CS 4, 44344 Bouguenais Cedex, France*
[3] *Institut P' UPR 3346, LEA UMR 6609, Bâtiment K 40, Avenue du Recteur Pineau, 86022 Poitiers Cedex, France*

Correspondence should be addressed to Alexis Billon, abillon@ulg.ac.be

Academic Editor: C. N. Wang

Industrial spaces are known to be very noisy working environment. This noise exposure can be uncomfortable, tiring, or even harmful, at worst. Industrial spaces have several characteristics: they are often huge flat volumes fitted with many obstacles and sound sources. Moreover, they are usually surrounded by rooms where low noise levels are required. The existing prediction tools can seldom model all these phenomena accurately. In this paper, a prediction model based on a diffusion equation is presented. The successive developments carried out to deal with the various propagating phenomena met in industrial spaces are shown. For each phenomenon, numerical or experimental examples are given to highlight the validity of this model. It is also shown that its computation load is very little in comparison to ray-tracing-based methods. In addition, this model can be used as a reliable and flexible tool to study the physics of the coupling between rooms. Finally, an application to a virtual factory is presented.

1. Introduction

In industrial spaces, workers are often exposed to intense sound fields. So, for their security and comfort, reliable acoustic predictions must be obtained. Moreover, these industrial spaces are often surrounded by other rooms, such as offices, where low noise levels are required.

Industrial spaces present a number of acoustical characteristics. Firstly, they are huge volumes and so can hardly be treated using methods solving the propagation equation [1, 2]. Their height is often small compared to their length and width, giving birth to nondiffuse reverberated sound fields discarding methods based on the classical theory of reverberation [3]. They often contain many obstacles (machines, stockpiles, benches, etc.) which scatter and absorb the propagating sound: their acoustic behaviour of the room, when fitted, is very different from that of the same room, when empty.

To solve this particular characteristic, various prediction models have been developed including analytical models [4–10], empirical models [11], and simplified models [12]. Numerical models have also been developed, based on the ray-tracing concept [13]. However, the applications of these different models remain limited due to their assumptions concerning the obstacles or the enclosing room. The most flexible and reliable model was developed by Hodgson [14], extending the ray-tracing model of Ondet and Barbry [13] to take into account diffuse reflections at walls.

The ray-tracing method can treat coupling of rooms through aperture, but they seldom take into account coupling via transmission through common walls. Moreover, some important reflection paths can be missed in configurations composed of multiple connected spaces [15]. In industrial spaces, the number of sound sources can also be high leading to prohibitive computation times to obtain reliable results using the ray-tracing method.

Picaut et al. [16] have proposed a model, first derived by Ollendorff [17], to describe the local acoustic energy density in rooms with perfectly diffuse reflecting walls. By using a physical analogy with the diffusion of particles in a medium containing spherical scattering objects, as presented by Morse and Feshbach [18], they showed that the acoustic energy density may be the solution of a diffusion equation. This diffusion problem can be then analytically solved for one-dimensional configuration [19]. More recently, Valeau et al. [20] have generalized this model to three-dimensional enclosures and proposed a numerical implementation of this model for room-acoustic predictions. This model was successively extended to take into account the acoustic specificities met in industrial spaces. The diffusion model is derived from the more computationally demanding transport theory which is used to model the propagation of sound in street canyon [21] or in long rooms [22, 23].

In Section 2, the basic equation of the diffusion model is introduced. Section 3 presents the extension of this model to mixed specular/diffuse reflections. The atmospheric attenuation is taken into account in Section 4. The coupling through aperture and through partition walls is presented in Sections 5 and 6, respectively. Section 7 shows how scattering obstacles modify the diffusion model. The diffusion model can then be used as a tool to investigate the coupling between rooms, as presented in Section 8. Finally, an application to a virtual factory is presented in Section 9. Along this paper, numerical and experimental comparisons of the diffusion model are showed and computation times are indicated.

2. Principle

2.1. Diffusion Equation. For a room of volume V and surface area S, the mean-free path of the room λ can be evaluated by the simple analytical relation [24]

$$\lambda = \frac{4V}{S}. \quad (1)$$

Following the physical analogy with the diffusion of particles in a scattering medium, the local acoustic energy density flux $\mathbf{J}(\mathbf{r}, t)$ can be approximated as the gradient of the acoustic energy density $w(\mathbf{r}, t)$:

$$\mathbf{J}(\mathbf{r}, t) = -D \operatorname{grad}(w(\mathbf{r}, t)), \quad (2)$$

where the variables $\mathbf{r}$ and t denote the position and time, respectively. D is the diffusion coefficient, and its analytical expression is directly taken from the theory of diffusion for particles in a scattering medium:

$$D = \frac{\lambda c}{3}, \quad (3)$$

where c is the speed of sound. This term takes the room morphology into account through its mean-free path.

Let us consider the case of a room containing an acoustic omnidirectional point source located at position $\mathbf{r}_s$ and with an output acoustic power P. It can be shown from (2) that the acoustic energy density is the solution of the following diffusion equation [20]:

$$\frac{\partial w(\mathbf{r}, t)}{\partial t} - D\nabla^2 w(\mathbf{r}, t) = P(\mathbf{r}, t) \text{ in } V. \quad (4)$$

In these equations, ∇^2 is the Laplace operator and V denotes the domain delimited by the room surfaces. In (4), the right-hand term is a source term which models the omnidirectional acoustic source in terms of power output and location [20]. This model assumes that the variations of energy density and energy flow remain small over one mean-free path [16, 18]. This assumption implies that the receiving location must be sufficiently far the sound source in both time and space for the energy to scatter sufficiently, as pointed out by Navarro et al. [25].

2.2. Boundary Conditions. Equation (4) expresses the temporal and spatial evolutions of the acoustic energy density throughout the room and must be completed by boundary conditions taking into account the absorption acoustic energy at the room's surfaces. A common way of describing energy exchanges at a boundary is by the use of a mixed boundary condition, in the following form:

$$\mathbf{J}(\mathbf{r}, t) \cdot \mathbf{n} = -D\frac{\partial w}{\partial \mathbf{n}} = hw(\mathbf{r}, t) \text{ on } S, \quad (5)$$

where h is an exchange coefficient, similar to a convection coefficient in heat transfer. Firstly, this expression of the exchange coefficient was proposed [20]:

$$h = \frac{c\alpha}{4}, \quad (6)$$

where α is the Sabine absorption coefficient. However, this boundary condition was limited to low absorption coefficients ($\alpha < 0.2$). To improve the obtained prediction for higher absorption coefficients, an exchange coefficient based on the Eyring absorption coefficient was later introduced [26, 27]:

$$h = -\frac{c\ln(1-\alpha)}{4}. \quad (7)$$

However, this expression becomes singular for $\alpha = 1$. Using an analogy with light diffusion, Navarro et al. [25] obtained the following expression of the exchange coefficient:

$$h = \frac{c\alpha}{2(2-\alpha)}. \quad (8)$$

Equations (4) and (8) can also be obtained through the spherical harmonics' expansion of a transport equation [28] or a radiative transfer model [25]. This last derivation permits to show that the diffusion model can be considered as a natural extension of the geometrical acoustics assumptions [25]. Previously, Valeau et al. [20] demonstrated that the diffusion model can also be seen as an extension of the classical theory of reverberation to nondiffuse sound fields.

In the diffusion model, the boundary conditions state the frequency band under consideration by the modelling, as in other geometrical models. Changing the evaluated frequency band comes to change the values of the absorption coefficients. Otherwise, the absorption is described locally: each surface is described by its own absorption. This more accurate description is an advantage over the classical theory of reverberation theory which considers only a mean absorption coefficient per room.

2.3. Room Acoustics Criteria. The most critical room-acoustics criterion in industrial space is the sound pressure level. The diffusion equation models only the reverberated part of the sound field and the direct sound field must be added at a later stage. Thus, the sound pressure level (SPL) at a location $\mathbf{r}$ can be obtained with [20]:

$$\mathrm{SPL}(\mathbf{r}) = 10 \times \log\left(\rho c\left(\frac{P}{4\pi d^2} + cw(\mathbf{r})\right) \times \frac{1}{P_{\mathrm{ref}}^2}\right), \quad (9)$$

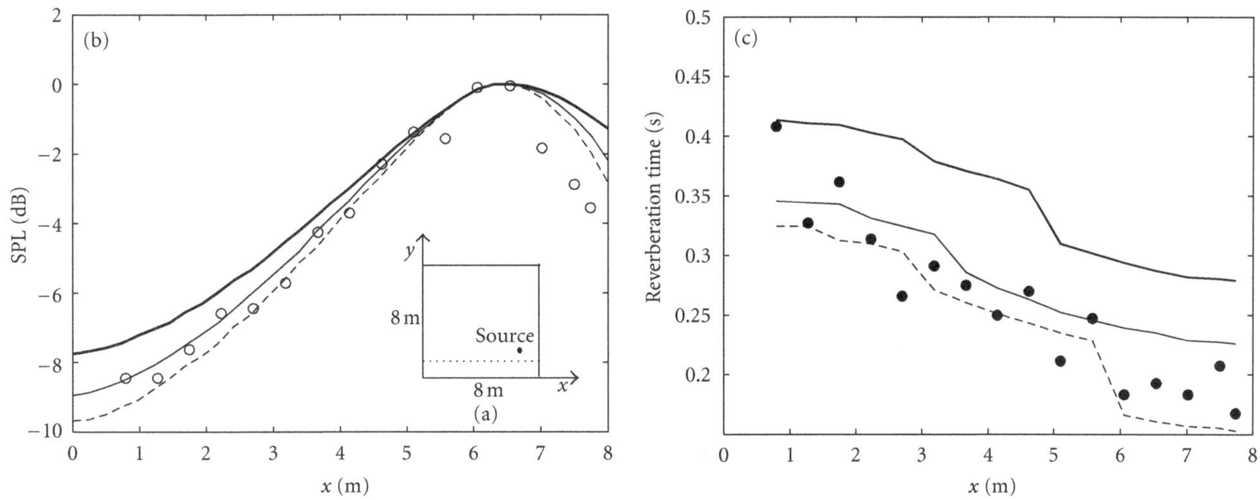

FIGURE 1: (a) Top view of the considered flat room. The black dot indicates the sound source, and the dotted line indicates the location of the measurements. (b) Evolution of the sound pressure level at the 1 kHz octave band. (c) Evolution of the reverberation time at the 1 kHz octave band [34]. (•) Experimental data, (—) diffusion model with (6) boundary condition, (- -) diffusion model with (7) boundary condition, and (–) diffusion model with (8) boundary condition.

where $d = |\mathbf{r} - \mathbf{r}_s|$ is the distance between the receiver and the source, ρ the air density, and $P_{\text{ref}} = 2 \times 10^{-5}$ Pa.

Most of the room-acoustics criteria are related to the temporal behaviour of the impulse response [29]. The diffusion model assumes that the energy is sufficiently scattered. In mixing rooms, the sound field becomes diffuse after about three mean-free time (a mean-free time is equal to λ/c, the time needed to travel a mean-free path) [30]. So the sound field ought to be sufficiently scattered at shorter times for diffusion model to be valid. Valeau et al. [20] state that only one mean-free time is necessary, whereas Xiang et al. [31] propose that two mean-free times are needed. A systematic comparison with a ray-tracing software in terms several room-acoustics criteria concludes that the sound decays evaluated by the diffusion model become valid after two mean-free times [32]. So the criteria based on the early part of the sound decay, such as the early decay time EDT, the definition D_{50}, or the clarity C_{80}, cannot be reliably predicted. On the other hand, the reverberation time evaluated from the backward integrated sound decay [31, 33] can be accurately evaluated as it will be shown in the following through several examples.

2.4. Resolution of the Diffusion Problem. For one-dimension configurations, Picaut et al. [19] obtained analytical solutions for the diffusion problem. Valeau et al. [20] proposed a numerical implementation using the finite element method, permitting to simulate more realistic three-dimension configurations. Using the diffusion model, the acoustic problem can be reduced to a classical conduction-convection heat transfer problem for which finite element softwares are many and well validated. Also, using unstructured mesh, even the most complex geometry can be treated. Moreover, the mesh can be rather coarse, with grid size about the mean-free path of the considered geometry [20]. So most problems imply only several thousands of degree of freedom and can be rapidly solved. For sound pressure levels, the stationary form of (4) can be solved and results are obtained in a few seconds for a given frequency. The sound decays are obtained in a few tens of seconds, solving the time-dependent form of (4). Moreover, the obtained results (SPL or sound decays) are evaluated throughout the whole calculation domain implying that no recalculation is needed if the receiver is moved.

2.5. Application to a Flat Room. The diffusion model associated to the three boundary conditions introduced in (6), (7), and (8) is compared to experimental data both in terms of sound pressure level and in terms of reverberation time for a $8 \times 8 \times 1.2\,\text{m}^3$ flat room (for more details, see [34]). The three models agree well with the experimental data in terms of SPL, with a maximum deviation of about 2 dB (Figure 1). A good agreement is also found in terms of reverberation times with the boundary conditions of (7) and (8). This example shows that reliable predictions of both sound levels and reverberation times can be obtained even for a configuration presenting a sound field far from being diffuse.

3. Mixed Specular/Diffuse Reflections

To obtain a diffusion equation, the reflections at walls were assumed to be perfectly diffuse. However, the reflections by walls are usually a combination of diffuse and specular reflections [35]. Valeau et al. [36] pointed out that increasing the value of the diffusion coefficient D permits better predictions when specular reflections. By systematically comparing the results of the diffusion model with a cone-tracing software, Foy et al. [37] proposed a modified diffusion constant D_m:

$$D_m = \kappa D, \tag{10}$$

where D is the theoretical diffusion constant given by (3) and κ is a function depending only on s, the scattering coefficient

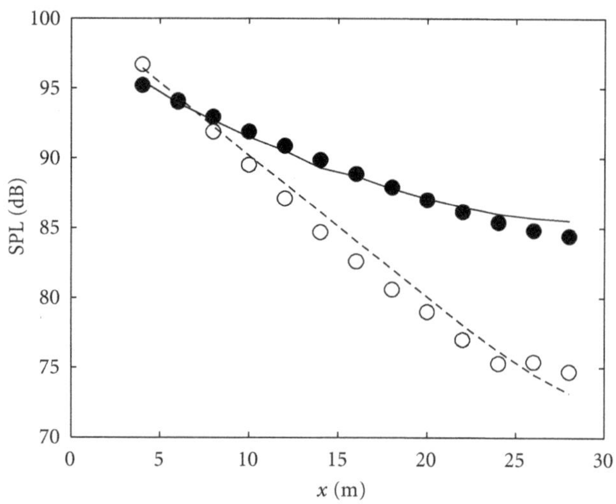

FIGURE 2: Evolution of the sound pressure (SPL) as a function of the receiver location in a centre axis of a long room of size $30 \times 2 \times 2$ m^3. (o) and (•) diffusion model, (- -) and (—) cone-tracing software, for $s = 0.2$ and $s = 1$, respectively.

of the surfaces [38]. κ is defined by the following empirical relation:

$$\kappa = -2.238 \ln(s) + 1.549 \quad \text{for } s > 0. \qquad (11)$$

It should be noted that, for $s = 1$ (fully diffuse reflections), the modified diffusion constant is not equal to the theoretical one ($\kappa \neq 1$) due to the empirical nature of the relation. Using this model, reliable results are obtained in terms of sound pressure levels whereas the reverberation times evaluated for $s < 0.4$ are erroneous [33]. As an example, Figure 2 plots the SPL for a $30 \times 2 \times 2$ m^3 long room with $s = 0.2$ and 1 showing the good agreement between the diffusion model and the cone-tracing software.

More recently, Visentin et al. [39] evaluated an estimated diffusion constant by comparing the acoustic energy intensity calculated by a particle-tracing and the diffusion model and obtained a spatially varying diffusion constant. In the vicinity of the source, the estimated diffusion constant is close to the one evaluated by (3) and it increases with the source distance [39]. This work introduces another more rigorous way to define a diffusion constant for specular reflecting surfaces.

4. Atmospheric Attenuation

The absorption of acoustical energy by the medium can have a significant impact on its propagation, particularly at high frequencies and for large rooms. An atmospheric attenuation term can be introduced in the diffusion model, leading to a modified expression of the diffusion constant [40]:

$$D' = D \times \frac{1}{1 + m\lambda}, \qquad (12)$$

where m is the coefficient of atmospheric attenuation in m^{-1} and D is the diffusion constant without atmospheric attenuation as defined in (3). However, the sound absorption being usually very small over a mean-free path (i.e., $m \ll 1$), one can consider thereafter that $D' \approx D$. The diffusion equation for the energy density w in the room, with a sound source term $P(\mathbf{r}, t)$, becomes [37]:

$$\frac{\partial}{\partial t} w(\mathbf{r}, t) - D\nabla^2 w(\mathbf{r}, t) + mcw(\mathbf{r}, t) = P(\mathbf{r}, t) \text{ in } V. \qquad (13)$$

In this equation, the attenuation term has the typical form of an absorption term within the diffusion framework. Expanding a radiative transfer model, Navarro et al. [25] end up with the same expressions. Figure 3 shows that a good agreement is obtained between the diffusion model and a cone-tracing software for a $30 \times 30 \times 3$ m^3 flat room both in terms of SPL and RT [40]. In this example, the atmospheric attenuation is set to 0 and 0.01 m^{-1} corresponding to a 7.2 kHz acoustic wave propagating in the air at 20°C and with 50% of relative humidity [38].

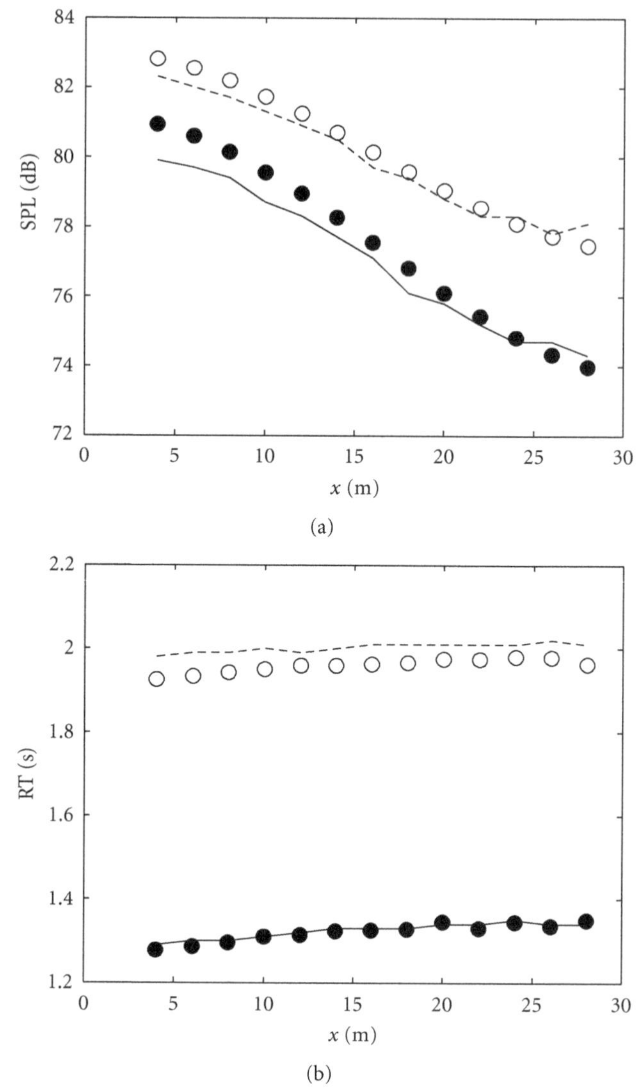

FIGURE 3: (a) Sound pressure level and (b) reverberation time as a function of the receiver location in a flat room of size (30×30×3) m^3 [40]: (o) and (•) diffusion model, (- -) and (—) cone-tracing software, without and with atmospheric attenuation, respectively.

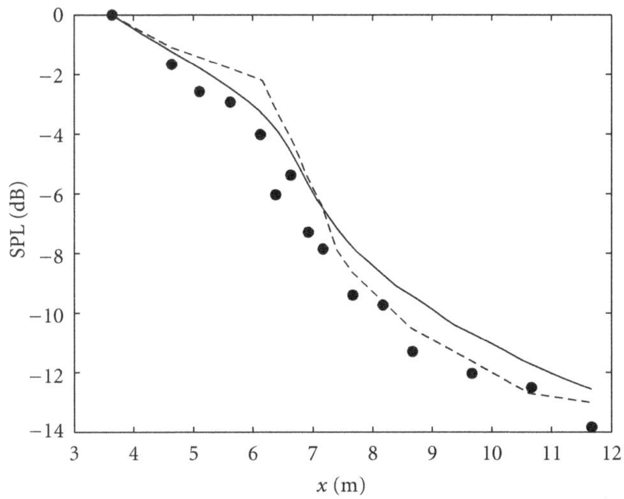

FIGURE 4: Evolution of the sound pressure level through the coupling aperture at the 1 kHz octave band [42]: (•) experimental data, (—) diffusion model, and (- -) cone-tracing software.

5. Coupling through an Aperture

Coupled volumes systems, composed of two or more spaces that are connected through acoustically transparent openings (i.e., a coupling aperture), have attracted considerable attention in architectural acoustics. In concert halls, these configurations are of a particular interest, allowing to conciliate a high clarity with a strong reverberance thanks to double-sloped decays [1, 41]. This configuration can also be found in various buildings or constructions such as industrial halls or office spaces.

For simulating the acoustics of two coupled rooms connected by an open aperture, a nonhomogeneous diffusion problem is to be solved where the computation domain V defined by the two coupled rooms is decomposed into two subvolumes, V_1 and V_2 [42]. Moreover, it is assumed that the value of the mean-free path in each room is unaffected by the coupling. The following set of equations can then be written [42]:

$$\begin{aligned} \frac{\partial w(\mathbf{r},t)}{\partial t} - D_1 \nabla^2 w(\mathbf{r},t) &= P(\mathbf{r},t) \text{ in } V_1, \\ \frac{\partial w(\mathbf{r},t)}{\partial t} - D_2 \nabla^2 w(\mathbf{r},t) &= 0 \text{ in } V_2, \end{aligned} \tag{14}$$

together with some boundary conditions. It should be pointed out that this method can be extended to an arbitrary number of coupled spaces without further assumptions or to an arbitrary of sound sources. Figure 4 plots results in the configuration of two coupled classrooms [42] showing the good agreement of the diffusion model with experimental data and a cone-tracing software in terms of SPL.

6. Coupling through a Partition Wall

Acoustical energy transferring from one room to another through partition walls can significantly deteriorate the building comfort. To evaluate this transfer, the standards are based on the theory of reverberation [43, 44] and their predictions are reliable as long as the assumptions of a diffuse sound field hold [3]. This energy transfer can be computed using the diffusion model. Firstly, the partition wall was assimilated as a surface source assuming a diffuse field in the room containing the source [20]. Later, the diffusion model was extended to solve the whole transmission problem and to consider nondiffuse sound fields on both sides of the partition wall [45]. Two diffusion equations must be considered, one for each enclosure:

$$\begin{aligned} -D_1 \nabla^2 w_1(\mathbf{r},t) &= P(\mathbf{r},t) \text{ in } V_1, \\ -D_2 \nabla^2 w_2(\mathbf{r},t) &= 0 \text{ in } V_2, \end{aligned} \tag{15}$$

where D_1 and D_2 are the diffusion constants of the source and adjacent rooms, respectively. Note that only the stationary problem is considered here as the sound pressure level is the main criterion of interest in acoustic transmission problem.

At the coupling area, the energy exchanges between the rooms must be expressed as well as its absorption at its surface. Thus, the following boundary condition is obtained on the source room's side with a normal exterior vector $\mathbf{n}_1$ [45]:

$$D_1 \frac{\partial w_1}{\partial \mathbf{n}_1} + h_1 w_1(\mathbf{r},t) = \frac{\tau c}{4} w_2. \tag{16}$$

The left hand term is similar to (5) and depicts the loss of energy form the source room to the partition wall. The right hand term stands for the energy transferred from the adjacent room to the source room through the partition wall as a function of the transmission loss R (or the transmission coefficient $\tau = 10^{-R/10}$). A similar boundary condition can be written on the adjacent room's side with a normal exterior vector $\mathbf{n}_2$:

$$D_2 \frac{\partial w_2}{\partial \mathbf{n}_2} + h_2 w(\mathbf{r},t) = \frac{\tau c}{4} w_1. \tag{17}$$

The right hand term describes here the energy transferred from the source room. Similarly to the coupling through apertures, the principle can be extended to an arbitrary number of coupled rooms and sound sources. Otherwise, it can be shown that the energy balance obtained using the diffusion model can be reduced to the one obtained using the theory of reverberation, if a diffuse sound field is assumed in each room [45]. Figure 5 depicts the difference of sound level between two coupled classrooms [45]. A good agreement is between the measurement and the diffusion model.

7. Diffusion by Scattering Objects

As stated in the introduction, the industrial halls are seldom empty and contain many obstacles that scatter and absorb the sound energy during its propagation. This scattering objects can be modelled statistically and characterised by their density n_f (i.e., the number of scattering objects per unit volume), their average scattering cross-section Q_f, and their absorption coefficient α_f [4, 5]. The scattering of sound by this obstacles can be then modelled by a diffusion process

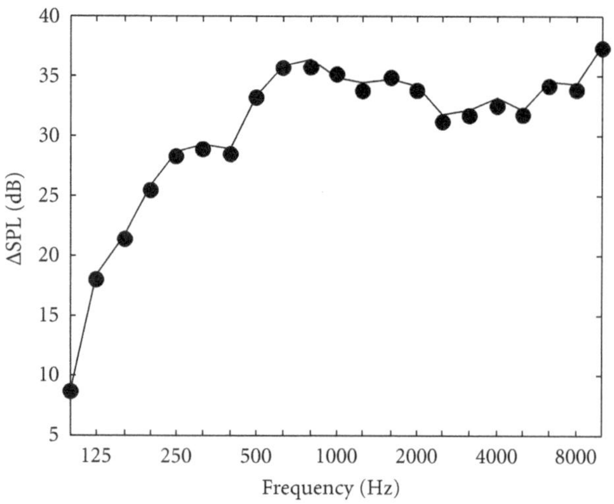

FIGURE 5: Sound pressure level difference by third octave band [45]: (•) experimental data and (—) diffusion model.

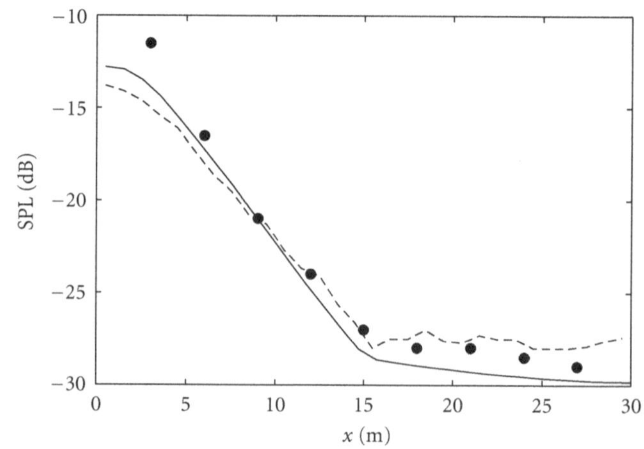

FIGURE 6: Sound pressure level distribution [36]: (•) experimental data [13], (—) diffusion model, and (- -) ray-tracing software.

characterised by a mean-free path $\lambda_f = 1/(n_f Q_f)$ and a diffusion constant $D_f = \lambda_f c/3$ [9]. The mean-free path of a sound particle travelling in a room characterised by a mean-free path λ_e (for empty) and the scatterers previously described can be then written [32]:

$$\lambda = \frac{\lambda_e \lambda_f}{\lambda_e + \lambda_f}. \tag{18}$$

The diffusion equation for the acoustic energy density in a room containing scattering obstacles is then [36]:

$$\frac{\partial w(\mathbf{r},t)}{\partial t} - Dt\nabla^2 w(\mathbf{r},t) + c\frac{\alpha_f}{\lambda_f} w(\mathbf{r},t) = P(\mathbf{r},t) \text{ in } V, \tag{19}$$

with

$$D_t = \frac{c\left(\lambda_e \lambda_f\right)}{3\left(\lambda_e + \lambda_f\right)} = \frac{D_e D_f}{D_e + D_f}, \tag{20}$$

where D_e is the diffusion constant in the empty room a defined in (3). Equation (20) describes the combination of two diffusion processes, one due to the walls and the other due to the scattering objects. The additional term in (19) takes into account the absorption of sound energy by the scattering objects. Figure 6 shows that the obtained agreement of the diffusion model with the experimental data [13] is very good, similar to the one obtained with a ray-tracing software [36].

8. Diffusion Model and Coupled Rooms

Due to concert halls applications, systems of rooms coupled through apertures coupled volumes systems have attracted considerable attention. The diffusion model was shown to be a flexible and reliable tool to investigate the behaviour of these systems. It was firstly validated by comparison with statistical theory and experimental data [42]. It reliably models the gradual transition both in terms of sound pressure level (see Figure 5) and in terms of sound decays through the aperture. Jing and Xiang [46] used the diffusion model to study the energy flow from one room to another as a function of the time. It permits to observe that, for double-sloped decays, the energy flows back from the adjacent room to the source room. This reversal of the energy flow is associated with a dip of the energy-flow decay. These phenomena are not observed when the sound decays are not double-sloped. Moreover, Jing and Xiang [28] showed that the double-sloped decays can be accurately modelled by comparison with scale model measurements. It permits to link the energy flow reversal and the dip observed in its magnitude to the switching of slope in the measured sound decay. Pu et al. [47] observed that the double-sloped decays also occur in the adjacent room and is not restricted to the source room.

9. Application to a Virtual Factory

To exhibit the interest of the diffusion model to real-life applications, a configuration, similar to a small factory, is presented in Figure 7 [48]. This building is composed of a $20 \times 25 \times 10\ \text{m}^3$ hall (A in Figure 7) connected through a $0.9 \times 2.1\ \text{m}^2$ aperture to a $20 \times 2.5 \times 2.5\ \text{m}^3$ corridor (B). The walls of the hall are in concrete ($\alpha = 0.03$), and the zone A′ ($9 \times 7 \times 5\ \text{m}^3$) is fitted with scattering objects with absorption $\alpha_f = 0.3$ and density $n_f = 0.25$. Two $5.9 \times 3.5 \times 2.5\ \text{m}^3$ offices (D and E) are connected through $0.9 \times 2.1\ \text{m}^2$ openings as well and a $8 \times 3.5 \times 2.5\ \text{m}^3$ workshop (E) through a $2 \times 2.1\ \text{m}^2$ opening. Four sound sources with different sound power level are located in the hall A: source 1 (120 dB) and sources 2, 3, and 4 (100 dB). Another 100 dB source is located in the workshop (5). The atmospheric sound attenuation is set to 0.005 dB/m.

In this study, two configurations of the corridor are compared. In the first configuration, the corridor is specularly reflecting ($s = 0.2$) and its absorption is homogeneous

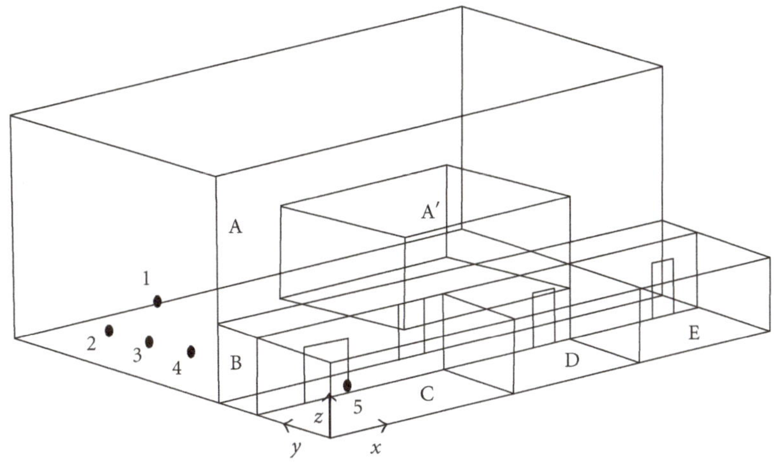

FIGURE 7: Sketch of the simulated building. A is an industrial hall where A′ represents a volume fitted with scattering objects. B is a corridor, C a small workshop, D and E two offices. The numbers 1–5 indicate the sound sources.

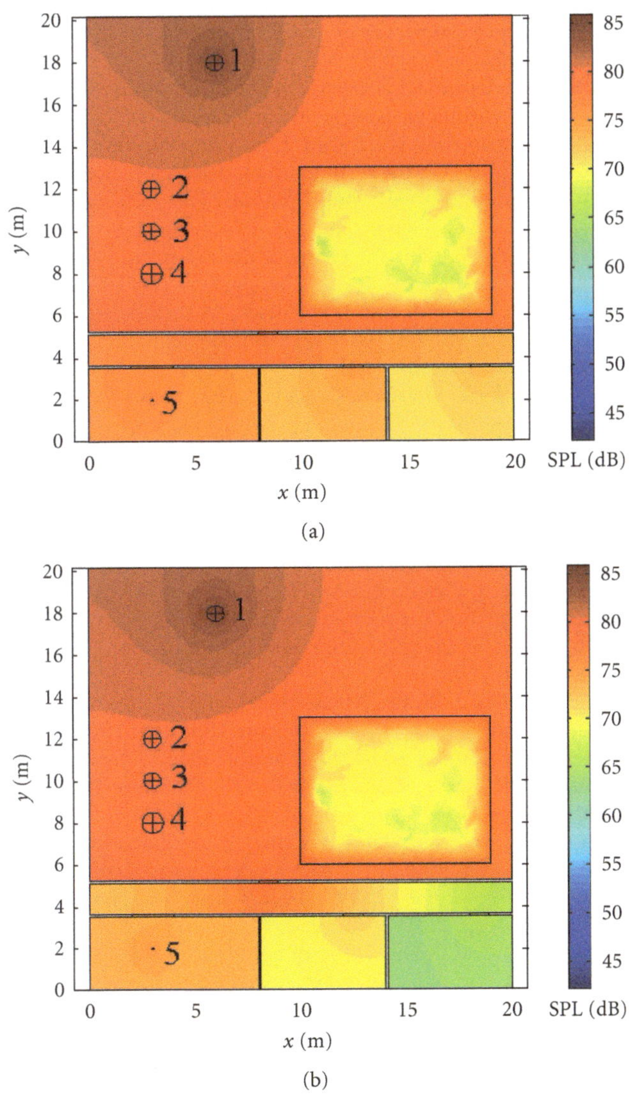

FIGURE 8: Color maps of the sound pressure level (SPL) of the reverberated sound field at 1.2 m: (a) nontreated configuration, (b) treated configuration.

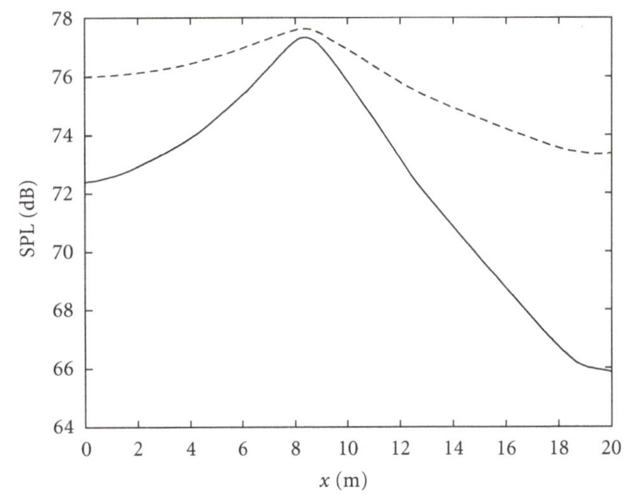

FIGURE 9: Sound pressure level (SPL) along the corridor at 1.2 m: (- -) nontreated corridor, (—) treated corridor.

($\alpha = 0.06$). In the second configuration, the corridor's surfaces are diffuse ($s = 1$) and the ceiling is absorbent ($\alpha = 0.6$).

The computation time is around 30 s for the steady state sound levels. Due to the geometrical complexity of the problem, a simulation carried out using a ray-tracing software would need a huge number of rays (and so a very long computing time to obtain) in order to obtain physically consistent results.

Figure 8 presents color maps of sound pressure level of the reverberated sound field at 1.2 m high for both configurations. In particular, one can observe the decrease of SPL in the fitting zone, in comparison with the SPL in the hall, due to additional absorption from the scattering objects. The effect of the treatment is also clearly effective in the connected rooms. For example, the SPL decreases of about 5 dB in room E. Figure 9, plotting the SPL along the corridor at 1.2 m high, allows one to observe more closely the effect of treatment. The increase of absorption and scattering raises the sound attenuation of more than 8 dB. The acoustics treatment in the corridor is also responsible for the decrease of the sound energy within the offices (D and E).

10. Conclusions

Industrial spaces often consist in huge, flat volumes fitted with a large number of diffracting objects and numerous sound sources. Moreover, they are often connected to other rooms (like offices) through common walls or apertures. These characteristics lead to complex acoustic problem to solve.

In this paper, a numerical model based on a diffusion equation was reviewed. It has been shown that the successive developments of this model succeed to solve most of the met difficulties of such acoustical spaces: nondiffuse sound field, atmospheric attenuation, coupling through apertures, transmission through partition walls, and sound absorption by scattering objects. This model can also be extended to an arbitrary number of sound sources or coupled spaces. Numerical and experimental comparisons were presented for each phenomenon exhibiting the reliability of the proposed model, both in terms of sound pressure levels and in terms of reverberation times. Moreover, the computation time has shown to remain very limited in regards to classical geometrical acoustics software. This model was thus used as a flexible and reliable tool to investigate the behaviour of the coupled-rooms, giving a new understanding of their behaviour. A real-life application to a virtual factory was presented, showing the capacity of the diffusion model in complex industrial spaces.

However, the limitations of the model were also pointed out. It cannot reliably model the early part of the sound decay and so cannot predict room-acoustics parameters such as definition, clarity, or early decay time. In spaces covered with specularly reflecting surfaces, the diffusion model permits to predict reliable sound pressure levels but is unable to obtain physically consistent reverberation times. The model also cannot predict the diffraction occurring around sound barriers. Those are several of the remaining challenges to be solved in the future.

Acknowledgment

The authors would like to thank the Agence de l'Environnement et de la Maîtrise de l'Énergie (ADEME) for providing financial support to this work.

References

[1] L. Cremer and H. H. Muller, *Principles and Applications of Room Acoustics*, vol. 1, Applied Science, London, UK, 1978.

[2] H. Kuttruff, *Room Acoustics*, Spon, London, UK, 4th edition, 1999.

[3] M. Hodgson, "When is diffuse-field theory applicable?" *Applied Acoustics*, vol. 49, no. 3, pp. 197–207, 1996.

[4] S. Jovivic, *Anleitung zur Vorausbestimmung des Schallpegels in Betriebsgebäuden*, Müller-BBM GmbH, München, Germany, 1979.

[5] K. H. Kuttruff, "Sound decay in reverberation chambers with diffusing elements," *Journal of the Acoustical Society of America*, vol. 69, no. 6, pp. 1716–1723, 1981.

[6] E. A. Lindqvist, "Sound attenuation in large factory spaces," *Acustica*, vol. 50, no. 5, pp. 313–328, 1982.

[7] M. Hodgson, *Physical and theoretical models as tools for the study of factory sound fields*, Ph.D. dissertation, University of Southampton, 1983.

[8] N. Auletta, "Contribution à l'étude de la propagation du son dans les locaux industriels encombrés—1ère partie (etude théorique)," *Revue d'Acoustique*, vol. 75, pp. 477–487, 1985.

[9] U. J. Kurze, "Scattering of sound in industrial spaces," *Journal of Sound and Vibration*, vol. 98, no. 3, pp. 349–364, 1985.

[10] P. Janeček, "A model for the sound energy distribution in work spaces based on the combination of direct and diffuse sound fields," *Acustica*, vol. 74, pp. 149–156, 1991.

[11] N. Heerema and M. Hodgson, "Empirical models for predicting noise levels, reverberation times and fitting densities in industrial workrooms," *Applied Acoustics*, vol. 57, no. 1, pp. 51–60, 1999.

[12] M. Hodgson, "Experimental evaluation of simplified models for predicting noise levels in industrial workrooms," *Journal of the Acoustical Society of America*, vol. 103, no. 4, pp. 1933–1939, 1998.

[13] A. M. Ondet and J. L. Barbry, "Modeling of sound propagation in fitted workshops using ray tracing," *Journal of the Acoustical Society of America*, vol. 85, no. 2, pp. 787–796, 1989.

[14] M. Hodgson, "On the accuracy of models for predicting sound propagation in fitted rooms," *Journal of the Acoustical Society of America*, vol. 88, no. 2, pp. 871–878, 1990.

[15] H. Lehnert, "Systematic errors of the ray-tracing algorithm," *Applied Acoustics*, vol. 38, no. 2–4, pp. 207–221, 1993.

[16] J. Picaut, L. Simon, and J. D. Polack, "A mathematical model of diffuse sound field based on a diffusion equation," *Acustica*, vol. 83, no. 4, pp. 614–621, 1997.

[17] F. Ollendorff, "Statistische raumakustik als diffusionsproblem," *Acustica*, vol. 21, no. 4, pp. 236–245, 1969.

[18] P. Morse and H. Feshbach, *Methods of Theoretical Physics*, McGraw-Hill, New York, NY, USA, 1953.

[19] J. Picaut, L. Simon, and J. D. Polack, "Sound field in long rooms with diffusely reflecting boundaries," *Applied Acoustics*, vol. 56, no. 4, pp. 217–240, 1999.

[20] V. Valeau, J. Picaut, and M. Hodgson, "On the use of a diffusion equation for room-acoustic prediction," *Journal of the Acoustical Society of America*, vol. 119, no. 3, pp. 1504–1513, 2006.

[21] T. Le Pollès, J. Picaut, M. Bérengier, and C. Bardos, "Sound field modeling in a street canyon with partially diffusely reflecting boundaries by the transport theory," *Journal of the Acoustical Society of America*, vol. 116, no. 5, pp. 2969–2983, 2004.

[22] Y. Jing, E. W. Larsen, and N. Xiang, "One-dimensional transport equation models for sound energy propagation in long spaces: theory," *Journal of the Acoustical Society of America*, vol. 127, no. 4, pp. 2312–2322, 2010.

[23] Y. Jing and N. Xiang, "One-dimensional transport equation models for sound energy propagation in long spaces: simulations and experiments," *Journal of the Acoustical Society of America*, vol. 127, no. 4, pp. 2323–2331, 2010.

[24] W. B. Joyce, "Sabine's reverberation time and ergodic auditorium," *Journal of the Acoustical Society of America*, vol. 58, no. 3, pp. 643–655, 1975.

[25] J. Navarro, F. Jacobsen, J. Escolano, and J. Lopez, "A theoretical approach to room acoustic model based on a radiative transfer

model," *Acta Acustica United with Acustica*, vol. 96, pp. 1078–1089, 2010.

[26] Y. Jing and N. Xiang, "A modified diffusion equation for room-acoustic predication (L)," *Journal of the Acoustical Society of America*, vol. 121, no. 6, pp. 3284–3287, 2007.

[27] A. Billon, J. Picaut, and A. Sakout, "Prediction of the reverberation time in high absorbent room using a modified-diffusion model," *Applied Acoustics*, vol. 69, no. 1, pp. 68–74, 2008.

[28] Y. Jing and N. Xiang, "Investigation on higher orders of spherical harmonics equations for efficient room-acoustics predictions," in *Proceedings of the 19th International Congress on Acoustics (ICA '07)*, pp. 1–4, Madrid, Spain, 2007.

[29] "Acoustics—measurement of the reverberation time of rooms with reference to other acoustical parameters," International Organization for Standardization Document No. ISO 3382, 1997.

[30] A. Billon and J.-J. Embrechts, "Numerical evidence of mixing in rooms using the free path temporal distribution," *Journal of the Acoustical Society of America*, vol. 130, no. 3, pp. 1381–1389, 2011.

[31] N. Xiang, Y. Jing, and A. C. Bockman, "Investigation of acoustically coupled enclosures using a diffusion-equation model," *Journal of the Acoustical Society of America*, vol. 126, no. 3, pp. 1187–1198, 2009.

[32] J. Escolano, J. M. Navarro, and J. J. López, "On the limitation of a diffusion equation model for acoustic predictions of rooms with homogeneous dimensions," *Journal of the Acoustical Society of America*, vol. 128, no. 4, pp. 1586–1589, 2010.

[33] M. R. Schroeder, "New method of measuring reverberation time," *Journal of the Acoustical Society of America*, vol. 37, no. 1, pp. 409–412, 1965.

[34] Y. Jing and N. Xiang, "On boundary conditions for the diffusion equation in roomacoustic prediction: theory, simulations, and experiments," *Journal of the Acoustical Society of America*, vol. 123, no. 1, pp. 145–153, 2008.

[35] M. Hodgson, "Evidence of diffuse surface reflections in rooms," *Journal of the Acoustical Society of America*, vol. 89, no. 2, pp. 765–771, 1991.

[36] V. Valeau, M. Hodgson, and J. Picaut, "A diffusion-based analogy for the prediction of sound fields in fitted rooms," *Acta Acustica United with Acustica*, vol. 93, no. 1, pp. 94–105, 2007.

[37] C. Foy, V. Valeau, A. Billon, J. Picaut, and A. Sakout, "An empirical diffusion model for acoustic prediction in rooms with mixed diffuse and specular reflections," *Acta Acustica United with Acustica*, vol. 95, no. 1, pp. 97–105, 2009.

[38] T. J. Cox, B. I. L. Dalenback, P. D'Antonio et al., "A tutorial on scattering and diffusion coefficients for room acoustic surfaces," *Acta Acustica United with Acustica*, vol. 92, no. 1, pp. 1–15, 2006.

[39] C. Visentin, V. Valeau, N. Prodi, and J. Picaut, "A numerical investigation of the sound intensity field in rooms by using diffusion theory and particle tracing," in *Proceedings of the 19th International Congress on Acoustics (ICA '10)*, pp. 1–6, Sydney, Australia, 2010.

[40] A. Billon, J. Picaut, C. Foy, V. Valeau, and A. Sakout, "Introducing atmospheric attenuation within a diffusion model for room-acoustic predictions (L)," *Journal of the Acoustical Society of America*, vol. 123, no. 6, pp. 4040–4043, 2008.

[41] J. E. Summers, R. R. Torres, and Y. Shimizu, "Statistical-acoustics models of energy decay in systems of coupled rooms and their relation to geometrical acoustics," *Journal of the Acoustical Society of America*, vol. 116, no. 2, pp. 958–969, 2004.

[42] A. Billon, V. Valeau, A. Sakout, and J. Picaut, "On the use of a diffusion model for acoustically coupled rooms," *Journal of the Acoustical Society of America*, vol. 120, no. 4, pp. 2043–2054, 2006.

[43] "Acoustics—measurement of sound insulation in buildings and of building elements—part 4: field measurements of airborne sound insulation between rooms," International Organization for Standardization Document No. ISO 140-4, 1998.

[44] "Building acoustics—estimation of acoustic performance of buildings from the performance of elements—part 1: airborne sound insulation between rooms," European Committee for Standardization Document No. EN-12354-1, 2000.

[45] A. Billon, C. Foy, J. Picaut, V. Valeau, and A. Sakout, "Modeling the sound transmission between rooms coupled through partition walls by using a diffusion model," *Journal of the Acoustical Society of America*, vol. 123, no. 6, pp. 4261–4271, 2008.

[46] Y. Jing and N. Xiang, "Visualizations of sound energy across coupled rooms using a diffusion equation model," *Journal of the Acoustical Society of America*, vol. 124, no. 6, pp. EL360–EL365, 2009.

[47] H. Pu, X. Qiu, and J. Wang, "Different sound decay patterns and energy feedback in coupled volumes," *Journal of the Acoustical Society of America*, vol. 129, no. 4, pp. 1972–1980, 2011.

[48] J. Picaut, V. Valeau, A. Billon, and A. Sakout, "Sound field modeling in architectural acoustics using a diffusion equation," in *Proceedings of the 20th International Conference on Noise*, pp. 1–8, Honolulu, Hawaii, USA, 2006.

18

Multiengine Speech Processing Using SNR Estimator in Variable Noisy Environments

Ahmad R. Abu-El-Quran,[1] Adrian D. C. Chan,[2] and Rafik A. Goubran[2]

[1] *Department of Enterprise and Higher Education, International Turnkey Systems, ITS Tower, Mubarak Al Kabeer Street, P.O. Box 26729, Safat, Kuwait City 13128, Kuwait*
[2] *Department of Systems and Computer Engineering, Carleton University, 1125 Colonel By Drive, Ottawa, ON, Canada K1S 5B6*

Correspondence should be addressed to Ahmad R. Abu-El-Quran, ahmad.rami@its.ws

Academic Editor: Akira Ikuta

We introduce a multiengine speech processing system that can detect the location and the type of audio signal in variable noisy environments. This system detects the location of the audio source using a microphone array; the system examines the audio first, determines if it is speech/nonspeech, then estimates the value of the signal to noise (SNR) using a Discrete-Valued SNR Estimator. Using this SNR value, instead of trying to adapt the speech signal to the speech processing system, we adapt the speech processing system to the surrounding environment of the captured speech signal. In this paper, we introduced the Discrete-Valued SNR Estimator and a multiengine classifier, using Multiengine Selection or Multiengine Weighted Fusion. Also we use the SI as example of the speech processing. The Discrete-Valued SNR Estimator achieves an accuracy of 98.4% in characterizing the environment's SNR. Compared to a conventional single engine SI system, the improvement in accuracy was as high as 9.0% and 10.0% for the Multiengine Selection and Multiengine Weighted Fusion, respectively.

1. Introduction

Speech processing systems, such as Speaker identification (SI) and Automatic Speech Recognition (ASR), have two operating modes: a training mode and a testing mode. The goal of the training mode is for the classifier to learn how to discern between different speech classes, using exemplars whose classes are known. The purpose of the testing mode is to classify speech data, from an unknown class, based on the training.

The performance of most speech processing systems is degraded severely when the training and the testing audio are captured in different acoustic environments (e.g., different levels of noise [1–4]). For example, training speech may be captured using a telephone handset in a noise-free environment, while the test speech is captured using a hands-free phone in a noisy environment. The conventional method to improve the performance of speech processing in noisy environments is to perform speech enhancement prior to speech processing [2–7]. In this method, training speech signals are obtained in a clean environment (i.e., no noise). The test speech signals, however, are obtained in environments that may contain noise. Denoising and/or speech enhancement methods may be applied to these signals to reduce the effects of the environment; however, often these methods enhances the speech signal audibility from the human perspective, but ignores the fact that the enhancements may remove or distort information in the speech signal, which may be essential to the speech processing.

In [2], the effect of five different speech enhancement techniques on a GMM-based ASR system was examined. These enhancement techniques were (1) perceptual wavelet adaptive denoising (PWAD) (TPS-PWAD), (2) VAD-based noise floor estimation (VAD-NFE), (3) minimum-mean-square-error-based two-dimensional spectral enhancement (MMSE-TDSE), (4) two-pass adaptive noise estimation (TPANE), and (5) combined two-pass spectral and PWAD. The performance of the speech enhancement techniques was evaluated using Perceptual Evaluation of Speech Quality-Mean Opinion Score (PESQ-MOS) [7, 8]. PESQ-MOS predicts the subjective quality of input speech by returning a score between 0 and 4.5; the higher the score, the cleaner the input speech (i.e., higher SNR). These enhancement techniques were used as preprocessing blocks to the ASR engine. The overall performance was evaluated based on the ASR classification accuracy using each enhancement technique.

The author of [2] noted the following.

(i) The speech enhancement techniques of the corrupted speech signal with white noise generally improved the classification accuracy of the ASR; however, this improvement is inconsistent for different SNR values for a single enhancement technique.

(ii) The MMSE-TDSE has the greatest average improvement in the speech quality PESQ-MOS score, over SNR range from 0 to 20 dB, an average improvement of 0.7 points. This improvement, however, did not translate into the best ASR classification accuracy.

(iii) At SNR = 20 dB, the MMSE-TDSE speech enhancement resulted in a decrease in ASR classification accuracy of −0.71%, relative to ASR without speech enhancement techniques.

(iv) The VAD-NFE achieved the best SNR improvement of 0.5 dB and 3.7 dB at 20 and 15 dB, respectively; however, this did not translate into the best ASR classification accuracy at these SNR values.

The results of [2] suggest that optimizing the quality of speech perception or SNR, using speech quality enhancement techniques, does not optimize speech processing performance. Indeed, speech quality enhancement techniques may actually degrade the speech processing performance instead of improving it.

In [5], the combination of four speech enhancement techniques have been combined to enhance the ASR in noisy environment. These speech enhancement techniques were (1) spectral subtraction with smoothing of time direction, (2) temporal domain SVD-based speech enhancements, (3) GMM-based speech estimation, and (4) KLT-based comp-filtering. In [5], two combination methods have been introduced to combine the speech enhancement techniques: selection of front-end processor and combination of results from multiple recognitions processors. The speech enhancement techniques are applied on the captured noisy speech. A combination of these speech enhancement techniques is combined using one of the two proposed combination methods. The overall performance was evaluated based on the ASR classification accuracy using the two combination techniques. The authors of [5] reported an ASR accuracy enhancement of 5.4% of the proposed combination methods; compared to best performing standalone speech enhancement technique. This combination has been achieved using a noise environment detector with a detection accuracy of 54%. Using an ideal noise environment detector with detection accuracy of 100% can enhance the ASR accuracy with only 1.9%, compared to the environment detector with detection accuracy of 54%. As indicated in [5] the main factor of enhancing the ASR accuracy is the proposed combination method using different speech enhancement techniques. Meanwhile the authors [5] did not discussed the fact indicated in [2] that speech enhancement techniques may actually degrade the speech processing performance instead of improving it.

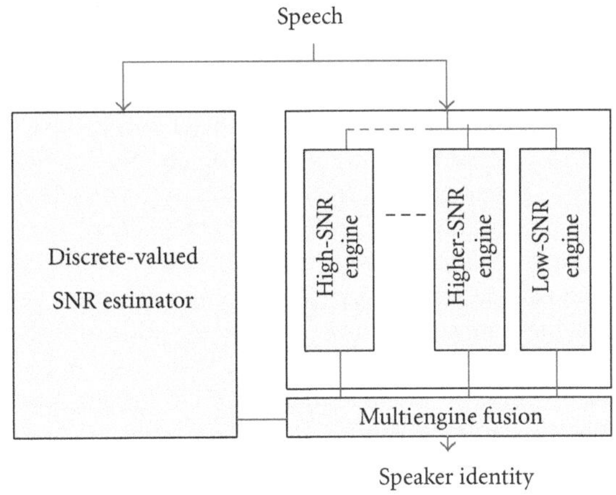

Figure 1: Multiengine SI system.

In this paper, we introduce a speech processing system that adapts to the surrounding environment of the speech signal, rather than trying to adapt the speech signal to the system; in this system, no speech enhancements are applied to the captured speech signal. This speech processing system characterizes its environment from the captured speech signal and adapts the speech processing to better suit the environment, enhancing the overall performance in variable environments. In this paper, speaker identification (SI) is used as the speech processing system.

The SI system proposed in this paper uses multiple SI engines, each trained in a noisy environment but with different SNR levels. Two multiengine fusion SI systems are proposed: Multiengine Selection and Multiengine Weighted Fusion (Figure 1). These proposed systems permit the intelligent fusion of the multiple SI engines using noise information of the surrounding environment. The noise information is determined by a Discrete-Valued SNR Estimator from the incoming speech signal. The proposed system enhances the SI performance, improving classification accuracy and robustness to different noise conditions.

The multiengine fusion system is being proposed for use in a novel security instrument using microphone array. This security instrument is capable of robustly determining the identity and location of a talker, even in a noisy environment. The instrument is also capable of identifying the audio type and location of nonspeech audio signals (e.g., footsteps, windows breaking, and cocktail noise) [9]. This instrument can be incorporated in a variety of applications, such as hands-free audio conferencing systems to perform voice-based security authentication for the conference users [10]. Also, this instrument can be utilized in smart homes and health monitoring for older adults that will support independent living and enable them to continue to age in their own homes [11]. Smart homes can benefit from this instrument by monitoring their occupants through analysis of the audio signals present in the home and recognition of certain sound sources. For instance, respiration, falls to the ground, and water leaking sounds could be detected. The proposed instrument could perform further analyses and make decisions according to the captured audio type, such as extracting the respiration rate, calling emergency help, or alerting the occupant that a plumber may be needed. The analysis of the audio data preserves the privacy of the users better than video monitoring.

Previously mentioned applications require high performance to have successful functionality. In addition, the audio processing system needs to be robust even in variable conditions, such as environments with different levels of noise. Poor instrument robustness would limit utility of these systems.

The remainder of the paper is organized as follows. In Section 2 of this paper, we describe the overall implementation of the proposed approach. This section introduces and discusses the speech/nonspeech classifier, the Discrete-Valued SNR Estimator, the multiengine SI system, and the two methods of multiengine fusion: Multiengine Selection and Multiengine Weighted Fusion. Performance of the multiengine SI system is evaluated using a library of speech signals in noisy environments. In Section 3, results of this performance evaluation are discussed and compared against the baseline method of SI. Major conclusions are summarized in Section 4.

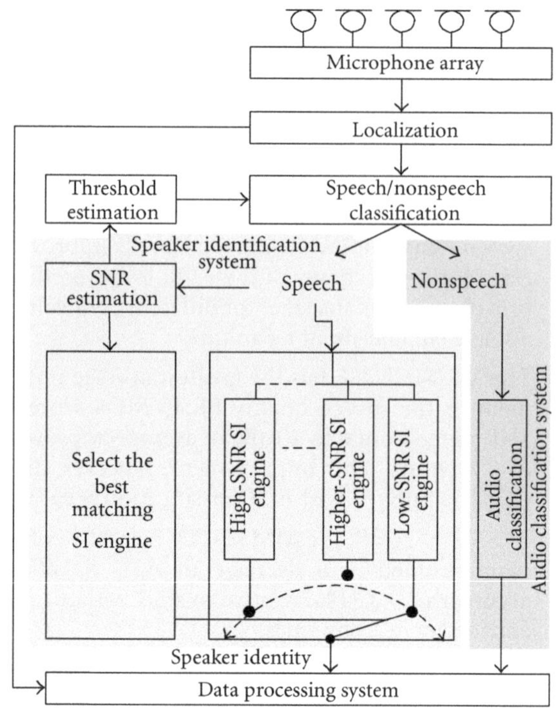

FIGURE 2: Block diagram of proposed system.

2. Proposed System

2.1. Instrument Overview. A block diagram of the overall instrument is shown in Figure 2. The microphone array captures the audio signal then sends the digitized audio signal to the localization block to determine the location of the sound source [12, 13] in addition to steering the microphone array to reduce undesired acoustic noise.

A speech/nonspeech classifier is used as a frontend to appropriately select the next stage in audio processing, which is either the SI system or the nonspeech audio classification system [9]. The speech/nonspeech classifier is also adapted to the environment, so that it is robust to varying levels of noise. In this paper, we utilize the Discrete-Valued SNR Estimator (Section 2.4) to adapt the speech/nonspeech classifier to the surrounding environment based on the estimated SNR value. If the detected signal is nonspeech, the audio classification system will determine the audio type of the detected signal (e.g., footsteps, widows breaking, and door opening sounds); otherwise, the speech signal will be sent to both the Discrete-Valued SNR Estimator and the multiengine SI, consisting of *E*SI engines to identify the talker. Each SI engine is trained in an environment with a specific noise condition. Using the SNR estimate value from the Discrete-Valued SNR Estimator, the system fuses the outputs of the *E*SI engines.

We are introduced to a new fusion method, Multiengine Weighted Fusion. Multiengine Selection selects the SI engine with training environments that best matches the surrounding environment; the multiengine SI system decision is equal to the decision associated with this single SI engine. The Multiengine Weighted Fusion system intelligently weights and combines the output decisions of the *E*SI engines; the higher the similarity of the testing environment to the engine's training environment, the higher the weighting that engine receives.

The resultant talker identity (or audio type for nonspeech audio), along with the location information, is sent to the data processing system for application specific tasks. For example, these data can be utilized in audio-video hands-free conferencing, where a video camera can be appropriately steered to talkers and not steered toward undesired audio sources.

2.2. Speech/Nonspeech Classification. In [9, 14], pitch ratio (PR) was used to perform speech/nonspeech classification. The PR algorithm calculates the presence of human pitch (*70 Hz < human pitch < 280 Hz*) in multiple small frames in certain audio segment duration as shown in Figure 3. The PR is calculated as follows:

$$\text{Pitch Ratio (PR)} = \frac{\text{NP}}{\text{NF}}, \tag{1}$$

where

$$\text{NF} = \left\lfloor \frac{(\text{SD} - \text{FS})/(1 - \text{OL})}{\text{FS}} \right\rfloor + 1, \tag{2}$$

where $\lfloor \bullet \rfloor$ is floor operator, NP is the number of frames that have human pitch, and NF is the total number of frames. This PR value is then compared to a certain threshold. The speech/nonspeech decision is made as follows:

$$\text{Decision} = \begin{cases} \text{Speech}, & \text{if PR} > \text{Threshold}, \\ \text{Nonspeech}, & \text{if PR} < \text{Threshold}. \end{cases} \tag{3}$$

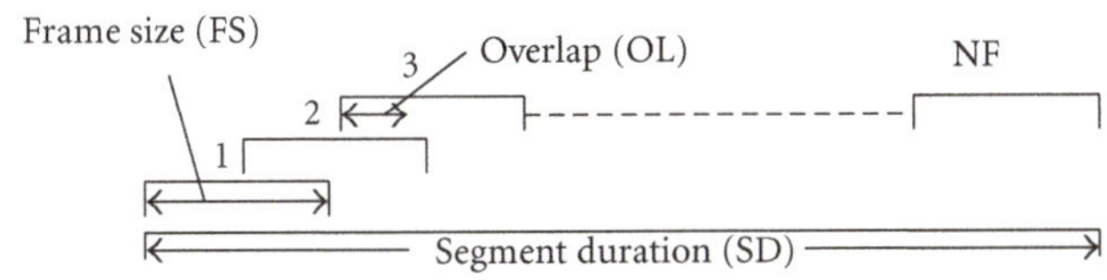

FIGURE 3: PR parameters.

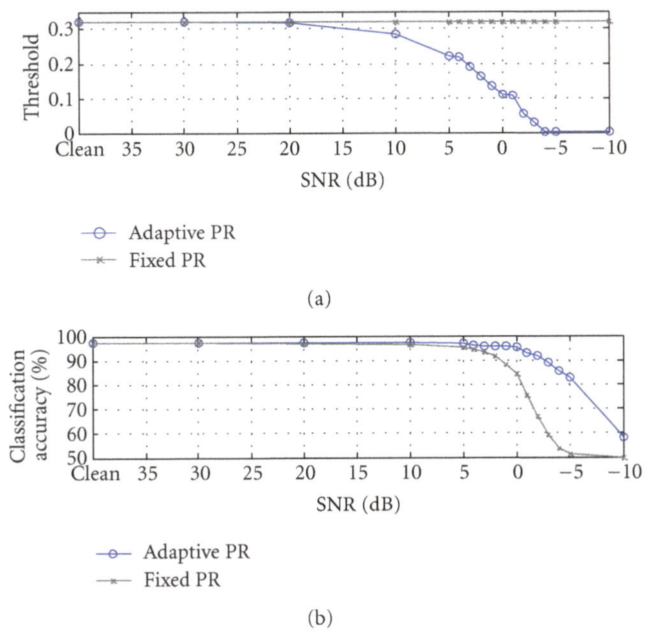

FIGURE 4: Relation between the PR threshold and SNR (sampling rate = 8 KHz, FS = 60 ms, SD = 450 ms, and OL = 0.85).

In [9], a threshold was empirically determined as a function of frame size (FS), segment duration (SD), and overlap (OL). Noise was shown to deteriorate the pitch period of sound. This phenomenon reduces the capability of the pitch detection algorithm to detect the presence of the pitch. To remedy this, the threshold of the PR algorithm is adapted to maximize the accuracy of the speech/nonspeech classifier. We have found that the optimal threshold (Figure 4(a)), the threshold corresponding to the maximum speech/nonspeech classification accuracy (Figure 4(b)), reduces as the SNR reduces. This adaptive speech/nonspeech algorithm was termed the adaptive pitch ratio (APR) algorithm [9]. Adaption is performed based on the SNR estimates from the Discrete-Valued SNR Estimator.

If the decision of the APR is nonspeech, further audio classification is performed to determine the audio type. Nonspeech audio classification will not be discussed here, as it is not the focus of this paper; however, readers can refer to [9] for more details. If the decision of the APR is speech, then an SI algorithm is used to identify the talker.

2.3. Discrete-Valued SNR Estimator. An SNR estimator that is computationally efficient with low complexity is desirable for this system, as it enables the selection of the SI engine in real time. Output of this estimator should be a discrete value that matches one of the training environments' SNRs. This estimator provides information about the noise in the surrounding environment by extracting features from the captured audio sounds emanating within this environment.

In this paper, the Discrete-Valued SNR Estimator is implemented using covariance matrix modeling [1, 15, 16]. The feature vector of the noisy environment is the mel-frequency spectrum coefficients (MFSCs) and the delta MFSC. The MFSCs are computed in a similar fashion as mel-frequency cepstrum coefficients (MFCCs) using a filter bank output but without the use of the discrete cosine transform (DCT) [17]. The DCT of the MFSC vector decorrelates the feature vectors' elements, which leads to a feature vector with low sensitivity to environment change in the audio signal and high sensitivity to the talker (i.e., dependent on the talker) [18]. In our case we desire a feature vector with high sensitivity to the noise but independent of the talker (i.e., provides information about the environments without being affected by who uttered the speech). Thus, the DCT stage has been removed from the feature vector generation process.

To implement the Discrete-Valued SNR Estimator, let X_e represent the covariance matrix of the M training feature vectors for environment $e\{\vec{x}_j,\ 1 \leq j \leq M\}$ and let Y represent the covariance matrix of the N test feature vectors $\{\vec{y}_i,\ 1 < i < N\}$ from an unknown environment. Feature vectors for both training and testing covariance matrices, have a length L; therefore, X_e and Y are matrices of size $L \times L$ and are computed as follows:

$$X_e = \frac{1}{M}\sum_{i=1}^{M}(\vec{x}_i - \overline{x})^T(\vec{x}_i - \overline{x}),$$
$$Y = \frac{1}{N}\sum_{j=1}^{N}\left(\vec{y}_j - \overline{y}\right)^T\left(\vec{y}_j - \overline{y}\right), \tag{4}$$

where vector $\overline{x}$ is the mean of the M training vectors $\vec{x}_i$ and vector $\overline{y}$ is the mean of the N test vectors $\vec{y}_j$. Using the following property:

$$\frac{1}{N}\sum_{j=1}^{N}\left(\vec{y}_j - \overline{y}\right)^T X_e^{-1}\left(\vec{y}_j - \overline{y}\right) = \operatorname{tr}\left(YX_e^{-1}\right). \tag{5}$$

The distance between training environment e and test environment can be calculated using the arithmetic harmonic sphericity measure as follows [1, 16, 18]:

$$d(e) = \log\left[\operatorname{tr}\left(YX_e^{-1}\right)\operatorname{tr}(X_eY^{-1})\right] - 2\log(L), \quad 1 \leq e \leq E, \tag{6}$$

where L is the feature vector length. The distance $d(e)$ represents the similarity between training environment e and the test environment; the smaller the distance, the more the similarity between the training and test environment. Then the distance vector is generated as follows:

$$D = [d(1)d(2)\cdots d(E)]^T, \quad d(e) \geq 0. \tag{7}$$

The proposed instrument only requires a rough estimation of the SNR value to set the threshold of the APR and to calculate the distance between the surrounding environment and the training environments to fuse the output of the multiengines. Other SNR estimators, such as SNR estimation using high-order statistics [19], have higher resolution (i.e., estimates SNR as continuous scale); meanwhile, they have higher complexity. The increased precision is not needed in the proposed instrument implementation, so the increased complexity can be avoided.

2.4. Speaker Identification Engines. Gaussian mixed models (GMM) are used to implement the individual SI engines. GMMs are used commonly for speaker identification [20, 21]. During training, the parameters of each talker GMM is determined from a sequence of M training vectors. First, the vectors are grouped into clusters using Linde et al.'s (LBG) clustering algorithm [22]. Second, the mean and variance of each cluster are then computed. During testing, the probability that a given talker uttered a certain test speech is computed from the GMMs. Using extracted N feature vectors from the test utterance, the SI engine computes a log-likelihood probability vector as follows:

$$L = [l(1)l(2)\cdots l(S)]^T, \tag{8}$$

where S is the number of elements in this vector which is equal to the numbers of the talkers. The SI engine selects the talker that uttered the test speech as the talker with the highest log-likelihood value. In this work, a 32-GMM is used as a classifier, with a feature vector of 16 MFCCs.

The accuracy of the SI engine can be degraded by a mismatch between the training and test speech. In the context of this work, this mismatch occurs when different levels of noise are present in the training and testing sets. In the proposed multiengine SI system, a set of ESI engines are used, each trained with different levels of noise. For the multiengine SI, E log-likelihood probability vectors are produced, one vector from each engine. Using the SNR estimate from the Discrete-Valued SNR Estimator multiengine fusion is performed.

2.5. Multiengine Selection. Each environment e in a set of E training environments is modeled by a single covariance matrix X_e. A test noisy utterance from an unknown environment is modeled by a covariance matrix Y. The distance $d(e)$ is computed for each training environment. The SI engine associated with the minimum distance is selected, and the talker identity determined by this engine is used as the SI system output (Figure 5). The engine selection is calculated as follows:

$$\hat{e} = \arg\min D, \quad 1 \le e \le E. \tag{9}$$

The estimated SNR value is the SNR of the selected training environment $\{\text{SNR} = \text{SNR}_{\hat{e}},\ 1 \le \hat{e} \le E\}$.

2.6. Multiengine Weighted Fusion. Performance of SI engines varies in different environments. This performance variation depends on the similarity between the test and the training environment. Using this fact, SI system can utilize the output decision of each engine and combine the decisions of these engines (Figure 6).

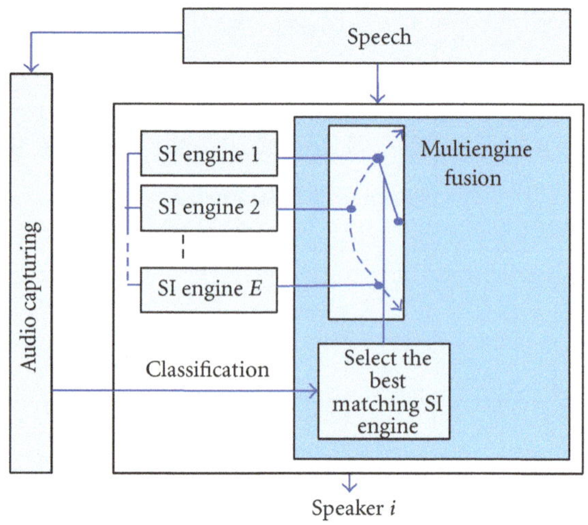

FIGURE 5: Multiengine speaker identification using the Engine Selection method.

Assume that we have S talkers t_i $(i \in (\Omega = \{1, 2, \ldots, S\}))$ and ESI engines trained in E environments. The conditional probability for each talker t_i generated from engine e is denoted as $p(t_i \mid \text{env}_e)$ and is computed using the following steps. Each SI engine that is trained in environment e generates a maximum log-likelihood vector L_e as follows:

$$L_e = [l_e(1)l_e(2)\cdots l_e(S)]^T, \tag{10}$$

where $l_e(i)$ is likelihood of the ith talker from the eth SI engine. These likelihoods are normalized as

$$p(t_i\text{env}_e) = \frac{l_e(i) - \min_j[l_e(j)]}{\sum_k \left(l_e(k) - \min_j[l_e(j)]\right)}. \tag{11}$$

This normalization ensures that the values of $p(t_i \mid \text{env}_e)$ have properties consistent with probabilities; that is,

$$\sum_{i=1}^{S} p(t_i \mid \text{env}_e) = 1, \quad p(t_i \mid \text{env}_e) \ge 0. \tag{12}$$

The probability that the surrounding environment is environment e is denoted as $p(\text{env}_e)$ and is computed as follows. First, the distance between the surrounding environment and the training environment is measured using the Discrete-Valued SNR Estimator. Then the distance vector is generated as in (7). The training environment e with the smallest distance $d(e)$ value has the highest similarity with the surrounding environment. The distances are normalized to the probability of the surrounding environments as follows:

$$p(\text{env}_e) = \frac{d(e) - \max_j[d(j)]}{\sum_k \left(d(k) - \max_j[d(j)]\right)}. \tag{13}$$

This normalization ensures the values of $p(\text{env}_e)$ have properties consistent with probabilities; that is:

$$\sum_{e=1}^{E} p(\text{env}_e) = 1, \quad p(\text{env}_e) \ge 0. \tag{14}$$

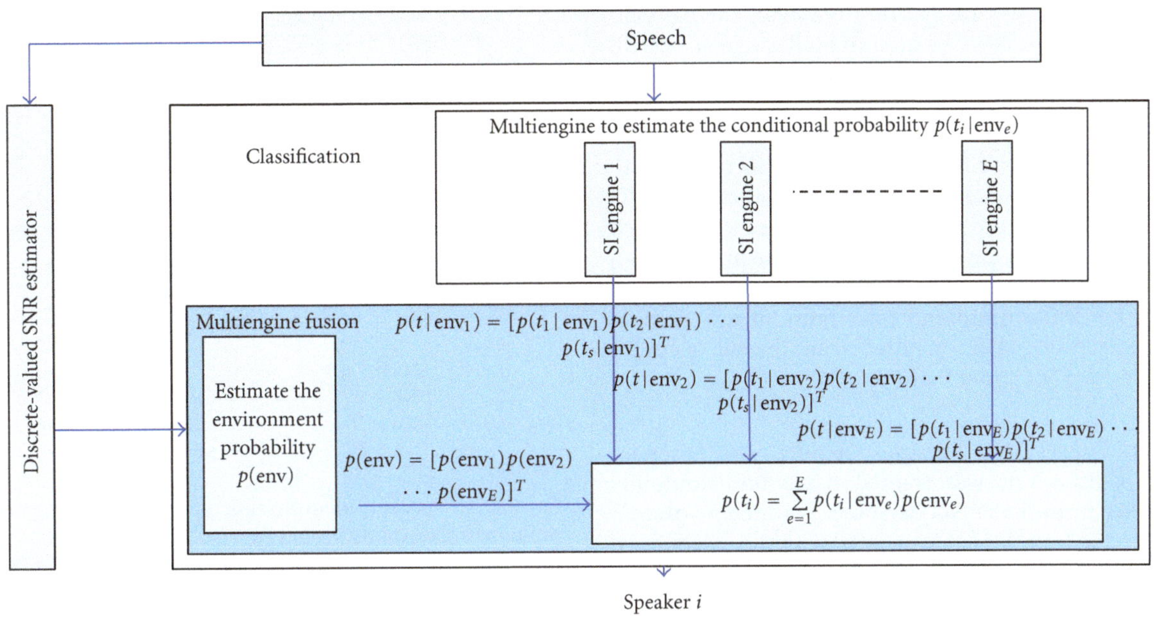

FIGURE 6: Multiengine speaker identification using the Weighted Fusion method.

The probability of talker t_i is computed for the SI system as follows:

$$p(t_i) = \sum_{e=1}^{E} p(t_i \mid \text{env}_e)p(\text{env}_e). \quad (15)$$

The decision of the Multiengine Weighted Fusion engines is the talker with the maximum probability. The Multiengine Weighted Fusion method generalizes the solution for the speaker identification systems in a variable environment.

3. Experiment and Results

3.1. Audio Database. A speech/nonspeech audio database containing a total of 1356 audio segments was used to train and test the speech/nonspeech classifier. Audio segments were 0.45 seconds in length, with a sampling rate of 8 kHz. The speech/nonspeech audio database contains nonspeech audio (e.g., windows breaking, wind, fan, and footsteps sounds) and speech audio. A full list of the audio types can be found in Table 1. Half of each audio type in the speech/nonspeech audio database was used in training the speech/nonspeech classifier, and the remaining half was used as a test set.

The KING speech database is used in training and testing the SI system. The speech/nonspeech portion of the system was trained using the previously mention speech/nonspeech audio database. The KING database consists of speech from 51 speakers; each speaker has 10 utterances with the exception of 2 speakers, which only have 5 utterances [23]. Data are sampled at a rate of 8 kHz, with 16-bit resolution, and utterance durations ranging from 30 to 60 seconds, which were all shortened to 15 seconds. The first 3 utterances of each speaker was used to form a training set (153 utterances), and the remaining utterances were used to form a test set (347 utterances). For the speech/nonspeech and KING databases, the training sets and test sets are mutually exclusive.

TABLE 1: Speech/nonspeech audio database.

Audio class	Number of Audio files
Speech	
Male	282
Female	282
Nonspeech	
Chair	20
Cocktail noise	20
Door	61
Drawer	20
Fan noise	169
Footsteps	41
Glass breaking	145
Rain	50
Thunder	40
Traffic	20
Water	20
Wind	186
Total	1356

A total of 6 levels of SNR (−10, 0, 10, 20, and 30 dB and a clean environment) were used for training, using additive white Gaussian noise to adjust the SNR. A total of 11 levels of SNR (−10, −5, 0, 5, 10, 15, 20, 25, 30, and 35 dB and a clean environment) were used for testing; 5 levels of SNR were different from the training SNR levels. Additive white Gaussian noise and pink noise were used to create the different levels of noise; therefore, while the SNR level may be the same in the training and test set, the type of noise may differ.

3.2. Speech/Nonspeech Classification Results. The speech/nonspeech classifier was tested over an SNR range of −10 dB to clean environment, using white and pink noise. The average performance using the pink and white noise of APR algorithm is shown in Table 2. The average classification performance with pink noise is similar to the performance with white noise. Also, the classification performance of KING database is similar to the speech in the speech/nonspeech audio database. Due to the deterioration of the pitch period, APR has a poor speech classification performance of 32.5% and 28.2% for white noise and pink noise at low SNR = −10 dB, respectively. APR maintains an overall accuracy above 93.0% for SNR above 0 dB.

3.3. Discrete-Valued SNR Estimator Results. The Discrete-Valued SNR Estimator was trained using one randomly selected talker from the KING database. Training involved 3 utterances in the training set, associated with this talker, with the 6 training SNR levels. Discrete-Valued SNR Estimator training and testing are conducted using a feature vector of 19 MFSCs and delta MFSCs. The minimum distance is used to classify test cases to one of the training environments as described in Section 2.4. Performance of the Discrete-Valued SNR Estimator is evaluated on the KING database test set, using the 11 testing SNR levels with white and pink noise.

Table 3 summarizes the results of the Discrete-Valued SNR Estimator classification. The total number of utterances for $j$thSNR value in the test set is equal to U_j, where $U_j = 347$ for $j = 1$ (clean environment) and $U_j = 2 \times 347 = 694$ for $j \neq 1$ (noisy environments). The elements of Table 3 Q_{ji} are the percentage of the total number of utterances U_j that were classified to the ith training SNR value.

The Discrete-Valued SNR Estimator output decision is assumed to be correct if the SNR of the selected training environment has the closest value to the SNR of the surrounding test environment or one of the adjacent environments in terms of SNR (these are set in bold in Table 3). For example if the surrounding environment has an SNR = 15 dB, then the output decision is assumed to be correct if the selected SNR = 10 or 20 dB (accuracy = 30.7% + 62.1% = 92.8%).

The SNR classification is not expected to require high precision, which justifies the inclusion of the adjacent SNR values in the definition of a correct decision. Overall, the average accuracy of the Discrete-Valued SNR Estimator is 98.4% overall accuracy.

3.4. Multiengine Selection Results. Six SI engines were trained using the SNR values of the training environments. Figure 7 plots the results of the 6 individual SI engines with two different noise types (white and pink noise) using the KING database. The maximum performance of a single SI occurs when the test data from a particular noisy environment is used with the SI engine trained in that environment. Also, the performance of SI engines trained in similar noise environments is better than those trained in environments that greatly differ.

The estimate of the SNR from the Discrete-Valued SNR Estimator is utilized to select the best performing SI engine

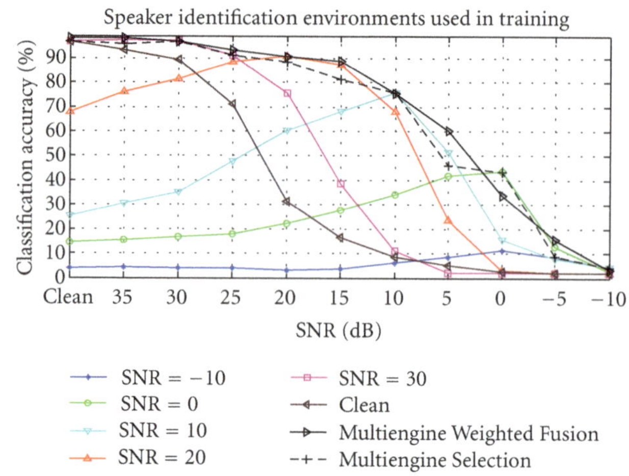

FIGURE 7: Speaker identification accuracy using the proposed multiengine methods.

for Multiengine Selection method. This method has average classification accuracy of 88.7% in noisy environments over an SNR range from 10 dB to clean environment. Compared to the accuracy using the SI engine trained in a clean environment, the improvement in accuracy was as high as 30.6%. Also the Multiengine Selection achieved an improvement in accuracy as high as 17% over an SNR range of −10 to 10 dB. Due to inaccuracies in the Discrete-Valued SNR estimator, the Multiengine Selection accuracy is slightly lower than the best individual SI engine at each SNR level. The Multiengine Selection maintains an accuracy above 75% for SNR above 10 dB, whereas the clean SI engine only does so for SNR above 30 dB.

3.5. Multiengine Weighted Fusion Results. Figure 7 plots the Multiengine Weighted Fusion SI engines' performance using the KING database. The estimate of the SNR from the Discrete-Valued SNR Estimator is utilized to weight and combine the SI decision of the 6 SI engines. Figure 7 shows that this Multiengine Weighted Fusion method has average classification accuracy of 91.3% in noisy environments over an SNR range of 10 dB to clean environment. The identification accuracy of the Multiengine Weighted Fusion method outperforms Multiengine Selection method (88.7%) and the identification accuracy using the SI engine trained in a clean environment (58.1%).

For SNR 5 ≥ dB, the classification accuracy of the Multiengine Weighted Fusion SI engines is superior to all individual SI engines. For example, at SNR = 15 dB, the engines trained at SNR = 10 and 20 dB have an accuracy of 68% and 87%, respectively, the Multiengine Selection classification accuracy has an accuracy of 81%, and the Multiengine Weighted Fusion has an accuracy of 89%. Naturally, the Multiengine Selection method cannot perform better than all individual SI engines and as previously mentioned is typically worse due to inaccuracies in the Discrete-Valued SNR estimator.

TABLE 2: Speech/nonspeech classification results (%).

Audio class	SNR (dB)					
	−10	0	10	20	30	Clean
White noise						
Nonspeech	84.0	95.2	98.1	98.3	98.6	98.6
Speech	32.0	93.2	96.8	97.3	97.3	97.3
KING	33.0	93.7	98.0	98.6	98.6	98.6
Average	58.2	94.3	97.7	98.1	98.3	98.3
Pink noise						
Nonspeech	87.0	94.0	97.5	97.8	98.4	98.6
Speech	27.1	90.0	96.0	95.9	95.9	96.7
KING	29.2	89.1	97.5	98.2	97.4	97.5
Average	57.5	91.7	97.1	97.8	97.9	97.9
Speech/nonspeech average	57.8	93.0	97.4	98	98.1	98.1

TABLE 3: Discrete-valued noise estimator classification results (%).

	Training env.					
Test env.	Clean	30	20	10	0	−10
Clean	**100.0**	0.0	0.0	0.0	0.0	0.0
35	**67.1**	**32.9**	0.0	0.0	0.0	0.0
30	**34.0**	**64.5**	**1.5**	0.0	0.0	0.0
25	10.1	**70.0**	**19.9**	0.0	0.0	0.0
20	0.0	**19.5**	**78**	**2.5**	0.0	0.0
15	0.0	7.2	**62.1**	**30.7**	0.0	0.0
10	0.0	0.0	**18.6**	**78.5**	**2.9**	0.0
5	0.0	0.0	0.0	**29.8**	**70.2**	0.0
0	0.0	0.0	0.0	**0.0**	**97.1**	**2.9**
−5	0.0	0.0	0.0	0.0	**2.9**	**97.1**
−10	0.0	0.0	0.0	0.0	**0.0**	**100**

The Multiengine Weighted Fusion method has higher complexity than the Multiengine Selection method, albeit only slightly higher complexity associated with (11), (13), and (15). Meanwhile, the Multiengine Weighted Fusion method has higher classification accuracy.

For variable environments with a known SNR set, the SI training environments can be matched with the testing environments. In cases, where the training and testing environments are the same (SNR = 30, 20, 10, 0, and −10 dB and clean environment), the Multiengine Selection method has almost similar accuracy as the Multiengine Weighted Fusion method. This makes the Multiengine Selection method perhaps a better solution because of its lower complexity; however, when the testing environments are different from the training environments (SNR = 35, 25, 15, 5, −5 dB), the classification performance enhancement of the Multiengine Weighted Fusion is 7.2% compared to the Multiengine Selection. This suggests that the Multiengine Weight Fusion is a more generalizable solution.

For high-SNR environments (SNR ≥ 30 dB), it seems there is no benefit for a multiengine system. Indeed, the engine trained at SNR = 30 dB has similar accuracy as the multiengine systems (Figure 7); however, it is clearly not robust as there is severe degradation in performance for SNR < 30 dB.

3.6. Speech/Nonspeech Classification with Multiengine SI. The overall adaptive system accuracy (with SNR estimator and APR) is compared with a nonadaptive system (clean SI engine and PR threshold = 0.32) (Figure 8) using the KING database. The main contributor of the accuracy enhancement of the introduced instrument at SNR above 10 dB is the Multiengine SI system (from Table 2 the performance of the APR is nearly perfect for SNR above 10 dB). The per formance curve of the Multiengine Selection, shown in Figure 8, has a large dip at SNR = 5 dB. For SNR below 10 dB the performance curve could be smoothed by decreasing the SNR spacing between SI engines.

The classification accuracy performance using a perfect SNR estimator with the APR and Multiengine Selection is shown in Figure 8. Using this SNR value, the best-performing SI engine and the APR threshold are selected. Compared to APR with Multiengine Weighted Fusion using the Discrete-Valued SNR Estimator, the improvement in accuracy is only 1.6% over SNR range clean to −10 dB. This

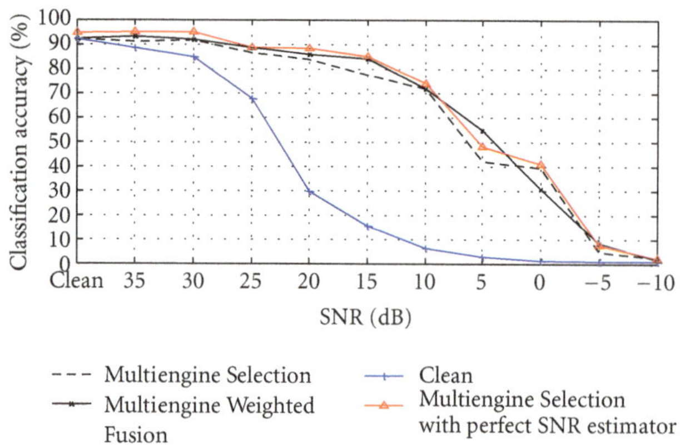

FIGURE 8: Overall accuracy of the introduced system.

result justifies the utility of the rough estimation of the SNR value.

4. Conclusion

This paper confirms that the closer the match between the training and testing environment the higher the SI accuracy. This paper proposed a security instrument that can detect location and identity of talker in noisy environment. To achieve this instrument, an adaptive multiengine speaker identification system and Discrete-Valued SNR Estimator are proposed. The Discrete-Valued SNR Estimator is also utilized to adapt the speech/nonspeech classifier. The proposed instrument is reliable and robust even in variable noise environments.

In the multiengine SI system, six speaker identification engines were utilized to accommodate five noisy environments in addition to clean environment. The SI system using the Multiengine Selection and Multiengine Weighted Fusion methods achieved average identification accuracies of 88.7% and 91.3%, respectively; this represents an enhancement of 30.6% and 33.2% over the SI engine trained in a clean environment, respectively, for SNR ≥ 10 dB. Also the Multiengine Selection and the Multiengine Weighted Fusion achieved an improvement in accuracy as high as 17% and 18.5% over an SNR range of −10 to 10 dB. While the Multiengine Weighted Fusion has high accuracy and reliability than the Multiengine Selection, the tradeoff is a slightly higher complexity.

In future work, this framework of this system (i.e., the combination of the Discrete-Value SNR Estimator and a Multiengine Fusion method) can be deployed in other applications, such as biomedical applications. An example of a biomedical application area that can benefit from the proposed instrument is hearing aid devices. Hearing aids could utilize the proposed instrument to customize their parameters to achieve an optimal hearing performance in variable environments. Modern hearing aids allow the user to set their internal equalizer or beam former variables to predefined settings by selecting the surrounding environment. The deployment of the proposed instrument in hearing aids can automate the selection of the predefined settings according to the surrounding environment.

Acknowledgment

The authors would like to acknowledge the funding from CITO, Mitel, NSERC, and OGS.

References

[1] H. Gish, "Robust discrimination in automatic speaker identification," in *IEEE International Conference on Acoustics, Speech, and Signal Processing: Speech Processing*, pp. 289–292, Albuquerque, NM, USA, April 1990.

[2] A. El-Solh, *Evaluation of speech enhancement techniques for speaker recognition in noisy environments*, M.S. thesis, Carleton University, 2006.

[3] A. El-Solh, A. Cuhadar, and R. A. Goubran, "Evaluation of speech enhancement techniques for speaker identification in noisy environments," in *9th IEEE International Symposium on Multimedia-Workshops*, pp. 235–239, Taichung, Taiwan, December 2007.

[4] J. Ming, T. J. Hazen, J. R. Glass, and D. A. Reynolds, "Robust speaker recognition in noisy conditions," *IEEE Transactions on Audio, Speech and Language Processing*, vol. 15, no. 5, pp. 1711–1723, 2007.

[5] N. Kitaoka, S. Hamaguchi, and S. Nakagawa, "Noisy speech recognition based on integration/selection of multiple noise suppression methods using noise GMMs," *IEICE Transactions on Information and Systems*, vol. E91-D, no. 3, pp. 411–421, 2008.

[6] R. Gomez and T. Kawahara, "Optimizing spectral subtraction and Wiener filtering for robust speech recognition in reverberant and noisy conditions," in *IEEE International Conference on Acoustics, Speech, and Signal Processing (ICASSP '10)*, pp. 4566–4569, Dallas, Tex, USA, March 2010.

[7] International Telecommunications Union (ITU), *Perceptual evaluation of speech quality (PESQ): An objective method for end-to-end speech quality assessment of narrow-band telephone networks and speech codecs*, Telecom Standardization Section, ITU-T, 2001.

[8] T. Yamada, M. Kumakura, and N. Kitawaki, "Performance estimation of speech recognition system under noise conditions using objective quality measures and artificial voice," *IEEE Transactions on Audio, Speech and Language Processing*, vol. 14, no. 6, pp. 2006–2013, 2006.

[9] A. R. Abu-El-Quran, R. A. Goubran, and A. D. C. Chan, "Security monitoring using microphone arrays and audio classification," *IEEE Transactions on Instrumentation and Measurement*, vol. 55, no. 4, pp. 1025–1032, 2006.

[10] S. Ntalampiras, I. Potamitis, and N. Fakotakis, "Probabilistic novelty detection for acoustic surveillance under real-world conditions," *IEEE Transactions on Multimedia*, vol. 13, no. 4, pp. 713–719, 2011.

[11] A. M. Vainio, M. Valtonen, and J. Vanhala, "Proactive fuzzy control and adaptation methods for smart homes," *IEEE Intelligent Systems*, vol. 23, no. 2, pp. 42–49, 2008.

[12] L. Wang, N. Kitaoka, and S. Nakagawa, "Robust distant speech recognition by combining multiple microphone-array

processing with position-dependent CMN," *EURASIP Journal on Applied Signal Processing*, vol. 2006, Article ID 95491, 11 pages, 2006.

[13] E. Mabande, A. Schad, and W. Kellermann, "Design of robust superdirective beamformers as a convex optimization problem," in *IEEE International Conference on Acoustics, Speech, and Signal Processing (ICASSP '09)*, pp. 77–80, Taipei, Taiwan, April 2009.

[14] A. R. Abu-El-Quran and R. A. Goubran, "Adaptive pitch-based speech detection for hands-free applications," in *IEEE International Conference on Acoustics, Speech, and Signal Processing (ICASSP '05)*, pp. 305–308, March 2005.

[15] F. Bimbot, I. Magrin-Chagnolleau, and L. Mathan, "Second-order statistical measures for text-independent speaker identification," *Speech Communication*, vol. 17, no. 1-2, pp. 177–192, 1995.

[16] R. D. Zilca, "Text-independent speaker verification using covariance modeling," *IEEE Signal Processing Letters*, vol. 8, no. 4, pp. 97–99, 2001.

[17] J. Pinquier, J. L. Rouas, and R. André-Obrecht, "A fusion study in speech/music classification," in *IEEE International Conference on Accoustics, Speech, and Signal Processing*, pp. 17–20, Hong Kong, April 2003.

[18] J. S. Gammal and R. A. Goubran, "Combating reverberation in speaker verification," in *Proceedings of the IEEE Instrumentation and Measurement Technology Conference (IMTC '05)*, pp. 687–690, Ottawa, Canada, May 2005.

[19] E. Nemer, R. Goubran, and S. Mahmoud, "SNR estimation of speech signals using subbands and fourth-order statistics," *IEEE Signal Processing Letters*, vol. 6, no. 7, pp. 171–174, 1999.

[20] D. A. Reynolds and R. C. Rose, "Robust text-independent speaker identification using Gaussian mixture speaker models," *IEEE Transactions on Speech and Audio Processing*, vol. 3, no. 1, pp. 72–83, 1995.

[21] P. J. Barger and S. Sridharan, "On the performance and use of speaker recognition systems for surveillance," in *IEEE International Conference on Video and Signal Based Surveillance (AVSS '06)*, Sydney, Australia, November 2006.

[22] Y. Linde, A. Buzo, and R. M. Gray, "An algorithm for vector quantizer design," *IEEE Transactions on Communications Systems*, vol. 28, no. 1, pp. 84–95, 1980.

[23] National Institute of Standards and Technology (NIST), "Brief Description of the KING Speech Data Base," 1992.

19

Transducer Field Imaging Using Acoustography

Jaswinder S. Sandhu,[1] Robert W. Schoonover,[2] Joshua I. Weber,[1] J. Tawiah,[1] Vitaliy Kunin,[1] and Mark A. Anastasio[2]

[1] *Santec Systems, Inc., 2924 Malmo Drive, Arlington Heights, IL 60005, USA*
[2] *Department of Biomedical Engineering, Washington University in St. Louis, St. Louis, MO 63130, USA*

Correspondence should be addressed to Jaswinder S. Sandhu, j-sandhu@santecsystems.com

Academic Editor: Jafar Saniie

A common current practice for transducer field mapping is to scan, point-by-point, a hydrophone element in a 2D raster at various distances from the transducer radiating surface. This approach is tedious, requiring hours of scanning time to generate full cross-sectional and/or axial field distributions. Moreover, the lateral resolution of the field distribution image is dependent on the indexing steps between data points. Acoustography is an imaging process in which an acousto-optical (AO) area sensor is employed to record the intensity of an ultrasound wavefield on a two-dimensional plane. This paper reports on the application of acoustography as a simple but practical method for assessing transducer field characteristics. A case study performed on a commercial transducer is reported, where the radiated fields are imaged using acoustography and compared to the corresponding quantities that are predicted numerically.

1. Introduction

An important aspect of transducer quality control is to assess the radiated ultrasonic field characteristics, which can be affected by factors such as piezoelectric element misalignment with respect to transducer housing, material inhomogeneity, lack of bonding, electrode design and contact placement, acoustic lens quality, and wear plate uniformity and adhesion, to mention a few. A common practice for assessing beam characteristics is to scan a small hydrophone element point-by-point in a plane perpendicular to the transducer's radiation axis [1, 2]. This generates a 2D cross-sectional field distribution at a given distance away from the transducer's radiating surface. To generate a 3D field distribution, 2D cross-sectional distributions are generated at increasing distances along the radiation axis. Hydrophone scanning is slow and tedious, requiring many hours of scanning time to generate the complete field characteristics of the transducer. Moreover, the lateral resolution of the hydrophone scanning method is dependent on the indexing steps between data points as well as the hydrophone element size.

Acoustography [3, 4] is a full-field ultrasonic imaging process wherein an acousto-optical (AO) area sensor is employed to convert ultrasound into a visual image in real time. The AO sensor converts ultrasound energy directly into a visual image by virtue of the inherent acousto-optic effect in our proprietary liquid crystal (LC) or "mesophase" material contained in the AO sensor [3–5]. Theoretical and experimental investigations into the dynamics of liquid crystal devices have been performed to assess the role of geometrical and liquid crystal properties in similar devices [6, 7]. Liquid crystal devices utilizing the AO effect have found utility in acoustic holography [8], medical imaging [9], optical devices [10], and the inspection of composites [11].

In this paper, we report on the application of acoustography for mapping transducer fields. Since it can yield high spatial-resolution recordings of a 2D cross-sectional field distribution without the need for mechanical scanning, acoustography represents a simple and practical method for this application. Experimental studies are conducted in which acoustography is employed to map the fields produced by a commercial transducer at a collection of distances along the transducer's radiation axis. These recordings are compared to estimates of the fields obtained by numerically solving the associated Rayleigh diffraction integral.

2. Background: Basic Principles of AO Sensors

In the AO sensor used in this work, the LC molecules are initially arranged to be parallel to each other but perpendicular to the confining substrates; confining substrates are acoustically transparent, but one of the substrates is also optically transparent. In this configuration of the molecules, the LC material behaves like a slab of a uniaxial, positive birefringent crystal with optic axis parallel to the LC molecular alignment direction ($\underline{n}$). For such an LC layer, the optical behavior (i.e., brightness change, T) under polarized light can be described by [12]:

$$T = \sin^2\left[\left(\frac{2\pi d}{\lambda}\right) \cdot \left(n_{\parallel} - n_{\perp}\right) \cdot \sin^2\theta(I)\right], \qquad (1)$$

where d is LC layer thickness; λ is wavelength of light; $n_{\parallel}$ is refractive index of LC along the optic axis ($\underline{n}$); $n_{\perp}$ is refractive index of LC perpendicular to optic axis ($\underline{n}$); $\theta(I)$ is ultrasonically induced tilt angle of the LC molecules that is a function of the ultrasonic intensity I.

In the absence of an ultrasound field, the molecular tilt angle, θ, is zero, and the AO sensor is uniformly dark across the field of view. However, when an area of the AO sensor is exposed to an ultrasonic field, the LC molecules experience a torque [13], which tilts the LC molecules in the ultrasonically exposed region of the AO sensor. As a result, the ultrasonically exposed area of the AO sensor appears bright. Figure 1 illustrates this process. The local induced brightness level of the AO sensor is related to the local ultrasonic intensity [13]; therefore, an ultrasonic intensity image can be created from the visual image of the AO sensor.

The lateral resolution in acoustography is very high because the visual image is created by the local interaction between the ultrasound field and the (~2 nanometer) LC molecules. In practice, the transverse resolution of this method is limited by the optical camera used to image the AO sensor. For this study, an optical system with 0.31 mm pixel size was chosen; however, higher-resolution optical cameras may be used when the application warrants it. The contrast resolution in acoustography depends on the AO sensor's acousto-optic transfer curve [6, 8], which is analogous to the characteristic curve of an X-ray detector. The AO transfer curve expresses the relationship between ultrasonic intensity and the corresponding optical density (brightness level).

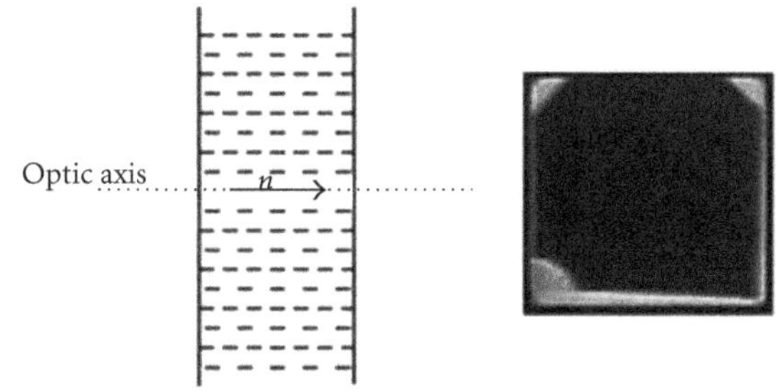

(a) AO sensor without ultrasound (full dark field image)

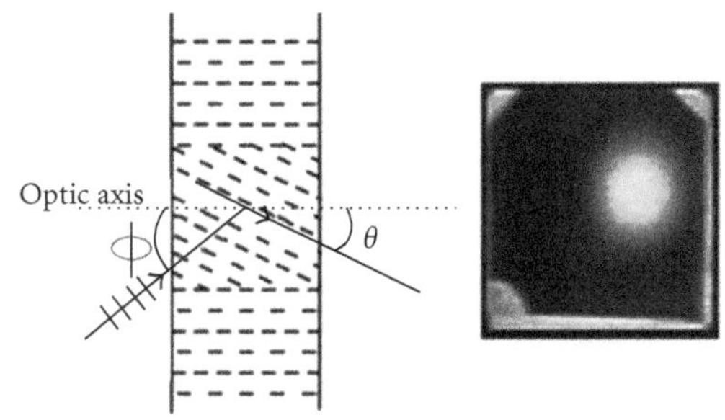

(b) AO sensor stimulated with ultrasound (area shown as bright spot)

Figure 1: In panel (a), a schematic of the AO sensor before insonification by an ultrasound field is shown alongside an image of the AO sensor, when the LC molecules are aligned along the optical axis. In panel (b), a schematic of the AO sensor after insonification by an ultrasound field is shown alongside an image of the AO sensor, when the LC molecules have been perturbed by an ultrasound field.

3. Methods

3.1. Imaging System. Figure 2 displays an overhead view of the acoustography imaging system. The system consists of a water-immersion tank with dimensions 36″(L) × 36″(W) × 12″(D); a 1″ diameter flat circular test transducer (GE Inspection Technologies) suspended in the water tank; a 5″ × 5″ area AO sensor; a digital video camera (Basler A311, not shown) housed in the Optics Box. The optical system has an effective pixel size of 0.31 mm × 0.31 mm. The test transducer was attached to a moveable slide, where the position of the movable slide could be set with an accuracy of ±1 mm with respect to the position of the AO sensor. The beam angle of the test transducer was set with an accuracy of better than 0.5°.

3.2. Experimental Studies. For a flat, circular transducer, such as the one used in this experiment, the ultrasonic intensity along the central axis of the transducer can be expressed as [14]

$$\frac{I(z)}{I(0)} = \sin^2\left[\frac{\pi}{\lambda}\left(\sqrt{r^2 + z^2} - z\right)\right], \qquad (2)$$

where $I(0)$ is the maximum intensity, $I(z)$ is the intensity at distance z from the transducer radiating surface, r is radius of the transducer radiating element, and λ is the acoustic wavelength in water. Two-dimensional images of the radiation pattern of the cylindrical transducer were made at four distances, as measured along the central axis of the transducer. Each measurement distance corresponded to either a minimum or maximum of the intensity along the axis (see Figure 3), or a point in the far-field of the transducer. The transition between the near-field and the far-field for this transducer was calculated to be at 35.3 cm.

The nominal frequency of the transducer was 3.5 MHz. However, the transducer was operated at 3.3 MHz to match the operating frequency of the AO sensor. The transducer beam angle was set at 21° with respect to the AO sensor

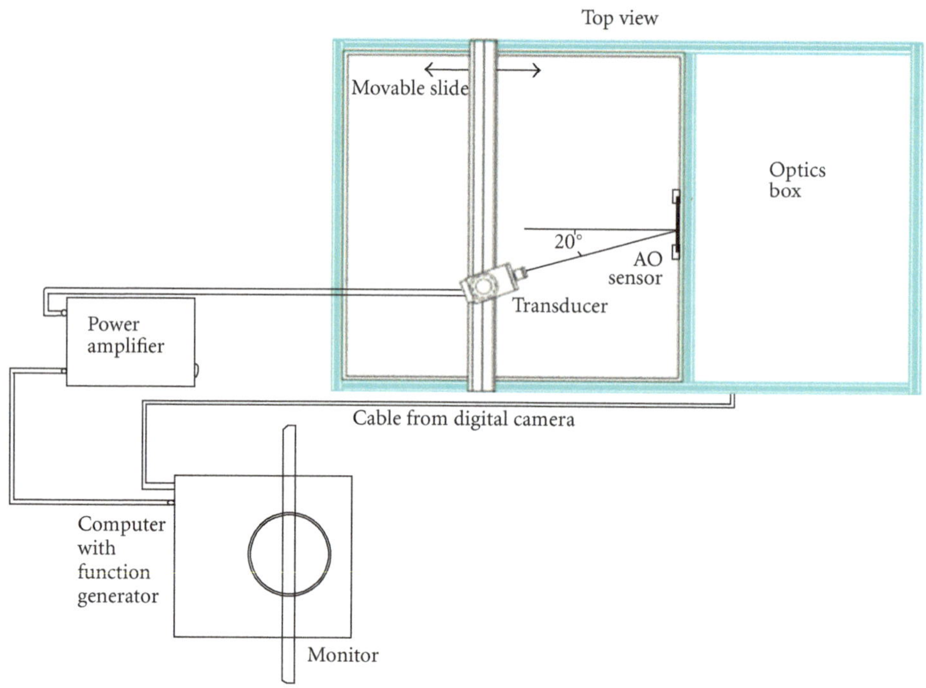

FIGURE 2: An overhead view of the acoustic imaging system.

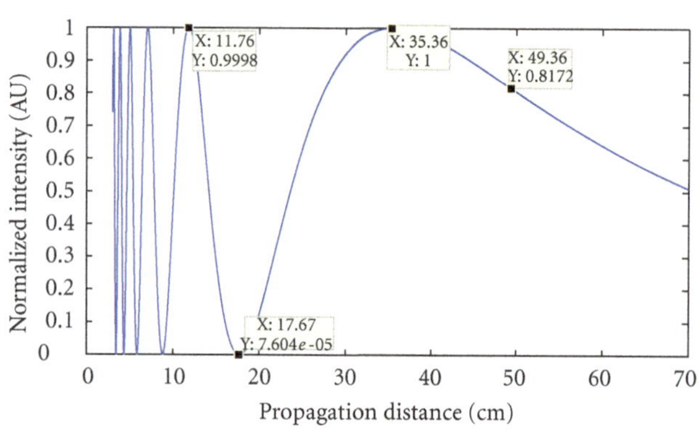

FIGURE 3: Intensity profile along the central radiation axis of a 1″ diameter transducer element operated at 3.3 MHz.

normal, to avoid reverberations between the AO sensor and the transducer's radiating surface. The transducer was energized using a continuous wave (CW) signal.

For each measurement distance, the intensity map on the AO sensor was captured after less than 10 seconds of insonification. Image capture and processing by the camera system was performed in 30 seconds. The AO sensor has a reset time (to realign the LC molecules) of less than 5 seconds. Thus, the ultrasound intensity at >22,00 distinct spatial positions (142 × 156 array) was measured in less than 35 seconds total.

3.3. Computer-Simulation Studies. Simulations of the ultrasonic field produced by a 1″ transducer were also performed to corroborate the images acquired in the experimental studies. The simulations, performed using the DREAM Toolbox for MATLAB [15], calculated the spatial impulse response of a 0.95″ diameter transducer on a plane rotated by 21° with respect to the central axis of the transducer, a distance z away (where z corresponds to the various measurement distances). The pixel size was chosen to be the same as the effective pixel size of the AO sensor + optical system: 0.31 mm × 0.31 mm. The choice of 0.95″ for the transducer diameter was chosen to best match the experimental data. The spatial impulse response was a 2D × 1D signal (two spatial dimensions to represent the measurement plane and one temporal dimension to represent the time-varying ultrasound field). The discrete Fourier transform of the resulting time series at each pixel in the measurement plane was computed by use of the fast Fourier transform algorithm. The element of the discrete Fourier transform, most closely corresponding to 3.3 MHz at that pixel, was selected. The resulting 2D data, corresponding to a monochromatic ultrasound field at 3.3 MHz on the measurement plane, were then squared to represent the ultrasound intensity. The simulations assume a perfectly flat transducer with specified diameter.

4. Experimental Results

4.1. Experimental and Simulated Images. The first column of Figure 4 displays the ultrasonic field maps recorded by use of the AO sensor in the experimental studies. Specifically, the images depict the field intensity distributions on planes located at distances z = 11.76 cm (panel (a)), 17.67 cm (panel (d)), 35.36 cm (panel (g)), and 49.36 cm (panel (j)) from the transducer face along the radiation axis.

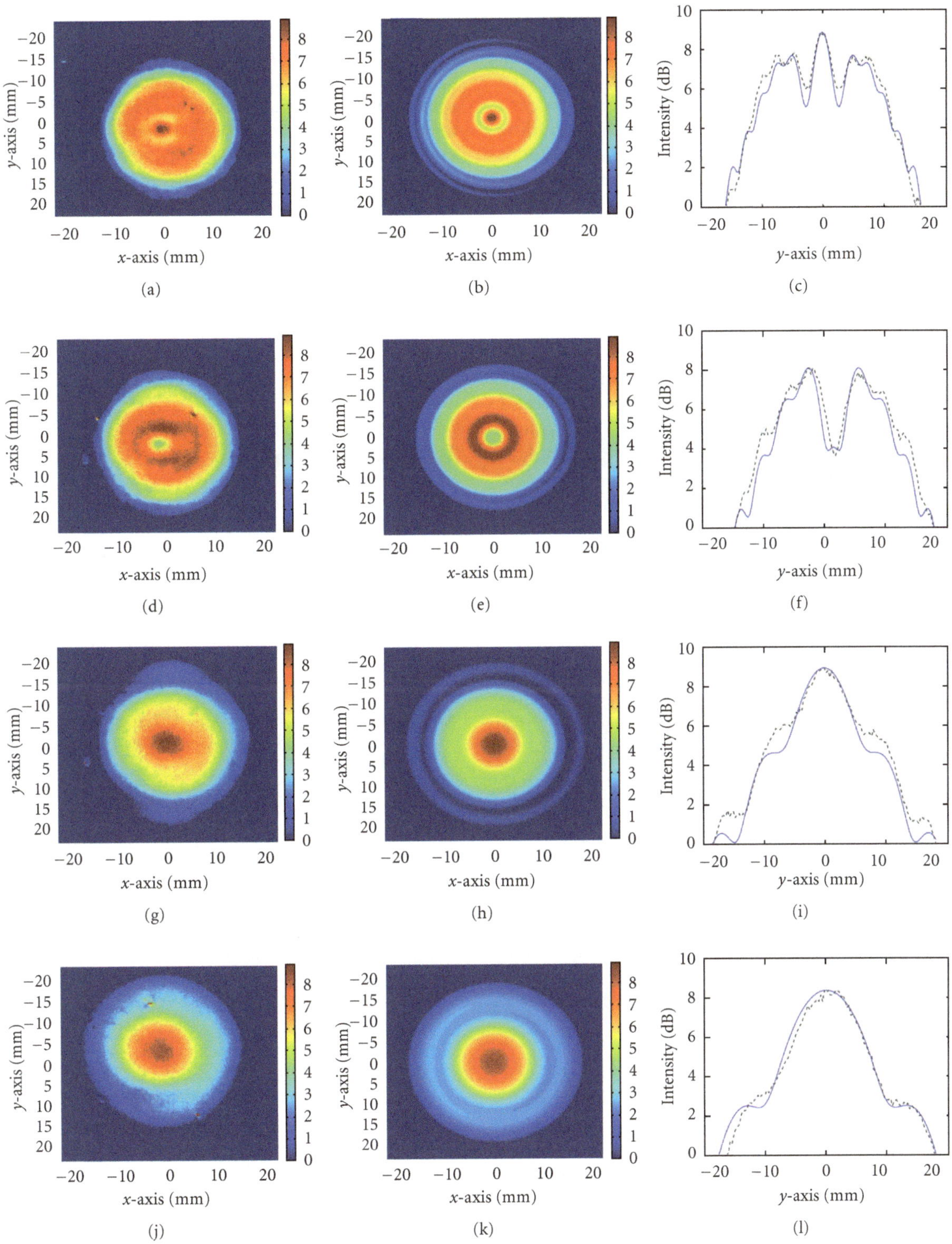

Figure 4: Images of the experimentally determined transducer radiation patterns are shown in the first column. In the second column, theoretically determined transducer radiation patterns are shown. Line plots through the center of the experimental data (black dashed lines) and the theoretical data (blue solid lines) are shown in the third column. The first row corresponds to a propagation distance of 11.76 cm (the location of an intensity maximum on the optical axis in the near zone). The second row corresponds to a propagation distance of 17.67 cm (the location of an intensity minimum on the optical axis in the near zone). The third row corresponds to a propagation distance of 35.36 cm (the near-zone/far-zone transition distance). The fourth row corresponds to a propagation distance of 49.36 cm (a distance in the far zone).

The distances were chosen based on the on-axis intensity profile in Figure 3. The ultrasound intensity in these images was displayed in decibels.

The corresponding simulated intensity images are displayed in the middle column of Figure 4. The simulated intensity data was scaled so that the peak intensity in each simulation matched the peak intensity in the corresponding experimental data. Plots along the y-axis ($x = 0$) of both the experimental (black dashed lines) and theoretical (blue solid lines) results are shown in panels (c), (f), (i), and (l) of Figure 4.

4.2. Discussion of Results. One notes that there is an overall qualitative agreement between the intensity patterns generated through simulation and those obtained with the AO sensor. Comparisons of the intensities in the first and second columns of Figure 4 show strong agreement in predicting the ring structure in the near-zone measurements (first and second rows). For example, in Figure 4(c), the experimental measurements illustrate the same peak-trough-shoulder structure as the theoretical results do for $z = 11.76$ cm. The experimental data also matches the theoretical model in predicting the depth of the trough ($y = 2$ mm) between the two intensity peaks ($y = -3$ mm and $y = 5$ mm) in Figure 4(f) ($z = 17.67$ mm) as well as the predicting the presence of the intensity shoulders at $y = 9$ mm and $y = 14$ mm. The height of the shoulders is slightly underestimated by the simulation results, however. In Figures 4(i) and 4(l), one notes that the simulation matches the experimental results in predicting the width of the radiation pattern as well as predicting the intensity shoulder at $y = \pm 10$ mm in panel (i). The simulation underestimates that shoulder height, however.

The quantitative agreement between the two results may be improved upon through better modeling of the dynamics of the ultrasound field in the AO sensor. The transmission of the ultrasound field through the substrate and into the LC layer of the AO sensor is not accounted for in this analysis. Because of the large speed-of-sound mismatch between the substrate layer (~5000 m/s) and water (~1500 m/s), the transmission of various components of the angular spectrum [16] of the ultrasound field can vary greatly. By accounting for the transmission of sound through the medium, better quantitative accuracy will be achieved.

5. Conclusion

The possibility of applying acoustography to provide a simple but practical tool for mapping ultrasonic fields radiated from ultrasonic transducers has been demonstrated. Since acoustography does not require elaborate mechanical scanning equipment, it is suitable for routine use for quality assessment of transducer fields. The latter capability may be particularly important for calibration of phased array systems, which are being used more and more for nondestructive testing. Other benefits from acoustography have also been demonstrated: each intensity map took <35 seconds compared to hours using current conventional scanning techniques and methods; the entire procedure is user friendly and can be performed by low-skilled operators; the method does not require calibration of the measurement device.

Future plans include a more indepth, head-to-head comparison and validation study of acoustography against the ball reflector method. This will remove any discrepancies that result from simulation modeling errors. In addition, refinement of the process will be pursued including automation, software development, and improved data analysis by obtaining the AO 2D cross-sectional fields and being able to generate true 3D axial profiles using frame-grabbing video output data. This is a distinct advantage over the convention ball reflector method, where the lateral resolution is primarily determined by the indexing step size. Fine indexing steps are required to achieve high resolution but that greatly increases the scan time required to map the ultrasound field. Further improvements in the image processing and analysis will also be pursued to account for the heterogeneous structure of the AO sensor.

Acknowledgment

R.W. Schoonover and M. A. Anastasio acknowledge the support from the National Institute of Health (NIH) awards EB010049 and EB009715.

References

[1] P. G. Kenny, J. J. Gruber, and J. M. Smith, "Ultrasonic transducer characterization," *Materials Evaluation*, vol. 45, no. 6, pp. 730–735, 1987.

[2] R. Lal and D. K. Das-Gupta, "Characterization of ultrasonic transducers," *IEEE transactions on electrical insulation*, vol. 24, no. 3, pp. 473–480, 1989.

[3] J. S. Sandhu, "Acoustography," in *Special Nondestructive Testing Methods, Nondestructive Testing Handbook*, P. O. Moore and P. McIntire, Eds., vol. 9, pp. 278–284, American Society for Nondestructive Testing, Columbus, Ohio, USA, 2nd edition, 1995.

[4] J. S. Sandhu, "Acoustography: a new imaging technique and its applications to nondestructive evaluation," *Materials Evaluation*, vol. 46, pp. 608–613, 1988.

[5] O. A. Kapustina, "Acoustooptical phenomena in liquid crystals," *Molecular Crystals & Liquid Crystals*, vol. 112, pp. 1–164, 1984.

[6] A. P. Malanoski, V. A. Greanya, B. T. Weslowski, M. S. Spector, J. V. Selinger, and R. Shashidhar, "Theory of the acoustic realignment of nematic liquid crystals," *Physical Review E*, vol. 69, no. 2, Article ID 021705, 2004.

[7] V. A. Greanya, A. P. Malanoski, B. T. Weslowski, M. S. Spector, and J. V. Selinger, "Dynamics of the acousto-optic effect in a nematic liquid crystal," *Liquid Crystals*, vol. 32, no. 7, pp. 933–941, 2005.

[8] G. L. Rodríguez, J. Weber, J. S. Sandhu, and M. A. Anastasio, "Feasibility study of complex wavefield retrieval in off-axis acoustic holography employing an acousto-optic sensor," *Ultrasonics*, vol. 51, no. 8, pp. 847–852, 2011.

[9] J. S. Sandhu, R. A. Schmidt, and P. J. La Rivìre, "Full-field acoustomammography using an acousto-optic sensor," *Medical Physics*, vol. 36, no. 6, pp. 2324–2327, 2009.

[10] Y. J. Liu, X. Ding, S. C. S. Lin, J. Shi, I. K. Chiang, and T. J. Huang, "Surface acoustic wave driven light shutters using

polymer-dispersed liquid crystals," *Advanced Materials*, vol. 23, no. 14, pp. 1656–1659, 2011.

[11] J. S. Sandhu, H. Wang, and W. J. Popek, "Acoustography for rapid ultrasonic inspection of composites," in *Nondestructive Evaluation of Materials and Composites*, Proceedings of SPIE, pp. 117–124, December 1996.

[12] M. Born and E. Wolf, *Principles of Optics*, Pergamon Press, New York, NY, USA, 1987.

[13] J. S. Sandhu, H. Wang, and W. J. Popek, "Liquid crystal based acoustic imaging," in *Liquid Crystal Materials, Devices and Flat Panel Displays*, R. Shashidhar and B. Gnade, Eds., vol. 3955 of *Proceedings of SPIE*, pp. 94–108, January 2000.

[14] L. E. Kinsler, A. R. Frey, A. B. Coppens, and J. V. Sanders, *Fundamentals of Acoustics*, Wiley, 1999.

[15] F. Lingvall, T. Olofsson, and T. Stepinski, "Synthetic aperture imaging using sources with finite aperture: deconvolution of the spatial impulse response," *Journal of the Acoustical Society of America*, vol. 114, no. 1, pp. 225–234, 2003.

[16] L. Mandel and E. Wolf, *Optical Coherence and Quantum Optics*, Cambridge University Press, 1995.

20

Advanced Applications for Underwater Acoustic Modeling

Paul C. Etter

Northrop Grumman Corporation, P.O. Box 1693, Baltimore, MD 21203, USA

Correspondence should be addressed to Paul C. Etter, paul.etter@ngc.com

Academic Editor: Jafar Saniie

Changes in the ocean soundscape have been driven by anthropogenic activity (e.g., naval-sonar systems, seismic-exploration activity, maritime shipping and windfarm development) and by natural factors (e.g., climate change and ocean acidification). New regulatory initiatives have placed additional restrictions on uses of sound in the ocean: mitigation of marine-mammal endangerment is now an integral consideration in acoustic-system design and operation. Modeling tools traditionally used in underwater acoustics have undergone a necessary transformation to respond to the rapidly changing requirements imposed by this new soundscape. Advanced modeling techniques now include forward and inverse applications, integrated-modeling approaches, nonintrusive measurements, and novel processing methods. A 32-year baseline inventory of modeling techniques has been updated to reflect these new developments including the basic mathematics and references to the key literature. Charts have been provided to guide soundscape practitioners to the most efficient modeling techniques for any given application.

1. Introduction

Over the past several decades, the soundscape of the marine environment has responded to changes in both natural and anthropogenic influences. A soundscape is a combination of sounds that characterize, or arise from, an ocean environment. The study of a soundscape is sometimes referred to as acoustic ecology. The idea of a soundscape refers to both the natural acoustic environment (consisting of natural sounds including animal vocalizations, the sounds of weather, and other natural elements) and anthropogenic sounds (created by humans) including sounds of mechanical origin associated with the use of industrial technology. The disruption of the natural acoustic environment results in noise pollution.

This paper is concerned with the underwater soundscape. The field of underwater acoustics enables us to observe quantitatively and predict the behavior of this soundscape and the response of the natural acoustic environment to noise pollution. Specifically, underwater acoustics entail the development and employment of acoustical methods to image underwater features, to communicate information via the oceanic waveguide, or to measure oceanic properties. In the present context, underwater acoustics encompasses both the science and the technology necessary to deploy functioning acoustical systems in support of naval and commercial operations.

Broadly defined, modeling is a method for organizing knowledge accumulated through observation or deduced from underlying principles. Modeling applications fall into two basic categories: prognostic and diagnostic. *Prognostic applications* include prediction and forecasting functions where future oceanic conditions or acoustic sensor performance must be anticipated. *Diagnostic applications* include system-design and analysis functions typically encountered in engineering tradeoff studies.

The challenges of managing the underwater soundscape are being met by enabling technologies and by emerging solutions. Throughout this paper, the utility of the available inventory of models is stressed and relevant examples from the recent literature are provided in support.

After a brief background in Section 1, the balance of this paper is divided into three main sections. Section 2 addresses evolving challenges. Section 3 discusses enabling technologies. Section 4 reviews emerging solutions. Finally, Section 5 summarizes the notable advances in underwater acoustic modeling that support management of the underwater soundscape.

2. Evolving Challenges

2.1. Background. The soundscape baseline is defined by ambient noise, which is the prevailing background of sound at a particular location in the ocean at a given time of the year. It does not include transient sounds such as the noise of nearby ships and marine organisms, or of passing rain showers. In practice, ambient noise excludes all forms of self-noise, such as the noise of current flow around the sonar. For sonar processing, however, it is the background of noise (including interfering sounds), typical of the time, location, and depth against which an acoustic signal must be detected.

2.2. Naval Operations in Coastal Environments

2.2.1. The Coastal Environment. Coastal environments are generally characterized by high spatial and temporal variabilities. When coupled with attendant acoustic spectral dependencies of the surface and bottom boundaries, these natural variabilities make coastal regions very complex acoustic environments. Specifically, changes in the temperature and salinity of coastal waters affect the refraction of sound in the water column. These refractive properties have a profound impact on the transmission of acoustic energy in a shallow-water waveguide with an irregular bottom and a statistically varying sea surface. Thus, accurate modeling and prediction of the acoustic environment is essential to understanding sonar performance in coastal oceans.

Physical processes controlling the hydrography of shelf waters often exhibit strong seasonal variations. Annual cycles of alongshore winds induce alternating periods of upwelling and downwelling. The presence of coastal jets and the frictional decay of deep-water eddies due to topographic interactions further complicate the dynamics of coastal regions. Episodic passages of meteorological fronts from continental interiors affect the thermal structure of the adjacent shelf waters through intense air-sea interactions. River outflows create strong salinity gradients along the adjacent coast. Variable bottom topographies and sediment compositions with their attendant spectral dependencies complicate acoustic bottom boundary conditions. At higher latitudes, ice formation complicates acoustic surface boundary conditions near the coast. Waves generated by local winds under fetch-limited conditions, together with swells originating from distant sources, conspire to complicate acoustic surface boundary conditions and also create noisy surf conditions. Marine life, which is often abundant in nutrient-rich coastal regions, can generate or scatter sound. Anthropogenic sources of noise are common in coastal seas including fixed sources such as drilling rigs and mobile sources such as merchant shipping and fishing vessels. Surface weather, including wind and rain, further contribute to the underwater noise field. Even noise from low-flying coastal aircraft can couple into the water column and add to the background noise field.

2.2.2. Littoral Operations. Over the past decade, naval mission requirements have shifted from open-ocean operations to shallow-water (or littoral) scenarios. For convenience, shallow water will be defined by water depths less than 200 meters. This has not been an easy transition for sonar technologists since sonar systems that were originally designed for operation in deep water seldom work optimally in coastal regions. This has also held true for modeling and simulation (M&S) technologies, which have undergone a redefinition and refocusing to support a new generation of multistatic naval systems that are intended to operate efficiently in littoral regions while still retaining a deepwater capability. Shallow-water geometries increase the importance of boundary interactions, which diminish acoustic energy through scattering and also complicate localization of diesel submarines and coastal mines due to multipath propagation. Moreover, the higher levels of interfering noises encountered in coastal regions combined with higher levels of boundary reverberation mask signals of interest. In advance of naval deployments, synoptic meteorological and oceanographic (METOC) measurements are often required in remote or hostile (i.e., harsh or heavily defended) coastal environments to forecast acoustic sensor performance. Coupled atmosphere-ocean-acoustic models could reduce the need for hazardous *in situ* data collection by numerically computing initial states for the embedded acoustic models.

In support of naval operations in littoral regions, acoustical oceanographers have employed ocean-acoustic models as adjunct tools that can be used to conduct rapid environmental assessments (REAs) in remote locations. Due to an increased awareness of the potential technological impacts on marine life, naval commanders and acoustical oceanographers must also be aware of new environmental regulations governing the acoustic emissions of their sonar systems.

In shallow water, interactions of the acoustic fields with the sea bed require an understanding of the sedimentary structure of the bottom to a level of detail that is usually not required in deep-water environments. In the forward-propagation case, this means that a significant amount of information is necessary to properly characterize the bottom boundary to ensure the generation of high-fidelity model outputs. This generally requires a good understanding of the physics of bottom-interacting acoustics in diverse ocean environments.

Sonar clutter, particularly in shallow-water environments, introduces false targets that change the statistics of the reverberation signal. Specifically, clutter increases the probability of false alarm for a given probability of detection. This is because clutter adds to the length of the tails of the reverberation-envelope PDF (probability distribution function), moving the statistics away from the Rayleigh canonical form. Clutter can be caused by target-like features, either natural or man-made, or by non-Gaussian distributions of the scatterers. Typically, high-bandwidth or highly directive systems (or both) have more problems with clutter since, as the size of the scattering patch is reduced, the PDF of the generally non-Gaussian scatterer distributions becomes resolved by the system [1].

2.2.3. Training Ranges. At issue here is operational naval training with active sonars. These high-power multistatic sonars have become more important in the face of improved diesel-electric submarine threats operating in complex coastal environments.

The US Navy has explored the environmental consequences of installing and operating an undersea warfare training range (USWTR) in conjunction with appropriate coordination and consultation with the National Marine Fisheries Service (NMFS) and in compliance with applicable laws and executive orders including the Marine Mammal Protection Act (MMPA), the Endangered Species Act (ESA), the National Environmental Policy Act (NEPA), and the Coastal Zone Management Act (CZMA).

2.2.4. Underwater Networks. Ocean-bottom sensor nodes are used for oceanographic data collection, pollution monitoring, offshore exploration, tactical surveillance applications, and rapid environmental assessments [2, 3]. Factors that determine the temporal and spatial variability of the acoustic channel also limit the available bandwidth of the ocean channel and make it dependent on range and frequency. Specifically, long-range systems (~10 km) have bandwidths of a few kilohertz while short-range systems (~0.1 km) have bandwidths on the order of a hundred kilohertz. A moored-buoy ocean observatory system comprising oceanographic sensors was linked by acoustic communications to retrieve data from sensors in the water column at ranges of approximately 3 km [4]; the observatory was deployed off Vancouver Island in the northeastern Pacific Ocean in May 2004 (for 13 months) to study the correlation of seismicity and fluid flow in a seep area along the Nootka fault.

Underwater networks consist of variable numbers of sensors and vehicles deployed in concert to perform collaborative monitoring tasks over a given area. Underwater sensor networks comprise nodes that communicate via acoustic waves over multiple wireless hops to perform collaborative tasks such as environmental monitoring, naval surveillance, and oceanic exploration. Nodes in underwater sensor networks are constrained by harsh physical environments. Data delivery schemes originally designed for terrestrial sensor networks are unsuitable for use in the underwater environment. Relatively few new schemes have been proposed for underwater use, and no single scheme has yet emerged as the *de facto* standard.

Underwater acoustic communications are influenced by spreading loss, noise, multipath discrimination, Doppler spread, and high and variable propagation delays. Moreover, underwater acoustic channels normally have low data rates and time-varying fading. These factors determine the temporal and spatial variability of the acoustic channel and make the available bandwidth of the ocean channel both limited and dependent on range and frequency. Challenges due to the presence of fading, multipath, and refractive properties of the sound channel necessitate the development of precise underwater-channel models. Some existing channel models are simplified and do not consider multipaths or fading. Multipath interference due to boundary reflection in shallow-water acoustic communications poses major obstacles to reliable high-speed underwater communication systems.

Cooperative transmission is a new wireless communication technique in which diversity gain can be achieved by utilizing relay nodes as virtual antennae. These transmission techniques have been investigated for underwater acoustic communications. First, the performance of several cooperative transmission schemes was studied in an underwater scenario. Second, by taking advantage of the relatively low propagation speed of sound in water, a new wave cooperative transmission scheme was designed in which the relay nodes amplified the signal received from the source node and then forwarded the signal immediately to the intended destination. The goal was to alter the multipath effect at the receiver. Third, the upper bound of performance was derived for the proposed wave cooperative transmission scheme. The simulation results showed that the proposed wave cooperative transmission had significant advantages over both the traditional direct transmission and the existing cooperative transmission schemes originally designed for radio wireless networks [5].

Localization algorithms are relevant to underwater sensor networks, but there are challenges in meeting requirements imposed by emerging applications for such networks in offshore engineering [6]. Localization algorithms can be broadly categorized into *range-based* and *range-free* schemes. Range-based schemes use precise distance or angle measurements to estimate the location of nodes in a network. Range-free schemes are simpler than range-based schemes, but they only provide a coarse estimate of a node's location.

Underwater networking is an enabling technology for the operation of autonomous underwater vehicles. In particular, *ad hoc* networks entail wireless communications for mobile hosts called nodes. In these networks, there is no fixed infrastructure. Mobile nodes that are within range communicate directly via wireless links, while those that are far apart rely on other nodes to relay messages as routers. Node mobility in an *ad hoc* network causes frequent changes of the network topology. Since *ad hoc* networks can be deployed rapidly with relatively low cost, they are attractive for military, emergency, commercial and scientific applications [2, 3]. A channel simulator was developed for testing the performance of unmanned undersea vehicle (UUV) communications [7].

2.2.5. Unmanned Underwater Vehicles. Autonomous underwater vehicles (AUVs), or unmanned undersea vehicles (UUVs), constitute part of a larger group of undersea systems known as unmanned underwater vehicles, a classification that includes nonautonomous remotely operated vehicles (ROVs) that are controlled and powered from the surface by an operator (or pilot) via an umbilical connection.

Underwater gliders actually constitute a new class of autonomous underwater vehicles that glide by controlling their buoyancy and attitude using internal actuators [8]. Gliders have useful applications in oceanographic sensing and data collection because of their low cost, autonomy, and capability for long-range extended-duration deployments. They serve as adjuncts to ship-based hydrographic casts, towed sensors, UUV/AUV and satellite-based sensors, but they also present challenges in communications common to all untethered subsurface sensors.

2.3. Marine Seismic Operations

2.3.1. Seismic Exploration. Marine seismic surveys are used to assess the location of hydrocarbon resources, including gas and oil. There are two acquisition methods: 2D and 3D. The 2D method tows a single seismic cable (or streamer) behind the seismic vessel, together with a single source. The reflections from the subsurface of the sea floor are assumed to lie directly below the path (sail line) of the vessel. In a 3D survey, groups of sail lines (or swathes) are used to acquire orthogonal or oblique lines relative to the acquisition direction. By utilizing more than one source together with many parallel streamers towed by the seismic vessel, the acquisition of many closely spaced subsurface 2D lines can be achieved by a single sail line. Computationally intensive processing is necessary to produce a 3D image of the subsurface of the sea floor. The source arrays are powered by high-pressure air that is compressed onboard the seismic vessel. These compressors are capable of recharging the airguns rapidly and continuously, enabling the airgun source arrays to be fired at approximately 10-second intervals for periods of up to 12 hours. Typical towing depths range from 4-5 meters for shallow high-resolution surveys or 8–10 meters for deeper penetration, lower-frequency targets in open waters. Typical source outputs are approximately 220 dB re 1 μPa/Hz at 1 m. Other types of seismic sources include water guns and marine vibrators.

2.3.2. Marine Mammal Impacts. In 2002, the International Association of Geophysical Contractors (IAGC) hosted an informal meeting to discuss future *Acoustics Research* relevant to seismic operations related to the effects of seismic exploration on sperm whales in the Gulf of Mexico [9]. The IAGC offered its support for sperm-whale research through the contribution of a seismic-source vessel for controlled-exposure experiments. In response, a proposed sperm whale seismic study (SWSS) was approved by the Minerals Management Service (MMS) in 2002. (The MMS is now the Bureau of Ocean Energy Management, Regulation and Enforcement, or BOEMRE.) In subsequent years, IAGC was joined by a number of oil and gas companies to form the Industry Research Funders Coalition (IRFC) that has continued to provide contributions in support of SWSS studies.

Long-term (monthly to seasonal) movements and distributions of sperm whales were studied using satellite-tracked radio telemetry tags (S-tags). Short-term (hours) diving and swimming behavior and vocalizations of sperm whales were examined using recoverable digital-recording acoustic tags (D-tags) that logged whale orientation (i.e., pitch, roll, heading) and depth, as well as the sounds made by the whale and received at the whale from the environment. Diving depths and movements were examined using 3D passive acoustic tracking techniques.

To examine potential changes in the behavior of sperm whales when subjected to seismic airgun sounds, controlled exposure experiments (CEEs) were conducted using the D-tags in conjunction with a seismic-source vessel. The location and level of airgun sounds delivered at the tagged sperm whales were controlled by the science team. These CEEs provided data on the immediate and short-term (hours) response of sperm whales to airgun sounds. Longer-term avoidance or displacement behaviors of sperm whales to seismic vessel airgun sounds were examined using location data from the S-tags and from proprietary commercial seismic shot data.

The 3D tracking method requires at least two widely separated hydrophones to obtain the horizontal range and depth of acoustically active sperm whales and would thus be suited for eventual use on a standard seismic vessel, where the passive acoustic arrays (streamers) can be over a kilometer long. Instead of relying on four hydrophones deployed as a three-dimensional array (which would be difficult to deploy and process), the method used here exploited surface multipath (or "ghosts") to reduce the number of required hydrophones to three and further permitted the phones to be deployed along a single towed cable. The horizontal separation between the widely-spaced hydrophones needed to be at least 200 m in order to obtain adequate range and depth resolution at 1 km horizontal ranges. The method did not require the use of multipath from the ocean bottom, but when such bottom returns were detected they could provide an independent confirmation of these tracking procedures [10].

2.4. Shipping Activity. Shipping lanes, a term used to indicate the general flow of merchant traffic between two ports, are routes that historically have been optimized for shortest distances and travel times, and which are modified to avoid extreme weather events [11]. Noise from distant shipping generally occupies the frequency band 20–500 Hz.

A comparison of time-series measurements of ocean ambient noise over two periods (1963–1965 and 1994–2001) revealed that noise levels from the latter period exceeded those of the earlier period by about 10 dB in the frequency ranges 20–80 Hz and 200–300 Hz, and by about 3 dB at 100 Hz [12]. The observed increase was attributed to increases in shipping. Ambient noise measurements collected at the same site but separated by an interval of nearly 40 years (1964–1966 and 2003–2004) revealed an average noise increase of 2.5–3 dB per decade in the frequency band 30–50 Hz [13, 14].

2.5. Windfarm Development. Wind power, as an alternative to fossil fuels, is plentiful, renewable, widely distributed, clean, and produces no greenhouse gas emissions during operation. A wind farm, which is a group of wind turbines in the same location used for production of electric power, may be located offshore. The installation of ocean wind farms requires medium water depths (<30 m) and construction logistics such as access to specialized vessels to install the turbines. Economic wind generators require wind speeds of 16 km/h or greater.

A concerted effort has been made by industry to minimize any undesirable effects relating to windfarm development and operation [15]. One potential effect of offshore windfarm development is the creation of underwater noise.

Knowing the length of time the marine environment is exposed to an underwater noise source is useful when assessing environmental effects. Measurements of offshore wind turbine noise showed low-frequency sound levels with a maximum of 153 dB re 1 μPa at 1 m at 16 Hz. These measurements were of individual turbines of a relatively low power (less than 1 MW). Despite the low-level and low-frequency nature of the sound, behavioral reactions of marine mammals have been observed in response to the reproduction of wind-turbine noise.

2.6. Ocean Acidification. Climate change also affects the ocean soundscape. The emission of carbon into the atmosphere through the effects of fossil-fuel combustion and industrial processes increases atmospheric concentrations of carbon dioxide (CO_2). Ocean acidification, which occurs when CO_2 in the atmosphere reacts with water to create carbonic acid (H_2CO_3), is increasing.

The attenuation of low-frequency sound in the sea is pH dependent; specifically, the higher the pH, the greater the attenuation. Thus, as the ocean becomes more acidic (lower pH) due to increasing CO_2 emissions, the attenuation will diminish and low-frequency sounds will propagate farther, making the ocean noisier.

Recent investigations modeled what effect the increasing acidity of the ocean would have on ambient-noise levels in shallow water in the presence of internal waves [16]. This model assumed an isotropic distribution of noise sources. Exploring a scenario typical of the East China Sea, the noise at 3 kHz was predicted to increase by 30%, or about one decibel, as the pH decreased from 8.0 to 7.4. These results are representative of other contemporaneous investigations into this matter.

3. Enabling Technologies

3.1. Background

3.1.1. Regulatory Initiatives. An examination of anthropogenic sound in a global context considered the need for new regulatory initiatives to deal with the conflicting uses of ocean space related to noise [17]. This study identified the existing legal, economic, and political barriers to the creation and implementation of a new international regime designed to manage anthropogenic noise in the ocean.

The Committee on Potential Impacts of Ambient Noise in the Ocean on Marine Mammals was charged by the Ocean Studies Board of the US National Research Council to assess the state of our knowledge of underwater noise and recommend research areas to assist in determining whether noise in the ocean adversely affects marine mammals [18]. One of the findings of this committee was that models describing ocean noise are better developed than are models describing marine mammal distribution, hearing, and behavior. The biggest challenge lies in integrating the two types of models. The National Research Council [19] also examined what constitutes *biologically significant* in the context of Level B harassment as used in the latest amendments to the US Marine Mammal Protection Act (MMPA). The MMPA separates harassment into two levels. Level A harassment is defined as "any act of pursuit, torment, or annoyance which has the potential to injure a marine mammal or marine mammal stock in the wild." Level B harassment is defined as "any act of pursuit, torment, or annoyance which has the potential to disturb a marine mammal or marine mammal stock in the wild by causing disruption of behavioral patterns, including, but not limited to, migration, breathing, nursing, breeding, feeding, or sheltering." The MMPA, enacted in 1972, was the first legislation that called for an ecosystem approach to natural-resource management and conservation; it specifically prohibited the *take* (i.e., hunting, killing, capture, and/or harassment) of marine mammals.

The ocean biogeographic information system (OBIS) is an on-line worldwide atlas for accessing, modeling and mapping marine biological data in a multidimensional geographic context [20]. Also see the website at http://www.iobis.org/home.

3.1.2. Modeling Uncertainty. Uncertainty has been defined as a quantitative measure of our lack of knowledge of the sound-speed field and boundary conditions constituting the waveguide information necessary for simulation of the acoustic field [21]. This uncertainty is distinct from any errors related to numerical solution of the wave equation. Existing methods typically solve a deterministic wave equation separately over many realizations, and the resulting set of pressure fields is then used to estimate statistical moments of the field. Proper sampling may involve the computation of thousands of realizations to ensure convergence of the statistics.

A study of the effects of uncertainty in the modeling of anthropogenic impacts suggested a precautionary approach to regulation [22]. It was further noted that due to the complex patterns of sound propagation encountered in diverse shelf regions, some marine mammals may not necessarily encounter the average sound exposure conditions predicted for any given seismic survey.

3.2. Numerical Modeling Techniques. Four types of models will be discussed: propagation, noise, reverberation, and sonar performance. The order of presentation will follow that indicated in Figure 1. This box-type format will be used throughout the text to present compact summaries of methods that have been developed in detail elsewhere [23].

3.2.1. Propagation Models. As sound propagates through the ocean, the effects of spreading and attenuation diminish its intensity. Spreading loss includes spherical and cylindrical spreading losses in addition to focusing effects. Attenuation loss includes losses due to absorption, leakage out of ducts, scattering, and diffraction. Propagation losses increase with increasing frequency due largely to the effects of absorption. Sound propagation is profoundly affected by the conditions of the surface and bottom boundaries of the ocean as well as by the variation of sound speed within the ocean volume.

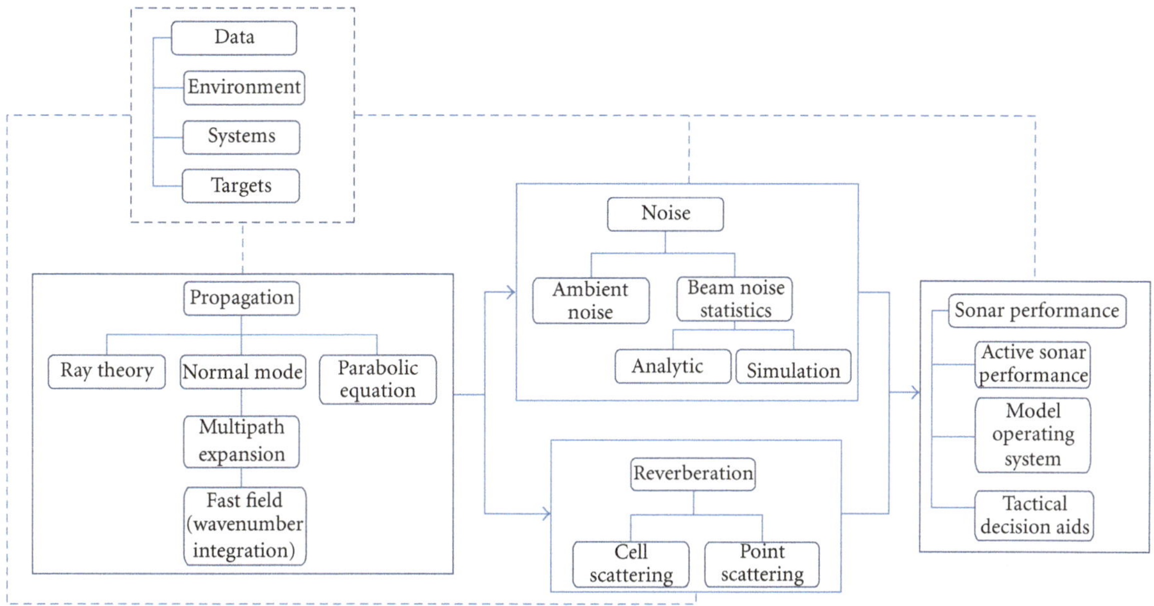

FIGURE 1: Flow of underwater acoustic modeling from propagation, through noise and reverberation, to sonar performance.

Sound-speed gradients introduce refractive effects that may focus or defocus the propagating acoustic energy.

Formulations of acoustic propagation models generally begin with the three-dimensional, time-dependent wave equation. For most applications, a simplified linear, hyperbolic, second-order, time-dependent partial differential equation is used:

$$\nabla^2 \Phi = \frac{1}{c^2} \frac{\partial^2 \Phi}{\partial t^2}, \quad (1)$$

where $\nabla^2 = (\partial^2/\partial x^2) + (\partial^2/\partial y^2) + (\partial^2/\partial z^2)$ is the Laplacian operator, Φ is the potential function, c is the speed of sound, and t is the time.

Subsequent simplifications incorporate a harmonic (single-frequency, continuous wave) solution in order to obtain the time-independent Helmholtz equation. Specifically, a harmonic solution is assumed for the potential function Φ:

$$\Phi = \phi e^{-i\omega t}, \quad (2)$$

where ϕ is the time-independent potential function, ω is the source frequency $(2\pi f)$, and f is the acoustic frequency. Then the wave equation (1) reduces to the Helmholtz equation:

$$\nabla^2 \phi + k^2 \phi = 0, \quad (3)$$

where $k = (\omega/c) = (2\pi/\lambda)$ is the wavenumber and λ is the wavelength. Equation (3) is referred to as the time-independent (or frequency-domain) wave equation.

Propagation models are integral to the higher-level modeling of noise, reverberation, and, ultimately, sonar performance. The graphic in Figure 1 shows the flow of sonar modeling from propagation, through noise and reverberation, through to sonar performance. Estimates of passive sonar performance would require the input of propagation and noise while active sonar performance would require inputs of both noise and reverberation.

Propagation models can be categorized into five distinct techniques [23].

(a) Ray-theoretical models calculate propagation loss on the basis of ray tracing.

(b) Normal-mode solutions are derived from an integral representation of the wave equation.

(c) Multipath expansion techniques expand the acoustic field integral representation of the wave equation in terms of an infinite set of integrals, each of which is associated with a particular ray-path family. Thus, each normal mode can then be associated with corresponding rays.

(d) In underwater acoustics, fast-field theory is also referred to as "wavenumber integration." In seismology, this approach is commonly referred to as the "reflectivity method" or "discrete-wavenumber method."

(e) The parabolic approximation approach replaces the elliptic reduced wave equation with a parabolic equation (PE). Use of the parabolic approximation in wave propagation problems can be traced back to the mid-1940s when it was first applied to long-range tropospheric radio wave propagation.

As shown in Figure 2, a further division can be made according to range-independent (1D, or depth-dependence only) or range-dependent environmental specifications, where environmental range-dependence can be 2D (depth and range) or 3D (depth, range, and azimuth). Since all five techniques are derived from the wave equation by restricting solutions to the frequency domain, the resulting models

- Frequency-domain solutions — $\nabla^2\phi + k^2\phi = 0$
  - Ray theory — $\phi = F(x,y,z)\, e^{iG(x,y,z)}$
  - Normal mode
  - Multipath expansion — $\phi = F(z) \cdot G(r)$ (normal mode, multipath expansion, fast field/wavenumber integration)
  - Fast field/wavenumber integration
  - Parabolic equation — $\phi = F(r,\theta,z) \cdot G(r)$
- Environmental range dependence
  - Range independent (1D) — $f(z)$
  - Range dependent (2D, 3D) — $f(z,r), f(z,r,\theta)$

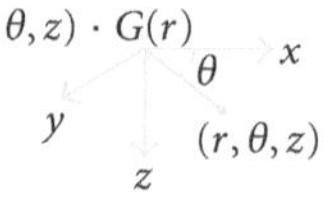

Figure 2: Organization of propagation models into five distinct techniques. A further division is made according to range-independent (1D) or range-dependent (2D or 3D) environmental specifications.

are appropriate for traditional sonar applications. (Solutions obtained in the time domain would be appropriate, e.g., for modeling shock propagation in the ocean.) Each of the five techniques has a unique domain of applicability that can be defined in terms of acoustic frequency and environmental complexity. These domains are determined by the assumptions that were invoked in deriving each solution. Hybrid formulations obtained by combining two or more different techniques are often developed to improve domain robustness.

Table 1 provides a summary of stand-alone propagation models. Superscript letters identify those models that have been added to the inventory since 2003. These letters refer to a brief summary and appropriate documentation. Model documentation can range from informal programming commentaries to journal articles to detailed technical reports containing a listing of the actual computer code. Corresponding information on the legacy models is provided in the 2003 baseline [23] and is not repeated here.

In Table 1: (*Propagation models*), (a) FeyRay was developed to accommodate the speed, fidelity, and implementation requirements of sonar trainers and simulators. It is a broadband, range-dependent, point-to-point propagation model optimized for computa-tional efficiency. FeyRay utilizes the Gaussian-beam approximation, which reduces the acoustic wave equation (a partial differential equation) to a more tractable system of ordinary differential equations [24–28], (b) PlaneRay provides a unique sorting and interpolation routine for efficient determination of a large number of eigenrays in range-dependent environments. No rays are traced into the bottom since bottom interaction is modeled by plane-wave reflection coefficients. The bottom structure is modeled as a fluid sediment layer over a solid half-space. This approach balances two conflicting requirements: ray tracing is valid for high frequencies while plane-wave reflection coefficients are valid for low frequencies where the sediment layers are thin compared with the acoustic wavelength [29–33], (c) PWRC is a ray-based model that performs geoacoustic inversions in range-dependent ocean waveguides. The pressure field is modeled approximately by separating the ocean propagation ray paths from the layered bottom interaction. The bottom interaction is included by using a full-wave description, making PWRC a hybrid model, in contrast to a full-ray theory approach that traces rays into the bottom layers. The field contribution from the bottom interactions partially includes beam-displacement effects associated with internally reflected or refracted returns from the sediment since the complex bottom reflection coefficients are obtained from a full-wave solution. This method is comparable in accuracy to normal mode and analytic solutions (in range-dependent environments) for frequencies >100 Hz [34], (d) Ray5, developed by Trond Jenserud at the Forsvarets forskningsinstitutt (Norwegian Defence Research Establishment), uses direct integration in a sound-speed field specified either analytically or by interpolation from measured data. The Ray5 program is well suited for ray-tracing calculations in acoustic fields described analytically. It needs the sound-speed values, their spatial derivatives and second derivatives at all points within the field. Hence, if these values can be given analytically as functions of the oceanographic and bathymetric parameters, no interpolation is necessary and the program reduces its computing time. Further developments in Ray5 have made it possible to calculate eigenrays. This allows phase information to be retained for a given frequency so that coherent pressure values can be summed for the rays arriving at the receiver. The actual pressure values for the individual rays are calculated by assuming the pressure distribution in the direction normal to the ray to have a Gaussian behavior (i.e., Gaussian beams). It is also possible to calculate the incoherent sound levels [35]. A separate report includes the MATLAB code for Ray5 [36], (e) RAYSON was developed by Semantic (France) to solve the Helmholtz equation using a ray-theoretic approximation for high frequencies. In a stratified (range-independent) environment, analytic solutions are obtained for ray paths that are portions of circles. In range-dependent environments, the ray equations are numerically integrated using a fourth-order Runge-Kutta method to propagate the rays along the range axis. The bottom composition can vary with range and the state of the sea surface can vary in time as well as in range. The software is coded in C++ and is available commercially [37–40], (f) XRAY combines ray tracing in a range-dependent water column with local full-field modeling of interactions with a seabed composed of multiple range-dependent layers of fluid

Table 1: Summary of underwater acoustic propagation models. Superscript letters identify those models that have been added to the inventory since 2003.

Technique	Range independent			Range dependent				
Ray theory	CAPARAY	PLRAY		ACCURAY	GRAB	LYCH	Pedersen	RAYWAVE
	FACT	RANGER		BELLHOP	GRASS	MEDUSA	PlaneRay[b]	RP-70
	FLIRT			Coherent DELTA	HARORAY	MIMIC	PWRC[c]	SHALFACT
	GAMARAY			FACTEX	HARPO	MPC	Ray5[d]	TRIMAIN
	ICERAY			FeyRay[a]	HARVEST	MPP	RAYSON[e]	XRAY[f]
Normal mode	AP-2/5	MODELAB	ORCA	ADIAB	COUPLE	KRAKEN	PROSIM	WKBZ
	BDRM	NEMESIS	POPP[g]	ASERT	CPMS	MOATL	SHAZAM	WRAP
	COMODE	NLNM	PROTEUS	ASTRAL	FELMODE	MOCTESUMA	SNAP/C-SNAP	3D ocean
	DODGE	NORMOD3	SHEAR2	CENTRO	IECM[h]	NAUTILUS	SWAMP[i]	
	FNMSS	NORM2L	Stickler	CMM3D	Kanabis	PROLOS	WEDGE	
Multipath Expansion	FAME	NEPBR				Integrated mode[j]		
	MULE	RAYMODE						
Fast field or wavenumber integration	FFP	OASES	SAFARI	CORE	RD-OASES	SAFRAN		
	Kutschale FFP	pulse FFP	SCOOTER	OASES-3D[k]	RDOASP			
	MSPFFP	RPRESS	SPARC	RDFFP	RDOAST			
Parabolic equation	Use single environmental specification			AMPE/CMPE	FEPES	MONM3D[m]	PE-FFRAME	Two-WayPE
				Cartesian 3DPE[l]	FOR3D	MOREPE	PESOGEN	ULETA
				CCUB/SPLN/CNP1	HAPE	NSPE[n]	PE-SSF (UMPE/MMPE)	UNIMOD
				corrected PE	HYPER	OS2IFD	RAM/RAMS/RAMGEO[p]	3DPE (NRL-1)
				DREP	IFD wide-angle	OWWE[o]	RMPE[q]	3DPE (NRL-2)
				FDHB3D	IMP3D	PAREQ	SNUPE	3DTDPA
				FEPE	LOGPE	PDPE	Spectral PE	3DWAPE[r]
				FEPE-CM	MaCh1	PECan	TDPE	

or solid materials [41], (g) POPP is a range-independent version of the PROLOS normal-mode propagation model [42], (h) IECM is a two-way coupled-mode formalism that provides an exact solution to the wave equation [43]. This model was used to establish a benchmark solution that is an exact numerical solution for the reverberation time series for an environment with a range-dependent, fine-scale rough bottom boundary that induces mode coupling and generates a scattered field. The solution includes scattering effects to all orders in that it sums the infinite series of forward and backward contributions at each range point and maintains energy conservation, (i) SWAMP is a range-dependent normal-mode model that contains closed-analytical forms of the vertical mode functions, which facilitate computation of one-way mode-coupling coefficients between adjacent range-independent regions by neglecting weak backscattering components [44, 45]. The model was used to understand the physics of pulse propagation in double-ducted shallow-water environments where precursors have been observed. The pulse temporal response is modeled using Fourier synthesis in the frequency domain. The model accounts for scattering events along the acoustic signal propagation path and has been extended to model acoustic pulse scattering by spherical elastic-shell targets in inhomogeneous waveguides within the T-matrix approach [46], (j) Integrated Mode extends the multipath expansion method to range-dependent environments [47]. This approach accounts for horizontal variations in bottom depth, bottom type, and sound speed using the stationary phase approximation, (k) OASES-3D target modeling framework is used to investigate scattering mechanisms of flush buried spherical shells under evanescent insonification [48–51], (l) The Cartesian 3D parabolic equation program implements a split-step Fourier algorithm with a wide-angle PE approximation, and is thus a 3D variant of the PE model of Thomson and Chapman [52]. The advantage of employing Cartesian coordinates in the numerical scheme is that the model resolution is uniform over the computational domain [53, 54], (m) MONM3D incorporates techniques that reduce the required number of model grid points. The concept of tessellation (i.e., covering the plane with a pattern in such a way as to leave no region uncovered) is used to optimize the radial grid density as a function of range, reducing the required number of grid points in the horizontal planes of the grid. The model marches the solution out in range along several radial propagation paths emanating from a source position. Tessellation, as implemented in MONM3D, allows the number of radial paths in the model grid to depend on range from the source. In addition, the model incorporates a higher-order azimuthal operator which allows a greater radial separation and reduces the required number of radial propagation paths [55], (n) NSPE, the Navy Standard PE model, consists of two methods of solving the acoustic parabolic wave equation: split-step Fourier parabolic equation model (SSFPE); and split-step Padé (finite-element) parabolic equation (SSPPE) known as RAM. [http://www.nrl.navy.mil/content.php?P=03REVIEW212], (o) OWWE [56] is based on the innovative one-way wave equation developed by Godin [57]. This equation was generalized by Godin to include the source terms and also to account for motion of the medium. The solutions of the differential OWWE are strictly energy conserving and reciprocal. The derivation presented for the multiterm Padé PE model is applicable to a broad class of finite-difference PE models, (p) RAMGEO is a version of RAM modified to handle sediment layers that are range dependent and parallel to the bathymetry [58], (q) RMPE is a ray-mode parabolic-equation solution that is expressed in terms of normal modes in the vertical direction and mode coefficients in the horizontal direction. The model is based on the beam-displacement ray-mode (BDRM) theory and the parabolic equation (PE) method. The BDRM theory is used to analyze the local normal modes. The PE method is used to solve the wave equations for mode coefficients [59], and (r) 3DWAPE incorporates higher-order finite-difference schemes to handle the azimuthal derivative term in a three-dimensional (3D) parabolic equation model [60]. Broadband pulse propagation problems were solved in a 3D waveguide using a Fourier synthesis of frequency-domain solutions (3DWAPE) in a penetrable wedge-shaped waveguide [61]. The 3DWAPE model includes a wide-angle paraxial approximation for the azimuthal component. This version of 3DWAPE was used to investigate broadband sound pulse propagation in two shallow-water waveguides: the 3D ASA benchmark wedge and the 3D Gaussian canyon [62].

The specific utility of these categories is further explained below. In applying ocean-acoustic propagation models, the analyst is normally faced with a decision matrix involving water depth (deep versus shallow), frequency (high versus low), and range-dependence (range-independent versus range-dependent ocean environments). The following assumptions and conditions were imposed in construction of Figure 3, which was originally adapted from F. B. Jensen (see [23]).

(1) Shallow water includes those water depths for which the sound can be expected to interact significantly with the sea floor. Typically, a maximum depth of 200 m is used to delimit shallow water regions. A more accurate definition of shallow water would be expressed in terms of water depth and acoustic wavelength [23].

(2) The threshold frequency of 500 Hz is somewhat arbitrary, but it does reflect the fact that above 500 Hz, many wave-theoretical models become computationally intensive. Also, below 500 Hz, the physics of some ray-theoretical models may become questionable due to restrictive assumptions.

(3) A solid circle indicates that the modeling approach is both applicable (physically) and practical (computationally). Distinctions based on speed of execution may change as progress is made in computational capabilities. A partial circle indicates that the modeling approach has some limitations in accuracy or in speed of execution. An open circle indicates that the modeling approach is neither applicable nor practical.

Model type	Applications							
	Shallow water				Deep water			
	Low frequency		High frequency		Low frequency		High frequency	
	RI	RD	RI	RD	RI	RD	RI	RD
Ray theory	○	○	◑	●	◑	◑	●	●
Normal mode	●	◑	●	◑	●	◑	◑	○
Multipath expansion	○	○	◑	◑	◑	◑	●	◑
Fast field	●	◑	●	◑	●	◑	◑	◑
Parabolic equation	◑	●	○	○	◑	●	◑	◑

Low frequency (<500 Hz) RI: range-independent environment
High frequency (>500 Hz) RD: range-dependent environment

● Modeling approach is both applicable (physically) and practical (computationally)

◑ Limitations in accuracy or in speed of execution

○ Neither applicable or practical

FIGURE 3: Domains of applicability of underwater acoustic propagation models.

To provide compact summaries, propagation models are arranged in categories reflecting the basic modeling technique employed (i.e., the five canonical approaches) as well as the ability of the model to handle environmental range dependence (Figure 3). Such factors define what is termed *domains of applicability*. Hybrid models occasionally compromise strict categorization, and some arbitrariness has been allowed in this classification process. The environmental range dependence considers variations in sound speed or bathymetry. Other parameters may be considered to be range dependent by some of the models, although they are not explicitly treated in this paper.

Figure 3 has been modified in two important respects relative to previous versions [23]. Specifically, a range-dependent capability has been added to the multipath-expansion and to the fast-field (or wavenumber integration) approaches. This change is warranted by the substantial progress made by modelers over the past several years.

Taken together, Figure 3 and Table 1 provide a useful mechanism for selecting a subset of candidate models once some preliminary information is available concerning the intended applications. Note that range-dependent models can also be used for range-independent environments by inserting a single environmental description to represent the entire horizontal range.

3.2.2. Noise Models. Noise is the prevailing, unwanted background of sound at a particular location in the ocean at a particular time. The local noise field is thus characterized by temporal, spatial, and spectral variabilities. The noise generated by natural or anthropogenic point sources is diminished through the effects of propagation to the sonar receiver.

Noise models can be segregated into two categories: ambient-noise models and beam-noise statistics models, as illustrated in Box 1. Ambient-noise models are applicable over a broad range of frequencies and consider noise originating from surface weather, biologics, shipping, and other commercial activities [63]. Beam-noise statistics models [64] predict the properties of low-frequency shipping noise using either analytic (deductive) or simulation (inductive) methods. Table 2 provides a summary of noise models. Superscript letters identify those models that have been added to the inventory since 2003. These letters refer to a brief summary and appropriate documentation. Model documentation can range from informal programming commentaries to journal articles to detailed technical reports containing a listing of the actual computer code. Corresponding information on the legacy models is provided in the 2003 baseline [23] and is not repeated here.

In Table 2: (*Noise models*), (a) ARAMIS consists of a number of FORTRAN, C++ and MATLAB codes that integrate US Navy standard databases with user-provided sonar system parameters to assess the performance of passive spatial processors [65], (b) DANM is the successor to ANDES. DANM predicts the azimuthal dependence of noise in the 25–5,000 Hz band [18]. The dynamic ambient noise model (DANM) provides a realistic simulation of the temporal noise field in which a passive receive array operates. The total noise field is obtained by separately calculating wind and shipping noise. The temporal variability of the noise field is simulated by moving merchant ships along major shipping lanes. Shipping databases provide seasonal information about shipping lanes between the world's major ports, as well as the type and number of ships that move in the lanes, (c) ISAAC uses a Gaussian ray-tracing approach to determine the acoustic ray paths between source and target, including those reflected from the sea surface and sea bed. The ray paths may refract with changes in bathymetry, water density, salinity, temperature, and sea-bed type. ISAAC

Ambient noise

Given: the directional noise intensity per unit solid angle $[N_s(\theta,\phi)]$.

(i) The horizontal noise directionality $[N(\phi)]$ is calculated from $[N_s(\theta,\phi)]$ as:

$$N(\phi) = \int_{-\pi/2}^{\pi/2} N_s(\theta,\phi)\cos\theta\, d\theta.$$

(ii) The vertical noise directionality $[N(\theta)]$ is calculated from $[N_s(\theta,\phi)]$ as:

$$N(\theta) = (1/2\pi)\int_0^{2\pi} N_s(\theta,\phi)\, d\phi.$$

(iii) The omnidirectional noise level (N) is then calculated as:

$$N = \int_0^{2\pi}\int_{-\pi/2}^{\pi/2} N_s(\theta,\phi)\cos\theta\, d\theta\, d\phi,$$

or

$$N = \int_0^{2\pi} N(\phi)\, d\phi.$$

(iv) The horizontal angle (ϕ) is measured positive clockwise from true North while the vertical angle (θ) is measured positive upward from the horizontal plane. No receiver beam patterns were convolved with the noise levels $N(\theta)$ and $N(\phi)$.

Beam-noise statistics

The averaged noise power at the beamformer output (Y) can be expressed as:

$$Y = \sum_{i=1}^{m}\sum_{j=1}^{n}\sum_{k=1}^{A_{ij}} S_{ijk}Z_{ijk}B_{ijk},$$

where m: number of routes in the basin; n: number of ship types; A_{ij}: number of ships of type j on route i (a random variable); S_{ijk}: source intensity of the kth ship of type j on route i (a random variable that is statistically independent of the source intensity of any other ship); Z_{ijk}: intensity transmission ratio from ship ijk to the receiving point; B_{ijk}: gain for a plane wave arriving at the array from ship ijk

Box 1: Organization of underwater acoustic noise models into two categories.

Table 2: Summary of underwater acoustic noise models. Superscript letters identify those models that have been added to the inventory since 2003.

Ambient noise	Beam-noise statistics
ANDES	*Analytic*
AMBENT	BBN shipping noise
ARAMIS[a]	BTL
CANARY	Sonobuoy noise
CNOISE	USI Array noise
DANES	
DANM[b]	*Simulation*
DINAMO	BEAMPL
DUNES	DSBN
FANM	NABTAM
ISAAC[c]	SIAM-I/II[e]
MONM[d]	
Normal mode ambient noise	
RANDI-I/II/III	

allows sensitive marine areas such as marine mammal locations, migratory routes and fisheries to be displayed in the GIS alongside acoustic propagation results. Noise impacts on individual species can be assessed by comparing the sound pressure levels generated from anthropogenic activities with sensitivity thresholds to perform environmental risk assessments. The system has been configured specifically for use by offshore industries, environmental agencies, regulators and others to help assess the environmental impact of underwater noise. The *dBht (species)* approach provides a measurement of sound that accounts for interspecies differences in hearing ability by passing the sound through a filter which mimics the hearing ability of the species. (The *dBht (species)* metric is a pan-specific metric incorporating the concept of "loudness" by using a frequency-weighted curve based on the species' hearing threshold as the reference unit for a dB scale. A large number of both field and controlled-laboratory measurements have been made of the avoidance of a range of idealized noises, using fish with greatly different hearing as a model. All data, irrespective of source or species, indicate a dependence of avoidance reaction on the dBht(species) level. The data indicate three regions: no reaction below 0 dBht (i.e., below the species' threshold of hearing), a cognitive avoidance region where increasing numbers of individuals will avoid the noise from 0 to 90 dBht, and instinctive reaction at and above 90 dBht where all animals will avoid the noise. This probabilistic model allows the behavioral impact of any noise source to be estimated [66].) This approach provides an indication of the noise level that will be received for the species at various distances from the noise source. These values can then be compared to published data to indicate distances at which a species will demonstrate a strong avoidance reaction, a temporary elevation of hearing threshold or a permanent elevation of hearing threshold [67], (d) MONM (marine operations noise model) incorporates a range-dependent, split-step parabolic equation acoustic model including a shear-wave computation capability. MONM has been used for precise estimation of noise produced by subsea construction noise, marine facilities operation, and seismic exploration, particularly in complex coastal regions. The core algorithm in MONM computes frequency-dependent acoustic transmission loss parameters along fans of radial tracks originating from each point in a specified set of source positions. The modeling is performed in individual one-third octave spectral bands covering frequencies from 10 Hz to several kHz, which covers the overlap between the auditory frequency range of marine mammals and the spectral region in which sound propagates significantly beyond the immediate vicinity of the source. The MONM software makes use of geo-referenced databases to automatically retrieve the bathymetry and acoustic-environmental parameters along each propagation traverse, and incorporates a tessellation algorithm that increases the angular density of modeling segments at greater ranges from a source to provide more computationally efficient coverage of the area of interest. The grid of transmission-loss values produced by the model for each source location is used to attenuate the spectral acoustic output levels of the corresponding noise source to generate absolute received sound levels at each grid point. These are then summed across frequencies to provide broadband levels. A further step of Cartesian resampling and summing of the received noise levels from all the sources in a modeling scenario yields the aggregate noise level for the entire operation on a regular grid from which contours can be drawn on a GIS map. The model can either generate contours at evenly spaced levels or draw boundaries representing biologically significant threshold levels [68], and (e) SIAM II (S.C. Wales, unpublished manuscript) was designed to provide many replications of surface-ship noise for horizontal array systems, particularly narrow-beam systems, but could also be used for omnidirectional systems. Its predecessor, SIAM I [69, 70], predicted ship-generated noise over the band 20–120 Hz by generating many replications so that ensemble statistics could be examined. A review report provides more details on SIAM (I/II) in addition to other legacy beam-noise statistics models [71]. Plotting packages for SIAM were described elsewhere [72].

3.2.3. Reverberation Models. Reverberation is sound that is scattered by the ocean boundaries (sea surface and sea floor) or by the volumetric inhomogeneities. Reverberation is produced by the sonar itself; therefore, the spectral characteristics are essentially the same as the transmitted sonar signal. The intensity of reverberation varies with the range of the scatterers (due to propagation loss) and also with the intensity of the transmitted signal.

Most traditional active sonars are configured in what is termed a monostatic geometry, meaning that the source and receiver are at the same position. In some sonar systems, however, the source and receiver are separated in range or depth, or both, in what is termed a bistatic configuration. Bistatic geometries are characterized by a triangle of source, target and receiver positions, and by their respective velocities. Such geometries are commonly employed in sonobuoy applications and also in active surveillance applications. Geometries involving multiple sources and receivers are termed multistatic.

Reverberation models can be categorized according to cell-scattering or point-scattering techniques (Box 2). Cell-scattering formulations divide the ocean into cells, where each cell contains a large number of uniformly distributed scatterers [73]. Point-scattering formulations assume a random distribution of (point) scatterers. Table 3 provides a summary of stand-alone reverberation models. Superscript letters identify those models that have been added to the inventory since 2003. These letters refer to a brief summary and appropriate documentation. Model documentation can range from informal programming commentaries to journal articles to detailed technical reports containing a listing of the actual computer code. Corresponding information on the legacy models is provided in the 2003 baseline [23] and is not repeated here.

In Table 3: (*Reverberation models*), (a) C-SNAP-REV computes reverberation using the C-SNAP range-dependent normal-mode model. Range dependence of the environment is treated as a one-way coupled-mode solution. Surface and bottom reverberation is obtained by integrating the received intensity over the area insonified by the emitted pulse and that contributes to the reverberation at a given time. An average sound speed is assumed for all the paths (i.e., no group-velocity dependence). The scattering is described by a mode-coupling matrix that is equivalent to the plane-wave scattering function evaluated at discrete angles corresponding to the modes. The incident and scattered mode angles are modified to take into account the local slope and are given by the local phase velocity and the sound speed at the interface. Coherent summation of mode contributions is used to correctly model the effects of deep-water convergent zones. The empirical scattering function is based on Lambert's rule. The model deals mainly with the monostatic case, though the technique is extendable to bistatic geometries. (Unpublished notes by Ellis, SACLANT Undersea Research Centre, La Spezia, Italy, various dates.) (b) HYREV is a high-frequency, monostatic reverberation model suitable for shallow-water environments. Arrival times and transmission losses from the source to scatterers are obtained from the appropriate eigenrays. The composite-roughness theory is used to predict the boundary scattering [74], (c) NOGRP first runs the normal-mode program POPP (a variant of PROLOS), which calls the normal mode subprogram MODES and writes out a binary file of mode information that is then read by the monostatic reverberation code. ROSELLA, an extension of NOGRP, is used to execute the reverberation calculations with beam patterns [75, 76], (d) PAREQ-REV is a range-dependent wave-theory model based on the parabolic approximation of the wave equation. The numerical method uses the split-step Fourier marching solution with automatic interpolation of environmental data with range. The code allows a choice of either the standard Tappert-Hardin parabolic equation or the wide-angle equation of Thomson-Chapman. Several choices of starting fields are provided, including a Gaussian source beam of varying width and tilt with respect to the horizontal. Reverberation from the ocean boundaries is computed using standard scattering laws: Lambert's rule for bottom backscatter, and either Chapman-Harris curves or Lambert's rule for sea-surface backscatter. The computational scheme uses reciprocity of propagation to compute the reverberation field for arbitrary receiver depths at the source range. (Unpublished notes by Schneider, SACLANT Undersea Research Centre,

Cell scattering:
Cell-scattering models assume:
(1) a homogeneous distribution of scatterers throughout the area or volume producing reverberation at any given time;
(2) a sufficiently high density of scatterers to ensure that a large number of scatterers occurs in an elemental volume (dV) or area (dA) at range r.
The projector beam pattern is denoted by $b(\theta, \phi)$ and the axial intensity at unit distance is I_0.
The receiver beam pattern is denoted by $b'(\theta, \phi)$ and s_v is the ratio of the intensity of the scattering produced by a unit volume at a distance of 1 m from the volume to the intensity of the incident sound wave.
The equivalent plane-wave volume reverberation level (RL_v) is

$$\mathrm{RL}_v = 10\log_{10}[(I_0/r^4)s_v \int_v b(\theta,\phi)b'(\theta,\phi)\,dV].$$

The plane-wave level of boundary reverberation is

$$\mathrm{RL}_b = 10\log_{10}[(I_0/r^4)s_b \int b(\theta,\phi)b'(\theta,\phi)\,dA].$$

Point scattering:
Point-scattering models are based on a statistical approach that assumes the scatterers are randomly distributed throughout the ocean. The echoes from each individual scatterer are then summed to compute the reverberation level.

Box 2: Organization of underwater acoustic reverberation models into two categories.

La Spezia, Italy, various dates.), (e) PERM-2D computes backscatter from range-dependent bathymetry in the oceanic waveguide [77]. This technique extends the approach of Collins and Evans [78] to problems involving small-scale and large-scale boundary roughness. The PERM-2D model subdivides the oceanic waveguide into range-independent regions and applies the single-scattering approximation (i.e., multiple forward and backward coupling in the scattered fields is neglected) to formulate a scattering problem for the reflected and transmitted pressure field at each range step. Forward-scattering loss, which can be significant at long ranges and very rough surfaces, is included in the solution. Wide-angle operators, which are accurate for the propagating and evanescent spectrum, are applied to yield stable and convergent iteration formulas for the reflected and transmitted fields [79]. Unlike perturbative methods that are restricted to small-roughness amplitudes, the PERM-2D model is valid for arbitrary roughness subject to the single-scattering approximation, (f) REVPA is a parabolic-equation reverberation model intended for shallow-water applications [80], (g) R-SNAP is a coherent monostatic reverberation model employing the range-dependent propagation model SNAP [81, 82], (h) ARTEMIS (adiabatic reverberation and target echo mode incoherent sum) is a general-purpose numerical model of bistatic target echo level and surface and bottom reverberation for bistatic arrangements in an arbitrary range-dependent environment with arbitrary sound-speed variation [83]. The model minimizes computation time while retaining a reasonably accurate power envelope. The approach is based on the adiabatic normal-mode approximation, but with the modal series treated as a continuum and with WKB mode amplitudes (excluding the oscillatory modulation). Outputs are three-dimensional and they can be presented in map form as target echo, reverberation or signal-to-reverberation ratio. Given sparse environmental data, the tradeoff between accuracy and speed is negotiated by intelligent interpolation. This is done by constructing quantities (functions of the desired variables, such as cycle distance, ray angles, etc.) that are more or less linear in space or in mode number. These are converted back to the original variables after linear interpolation, (i) BISTAR is a bistatic, range-dependent reverberation model based on the method of coupled normal modes. The environment is discretized into range-independent segments and the outward propagating field is coupled at the interfaces under the single-scatter hypothesis. The propagation theory and implementation are those of C-SNAP. In order to implement the reverberation prediction, the CW field estimate of C-SNAP has been augmented with a narrowband time-series estimator. The time-domain estimates for the range-dependent environment are available separately and are also used to obtain the reverberation estimates. The scattering process itself can be modeled either as a parametric scattering strength such as Lambert's law or via perturbation theory. The model includes coherent propagation to and from the scattering patch [84–87]. (Also, unpublished notes by K.D. LePage, SACLANT Undersea Research Centre, La Spezia, Italy, various dates.), (j) MURAL, the multistatic reverberation algorithm, supports trainer development, and can be used with any propagation model that produces range-sampled grids of transmission loss, travel time, launch and grazing angles. MURAL contains functions that calculate propagation, scattering and beam patterns. The operation of MURAL couples the algorithm controls to the requested resolution of the prediction with the goal of self-optimizing its performance for the requested resolution [88], (k) S-SCARAB is a range-independent raytracing model for calculating both forward propagating and reverberant acoustic fields. Features include improved algorithms to calculate the coherent forward propagating field, the inclusion of upward refracting rays in the sediment, definition of bistatic source-receiver geometries, and the possibility of specifying 3D beam patterns of the source and receivers. The forward and back propagating acoustic fields are calculated by tracing rays both in the water column and sediment. Only contributions from the ocean bottom are considered in calculation of the reverberation. The local scattering properties of the seabed are described by known power-law expressions for both the interfaces of stratified sediment layers and volume inhomogeneities in the sediment. These scattering kernels

TABLE 3: Summary of underwater acoustic reverberation models. Superscript letters identify those models that have been added to the inventory since 2003.

Cell scattering	
Monostatic	Bistatic/multistatic
C-SNAP-REV[a]	ARTEMIS[h]
DOP	BAM
EIGEN/REVERB	BiKR
HYREV[b]	BiRASP
MAM	BISAPP
NOGRP/ROSELLA[c]	BISSM
PAREQ-REV[d]	BISTAR[i]
PEREV	MURAL[j]
PERM-2D[e]	OGOPOGO
REVMOD	RASP
REVPA[f]	RUMBLE
REVSIM	S-SCARAB[k]
R-SNAP[g]	
TENAR	
Point scattering	
Monostatic	Bistatic/multistatic
REVGEN	BORIS-SSA[m]
RITSHPA[l]	Under-ice reverberation Simulation

are dependent on physical descriptors of the bottom such as seabed roughness. S-SCARAB is computationally efficient compared to other reverberation tools based on normal-mode or parabolic equation approaches, particularly at higher frequencies [89], (l) RITSHPA is a reverberation module intended for use in high-resolution, wideband sonar simulators [90] in which reverberation is stochastic (non-Rayleigh) and follows the K-distribution [91]. Hybrid multipaths, in which the return path is different from the transmit path, are also considered. RITSHPA assumes spherical propagation loss (i.e., isovelocity water column) and uses widely known formulas to compute reflection and scattering at boundaries, and (m) BORIS-SSA (bottom reverberation from inhomogeneities and surfaces small-slope approximation) simulates time series resulting from acoustic scattering off various seafloor types involving various source-receiver geometries [92–96]. This package is an upgrade of BORIS-3D. The model parameters characterize the sonar directivity and pulse shape, the geometrical configuration of the scattering problem and the geophysical characteristics of the seafloor, the sea surface, or other surfaces. These surfaces can have various statistical behaviors or can be obtained from deterministic data based on measured surface heights.

3.2.4. Sonar Performance Models. Sonar performance models combine environmental models, propagation models, noise models, reverberation models, and appropriate signal-processing models to solve the sonar equations. A further segregation can be made according to source/receiver geometry into monostatic and bistatic categories (Box 3). The monostatic formulation of the sonar equations follows a traditional format [73], while the bistatic formulation is adapted from H. Cox (see [23]). The performance of passive sonars (i.e., those that detect sound emitted from a target of interest) could be modeled using the appropriate environmental descriptors together with suitable propagation-loss and noise models. The performance of active sonars (i.e., those that transmit an interrogation signal and then detect the echo returned from a target of interest) could be modeled similarly with the addition of suitable reverberation models.

Sonar performance models can be further categorized as active sonar models, model operating systems and tactical decision aids. Model-operating systems provide a framework for the direct linkage of data-management software with computer-implemented codes of acoustic models, thus facilitating the construction of versatile simulation capabilities. Model-operating systems are further distinguished from stand-alone active sonar performance models by virtue of their ability to conduct sensitivity analyses by computing components of the active-sonar equation using alternative solution techniques. Since sonar model operating systems normally utilize existing ocean-acoustic models and standard oceanographic databases, these systems are unique only in the sense of the number and types of models and databases included, and the particular architectures, graphical user interfaces (GUIs), and other features employed. Tactical decision aids represent a form of engagement-level simulation that blends environmental information with tactical rules. These decision aids guide system operators and scene commanders in planning missions and allocating resources by exploiting knowledge of the operating environment. Table 4 provides a summary of sonar performance models. Superscript letters identify those models that have been added to the inventory since 2003. These letters refer to a brief summary and appropriate documentation. Model documentation can range from informal programming commentaries to journal articles to detailed technical reports containing a listing of the actual computer code. Corresponding information on the legacy models is provided in the 2003 baseline [23] and is not repeated here.

In Table 4: (*Sonar performance models*), (a) ESPRESSO is a minehunting sonar performance assessment tool developed as a NATO standard for interfacing with NATO planning and evaluation tactical decision aids [97]. It uses BELLHOP as a propagation submodel, which has also been modified to calculate beam-based, high-frequency reverberation [98]. Espresso exists in two versions: one intended for scientific use and the other for military use. The scientific version of Espresso provides greater flexibility than the military version, including the ability to select sub-models and view the results of any sub-model. A user guide addresses the user interface for Espresso and describes the underlying software models and data output options available within Espresso [99]. There is a separate user guide for the military version, Espresso (m), which provides greater tailoring of the user interface, including the ability to customize parameters [100]; (b) LYBIN is a range-dependent, ray-theoretical model developed by Svein Mjølsnes of the Norwegian Defence Logistic Organization [101–103]. Range-dependent environmental inputs include bottom type and topography, volume backscatter, sound speed, temperature, salinity, wind speed, and wave height. Choices of

Active sonar equations (monostatic)

(i) Noise background

$$\mathrm{SL} - 2\mathrm{TL} + \mathrm{TS} = \mathrm{NL} - \mathrm{DI} + \mathrm{RD}_N.$$

(ii) Reverberation background

$$\mathrm{SL} - 2\mathrm{TL} + \mathrm{TS} = \mathrm{RL} + \mathrm{RD}_R.$$

Passive sonar equation

$$\mathrm{SL} - \mathrm{TL} = \mathrm{NL} - \mathrm{DI} + \mathrm{RD}.$$

SL: source level; TL: transmission loss; TS: target strength; NL: noise level; DI: receiving directivity index;

RL: reverberation level; RD: recognition differential.

Active sonar equations (bistatic)

The signal excess (SE) can be represented as:

(i) $\mathrm{SE} = \mathrm{ESL} - \mathrm{TL}_1 - \mathrm{TL}_2 - [(N_0 - \mathrm{AG}_N) \bigoplus R_0] + \mathrm{TS} - \Lambda - L$,

the energy source level (ESL) is related to the intensity source level (SL) as:

(ii) $\mathrm{ESL} = \mathrm{SL} + 10\log_{10} T$

where T is the duration of the transmitted pulse,

The echo energy level (EEL) received from the target at a hydrophone on the receiver array is then:

(iii) $\mathrm{EEL} = \mathrm{ESL} - \mathrm{TL}_1 - \mathrm{TL}_2 + \mathrm{TS}$,

where TS is the target strength, N_0 is the noise spectral level, R_0 represents the reverberation spectral level, AG_N is the array gain against noise, Λ is the threshold on the signal-to-noise ratio (SNR) required for detection,

L is a loss term to account for time spreading and system losses,

$\bigoplus$ represents power summation, TL_1 is the transmission loss from source (S) to target (T),

and TL_2 is the transmission loss from target (T) to receiver (R).

Box 3: Sonar performance models are based upon the sonar equations, which are the basic building blocks for both monostatic and bistatic sonar geometries.

calculation outputs include ray trace, transmission loss, reverberation (surface, volume, and bottom), noise, signal excess, probability of detection, travel time, and impulse response. The transmission-loss module was evaluated by NURC [104]. LYBIN is available commercially from the Forsvarets forskningsinstitutt (FFI), (c) MOC3D [105] is a 3D model developed from the 2D model MOCASSIN [106]. MOC3D was used to investigate the importance of out-of-plane sound propagation in a shallow-water experiment in the Florida Straits [107], (d) MODRAY was developed in conjunction with DSTO (Australia) to simulate the propagation of sound through the underwater environment [108, 109]. MODRAY uses classical ray-tracing theory to produce sound-pressure time series at one or more receivers. The marine environment is range-independent. Seafloor composition can be specified, the sound-speed profile can be arbitrary, noise includes wind, rain, biological and shipping sources, and scattering by marine organisms is included. MODRAY can model an arbitrary number of sound sources, reflectors and receivers stationed on moving platforms. MODRAY has been used extensively to model the effectiveness of underwater communications algorithms [110, 111], (e) SUPREMO is a multistatic sonar performance model that includes propagation, target echoes, reverberation and noise, all plotted as a function of delay time and bearing using an equivalent map projection. A modular propagation section makes it possible to separate the effects of propagation modeling (e.g., theoretical basis and range-dependence) from those of scattering computations (e.g., computational efficiency, bottom slope and scattering law). Special attention is paid to problems of interference from multiple sources firing in sequence, target aspect dependence from multiple receivers, mixed FM and CW, mismatched source and receivers, multiple displays (one for each bistatic pair), and presentation of results [82, 112–115]. Predictions of acoustic reverberation and target echo intensity made by the SUPREMO sonar performance model were compared with measured data gathered in the Malta Plateau region of the Mediterranean Sea. The observed model-measurement agreement demonstrated the suitability of SUPREMO for use with an environmentally adaptive, low-frequency, active sonar system [116]. Version 2.0 of SUPREMO has been documented [117, 118], (f) SWAMI models range-and-azimuth dependence (N × 2D) via adiabatic modes [119].

Table 4: Summary of sonar performance models. Superscript letters identify those models that have been added to the inventory since 2003.

Active sonar models	
Active RAYMODE	MINERAY
ALMOST	MOCASSIN
ASPM	MOC3D[c]
CASTAR	MODRAY[d]
CONGRATS	MSASM
ESPRESSO[a]	NISSM-II
GASS	SEARAY
HODGSON	SONAR
INSIGHT	SST
INSTANT	SUPREMO[e]
LIRA	SWAMI/DMOS[f]
LORA	SWAT
LYBIN[b]	UAIM[g]
Model operating systems	
CAAM	GSM - Bistatic
CALYPSO[h]	HydroCAM
CASS	PRISM
ESME/NEMO[i]	SPPS[j]
Tactical decision aids	
ASPECT SPPS[k]	
IMAT	
NECTA	

SWAMI has been used to support the towed integrated active-passive sonar (TIAPS), which was developed as a technology demonstrator for the Canadian Forces. Software development utilized a system test bed (STB) comprising a collection of scaleable, portable, and reusable components for constructing sensor-based applications. The toolset contains modules to produce predictions of transmission loss, reverberation (MONOGO), signal excess, and probability of detection. SWAMI includes the capability to model various source and receiver configurations including omnidirectional arrays, line-arrays (both horizontal and vertical) and volumetric arrays [120]. DMOS (DRDC Atlantic Model Operating System) is an evolution of the SWAMI suite of programs that enables a user to model transmission loss, reverberation, signal excess, and probability of detection for active sonars [121]. Originally, the suite was based on normal-mode theory (PMODES); however, normal-mode theory is best suited to shallow water and low acoustic frequencies, and users occasionally need to model reverberation and other parameters under other conditions. DMOS was enhanced to include a Gaussian-beam acoustic propagation model, BELLHOP, as an alternate propagation engine. A DRDC-extended version that included a range-dependent capability was chosen. DMOS may now be used to model both active and passive sonars in shallow or deep water; (g) UAIM is a system of computer algorithms designed to predict multibeam sonar performance and image the effects of variations in bathymetry, clutter objects, and bottom type, particularly in complex shallow-water environments [122]. The model addresses perceived weaknesses in existing mine-countermeasure (MCM) sonar performance models. The propagation code in UAIM was derived from RASP while the reverberation and signal excess codes were derived from GSM. Modifications were made to these codes to accommodate high-frequency range-dependent applications. A utility package called SoundGuide assists the operator in creating the large input file, executes UAIM and plots the resulting data, (h) CALYPPSO (ΚΑΛΥΨΩ) is an integrated computer environment that was developed for the analysis of underwater acoustic detection systems [123]. The communication language is Greek. It can treat passive detection, broadband, or LOFAR, as well as active detection, monostatic or multistatic. The system contains compact databases for environmental (coastline, bathymetry, oceanographic, geological) and operational (system parameters, target characteristics) data. Acoustic propagation calculations are performed using normal-mode, parabolic approximation, and ray-theoretical codes supporting broadband calculations in range-dependent environments. The results include transmission loss, reverberation levels, detection thresholds and probabilities of detection for a variety of user-defined operational scenarios. (This work was supported by the Greek MOD.) (i) ESME (effects of sound on the marine environment) is a multidisciplinary research and development effort to explore the interactions between anthropogenic sounds, the acoustic environment and marine mammals [124, 125]. The "ESME workbench" models the entire sound path including the sound sources (impulsive or explosive), the medium (water column and seafloor), and the temporary threshold shift (TTS) models of the marine mammals. (TTS refers to a temporary increase in the threshold of hearing, i.e., the minimum intensity needed to hear a sound at a specific frequency, but which returns to its preexposure level over time.) The goal is to predict impacts of anthropogenic sounds on marine mammals. This entails three elements: (1) accurate estimates of the sound field in the ocean; (2) accurate estimates of the cumulative sound exposure of the marine mammals; (3) reliable predictions of the incidence of TTS for the species of interest given the estimated cumulative exposure. The flexibility and computational efficiency of the ESME Workbench will be enhanced by merging the NEMO (NUWC exposure model) and ESME approaches into the "One Navy Model," which is intended to serve as the standard simulation system for use in predicting impacts of anthropogenic sound sources on marine life for environmental compliance purposes, (j) SPPS, the sonar performance prediction system (developed in Germany), evaluates the performance of sonar systems in various naval-warfare scenarios and assists in sonar-system design efforts. The model accommodates active and passive sonar detection problems involving both broadband and narrowband signals using a variety of sonar antennas. The integrated ocean-acoustic propagation model spans the frequency range 10 Hz to 1 MHz by using a hybrid combination of parabolic-equation, coupled-mode, and ray propagation submodels. Environmental databases describe the ocean environment, targets, platforms and sonar-system characteristics. The reverberation model predicts volume, surface and bottom components. Noise sources include ambient, biological, rain, shipping, ice, seismic exploration, and self-noise [126], and (k) ASPECT, the active system performance estimate computer tool, is a multistatic tactical decision aid (TDA). It computes estimates of system performance for active underwater acoustic sensors. ASPECT was originally designed to satisfy the requirements of mission planning software for the IEER (improved extended echo ranging) system. However, virtually any multistatic or monostatic active acoustic system can be modeled using this software package. ASPECT uses the FAME (fast multipath expansion) model for range-independent calculations of transmission loss and ASPM (active system performance model) for range-dependent transmission loss and reverberation. These computations are then fed into MSASM, which is capable of estimating the performance of active sonar systems for multiple sources, receivers, and targets. It simulates target motion including such features as normal and uniform probability distribution, various speed, and course distributions, as well as target evasion modeling to a limited degree. ASPECT version 2.0 (and beyond) supersedes the MSASM Interactive Execution and Optimization System (MINEOS), which was the first version of mission planning software developed for the IEER program. ASPECT also exists as a standalone version [127].

3.3. Inversion Techniques

3.3.1. Adjunct Tools for Oceanographers. Inverse acoustic sensing techniques presently constitute adjuncts to direct measurement methods. However, the application of inverse acoustic sensing techniques to dynamical studies of the oceans' boundaries and interior show great promise for three reasons. First, such data can be used to establish comparative baselines for other remote sensors, such as satellites, by providing synoptic portraitures of the interior oceans together with concurrent ground-truth data at the sea surface. Second, inverse acoustic techniques often afford useful insights into a broad class of oceanic phenomena since their successful employment relies heavily on the use of numerical models first to understand the role of the oceans as an acoustic medium. Third, inverse data provide estimates of spatially integrated and temporally averaged oceanic conditions that are not readily available from traditional oceanographic sensors.

Recently, acoustical oceanographers have employed underwater-acoustic models as adjunct tools for inverse-sensing techniques that can be used to obtain synoptic portraitures of large ocean areas or to monitor long-term variations in the ocean.

Useful information about the ocean can be derived from both forward and inverse applications of underwater sound. Direct (or forward) methods include traditional sonar applications. Inverse methods extract information from direct measurements of the physical properties of the ocean. These inverse methods combine the direct physical measurements with theoretical models of underwater acoustics. The objective is to estimate detailed underwater-acoustic fields from sparse physical measurements using the theoretical models as guides. Inversions of controlled-source, noise, and reverberation fields have been performed successfully (Box 4).

Inverse sensing techniques that employ acoustics have been used in several subdisciplines of geophysics including seismology, meteorology, and oceanography. In oceanography, inverse acoustic data provide estimates of spatially integrated and temporally averaged oceanic conditions that are not readily available from a traditional constellation of point sensors. Forward (or direct) models are important in solving inverse problems.

3.3.2. Propagation (Acoustic Field Measurements). Acoustic propagation characteristics in the deep oceans are determined largely by the refractive properties of the water column and, to a lesser extent, by the surface and bottom boundary conditions. Propagation measurements can be used to infer bulk properties of the water column such as temperature, sound speed, density, and currents. In shallow-ocean areas, where propagation characteristics can be strongly affected by the bottom boundary, propagation measurements can be used to infer properties of the sea floor such as composition and scattering characteristics. A wide spectrum of inverse problems has been addressed in underwater acoustics including estimation of geoacoustic parameters, acoustic thermometry, and shallow-water characterization. Inverse acoustic sensing methods utilizing the propagation characteristics of the oceans include matched field processing, ocean acoustic tomography and deductive geoacoustic inversion [23]. A new technique known as a time reversal mirror (TRM) uses inverse methods to refocus received signals back to the source.

Range-dependent ray-theoretical models are typically used for deep-water applications involving tomographic experiments as well as for high-frequency experiments in shallow water. Range-dependent wave-theoretical models based on normal-mode, fast-field (wavenumber integration), or parabolic-equation approaches are preferred for low-frequency experiments in shallow water.

3.3.3. Noise. The ambient noise field in the oceans is described by the spectral, spatial, and temporal characteristics of sound generated by both natural and industrial sources. Measurements of these characteristics can provide useful information regarding the nature of the noise sources themselves as well as physical features within the oceans. Examples of inverse applications of the noise field include wind speed determination, rainfall measurements, object imaging (*acoustic daylight*), and geoacoustic inversion.

3.3.4. Reverberation. The reverberation field in the oceans is the product of acoustic scattering by the surface and bottom boundaries and by inhomogeneities within the oceans. The utility of the reverberation field as an inverse sensing technique is analogous to that of the ambient noise field. For example, the reverberation field can be inverted to

Propagation
(i) Matched field processing:
(a) source localization,
(b) marine environment characterization.
(ii) Ocean acoustic tomography:
(a) density field (eddies, currents),
(b) temperature (climate monitoring).
(iii) Deductive geoacoustic inversion:
(a) sediment parameters,
(b) sea-floor scattering characteristics.
(iv) Time reversal mirror (TRM):
(a) signal refocusing.
Noise
(i) Field inversion:
(a) wind speeds,
(b) rainfall rates.
(ii) Acoustic daylight:
(a) object imaging.
(iii) Geoacoustic inversion:
(a) seabed acoustics.
Reverberation
(i) Field inversion:
(a) sea-floor imaging.
(ii) TRM nulling:
(a) reverberation attenuation.

Box 4: Summary of inverse ocean acoustic sensing techniques.

image the sea floor. A new development uses time reversal mirror (TRM) methods to attenuate reverberant returns.

4. Emerging Solutions

4.1. Background. This section addresses advanced processing methods in addition to novel methods by which to maximize the interpretation, and thus utility, of archival and experimental data.

4.2. Advanced Integrated Modeling Approaches

4.2.1. Acoustic Integration Model. The acoustic integration model (AIM) combines a movement simulator with underwater acoustic models to predict (and thus minimize and mitigate) the potential effect of sound on marine mammals [128]. Simulated sound sources and animals are programmed to move in location and depth over time in a realistic fashion. AIM is a Monte Carlo statistical model based on a whale movement and tracking model and an underwater acoustic backscattering model for a moving source. Currently, AIM incorporates the BELLHOP and PE propagation models in addition to the ETOP05 bathymetry and GDEM sound-speed profile databases. The ETOP05 data set is the highest resolution topographic data set with global coverage that is publicly available, with elevations posted every 5 arc minutes (approximately 10 km) for all land and sea floor surfaces. ETOP05 is distributed without restriction by the National Oceanic and Atmospheric Administration (NOAA) through its National Geophysical Data Center (NGDC).

4.2.2. Marine Mammal Movement Models. A method has been developed for modeling marine mammal movement and behavior (3MB) for use in environmental impact assessments [129]. Estimating the impact of anthropogenic sound on marine animals entails consideration of animal location and behavior at the time of sound exposure. The ESME model (see Table 4) incorporates 3MB to provide fine-scale control over simulated marine-animal (animat) movement and behavior. Control over the animats is scaleable to the information available regarding the species of concern. Movement and behavior are stochastically determined by sampling from distributions describing rates of movement in the horizontal and vertical planes, direction of travel, time at the surface between dives, time at depth, and time in and transition between behavioral states. Influence of behavior over each of the other distributions is also permitted.

As part of the NATO Undersea Research Centre (NURC) broadband environmentally adaptive concept, the SUPREMO sonar performance model was incorporated within a feedback mechanism for sonar-parameter optimization for any specified ocean environment [130]. The feasibility of using active sonars to monitor the movement of marine mammals was investigated in order to recommend stand-off distances within which an acoustic source should not be deployed [131].

4.2.3. Energy Flux Models. Some applications, such as work relating to acoustic impacts on marine mammals, do not require extremely high-fidelity model outputs. Transmission losses averaged over depth, for example, are often adequate.

An approach referred to as energy-flux models [132] is useful for rapid calculation of transmission loss where the propagation conditions are dominated by numerous boundary-reflected multipaths and when only the coarse characteristics of the acoustic field are needed.

In specific configurations, especially at long ranges in shallow-water environments, the transmitted field can be viewed as being composed of many paths propagating by successive reflections from the surface and bottom boundaries. Here, the acoustic energy will remain trapped between these two boundaries. Furthermore, if the acoustic frequency is high enough that the field oscillations can be considered to be random, then an average intensity can be calculated using simple algebraic formulas [23].

This concept can be extended to ocean environments where the sound speed is not constant, or where there are slight losses at the boundaries. In such cases, the transmitted field cannot be taken as a volumetric average. Rather, it has to be decomposed into its angular components and the cyclic characteristics of the various beams must be detailed [133].

This approach is very useful for predicting incoherent transmission losses in cases where there are a large number of modes (≥ 6). These flux models are fast because there is no requirement to find modes or eigenrays. They can also lead to very simple and intuitive expressions. This method has been extended for reverberation [134].

4.2.4. Waveguide Invariant Models. In weakly range-dependent ocean environments, it has been observed that the acoustic wavefield can be described in terms of ray-based or mode-based approaches [135, 136]. Using ray-based descriptions, distributions of the ray amplitude and phase (travel time) are largely controlled by the so-called stability parameter (α). Alternatively, using mode-based descriptions in either purely range-independent environments or in range-independent environments upon which weak (range-dependent) perturbations have been superimposed, many wavefield properties are controlled by the so-called waveguide invariant (β). When β is evaluated using asymptotic mode-theoretical results (in a stratified environment), it was found that $\beta = \alpha$, which is consistent with the well-known ray-mode duality. In weakly range-dependent environments, however, α was found to control ray stability and several measures of travel-time dispersion while β controlled the spread of modal-group delays due to mode coupling.

In essence, the waveguide invariant (β) summarizes the pattern of constructive and destructive interference between acoustic modes propagating in the ocean waveguide. It manifests itself as interference fringes (or striations) in a plot of frequency versus source-receiver separation. The waveguide invariant summarizes in a single scalar parameter the dispersive propagation characteristics in a waveguide.

4.3. Advanced Nonintrusive Measurement Approaches

4.3.1. Acoustic Transmission Options. In *acoustic probing*, transmit signals can be adapted according to *in situ* estimates of channel conditions. Waveform characteristics could be set using returns from the previous transmission as an indicator of channel response. Rapidly modifying processing parameters by analyzing previously received signals addresses the problem of fluctuating acoustic channel response owing to variations in environmental factors. Such optimal matching to the estimated channel response would improve target-echo stability [137].

Since sound level, frequency, and duration are critical parameters in causing physical damage to marine life, transmitting over a wider frequency spectrum using spread-spectrum techniques or noise-like signals could avoid tissue damage by reducing the time spent (or energy transmitted) in particular frequency bands. Spread-spectrum techniques use a pseudorandom spreading code to produce a wide-band source signature. Signals must possess high range and Doppler resolution and minimize cross-talk between different users.

Noise-like signals emulate ambient noise by covertly adapting to the fluctuating ambient noise field. Designers could model pseudo-random signals based on bioacoustic noise, surf noise (when high-frequency broadband transients should be useful [138]) or shipping noise (when near CW would be more useful).

Advances in sonar technologies have rendered modern sonar systems useful for *in situ* measurements of the ambient marine environment. For example, through-the-sensor measurements of the ocean impulse response have enabled modern sonars to perform collateral functions as *tactical environmental processors*.

A recent *through-the-sensor* remote-sensing technique for acoustic-parameter estimation has been referred to as SABLE (sonar active boundary loss estimation) [139]. Bottom scattering strengths were derived from active hull-mounted naval sonar data by using eigenray paths modeled by CASS and GRAB to associate sonar-signal attributes with specific propagation paths.

4.3.2. Collision Avoidance. Cetaceans are prone to collisions with fast-moving vessels. In areas of high cetacean and vessel densities, the sperm whale (*Physeter macrocephalus*) is of great concern. Sperm whales are highly vocal and can be localized with passive sonar; however, when at or near the surface (i.e., when they are at greatest risk of collision), they tend to stop vocalizing. Techniques employing active sonars have proved to be inefficient due to short detection ranges and high closing speeds. The efficiency of a passive-sonar solution was evaluated that used vocalizing whale clicks (at depth) as acoustic sources to detect silent whales [140]. This solution could serve as a noninvasive complement to a more complex passive localization and collision-avoidance system. A wideband N $\times$ 2D (range-and-azimuth-dependent) ray model was used to simulate a passive solution comprising an arbitrary number of active acoustic sources (vocalizing whales), an illuminated object (silent whale) and a passive-sonar receiver, all positioned in three-dimensional space with arbitrary bathymetry. Both curved-line and straight-line ray solutions were implemented, with the latter providing greater computational speeds at the expense of temporal and angular fidelity. The simulation recreated the resultant mixture of direct, reverberated and target-backscattered signals arriving at the receiver for any array configuration, any number of sources and one target. In the vicinity of the Canary Islands, the simulations demonstrated the applicability of the concept with a maximum detection range on the order of 1 km.

4.4. Advanced Processing Methods

4.4.1. Adjoint Modeling. The principal application of adjoint models is sensitivity analysis. When quantitative estimates of sensitivity are desired, a mathematical model of the phenomenon or relationship is required. While models have been used to assess the impacts of perturbations and thus estimate sensitivity, a more efficient approach is to use the model's adjoint to determine optimal solutions. The adjoint operates backward in the sense that it determines a gradient with respect to input from a gradient with respect to output. In a temporally continuous model, this would appear as integration backwards in time. If there are no numerical instabilities associated with irreversible processes in the tangent linear model acting forward in time, there will be none in the adjoint acting backward in time (a tangent linear model provides a first-order approximation to the evolution of perturbations in a nonlinear forecast trajectory). The greatest limitation to the application of adjoints is that the results are useful only when the linearized approximation is valid. The adjoint operator (matrix transpose) back-projects information from data to the underlying model. Geophysical modeling calculations generally use linear operators that predict data from models. The usual task is to find the inverse of these calculations, that is, to find models (or make maps) from the data. The adjoint operator tolerates imperfections in the data and does not demand that the data provide full information.

The concept of adjoint modeling was introduced in shallow water acoustics for solving inverse problems [141]. In related work, the concept of backpropagation was reviewed in the context of adjoint modeling [142]. The different implementations of this concept were compared and discussed in the framework of experimental acoustic inversion in shallow water with application to source localization, ocean acoustic tomography, geoacoustic inversion and underwater communications. Well-established inversion (or focalization) methods based on matched-field processing, model-based matched filter and time-reversal mirror are related to lesser known techniques such as acoustic retrogation and other variants of backpropagation. In contrast to the latter, adjoint-based variational inversion approaches make use of the adjoint of a forward model to backpropagate the model-data mismatch at the receiver toward the source.

An adjoint model was derived from a forward-propagation model (e.g., normal mode or parabolic equation) to propagate data-model misfits at the observation point back through the medium to the site of those medium perturbations that were not accounted for in the forward model and which gave rise to the observed data-model misfits [143]. This property makes adjoint models attractive for use in acoustic inversion experiments.

4.4.2. Stochastic Resonance. Stochastic resonance refers to a phenomenon that is manifested in nonlinear systems whereby weak signals can be amplified in the presence of noise [144]. Three components are necessary: a threshold, a periodic signal and a source of noise. The response of the system undergoes resonance-like behavior as the noise level is varied [145]. Stochastic resonance was applied to enhance the detection of target signals masked by shallow-water reverberation [146]. Specifically, parameter-induced stochastic resonance was used to tune system parameters to recover the spatial signal that was corrupted by Gaussian noise. This method also has applicability when processing spatial signals that are corrupted by K-distributed envelope noise (a standard model for radar clutter).

4.4.3. Optimization Techniques. Ant colony optimization (ACO) has elements in common with genetic algorithms [147]. Both are population-based algorithms that search a discrete space and provide uncertainty analyses. The main difference is the mechanism that handles and recombines components of better candidate solutions (i.e., ant-pheromones trails versus genetic operators). Ant colony optimization is further distinguished by having a form of memory (the ant pheromone trails), while genetic algorithms are without memory. Specifically, when the pheromones evaporate, identifiers of paths with above-average quality fade out. Thus, high rates of evaporation mean that only recent information can be retrieved (typical for short-term memory); alternatively, low rates of evaporation allow recollection of much older information (typical for long-term memory). When applied to inversion, the world of the ants acts as an analogy for the geoacoustic environment.

An efficient and reliable method was proposed for performing the inversion of a neural-network underwater acoustic model to obtain sea-floor parameters [148]. Two different versions of a modified particle-swarm optimization were used: two-step gradient approximation and hierarchical cluster-based approximation. Both approaches worked well.

Tabu search has traditionally been applied to combinational optimization problems [149, 150]. The tabu search begins by marching to a local minimum. This approach avoids entrapment in cycles by forbidding (tabu), or penalizing, moves that take the solution in the next iteration to points in the solution space previously visited.

4.4.4. Chaos. In the presence of weak fluctuations in the sound-speed field, it has been observed that ray trajectories in ocean waveguides exhibit a chaotic behavior in which the travel times along eigenrays form compact clusters, the center of which is close to the arrival time of an unperturbed ray with similar geometry [151]. The deep-water acoustic waveguide was modeled by an unperturbed sound-speed profile $[c_0(z)]$ combined with weak fluctuations of sound speed $[\delta c(r,z)]$ caused by a random field of internal waves conforming to the Garrett-Munk spectrum. Formally, ray equations in a random inhomogeneous medium can be considered as stochastic equations whose parameters are determined by a random function δc. Such nonlinear equations can be solved in the framework of ordinary perturbation theory, but only for short ranges. Instead, the ray structure was analyzed by using the Hamiltonian formalism in terms of action-angle canonical variables whose variation was small under conditions of weak perturbations. It was estimated that ray arrival times at a distance r occupied an interval whose average width increased in proportion with $r^{1.5}$, in agreement with previous investigators. Moreover, for $r < 10^3$ km, the ray trajectories forming the cluster deviated only slightly from the unperturbed ray trajectory and the cluster width grew in proportion with $r^{0.5}$. For $r > 10^3$ km, the width of the cluster grew in proportion with r^2.

Chaotic and stochastic nonlinear ray dynamics can occur in underwater-acoustic waveguides with longitudinal variations in the sound speed caused by internal waves [152]. This situation was investigated using a model of a "frozen" medium in which the temporal variations in the environment were neglected and only the spatial variations due to the comparatively small propagation time of sound in the ocean were considered. Coherent ray clusters were observed in which large fans of rays with close initial conditions preserved close current dynamical characteristics over long distances. The cluster structure could be considered to consist of statistical and coherent parts. Rays belonging to the statistical part propagated in the same areas of phase space with the same value of the Lagrangian, but did not correlate with each other and demonstrated exponential sensitivity to initial conditions. Rays belonging to the coherent part did not demonstrate sensitivity to initial conditions. This coherent clusterization might be a useful property for acoustic tomography in terms of determining spatio-temporal variations in the hydrological environment under conditions of ray chaos [153].

5. Summary

This paper has reviewed changes in the ocean soundscape, changes that have been driven by both anthropogenic activity and natural factors. New regulatory initiatives have placed additional restrictions on uses of sound in the ocean, and mitigation of marine-mammal endangerment is now an integral consideration in acoustic-system design and operation. Emphasis has been placed on leveraging advanced modeling methods to solve emerging scientific and engineering challenges in this field.

Naval sonars have been redesigned as high-power multistatic systems capable of operating in complex coastal environments against quiet submarines. Littoral training ranges to support this new generation of sonar systems must comply with more stringent environmental regulations. Such compliance is often assessed using advanced modeling techniques coupled with *in situ* data collection. Oceanographic data collection is supported by underwater networks consisting of sensors and vehicles deployed in concert. These networks employ nodes that communicate via acoustic channels that have low data rates and time-varying fading.

Seismic exploration has ventured into deeper waters in search of diminishing energy resources using source arrays that are powered by high-pressure air. Controlled-exposure experiments have studied changes in the behavior of nearby whales due to these powerful sources; however, modeling uncertainty complicates the regulatory interpretation of average sound-exposure conditions.

Increased merchant shipping activity has had a measureable impact on the ocean soundscape over the past four decades: the average noise has increased by 2.5–3 dB per decade in the frequency band 30–50 Hz. Moreover, the growing number of wind farms in coastal waters of medium depth (<30 m) has added to the soundscape at low frequencies and resulted in behavioral reactions of nearby marine mammals.

Climate change has also affected the ocean soundscape. The emission of carbon into the atmosphere through the effects of fossil-fuel combustion and industrial processes has increased atmospheric concentrations of carbon dioxide (CO_2). Ocean acidification, which occurs when CO_2 in the atmosphere reacts with water to create carbonic acid (H_2CO_3), has increased. The attenuation of low-frequency sound in the sea is pH dependent; specifically, the higher the pH, the greater the attenuation. Thus, as the ocean has become more acidic (lower pH) due to increasing CO_2 emissions, the attenuation has diminished and low-frequency sounds propagate farther, making the ocean noisier. Recent investigations have estimated that the noise at 3 kHz would increase by 30% if the pH decreased from 8.0 to 7.4.

Modeling tools traditionally used in underwater acoustics have undergone a necessary transformation; these tools include propagation, noise, reverberation, and sonar-performance models. Advanced modeling techniques now include forward and inverse applications, integrated-modeling approaches, nonintrusive measurements and novel processing methods. The flow of the sonar modeling processes was used to demonstrate advances in modeling technology relevant to assessments of the ocean soundscape. A 32-year baseline inventory of modeling techniques was updated with the latest developments, including basic mathematics and references to the key literature, to guide soundscape practitioners to the most efficient modeling techniques for any given application.

Advanced processing methods included integrated modeling approaches that combine marine-mammal movement simulators with underwater acoustic models to predict (and thus minimize and mitigate) the potential effect of sound on marine mammals. Advances have been achieved using energy-flux and waveguide-invariant techniques that can simplify interpretation of channel models. Nonintrusive measurement approaches included new acoustic transmission options to minimize marine-mammal impacts. Collision avoidance techniques have proved useful in areas of high cetacean and vessel density. Finally, adjoint modeling, stochastic resonance, optimization techniques, and chaos were shown to facilitate understanding of advance underwater acoustic modeling results.

Abbreviations

1D:	One-dimensional
2D:	Two-dimensional
3D:	Three Dimensional
3DWAPE:	3D wide-angle parabolic equation
3MB:	Marine mammal movement and behavior
ACM:	Association for computing machinery
ACO:	Ant colony optimization
AIM:	Acoustic integration model
ARAMIS:	Array response advanced modal integrated simulator
ARTEMIS:	Adiabatic reverberation and target echo mode incoherent sum
ASA:	Acoustical Society of America
ASPECT:	Active system performance estimate computer tool
AUV:	Autonomous underwater vehicle
BISTAR:	Incoherent bistatic reverberation for range-dependent environments

BLM: Bureau of Land Management (now BOEMRE)
BMT: British Maritime Technology Ltd
BOEMRE: Bureau of Ocean Energy Management, Regulation and Enforcement (formerly BLM)
BORIS-SSA: Bottom reverberation from inhomogeneities and surfaces-small-slope approximation
CALYPSO: Model operating system
Cartesian 3DPE: 3DPE Employing Cartesian Coordinates in the Numerical Scheme
CEE: Controlled exposure experiments
CiSE: Computational intelligence and software engineering
CMM: Caractérisation du Milieu Marin
CORDA: Centre for Operational Research and Defence Analysis
COWRIE: Collaborative Offshore Wind Energy Research Into the Environment
C-SNAP-REV: Reverberation using the C-SNAP Normal-Mode Model
CW: Continuous wave
CZMA: Coastal zone Management Act
DANM: Dynamic ambient noise model
dB: Decibel
DMOS: DRDC Atlantic model operating system
DRDC: Defence Research and Development Canada
DREA: Defence Research Establishment Atlantic (Canada)
DSTO: Defence Science and Technology Organisation (Australia)
EIA: Environmental impact assessment
ESA: Endangered Species Act
ESME: Effects of sound on the marine environment
ESPRESSO: Extensible performance and evaluation suite for sonar
ETOP05: Earth topography five-minute grid
FeyRay: Broadband, range-dependent Gaussian-beam propagation model
FFI: Forsvarets forskningsinstitutt (Norwegian Defence Research Establishment)
FFP: Fast field program
FM: Frequency modulation
FORTRAN: Formula translation
GDEM: Generalized digital environmental model
GIS: Geographic information system
GmbH: Gesellschaft mit beschränkter Haftung
GOATS: Generic Ocean Array Technology Sonar
GUI: Graphical User Interface
HYREV: HanYang University Reverberation Model
IAGC: International Association of Geophysical Contractors
IARIA: International Academy, Research, and Industry Association
ICC: International Conference on Communications
IECM: Integral equation coupled-mode
IEEE: Institute of Electrical and Electronics Engineers
IEER: Improved extended echo ranging
Integrated Mode: Multipath expansion method extended to range-dependent environments
IOMEDEX: Ionian Sea/Mediterranean Sea Exercise (part of LRAPP)
IRFC: Industry Research Funders Coalition
ISAAC: Impact on Species from Anthropogenic Acoustic Channels
JASA: Journal of the Acoustical Society of America
LRAPP: Long Range Acoustic Propagation Project
LSMS: Life system modeling and simulation
LYBIN: Range-dependent ray-theoretical propagation model
MATLAB: Matrix Laboratory (Mathworks, Inc.)
MCM: Mine countermeasures
METOC: Meteorology and oceanography
MINEOS: MSASM interactive execution and optimization system
MMPA: Marine Mammal Protection Act
MMS: Minerals Management Service
MOC3D: 3D Version of MOCASSIN
MOCASSIN: Monte Carlo Schall-Strahlen Intensitäten (Monte Carlo Sound Ray Intensities)
MOD: Ministry of Defence
MODRAY: Maritime Operations Division Ray-tracer
MONM: Marine Operations Noise Model
MONM3D: 3D PE Model for MONM
MONOGO: Reverberation Module in SWAMI
M&S: Modeling and Simulation
MTS: Marine Technology Society
MURAL: Multistatic reverberation algorithm
N × 2D: Quasi-3D modeling approach
NATO: North Atlantic Treaty Organization
NEMO: NUWC Exposure Model
NEPA: National Environmental Policy Act
NGDC: National Geophysical Data Center
NMFS: National Marine Fisheries Service
NOAA: National Oceanic and Atmospheric Administration
NOGRP: Fast normal mode (monostatic) reverberation model
NSPE: Navy Standard Parabolic Equation
NTNU: Norwegian University of Science and Technology
NURC: NATO Undersea Research Centre (formerly SACLANTCEN)

NUSC: Naval Underwater Systems Center (now NUWC)
NUWC: Naval Undersea Warfare Center (formerly NUSC)
OASES: Ocean acoustics and seismic exploration synthesis
OASES-3D: Version of OASES incorporating 3D scattering effects
OBIS: Ocean biogeographic information system
OCS: Outer continental shelf
ONR: Office of Naval Research
OWWE: One-way wave equation
PAREQ: Parabolic equation model
PAREQ-REV: Reverberation using the PAREQ parabolic equation model
PDF: Probability density (or distribution) function
PDPE: Pseudo-differential PE
PE: Parabolic equation
PERM-2D: Reverberation model
PlaneRay: Eigenray Model for range-dependent environments
PMODES: Range-independent normal-mode transmission-loss module in SWAMI
POPP: Range-independent version of PROLOS
PWRC: Plane wave reflection coefficient
RAM: Range-dependent acoustic model
RAMGEO: RAM modified for range-dependent sediment layers
Ray5: Range-dependent ray model
RAYSON: Range-dependent ray-theoretic model for high frequencies
RDS: Rapidly deployable system
REA: Rapid environmental assessment
REVPA: Parabolic-equation reverberation model
RITSHPA: Reverberation including the statistics of hybrid path arrivals
RMPE: Ray-Mode Parabolic Equation Model
ROSELLA: Extension of NOGRP to handle beam patterns
ROV: Remotely operated vehicle
R-SNAP: Coherent monostatic reverberation model
SABLE: Sonar active boundary loss estimation
SACLANT: Supreme Allied Commander, Atlantic
SACLANTCEN: SACLANT Undersea Research Centre
SIAM: Simulated ambient noise model
SIGBED: Special Interest Group on Embedded Systems (ACM)
SPPS: Sonar performance prediction system
S-SCARAB: SACLANTCEN-scattering reverberation and backscatter
SSF: Split-step Fourier
SSFEPE: SSF PE Model
SSPPE: Split-step Padé PE model
STB: System test bed
SUPREMO: Multistatic sonar model
SWAMI: Shallow Water Active-Sonar Modelling Initiative
SWAMP: Shallow Water Acoustic Modal Propagation
SWSS: Sperm Whale Seismic Study
TDA: Tactical decision aid
TIAPS: Towed integrated active-passive sonar
TRM: Time-reversal mirror
TTS: Temporary threshold shift
UAIM: Underwater acoustic imaging model
UDT: Undersea defence technology
USWTR: Undersea warfare training range
UUV: Unmanned undersea vehicle
WHOI: Woods Hole Oceanographic Institution
WKB: Wentzel, Kramers, and Brillouin
WUWNet: Workshop on Underwater Networks
XRAY: Range-dependent ray theoretical model combined with full-field modeling of seabed interactions.

References

[1] K. D. LePage, P. Neumann, and C. W. Holland, "Broad-band time domain modeling of sonar clutter in range dependent waveguides," in *in Proceedings of the MTS/IEEE Oceans 2006 Conference (OCEANS'06)*, Boston, Mass, USA, September 2006.

[2] I. F. Akyildiz, D. Pompili, and T. Melodia, "Challenges for efficient communication in underwater acoustic sensor networks," *ACM SIGBED Review*, vol. 1, no. 2, pp. 3–8, 2004.

[3] I. F. Akyildiz, D. Pompili, and T. Melodia, "Underwater acoustic sensor networks: research challenges," *Ad Hoc Networks*, vol. 3, no. 3, pp. 257–279, 2005.

[4] D. Frye, L. Freitag, R. Detrick et al., "An acoustically linked moored-buoy ocean observatory," *EOS, Transactions, American Geophysical Union*, vol. 87, no. 22, pp. 213–218, 2006.

[5] Z. Han, Y. L. Sun, and H. Shi, "Cooperative transmission for underwater acoustic communications," in *Proceedings of the IEEE International Conference on Communications, (ICC '08)*, pp. 2028–2032, Beijing, China, 2008.

[6] V. Chandrasekhar, W. K. G. Seah, Y. S. Choo, and H. V. Ee, "Localization in underwater sensor networks—survey and challenges," in *Proceedings of the 1st ACM International Workshop on Underwater Networks (WUWNet'06)*, pp. 33–40, Los Angeles, Calif, USA, September 2006.

[7] J. Wang, P. Cai, and D. Yuan, "An underwater acoustic channel simulator for UUV communication performance testing," in *Proceedings of the IEEE International Conference on Information and Automation*, pp. 2286–2290, Harbin, China, 2010.

[8] J. G. Graver, *Underwater gliders: dynamics, control and design*, PhD dissertation, Department of Mechanical and Aerospace Engineering, Princeton University, Princeton, NJ, USA, 2005.

[9] A. Jochens, D. Biggs, D. Engelhaupt et al., "Sperm whale seismic study in the Gulf of Mexico," Summary Report, 2002–2004 OCS Study MMS 2006-034, US Department of the Interior, Minerals Management Service, Gulf of Mexico OCS Region, New Orleans, La, USA, 2006.

[10] A. Thode, D. K. Mellinger, S. Stienessen, A. Martinez, and K. Mullin, "Depth-dependent acoustic features of diving sperm whales (*Physeter macrocephalus*) in the Gulf of Mexico," *Journal of the Acoustical Society of America*, vol. 112, no. 1, pp. 308–321, 2002.

[11] W. M. Carey and R. B. Evans, *Ocean Ambient Noise: Measurement and Theory*, Springer, New York, NY, USA, 2011.

[12] R. K. Andrew, B. M. Howe, J. A. Mercer, and M. A. Dzieciuch, "Ocean ambient sound: comparing the 1960s with the 1990s for a receiver off the California coast," *Acoustic Research Letters Online*, vol. 3, pp. 65–70, 2002.

[13] M. A. McDonald, J. A. Hildebrand, and S. M. Wiggins, "Increases in deep ocean ambient noise in the Northeast Pacific west of San Nicolas Island, California," *Journal of the Acoustical Society of America*, vol. 120, no. 2, pp. 711–718, 2006.

[14] M. A. McDonald, J. A. Hildebrand, S. M. Wiggins, and D. Ross, "A 50 year comparison of ambient ocean noise near San Clemente Island: a bathymetrically complex coastal region off Southern California," *Journal of the Acoustical Society of America*, vol. 124, no. 4, pp. 1985–1992, 2008.

[15] J. Nedwell and D. Howell, "A review of offshore windfarm related underwater noise sources," Report 544R0308, Subacoustech Ltd, commissioned by COWRIE (Collaborative Offshore Wind Research Into the Environment), Hampshire, UK, 2004.

[16] D. Rouseff and D. Tang, "Internal waves as a proposed mechanism for increasing ambient noise in an increasingly acidic ocean," *Journal of the Acoustical Society of America*, vol. 127, no. 6, pp. EL235–EL239, 2010.

[17] E. McCarthy, *International Regulation of Underwater Sound: Establishing Rules and Standards to Address Ocean Noise Pollution*, Kluwer Academic Publishers, Dordrecht, The Netherlands, 2004.

[18] National Research Council, *Ocean Noise and Marine Mammals*, The National Academies Press, Washington, DC, USA, 2003.

[19] National Research Council, *Marine Mammal Populations and Ocean Noise: Determining When Noise Causes Biologically Significant Effects*, The National Academies Press, Washington, DC, USA, 2005.

[20] J. F. Grassle, "The ocean biogeographic information system (OBIS): an on-line, worldwide atlas for accessing, modeling and mapping marine biological data in a multidimensional geographic context," *Oceanography*, vol. 13, no. 3, pp. 5–7, 2000.

[21] S. Finette, "Embedding uncertainty into ocean acoustic propagation models," *Journal of the Acoustical Society of America*, vol. 117, no. 3 I, pp. 997–1000, 2005.

[22] J. W. Lawson, "The use of sound propagation models to determine safe distances from a seismic sound energy source," Research Document 2009/060, Department of Fisheries and Oceans, Canadian Science Advisory Secretariat, 2009.

[23] P. C. Etter, *Underwater Acoustic Modeling and Simulation*, Spon Press, London, UK, 3rd edition, 2003.

[24] R.J. Howard, T. Foreman, and D. Clark, "Architectural and design considerations in propagation model selection and design, simulation interoperability standards organization," in *Proceedings of the Simulation Interoperability Workshop*, no. 00S-SIW-054, 2000.

[25] F. C. Newman, A. C. Biondo, M. D. Mandelberg, C. C. Matthews, and J. R. Rottier, "Enhancing realism in computer simulations: environmental effects," *Johns Hopkins APL Technical Digest*, vol. 23, no. 4, pp. 443–453, 2002.

[26] J. Collins, T. Foreman, and D. Speicher, "A new design for a sonar environmental effects server," in *Proceedings of the Simulation Interoperability Workshop*, no. 02F-SIW-114, Simulation Interoperability Standards Organization, 2002.

[27] T. Foreman and D. Speicher, "Expanded sonar environmental acoustic effects server capabilities for fleet battle experiments," in *Proceedings of the Simulation Interoperability Workshop*, no. 03F-SIW-062, Simulation Interoperability Standards Organization, 2003.

[28] J. B. Collins and C. G. Scannell, "Natural environment data services in distributed modeling and simulation," in *Net-Centric Approaches to Intelligence and National Security*, R. Ladner and F. E. Petry, Eds., pp. 149–174, Springer, New York, NY, USA, 2005.

[29] J. M. Hovem, "PlaneRay: an acoustic underwater propagation model based on ray tracing and plane-wave reflection coefficients," FFI-rapport 2008/00610, Forsvarets forskningsinstitutt/Norwegian Defence Research Establishment (FFI), 2008.

[30] J. M. Hovem, S. Yan, X. Bao, and H. Dong, "Modeling underwater communication links," in *Proceedings of the 2nd International Conference on Sensor Technologies and Applications (SENSORCOMM '08)*, pp. 679–686, IEEE Computer Society, 2008.

[31] J. M. Hovem and H. Dong, "PlaneRay: an underwater acoustic propagation model using ray tracing and plane wave reflection coefficients," *Journal of the Acoustical Society of America*, vol. 120, p. 3221, 2006.

[32] J. M. Hovem and D. P. Knobles, "A range-dependent propagation model based on a combination of ray theory and plane-wave reflection coefficients," *Journal of the Acoustical Society of America*, vol. 112, p. 2393, 2002.

[33] J. M. Hovem and D. P. Knobles, "A range dependent propagation model based on a combination of ray theory and plane wave reflection coefficients," in *Proceedings of the 10th International Congress on Sound and Vibration*, pp. 2593–2600, Stockholm, Sweden, July 2003.

[34] S. A. Stotts, D. P. Knobles, R. A. Koch, D. E. Grant, K. C. Focke, and A. J. Cook, "Geoacoustic inversion in range-dependent ocean environments using a plane wave reflection coefficient approach," *Journal of the Acoustical Society of America*, vol. 115, no. 3, pp. 1078–1102, 2004.

[35] U. Kristiansen, "Sound propagation in regions of variable oceanography—a summary of student work at NTNU," FFI-rapport 2010/00812, Norwegian Defence Research Establishment, 2010.

[36] H.S. Olsen, "Lydutbredelse i havområder med avstandsavhengig oseanografi," Norges teknisk-naturvitenskapelige universitet, Institutt for elektronikk og telekommunikasjon, Trondheim, Norway, 2008.

[37] T. S. Semantic, "RAYSON: underwater acoustic rays software," Product description literature, Semantic TS , Sanary, France, 2002.

[38] C. Viala, C. Noël, and G. Lapierre, "RAYSON: a real time underwater communication simulator and performance estimator," in *Proceedings of the 7th French Workshop on Underwater Acoustics*, Brest, France, 2004.

[39] C. Viala, C. Noël, Y. Stéphan, and M. Asch, "Real-time geoacoustic inversion of large band signals," in *Proceedings of the IEEE Oceans 2005 Europe Conference*, pp. 1392–1395, Brest, France, 2005.

[40] C. Viala, C. Noël, Y. Stéphan, and M. Asch, "Real-time geoacoustic inversion of broad band signals in deep water,"

in *Proceedings of the Caractérisation du Milieu Marin (CMM'06)*, Brest, France, 2006.

[41] E. Svensson, I. Karasalo, and J.-P. Hermand, "Hybrid raytrace modelling of an underwater acoustics communication channel," in *Proceedings of the 7th European Conference on Underwater Acoustics*, pp. 1211–1216, Delft, The Netherlands, 2004.

[42] D. D. Ellis, "A two-ended shooting technique for calculating normal modes in underwater acoustic propagation," DREA Report 85/105, Defence Research Establishment Atlantic, Nova Scotia, Canada, 1985.

[43] S. A. Stotts, D. P. Knobles, and R.A. Koch, "Scattering in a Pekeris waveguide from a rough bottom using a two-way coupled mode approach," *Journal of the Acoustical Society of America*, vol. 129, no. 5, pp. EL172–EL178, 2011.

[44] N. A. Sidorovskaia, "A new normal mode program SWAMP as a tool for modeling scattering effects in oceanic waveguides," in *Modelling and Experimental Measurements in Acoustics III*, D. Almorza, C. A. Brebbia, and R. Hernandez, Eds., pp. 267–275, WIT Press, Southampton, UK, 2003.

[45] N. A. Sidorovskaia, "Systematic studies of pulse propagation in ducted oceanic waveguides in normal mode representation," *The European Physical Journal, Applied Physics*, vol. 25, no. 2, pp. 113–131, 2004.

[46] I. M. Tamendarov and N. A. Sidorovskaia, "Unified acoustic model for simulating propagation and scattering effects in oceanic waveguides," *Journal of the Acoustical Society of America*, vol. 116, p. 2527, 2004.

[47] C. A. Clark, "Acoustic wave propagation in horizontally variable media," *IEEE Journal of Oceanic Engineering*, vol. 30, no. 1, pp. 188–197, 2005.

[48] H. Schmidt, A. Maguer, E. Bovio et al., "Generic oceanographic array technologies (GOATS) '98 – Bistatic seabed scattering measurements using autonomous underwater vehicles," Report SR-302, SACLANT Undersea Research Centre, La Spezia, Italy, 1998.

[49] H. Schmidt, I. Veljkovic, and M. Zampolli, "Bistatic scattering from buried targets in shallow water—experiment and modelling," in *Proceedings of the 7th European Conference on Underwater Acoustics*, Delft, The Netherlands, 2004.

[50] H. Schmidt, "Bistatic scattering from buried targets in shallow water," in *Proceedings of the Autonomous Underwater Vehicle and Ocean Modelling Networks (GOATS'00)*, E. Bovio, R. Tyce, and H. Schmidt, Eds., no. CP-46, SACLANT Undersea Research Centre, 2001.

[51] H. Schmidt, "OASES Version 3.1 user guide and reference manual," Department of Ocean Engineering, Massachusetts Institute of Technology, Cambridge, Mass, USA, 2004.

[52] D. J. Thomson and N. R. Chapman, "WIDE-angle split-step algorithm for the parabolic equation," *Journal of the Acoustical Society of America*, vol. 74, no. 6, pp. 1848–1854, 1983.

[53] T. F. Duda, "Initial results from a Cartesian three-dimensional parabolic equation acoustical propagation code," Tech. Rep. WHOI-2006-14, Woods Hole Oceanographic Institution, Woods Hole, Mass, USA, 2006.

[54] L. Y. S. Chiu, Y.-T. Lin, C.-F. Chen, T. F. Duda, and B. Calder, "Focused sound from three-dimensional sound propagation effects over a submarine canyon," *Journal of the Acoustical Society of America*, vol. 129, no. 6, pp. EL260–EL266, 2011.

[55] M. E. Austin and N. R. Chapman, "The use of tessellation in three-dimensional parabolic equation modeling," *Journal of Computational Acoustics*, vol. 19, pp. 221–239, 2011.

[56] D. Mikhin, "Exact discrete nonlocal boundary conditions for high-order Padé parabolic equations," *Journal of the Acoustical Society of America*, vol. 116, no. 5, pp. 2864–2875, 2004.

[57] O. A. Godin, "Reciprocity and energy conservation within the parabolic approximation," *Wave Motion*, vol. 29, no. 2, pp. 175–194, 1999.

[58] L. T. Fialkowski, J. F. Lingevitch, J. S. Perkins, D. K. Dacol, and M. D. Collins, "Geoacoustic inversion using a rotated coordinate system and simulated annealing," *IEEE Journal of Oceanic Engineering*, vol. 28, no. 3, pp. 370–379, 2003.

[59] L. Zhang, L. Da, and Y. Zhou, "A new ocean acoustic model in computational oceanographic physics," in *in Proceedings of the International Conference on Computational Intelligence and Software Engineering, CiSE'09*, Wuhan, China, December 2009.

[60] F. Sturm and J. A. Fawcett, "On the use of higher-order azimuthal schemes in 3-D PE modeling," *Journal of the Acoustical Society of America*, vol. 113, no. 6, pp. 3134–3145, 2003.

[61] F. Sturm, "Examination of signal dispersion in a 3-D wedge-shaped waveguide using 3DWAPE," *Acta Acustica United with Acustica*, vol. 88, no. 5, pp. 714–717, 2002.

[62] F. Sturm, "Numerical study of broadband sound pulse propagation in three-dimensional oceanic waveguides," *Journal of the Acoustical Society of America*, vol. 117, no. 3 I, pp. 1058–1079, 2005.

[63] W. Renner, "User's guide for the ANDES model," Tech. Rep. version 4.2, Science Applications International Corp., 1995.

[64] M. Moll, R. M. Zeskind, and W. L. Scott, "An algorithm for beam noise prediction," Report 3653, Bolt, Beranek and Newman, 1979.

[65] J. Arvelo, "Array response advanced modal integrated simulator (ARAMIS) user's guide, STX-98-010," Johns Hopkins Applied Physics Laboratory, Laurel, Md, USA, 1998.

[66] J. R. Nedwell, J. Lovell, and A. W. H. Turnpenny, "Experimental validation of a species-specific behavioral impact metric for underwater noise," *Journal of the Acoustical Society of America*, vol. 118, p. 2019, 2005.

[67] BMT Cordah EIA and Environmental Services, *Offshore Renewables. Noise Assessment and Modelling, Product Description Literature*, Aberdeen, UK, 2010.

[68] M. H. Laurinolli, C. D. S. Tollefsen, S. A. Carr, and S. P. Turner, "Assessment of the effects of underwater noise from the proposed Neptune LNG project. Part (3): noise sources of the Neptune project and propagation modeling of underwater noise," LGL Report TA4200-3, JASCO Research Ltd for LGL Ltd, Ontario, Canada, 2005.

[69] S. W. Marshall and J. J. Cornyn, "Ambient-noise prediction. Vol. 1–Model of low-frequency ambient sea noise," Report 7755, Naval Research Laboratory, Washington, DC, USA, 1974.

[70] S. W. Marshall and J. J. Cornyn, "Ambient-noise prediction. Vol. 2–Model evaluation with IOMEDEX data," Report 7756, Naval Research Laboratory, Washington, DC, USA, 1974.

[71] Science Applications Inc., "Review of models of beam-noise statistics," SAI-78-696-WA, McLean, Va, USA, 1977.

[72] D. Long, "FANM/SIAM noise plot program," Ocean Data Systems Inc., Rockville, Md, USA, 1974.

[73] R. J. Urick, *Principles of Underwater Sound*, McGraw-Hill, New York, NY, USA, 3rd edition, 1983.

[74] J. W. Choi, K.-S. Yoon, J. Na, J.-S. Park, and Y.N. Na, "Shallow water high-frequency reverberation model," *Journal of the Acoustical Society of Korea*, vol. 21, pp. 671–678, 2002 (Korean).

[75] D.D. Ellis, *Measurements and Analysis of Reverberation and Clutter Data*, DRDC Atlantic ECR 2007-065, Defence Research and Development Canada, Atlantic, 2007.

[76] D. D. Ellis, "Normal-mode models OGOPOGO and NOGRP applied to the 2006 ONR reverberation modeling workshop

problems," DRDC Atlantic TM 2006-289, Defence Research and Development Canada, Atlantic, 2008.
[77] J. F. Lingevitch, "A parabolic equation method for modeling rough interface reverberation," in *Proceedings of the International Symposium on Underwater Reverberation and Clutter*, P. L. Nielsen, C. H. Harrison, and J.-C. LeGac, Eds., Lerici, Italy, 2008.
[78] M. D. Collins and R. B. Evans, "A two-way parabolic equation for acoustic backscattering in the ocean," *Journal of the Acoustical Society of America*, vol. 91, no. 3, pp. 1357–1368, 1992.
[79] F. A. Milinazzo, C. A. Zala, and G. H. Brooke, "Rational square-root approximations for parabolic equation algorithms," *Journal of the Acoustical Society of America*, vol. 101, no. 2, pp. 760–766, 1997.
[80] G. Bouchage and K. D. LePage, "A shallow-water reverberation PE model," *Acta Acustica United with Acustica*, vol. 88, no. 5, pp. 638–641, 2002.
[81] K. LePage, "Monostatic reverberation in range dependent waveguides: the R-SNAP model," Report SR-363, SACLANT Undersea Research Centre, La Spezia, Italy, 2003.
[82] M. K. Prior, C. H. Harrison, and K. D. LePage, "Reverberation comparisons between RSNAP, SUPREMO and analytical solutions," Report SR-361, SACLANT Undersea Research Centre, La Spezia, Italy, 2002.
[83] C. Harrison, "ARTEMIS: a fast general environment reverberation model," in *Proceedings of the International Symposium on Underwater Reverberation and Clutter*, P. L. Nielsen and C. H. Harrison and J.-C. LeGac, Eds., p. 2008, Lerici, Italy, 2008.
[84] K. LePage, "Bottom reverberation in shallow water: coherent properties as a function of bandwidth, waveguide characteristics, and scatterer distributions," *Journal of the Acoustical Society of America*, vol. 106, no. 6, pp. 3240–3254, 1999.
[85] K. D. LePage, "Bistatic reverberation modeling for range-dependent waveguides," *Journal of the Acoustical Society of America*, vol. 112, pp. 2253–2254, 2002.
[86] K. D. LePage and C. H. Harrison, "Bistatic reverberation benchmarking exercise: BiStaR versus analytic formulas," *Journal of the Acoustical Society of America*, vol. 113, pp. 2333–2334, 2003.
[87] K. D. LePage and C. Harrison, "Effects of refraction on the prediction of bistatic reverberation in range dependent shallow water waveguides," *Journal of the Acoustical Society of America*, vol. 114, p. 2302, 2003.
[88] D. M. Fromm, "A computationally efficient multistatic reverberation algorithm," *Journal of the Acoustical Society of America*, vol. 129, p. 2631, 2011.
[89] V. Marconi, P. L. Nielsen, and C. Holland, "S-SCARAB: SACLANTCEN-Scattering reverberation and backscatter user guide and reference manual, NURC/SM-418," NATO Undersea Research Centre, La Spezia, Italy, 2004.
[90] F.-P. A. Lam, N. J. Konijnendijk, J. Groen, and D. G. Simons, "Non-Rayleigh wideband sonar reverberation modeling including hybrid multi-paths," in *Proceedings of the The Oceans'06 MTS/IEEE-Boston Conference*, Boston, Mass, USA, 2006.
[91] D. A. Abraham and A. P. Lyons, "Simulation of non-Rayleigh reverberation and clutter," *IEEE Journal of Oceanic Engineering*, vol. 29, no. 2, pp. 362–374, 2004.
[92] E. Pouliquen, A. P. Lyons, and N. G. Pace, "The Helmholtz-Kirchhoff approach to modeling penetration of acoustic waves into rough seabeds," *Journal of the Acoustical Society of America*, vol. 104, p. 1762, 1998.
[93] G. Canepa and C. Berron, "Characterization of seafloor geo-acoustic properties from multibeam data," in *Proceedings of the MTS/IEEE Oceans Conference*, Boston, Mass, USA, 2006.
[94] R. J. Soukup, G. Canepa, H. J. Simpson, J. E. Summers, and R. F. Gragg, "Small-slope simulation of acoustic backscatter from a physical model of an elastic ocean bottom," *Journal of the Acoustical Society of America*, vol. 122, no. 5, pp. 2551–2559, 2007.
[95] P. Staelens, *Defining and modeling the limits of high-resolution underwater acoustic imaging,*, Doctoral dissertation, Universiteit Gent, Belgium, 2009.
[96] P. Blondel and N. G. Pace, "Bistatic sonars: sea trials, laboratory experiments and future surveys," *Archives of Acoustics*, vol. 34, no. 1, pp. 95–109, 2009.
[97] G. L. Davies, "Project 03F–1: phase I completion report," Memo SM-399, SACLANT Undersea Research Centre, La Spezia, Italy, 2002.
[98] M. Meyer and G. L. Davies, "A beam-based high-frequency reverberation model," Report SR-365, SACLANT Undersea Research Centre, La Spezia, Italy, 2002.
[99] G. L. Davies and E. P. Signell, "Espresso—scientific user guide," Report NURC-SP-2006-003, NATO Undersea Research Centre, La Spezia, Italy, 2006.
[100] G. L. Davies and E. P. Signell, "Espresso(m)—user guide," Report NURC-SP-2006-002, NATO Undersea Research Centre, La Spezia, Italy, 2006.
[101] S. Mjølsnes, *LYBIN SGP-180(C)—Model Description*, The Royal Norwegian Navy Materiel Command, Bergen, Norway, 2000.
[102] K.T. Hjelmervik and G.H. Sandsmark, "In ocean evaluation of low frequency active sonar systems," in *Proceedings of the Acoustics '08 Conference*, pp. 2839–2843, Paris, France, 2008.
[103] E. Dombestein, A. Gjersøe, and M. Bosseng, "LybinCom 6.0—description of the binary interface," FFI-rapport 2009/02267, Forsvarets forskningsinstitutt/Norwegian Defence Research Establishment, 2010.
[104] C. M. Ferla, C. Isoppo, G. Martinelli, and F. B. Jensen, "Performance assessment of the LYBIN-2.0 propagation-loss model," Memo SM-384, SACLANT Undersea Research Centre, La Spezia, Italy, 2001.
[105] S. M. Ivansson, "Stochastic ray-trace computations of transmission loss and reverberation in 3-D range-dependent environments," in *Proceedings of the 8th European Conference on Underwater Acoustics*, pp. 131–136, Carvoeiro, Portugal, 2006.
[106] H. G. Schneider, "MOCASSIN: sound propagation and sonar range prediction model for shallow water environments. User's guide," Tech. Rep. 1990–9, Forschungsanstalt der Bundeswehr für Wasserschall- und Geophysik, Kiel, Germany, 1990.
[107] F. Sturm, S. Ivansson, Y.-M. Jiang, and N. R. Chapman, "Numerical investigation of out-of-plane sound propagation in a shallow water experiment," *Journal of the Acoustical Society of America*, vol. 124, no. 6, pp. EL341–EL346, 2008.
[108] K. Frith, "MODRAY underwater sound simulator user manual, Version 5.0," Australia Department of Defence, Defence Science and Technology Organisation, Maritime Operations Division, Aeronautical and Maritime Research Laboratory, Edinburgh, South Australia, 2003.
[109] Ebor Computing, "MODRAY, product description literature," Mile End, South Australia, 2005.
[110] R. Alksne, "Rapidly deployable systems (RDS) underwater acoustic telemetry trials report," Tech. Rep. DSTO-TN-0259, Australia Department of Defence, Defence Science and

Technology Organisation, Maritime Operations Division, Aeronautical and Maritime Research Laboratory, Victoria, Australia, 2000.
[111] D. McCammon, "Active sonar modelling with emphasis on sonar stimulators," Report CR 2004-130 Defence Research and Development Canada—Atlantic, 2004.
[112] A. Baldacci and C. H. Harrison, "SUPREMO prototype version 01 user's guide, M-141," SACLANT Undersea Research Centre, La Spezia, Italy, 2002.
[113] C. H. Harrison, "Scattering strength uncertainty," *Journal of the Acoustical Society of America*, vol. 112, p. 2253, 2002.
[114] C. H. Harrison, "SUPREMO: a multistatic sonar performance model, Memo SM-396," SACLANT Undersea Research Centre, La Spezia, Italy, 2002.
[115] C. H. Harrison, "Closed-form expressions for ocean reverberation and signal excess with mode stripping and Lambert's law," *Journal of the Acoustical Society of America*, vol. 114, no. 5, pp. 2744–2756, 2003.
[116] M. K. Prior and A. Baldacci, "Comparison between predictions made by the SUPREMO sonar performance model and measured data," Report SR-429, SACLANT Undersea Research Centre, La Spezia, Italy, 2005.
[117] M. K. Prior, "SUPREMO v.2.0 user guide," Report NURC-FR-2007-009, NATO Undersea Research Centre, La Spezia, Italy, 2007.
[118] M. K. Prior, "SUPREMO v.2.0 program description," Report NURC-FR-2007-011, NATO Undersea Research Centre, La Spezia, Italy, 2007.
[119] J. A. Theriault and D. D. Ellis, "Shallow-water low-frequency active sonar modelling issues," in *Proceedings of the MTS/IEEE Oceans '97 Conference*, p. 678, 672, Halifax, Canada, 1997.
[120] J. A. Theriault, S. Pecknold, N. Collison, and C. Calnan, "Modelling multibeam reverberation with an N × 2D model," in *Proceedings of the 8th European Conference on Underwater Acoustics*, Carvoeiro, Portugal, 2006.
[121] C. Calnan, "DMOS—Bellhop Extension," Tech. Rep. CR 2006-005, DRDC Atlantic, Defence Research and Development Canada—Atlantic, 2006.
[122] M. D. Wagstaff, R. W. Meredith, and H. A. Terrill-Stolper, "Multi-beam high frequency imaging in a range-dependent environment," in *Proceedings of the MTS/IEEE Oceans '97 Conference*, pp. 927–931, Halifax, Canada, 1997.
[123] M. A. Kalogerakis, E. K. Skarsoulis, J. S. Papadakis et al., "An integrated computer system for underwater acoustic detection analysis," in *Proceedings of the Undersea Defence Technology Conference, (UDT'04)*, Nice, France, 2004.
[124] H. J. Shyu and R. Hillson, "A software workbench for estimating the effects of cumulative sound exposure in marine mammals," *IEEE Journal of Oceanic Engineering*, vol. 31, no. 1, pp. 8–21, 2006.
[125] M. Siderius and M. B. Porter, "Modeling techniques for marine-mammal risk assessment," *IEEE Journal of Oceanic Engineering*, vol. 31, no. 1, pp. 49–60, 2006.
[126] "Applied Radar & Sonar Technologies GmbH, Sonar performance prediction system (SPPS)," product description literature, Bremerhaven, Germany, 2002.
[127] Applied Hydro-Acoustics Research Inc., "ASPECT user's guide," Version 2.0.1 Build 664, Centreville, VA, USA, 2001.
[128] A. S. Frankel, W. T. Ellison, and J. Buchanan, "Application of the Acoustic Integration Model (AIM) to predict and minimize environmental impacts," in *in Proceedings of the MTS/IEEE Oceans '02 Conference*, pp. 1438–1443, Biloxi, Miss, USA, October 2002.
[129] D. S. Houser, "A method for modeling marine mammal movement and behavior for environmental impact assessment," *IEEE Journal of Oceanic Engineering*, vol. 31, no. 1, pp. 76–81, 2006.
[130] G. Haralabus and A. Baldacci, "Impact of seafloor and water column parameter variations on signal-to-background ratio estimates—a sensitivity analysis based on sonar performance modeling," in *Proceedings of the 8th European Conference on Underwater Acoustics, 2006*, Carvoeiro, Portugal.
[131] P. D. Ward, P. Varley, and T. Clarke, "Detecting marine mammals using active sonar—a theoretical consideration," in *Proceedings of the Institute of Acoustics*, vol. 26, part 6, no. Symposium on Bio-Sonar Systems and Bioacoustics, 2004.
[132] D. E. Weston, "Acoustic flux formulas for range-dependent ocean ducts," *Journal of the Acoustical Society of America*, vol. 68, no. 1, pp. 269–281, 1980.
[133] X. Lurton, *An Introduction to Underwater Acoustics: Principles and Applications*, Springer, New York, NY, USA, 2002.
[134] C. W. Holland, "Propagation in a waveguide with range-dependent seabed properties," *Journal of the Acoustical Society of America*, vol. 128, no. 5, pp. 2596–2609, 2010.
[135] M. G. Brown, F. J. Beron-Vera, I. Rypina, and I. A. Udovydchenkov, "Rays, modes, wavefield structure, and wavefield stability," *Journal of the Acoustical Society of America*, vol. 117, no. 3, pp. 1607–1610, 2005.
[136] K. L. Cockrell and H. Schmidt, "A relationship between the waveguide invariant and wavenumber integration," *Journal of the Acoustical Society of America*, vol. 128, no. 1, pp. EL63–EL68, 2010.
[137] R. McHugh, D. McLaren, M. Wilson, and R. Dunbar, "The underwater environment—a fluctuating acoustic medium rich in marine life. Implications for active military sonar," *Acta Acustica United with Acustica*, vol. 91, no. 1, pp. 51–60, 2005.
[138] G. B. Deane and M. D. Stokes, "Model calculations of the underwater noise of breaking waves and comparison with experiment," *Journal of the Acoustical Society of America*, vol. 127, no. 6, pp. 3394–3410, 2010.
[139] W. E. Brown and M. L. Barlett, "Midfrequency 'through-the-sensor' scattering measurements: a new approach," *IEEE Journal of Oceanic Engineering*, vol. 30, no. 4, pp. 733–747, 2005.
[140] E. Delory, M. André, J. L. Navarro Mesa, and M. Van Der Schaar, "On the possibility of detecting surfacing sperm whales at risk of collision using others' foraging clicks," *Journal of the Marine Biological Association of the United Kingdom*, vol. 87, no. 1, pp. 47–58, 2007.
[141] J.-P. Hermand, M. Meyer, M. Asch, and M. Berrada, "Adjoint-based acoustic inversion for the physical characterization of a shallow water environment," *Journal of the Acoustical Society of America*, vol. 119, no. 6, pp. 3860–3871, 2006.
[142] M. Meyer and J.-P. Hermand, "Backpropagation techniques in ocean acoustic inversion: time reversal, retrogation and adjoint modeling—a review," in *Acoustic Sensing Techniques for the Shallow Water Environment: Inversion methods and experiments*, A. Caiti, N. R. Chapman, J.-P. Hermand, and S.M. Jesus, Eds., pp. 29–46, Springer, New York, NY, USA, 2006.
[143] P. Hursky, M. B. Porter, B. D. Cornuelle, W. S. Hodgkiss, and W. A. Kuperman, "Adjoint modeling for acoustic inversion," *Journal of the Acoustical Society of America*, vol. 115, no. 2, pp. 607–619, 2004.

[144] L. Gammaitoni, P. Hänggi, P. Jung, and F. Marchesoni, "Stochastic resonance," *Reviews of Modern Physics*, vol. 70, no. 1, pp. 223–287, 1998.

[145] H. Zhang, B. Xu, Z.-P. Jiang, and X. Wu, "Target detection in shallow-water reverberation based on parameter-induced stochastic resonance," *Journal of Physics A*, vol. 41, no. 10, Article ID 105003, 2008.

[146] B. Xu, L. Zeng, and J. Li, "Application of stochastic resonance in target detection in shallow-water reverberation," *Journal of Sound and Vibration*, vol. 303, no. 1-2, pp. 255–263, 2007.

[147] A. V. van Leijen, *The sound of sediments: acoustic sensing in uncertain environments*, PhD dissertation, Universiteit van Amsterdam, The Netherlands, 2010.

[148] B. B. Thompson, R. J. Marks, M. A. El-Sharkawi, W. J. Fox, and R. T. Miyamoto, "Inversion of neural network underwater acoustic model for estimation of bottom parameters using modified particle swarm optimizers," in *in Proceedings of the IEEE International Joint Conference on Neural Networks*, pp. 1301–1306, Portland, Ore, USA, July 2003.

[149] Z.-H. Michalopoulou and U. Ghosh-Dastidar, "Tabu for matched-field source localization and geoacoustic inversion," *Journal of the Acoustical Society of America*, vol. 115, no. 1, pp. 135–145, 2004.

[150] F. Glover and M. Laguna, *Tabu Search*, Kluwer Academic Publishers, Dordrecht, The Netherlands, 1997.

[151] A. L. Virovlyansky, "Signal travel times along chaotic rays in long-range sound propagation in the ocean," *Acoustical Physics*, vol. 51, no. 3, pp. 271–281, 2005.

[152] D. V. Makarov, M. Y. Uleysky, and S. V. Prants, "Ray chaos and ray clustering in an ocean waveguide," *Chaos*, vol. 14, no. 1, pp. 79–95, 2004.

[153] D. Makarov, S. Prants, A. Virovlyansky, and G. Zaslavsky, *Ray and Wave Chaos in Ocean Acoustics: Chaos in Waveguides*, World Scientific Publishing, Singapore, 2010.

Permissions

The contributors of this book come from diverse backgrounds, making this book a truly international effort. This book will bring forth new frontiers with its revolutionizing research information and detailed analysis of the nascent developments around the world.

We would like to thank all the contributing authors for lending their expertise to make the book truly unique. They have played a crucial role in the development of this book. Without their invaluable contributions this book wouldn't have been possible. They have made vital efforts to compile up to date information on the varied aspects of this subject to make this book a valuable addition to the collection of many professionals and students.

This book was conceptualized with the vision of imparting up-to-date information and advanced data in this field. To ensure the same, a matchless editorial board was set up. Every individual on the board went through rigorous rounds of assessment to prove their worth. After which they invested a large part of their time researching and compiling the most relevant data for our readers. Conferences and sessions were held from time to time between the editorial board and the contributing authors to present the data in the most comprehensible form. The editorial team has worked tirelessly to provide valuable and valid information to help people across the globe.

Every chapter published in this book has been scrutinized by our experts. Their significance has been extensively debated. The topics covered herein carry significant findings which will fuel the growth of the discipline. They may even be implemented as practical applications or may be referred to as a beginning point for another development. Chapters in this book were first published by Hindawi Publishing Corporation; hereby published with permission under the Creative Commons Attribution License or equivalent.

The editorial board has been involved in producing this book since its inception. They have spent rigorous hours researching and exploring the diverse topics which have resulted in the successful publishing of this book. They have passed on their knowledge of decades through this book. To expedite this challenging task, the publisher supported the team at every step. A small team of assistant editors was also appointed to further simplify the editing procedure and attain best results for the readers.

Our editorial team has been hand-picked from every corner of the world. Their multi-ethnicity adds dynamic inputs to the discussions which result in innovative outcomes. These outcomes are then further discussed with the researchers and contributors who give their valuable feedback and opinion regarding the same. The feedback is then collaborated with the researches and they are edited in a comprehensive manner to aid the understanding of the subject.

Apart from the editorial board, the designing team has also invested a significant amount of their time in understanding the subject and creating the most relevant covers. They scrutinized every image to scout for the most suitable representation of the subject and create an appropriate cover for the book.

The publishing team has been involved in this book since its early stages. They were actively engaged in every process, be it collecting the data, connecting with the contributors or procuring relevant information. The team has been an ardent support to the editorial, designing and production team. Their endless efforts to recruit the best for this project, has resulted in the accomplishment of this book. They are a veteran in the field of academics and their pool of knowledge is as vast as their experience in printing. Their expertise and guidance has proved useful at every step. Their uncompromising quality standards have made this book an exceptional effort. Their encouragement from time to time has been an inspiration for everyone.

The publisher and the editorial board hope that this book will prove to be a valuable piece of knowledge for researchers, students, practitioners and scholars across the globe.

List of Contributors

Satoru Gotoand and Mikio Tohyama
Wave Science Study, 2-7-8 Fujigaya, Kugenuma, Fujisawa-shi, Kanagawa 251-0031, Japan

Yoshinori Takahashi
Tokyo Metropolitan College of Industrial Technology, 8-17-1 Minami Senju, Arakawa-ku, Tokyo 116-0003, Japan

Minoru Sasaki and Takeshi Ueda
Department of Human and Information Systems Engineering, Gifu University, 1-1 Yanagido, Gifu 501-1193, Japan

Yshihiro Inoue
Department of Mechanical and Systems Engineering, Gifu University, 1-1 Yanagido, Gifu 501-1193, Japan

Wayne J. Book
School of Mechanical Engineering, Georgia Institute of Technology Atlanta, J. E. Love Jr. Manufacturing Building, 771 Ferst Drive, GA 30332-0405, USA

P. R. Dash, S. K. Sarangi and P. K. Pradhan
Mechanical Engineering Department, VSSUT Burla, Orissa 768018, India

K. Ray
Mechanical Engineering Department, IIT Kharagpur, Kharagpur 721302, India

Hyung Tae Kim, Cheol Ho Kim, Sung Bok Kang, Kang Won Lee and Hyun Hee Han
Manufacturing System Division, KITECH, 35-3 Hongcheon, Ipjang, Chungnam, Cheonan, Republic of Korea

Jae Ho Baek
RMS Technology, Techpia Building, 774 Dujeonng, Seobuk,Chungnam, Cheonan, Republic of Korea

Alberto Cavallo, Giuseppe De Maria, Ciro Natale and Salvatore Pirozzi
Dipartimento di Ingegneria dell'Informazione, Seconda Universita degli Studi di Napoli, Via Roma 29, 81031 Aversa, Italy

Y. M. Parulekar and G. R. Reddy
Reactor Safety Division, Bhabha Atomic Research Centre, Mumbai 400085, India

Mothanna Y. Abd and Nawal A. A. Jalil
Department of Mechanical and Manufacturing Engineering, Faculty of Engineering, Universiti Putra Malaysia, 43400 Serdang, Selangor, Malaysia

Azma Putra
Department of Structure and Materials, Faculty of Mechanical Engineering, Universiti Teknikal Malaysia Melaka, Hang Tuah Jaya, 76100 Durian Tunggal, Melaka, Malaysia

Jamaludin Noorzaei
Department of Civil Engineering, Faculty of Engineering, Universiti Putra Malaysia, 43400 Serdang, Selangor, Malaysia

Shahin S. Nudehi
Mechanical Engineering Department, Valparaiso University, Valparaiso, IN 46383, USA

Ghalib R. Ibrahim
Mechanical Engineering Department, College of Engineering, University of Anbar, Anbar, Iraq

A. Albarbar
Advanced Industrial Diagnostic Centre, School of Engineering, Manchester Metropolitan University, Manchester M1 5GD, UK

S. H. Gawande and L. G. Navale
Department of Mechanical Engineering, M. E. Society's College of Engineering, Pune-411 001, Maharashtra, India

M. R. Nandgaonkar
Department of Mechanical Engineering, Government College of Engineering, Pune-411 005, Maharashtra, India

D. S. Butala and S. Kunamalla
Research and Development, Greaves Cotton Limited, Pune-411 019, Maharashtra, India

R. K. Giridhar and S. G. Barad
Vibration Engineering Group, Gas Turbine Research Establishment, CV Raman Nagar, Bangalore 560093, India

P. V. Ramaiah and G. Krishnaiah
Mechanical Engineering Department, Sri Venkateswara University, Tirupati 517502, India

A. G. Sheard
Flakt Woods Limited, Axial Way, Colchester, Essex CO4 5ZD, UK

C. P. Vyasarayani
Department of Mechanical Engineering, Indian Institute of Technology Hyderabad, Ordnance Factory Estate, Yeddumailaram 502205, India

Sukhpreet Singh Sandhu
MSC Software, 101 North Brand Boulevard, Suite 900 Glendale, Santa Ana, CA 91203, USA

John McPhee
Systems Design Engineering, University of Waterloo, Waterloo, ON, Canada

Celine Quinsac, Adrian Basarab and Denis Kouame
University of Toulouse, IRIT UMR 5505 CNRS, 31062 Toulouse Cedex 9, France

Liz G. Nallim, Facundo J. Bellomo and Ricardo D. Quinteros
INIQUI-CONICET, Facultad de Ingenierıa, Universidad Nacional de Salta, Avenue Bolivia 5150, 4400 Salta, Argentina

Sergio Oller
International Center for Numerical Method in Engineering, (CIMNE) Technical University of Catalonia-Barcelona Tech (UPC), Edif. C1, Campus Nord, Jordi Girona 1-3, 08034 Barcelona, Spain

Ennes Sarradj
Department of Technical Acoustics, Brandenburg University of Technology, Siemens-Halske-Ring 14, 03046 Cottbus, Germany

Alexis Billon and Anas Sakout
LEPTIAB, University of La Rochelle, 17042 La Rochelle, France

Judicaël Picaut
Departement IM UR EASE, Ifsttar Nantes, Route de Bouaye, CS 4, 44344 Bouguenais Cedex, France

Vincent Valeau
Institut PUPR 3346, LEA UMR 6609, Batiment K 40, Avenue du Recteur Pineau, 86022 Poitiers Cedex, France

Ahmad R. Abu-El-Quran
Department of Enterprise and Higher Education, International Turnkey Systems, ITS Tower, Mubarak Al Kabeer Street, P.O. Box 26729, Safat, Kuwait City 13128, Kuwait

Adrian D. C. Chan and Rafik A. Goubran
Department of Systems and Computer Engineering, Carleton University, 1125 Colonel By Drive, Ottawa, ON, Canada

Jaswinder S. Sandhu, Joshua I. Weber, J. Tawiah and Vitaliy Kunin
Santec Systems, Inc., 2924 Malmo Drive, Arlington Heights, IL 60005, USA

Robert W. Schoonover and Mark A. Anastasio
Department of Biomedical Engineering, Washington University in St. Louis, St. Louis, MO 63130, USA

Paul C. Etter
Northrop Grumman Corporation, P.O. Box 1693, Baltimore, MD 21203, USA